Seino van Breugel
**A Dictionary of Atong**

# Pacific Linguistics

**Managing editor**
Alexander Adelaar

**Editorial board members**
Wayan Arka
Danielle Barth
Don Daniels
Nicholas Evans
Gwendolyn Hyslop
David Nash
Bruno Olsson
Bill Palmer
Andrew Pawley
Malcolm Ross
Dineke Schokkin
Jane Simpson

## Volume 664

Seino van Breugel

# A Dictionary of Atong

A Tibeto-Burman Language of Northeast India and Bangladesh

**DE GRUYTER**
MOUTON

ISBN 978-3-11-111473-6
e-ISBN (PDF) 978-3-11-071800-3
e-ISBN (EPUB) 978-3-11-071807-2
ISSN 1448-8310

**Library of Congress Control Number: 2020946081**

**Bibliographic information published by the Deutsche Nationalbibliothek**
The Deutsche Nationalbibliothek lists this publication in the Deutsche Nationalbibliografie;
detailed bibliographic data are available on the Internet at http://dnb.dnb.de.

© 2022 Walter de Gruyter GmbH, Berlin/Boston
This volume is text- and page-identical with the hardback published in 2021.
Typesetting: Integra Software Services Pvt. Ltd.
Printing and binding: CPI books GmbH, Leck
Photo credit: Seino van Breugel

www.degruyter.com

# Acknowledgements

I wish to thank all Atong speakers who have provided me with hospitality, help and information, which made it possible for me to make this dictionary. I am particularly thankful for the hospitality of the people of the villages of Badri Maidugythym and Siju: the family of Mr Sushil S. Marak (Nisawa•) and his wife Mrs Kelbish M. Sangma (Nisajyw•); Mrs Latith M. Sangma and Ms Janita M. Sangma; Mr Plindar R. Marak, Mr Peslar R. Marak and his wife Mrs Golaphy R. Sangma, their daughter Miss Jupina R. Sangma, their son Mr Stephen R. Sangma and the rest of their large family; Mrs Elsina R. Sangma (Aby Fernandajyw•), her mother Mrs Monjila R. Sangma, her brothers Mr Dilseng R. Sangma and Mr Gostar R. Sangma, Mr Jomuna R. Sangma, his wife Mrs D. Shira, their daughters Ms Radia D. Shira and Ms Dambe D. Shira, their son Mr Haiwash D. Shira and their nephew Mr Bairyck D. Sira, and Mr Samrat N. Marak.

Many of the words and sentences in this dictionary are taken from the texts I recorded during my stay with the Atongs on various occasions between 2004 and 2014. I want to thank all the authors of the stories I recorded: Ms Dorina A. Sangma, Mr Todan M. Sangma, Mr Ranus M Sangma, Mr Tonton M. Sangma and his parents, Mr Renjen S. Marak, Mr Thangring (Rangsewa•) M. Sangma, Mr Dalcheng M. Sangma, Mr Miksrang J. Momin, Mr Negverson M. Sangma, Mr Genda R. Marak, Mr Samrat N. Marak, Ms Monjilla R. Sangma, Mr Johan A. Sangma, Mr Nikseng S. Marak, Ms Janita M. Sangma, Mr Kempai A. Sangma, Ms Jamila M. Sangma, Mr, Derus R. Marak, Mr Sandish M. Sangma, Mr Dilseng R. Sangma, Mr Kiubirth M. Sangma, Mr Cheng M. Sangma, Mr Jentibirth M. Sangma, Mr Silcheng M. Sangma, Mr Wilseng S. Marak and Mr Aristo J. Monin.

When I collected the texts, they were first recorded on tape. They had to be written down, and I could not have done that by myself. Therefore, my special thanks go to Mr Sandish M. Sangma, Mr Salseng R. Sangma, Mr Ranus M. Sangma, Mr Tonton M. Sangma, Mr Bitter M. Sangma, Mr Inden R. Sangma, Mr Plindar R. Marak, Mr Shyam R. Marak and Mr Samrat N. Marak for their help in writing down and translating the stories. Their enthusiasm and devotion are truly remarkable!

Many words and sentences in this dictionary come from notes I made while I interacted with Atong speakers, and learned their language. This dictionary also contains some words that where intentionally added to the dictionary by a variety of speakers, all of whom I thank for their contributions. My thanks also go out to everyone who commented on my queries on Facebook.

I also extend my gratitude to the anonymous reviewers of this work, as well as to to Sander Adelaar, for the their valuable, insightful and welcome comments. Any shortcomings that may still persist in the text are entirely my own.

Finally, I thank the Atongs for their kindness, friendship, hospitality, care and patience while I was learning their language and culture. You have conquered my heart and mind, and I will never forget you.

Mythelbiok!
Seino van Breugel

# Contents

**Acknowledgements** —— V

**List of Tables and Maps** —— XIII

**Prologue** —— XV
    Introduction —— XV
    Atong and its Speakers —— XVI
    A note on Atong orthography —— XVIII
    Abbreviations of grammatical categories —— XXIII
    List of abbreviations of types of headword —— XXV
    Labels for the semantic domains of nouns in the Atong-English dictionary —— XXVII
    Symbols —— XXVIII

**PART 1: ATONG – ENGLISH DICTIONARY** —— 1
    What do we see in the Atong-English dictionary? —— 1
    Atong-English Dictionary —— 11

**PART 2: ENGLISH – ATONG DICTIONARY** —— 135
    What do we see in the English-Atong dictionary? —— 135
    How to use the English-Atong dictionary —— 137
    English-Atong Dictionary —— 140

**PART 3: SEMANTIC LEXICA** —— 185

| | | |
|---|---|---|
| 1 | Days of the week —— 185 | |
| 2 | Months of the year —— 186 | |
| 3 | Lexicon of kinship terms: Atong – English —— 187 | |
| 4 | Lexicon of kinship terms: English – Atong —— 191 | |
| 5 | Semantic lexicon of verbs and nouns —— 194 | |
| 5.1 | Nature and natural phenomena —— 194 | |
| 5.1.1 | Heavenly bodies —— 194 | |
| 5.1.2 | Parts of the day —— 194 | |
| 5.1.3 | Seasons —— 194 | |
| 5.1.4 | Weather —— 195 | |
| 5.1.5 | Fire and related words —— 195 | |
| 5.1.6 | Other natural phenomena —— 195 | |
| 5.2 | The earth, soil, products of the earth —— 195 | |
| 5.2.1 | Stones and rocks —— 195 | |

| | | |
|---|---|---|
| 5.2.2 | Precious stones —— **196** | |
| 5.2.3 | Metals and minerals —— **196** | |
| 5.2.4 | Water bodies and aquatic phenomena —— **196** | |
| 5.2.5 | Dirt, filth —— **196** | |
| 5.3 | Physical processes —— **196** | |
| 5.4 | Physical development —— **197** | |
| 5.5 | Humans —— **197** | |
| 5.5.1 | Human body parts —— **197** | |
| 5.5.2 | Products of the human body —— **199** | |
| 5.5.3 | Bodily functions —— **199** | |
| 5.5.4 | Afflictions of the human body —— **199** | |
| 5.5.5 | Physical sensations —— **200** | |
| 5.5.6 | Diseases and infections —— **200** | |
| 5.5.7 | Emotions and psychological feelings and states —— **200** | |
| 5.6 | Human behaviour —— **201** | |
| 5.6.1 | General behaviour —— **201** | |
| 5.6.2 | Child bearing and raising —— **203** | |
| 5.6.3 | Hunting and fishing —— **203** | |
| 5.6.4 | Killing —— **204** | |
| 5.6.5 | Death —— **204** | |
| 5.6.6 | Agriculture —— **204** | |
| 5.6.7 | Household chores —— **205** | |
| 5.6.8 | Religion —— **205** | |
| 5.6.9 | Games, toys and jokes —— **205** | |
| 5.6.10 | Festivals, ceremonies and events —— **205** | |
| 5.6.11 | Things humans and/or animals undergo —— **206** | |
| 5.7 | Language —— **206** | |
| 5.8 | Food and cooking, eating and drinking —— **207** | |
| 5.8.1 | Food items and ingredients used for food —— **207** | |
| 5.8.2 | Kitchen furniture —— **208** | |
| 5.8.3 | Cooking —— **208** | |
| 5.8.4 | Cooking and eating utensils —— **208** | |
| 5.8.5 | Eating and drinking —— **209** | |
| 5.9 | Cognition —— **209** | |
| 5.10 | Perception —— **209** | |
| 5.11 | People —— **210** | |
| 5.11.1 | Men, women and stages in life —— **210** | |
| 5.11.2 | Relationships between people —— **210** | |
| 5.11.3 | Profession/Function —— **210** | |
| 5.11.4 | Religious people —— **211** | |

| | | |
|---|---|---|
| 5.11.5 | Looks and afflictions —— **211** | |
| 5.11.6 | Persons with negative characteristics —— **212** | |
| 5.11.7 | Babies, children —— **212** | |
| 5.11.8 | Dead people —— **212** | |
| 5.11.9 | Nationality/Ethnicity —— **212** | |
| 5.11.10 | Institutions —— **213** | |
| 5.11.11 | Groups of people —— **213** | |
| 5.11.12 | Stages in life and death —— **213** | |
| 5.12 | Human products —— **213** | |
| 5.12.1 | Clothes —— **213** | |
| 5.12.2 | Jewellery and makeup —— **214** | |
| 5.12.3 | Tools —— **214** | |
| 5.12.4 | Weapons —— **215** | |
| 5.12.5 | Instruments and related words —— **215** | |
| 5.12.6 | Baskets and other receptacles —— **216** | |
| 5.12.7 | Bamboo mats —— **216** | |
| 5.12.8 | Business, trade and money —— **217** | |
| 5.12.9 | Electronics —— **217** | |
| 5.12.10 | Transportation and vehicles —— **217** | |
| 5.12.11 | Building and creating —— **218** | |
| 5.12.12 | Buildings and other man-made structures —— **218** | |
| 5.12.13 | House structure —— **218** | |
| 5.12.14 | Furniture —— **220** | |
| 5.12.15 | Other man-made products —— **220** | |
| 5.12.16 | Materials and substances —— **220** | |
| 5.13 | The supernatural —— **220** | |
| 5.13.1 | Supernatural beings —— **220** | |
| 5.13.2 | Dreams, magic and supernatural practices and experiences —— **221** | |
| 5.14 | Animals —— **221** | |
| 5.14.1 | Amphibians —— **221** | |
| 5.14.2 | Arthropods (spiders, scorpions, ticks, fleas, lice, centipedes, millipedes) —— **222** | |
| 5.14.3 | Birds —— **222** | |
| 5.14.4 | Fish —— **223** | |
| 5.14.5 | Insects —— **223** | |
| 5.14.6 | Mamals —— **224** | |
| 5.14.7 | Reptiles —— **225** | |
| 5.14.8 | Snails and clams —— **225** | |
| 5.14.9 | Worms and leeches —— **226** | |

| | | |
|---|---|---|
| 5.14.10 | Animal body parts and products —— **226** | |
| 5.14.11 | Animal behaviour —— **226** | |
| 5.14.12 | Animal dwellings —— **227** | |
| 5.15 | Plants and trees —— **227** | |
| 5.15.1 | Plants —— **227** | |
| 5.15.2 | Fruits, vegetables and leafy greens —— **228** | |
| 5.15.3 | Beans —— **229** | |
| 5.15.4 | Trees —— **229** | |
| 5.15.5 | Lianas —— **230** | |
| 5.15.6 | Creepers —— **230** | |
| 5.15.7 | Fungi and algae —— **230** | |
| 5.15.8 | Tubers, other edible roots and onions —— **230** | |
| 5.15.9 | Grasses —— **230** | |
| 5.15.10 | Bamboo —— **230** | |
| 5.15.11 | Bamboo parts —— **231** | |
| 5.15.12 | Rice —— **231** | |
| 5.15.13 | Parts of plants and trees —— **231** | |
| 5.15.14 | Seeds —— **232** | |
| 5.15.15 | Plant substances —— **232** | |
| 5.15.16 | Verbs pertaining to trees, plants and fruit —— **232** | |
| 5.16 | Places, spaces, position, direction —— **233** | |
| 5.16.1 | Places and spaces —— **233** | |
| 5.16.2 | Position and positioning —— **234** | |
| 5.16.3 | Directions of the compass —— **235** | |
| 5.16.4 | Toponyms —— **235** | |
| 5.17 | Shape —— **237** | |
| 5.18 | Time —— **237** | |
| 5.19 | Manipulation —— **238** | |
| 5.20 | Movement —— **241** | |
| 5.21 | Involuntary movement or change of state —— **243** | |
| 5.22 | Order —— **243** | |
| 5.23 | Phase —— **243** | |
| 5.24 | Noise and sound —— **244** | |
| 5.25 | Abstract —— **244** | |

**PART 4: GRAMMATICAL LEXICA —— 246**

| | |
|---|---|
| 1 | Lists of adjectives of Type 1 and stative verbs —— **246** |
| 2 | List of adjectives of Type 2 —— **249** |
| 3 | List of adverbs —— **251** |
| 4 | Lists of classifiers and volume words —— **252** |

| | | |
|---|---|---|
| 5 | List of collocations —— 256 | |
| 6 | List of demonstratives —— 259 | |
| 7 | List of discourse connectives —— 260 | |
| 8 | List of event specifiers —— 261 | |
| 9 | List of grammatical categories found in Atong and the morphemes associated with them —— 263 | |
| 10 | List of idiomatic expressions —— 265 | |
| 11 | List of ideophones —— 266 | |
| 12 | List of indefinite Proforms —— 268 | |
| 13 | List of interjections —— 269 | |
| 14 | List of interrogatives —— 271 | |
| 15 | List of particles —— 272 | |
| 16 | List of personal pronouns —— 273 | |
| 17 | List of postpositions —— 274 | |
| 18 | List of proclauses —— 275 | |
| 19 | Lists of quantifiers —— 276 | |
| 20 | List of time words —— 282 | |
| 21 | List of intransitive and transitive verbal pairs —— 283 | |
| 22 | List of verbs of emotion and interaction —— 284 | |

**PART 5: COMPENDIUM OF ATONG GRAMMAR —— 285**

| | |
|---|---|
| 1 | Overview of the word classes of Atong —— 285 |
| 2 | Adjectives —— 286 |
| 3 | Adverbs —— 288 |
| 4 | Classifiers and volume words —— 288 |
| 4.1 | Classifiers —— 288 |
| 4.2 | Volume words —— 293 |
| 5 | Collocations —— 294 |
| 6 | Conjunctions —— 296 |
| 7 | Demonstratives —— 296 |
| 8 | Discourse connectives —— 298 |
| 9 | Event specifiers —— 298 |
| 10 | Idiomatic expressions —— 299 |
| 11 | Ideophones —— 299 |
| 12 | Indefinite proforms —— 300 |
| 13 | Interjections —— 300 |
| 14 | Interrogatives —— 300 |
| 15 | Nouns and verbs —— 301 |
| 16 | Particles —— 303 |
| 17 | Personal pronouns —— 304 |

| 18 | Postpositions —— **305** |
|---|---|
| 19 | Proclauses —— **305** |
| 20 | Quantifiers —— **306** |
| 20.1 | Numerals —— **306** |
| 20.2 | Non-numeral quantifiers —— **308** |
| 21 | Time words —— **308** |
| 22 | Verbs of emotion and interaction —— **310** |

**Appendix of Photos** —— **311**

**References** —— **375**

**Index** —— **377**

# List of Tables and Maps

| | | |
|---|---|---|
| Table 1 | The Atong alphabet with example words and English glosses —— | **XVIII** |
| Table 2 | Example of the ordering of aspirated stops in the dictionary —— | **XIX** |
| Table 3 | Atong phonemes (in IPA) and their corresponding letters of the Atong alphabet —— | **XX** |
| Table 4 | Pairs of words with and without *raka* —— | **XXI** |
| Table 5 | Type 1 Adjectives —— | 246 |
| Table 6 | Stative verbs not yet identified as Type 1 Adjectives —— | 248 |
| Table 7 | Type 2 Adjectives —— | 249 |
| Table 8 | Adverbs sorted by denotation or function —— | 251 |
| Table 9 | Classifiers —— | 252 |
| Table 10 | Volume words —— | 255 |
| Table 11 | Synonymous collocations —— | 256 |
| Table 12 | Associative collocations —— | 257 |
| Table 13 | Decorative collocations where only the second member is a decorative word —— | 257 |
| Table 14 | Decorative collocations where both members are decorative words —— | 258 |
| Table 15 | Demonstratives —— | 259 |
| Table 16 | Discourse connectives —— | 260 |
| Table 17 | Event specifiers —— | 261 |
| Table 18 | Grammatical categories and their morphemes —— | 263 |
| Table 19 | Idiomatic expressions —— | 265 |
| Table 20 | Ideophones —— | 266 |
| Table 21 | Indefinite proforms —— | 268 |
| Table 22 | Interjections —— | 269 |
| Table 23 | Interrogatives —— | 271 |
| Table 24 | Particles —— | 272 |
| Table 25 | Personal pronouns —— | 273 |
| Table 26 | Positions with the markers of their complements —— | 274 |
| Table 27 | Proclauses —— | 275 |
| Table 28 | Atong numerals —— | 276 |
| Table 29 | Numerals borrowed from English —— | 280 |
| Table 30 | Numerals borrowed from Hindi —— | 281 |
| Table 31 | Non-numeral quantifiers —— | 281 |
| Table 32 | Time words —— | 282 |
| Table 33 | Intransitive and transitive verb pairs —— | 283 |
| Table 34 | Verbs of emotion and interaction —— | 284 |

| Table 35 | The predicative word classes —— **285** |
| Table 36 | The non-predicative word classes —— **286** |
| Table 37 | The semantic categories found in Type 1 and Type 2 adjectives —— **286** |

| Map 1 | The location of the state of Meghalaya within India —— **XVII** |
| Map 2 | Rough indication of the Atong speaking area in Meghalaya —— **XVIII** |

# PROLOGUE

## Introduction

This is a scholarly dictionary of Atong, which means that it is contains a lot of information which is meant for academics in universities. However, this dictionary can also be used by anyone who just wants to translate a word from Atong into English, or from English into Atong.

In PART 1 of this book, the reader finds the third edition of the Atong-English Dictionary, with more than 3900 entry words. The photos in the appendix serve as illustrations of some of the words in the dictionary. PART 2 presents the first edition of the English-Atong dictionary with almost 3000 entries.

The Atong lexicon and all of its suffixes and enclitics can be organised and hence probed in other ways than in a dictionary. PART 3 contains semantic lexica, i.e. lists of words organised according to their meanings. PART 4 presents lists of lexemes and morphemes grouped together based on grammatical criteria. In this part, the reader finds lists of the members of different Atong word classes, collocations and idiomatic expressions.

PART 5 is a compendium of Atong grammar, providing cursory information about the word classes, collocations and idiomatic expressions in Atong. Wherever possible, reference is made to *A grammar of Atong* (van Breugel 2014) and other relevant sources, so as to keep the amount of information in this volume succinct.

In this dictionary, mistakes discovered in previous editions (van Breugel 2009b, 2015c) have been corrected. These mistakes include spelling mistakes, mistakes in the shades of meaning of several words, mistakes in word-class assignment and others. Spelling mistakes were corrected in the field, as well as by comparing occurrences of the same word in different recorded texts and in field notes (see van Breugel 2014: 33–36). When necessary, native Atong speakers were contacted through social media to consult the author in times when fieldwork was not possible.

Moreover, fieldwork conducted in the years since the previous editions, as well as the preparation of van Breugel (2019), brought to light some previously unnoticed variation in the pronunciation of a number of words. These differences in pronunciation, reflected in the spelling, were added to the dictionary as variations of existing entry words. Also, new entries, sub-entries and examples have been added. This edition is therefore bigger and better than the previous ones.

The Atong words and sentences in this dictionary are spelled with the Atong alphabet, using the Atong spelling rules. The spelling rules of Atong are explained in the *Atong Spelling Guide* (van Breugel 2015a). However, a short note on the Atong orthography (spelling system) is given in this prologue. This note is followed by lists of the abbreviations and symbols used in this entire volume.

The grammatical terminology used in the Atong-English dictionary, the Atong word classes, and the meaning or function of Atong suffixes and enclitics are all explained and discussed in detail in van Breugel (2014), which presents a grammatical description of all major areas of the grammar of the language. This dictionary contributes to a more detailed documentation and description of the Atong language, adding to the existing corpus of works in or about Atong. These works are summed up in the list of references at the end of this book, together with the other sources referred to in this volume.

This book does not contain all the words of the Atong language. The words in this dictionary were mainly recorded in the villages of Badri Maidugythym and Siju, in the South Garo Hills District of Meghalaya. The author is aware that Atong is spoken in different ways in different places; therefore, this dictionary does not represent Atong as a whole. Thus, when an Atong speaker searches this dictionary and does not find a word from his or her variety of Atong, it does not mean that that word does not exist. When a word cannot be found, it means that it was not recorded by the author. The author hopes that future fieldwork will make it possible for him, or someone else, to add to this dictionary many more words from many different places where Atong is spoken in India, and across the border in Bangladesh. Finally, if any mistakes still remain in this book, please forgive the author.

## Atong and its Speakers

> The Atongs belong to the Garo Tribe, but many Atongs do not use Standard Garo to communicate in their daily lives. Instead, they use Atong, a speech variety closely related to Garo, but nonetheless quite different. Due to widespread bilingualism amongst the Atongs, mutual intelligibility between Atong and Garo is a one way street. This means that those who do not speak Atong will not understand it, but an Atong speaker will most probably understand Standard Garo. Just like any other form of speech, Atong is a valuable means of communication for the people who speak it.

Atong is spoken in Meghalaya State in Northeast India and adjacent areas in Bangladesh.[1] The variety of Atong recorded in this book is the variety spoken in India. Eberhard et al. (2020) states a population of 4,600 Atongs in India, with

---

[1] Atong's identifying codes are: ISO 639-3aot.

the addition that the total number of "users in all countries" is 10,000; however, no official numbers are available. Map 1 shows the position of Meghalaya within India, and Map 2 gives a rough indication of the area where Atong is spoken in Meghalaya. The highest concentration of speakers can be found in the South Garo Hills Districts and the western part of the Khasi Hills. Atong belongs to the Tibeto-Burman language family, just like its closest linguistic relatives Koch, Ruga and Rabha, as well as the different dialects of Garo (see Burling 2003a: 175–177).

**Map 1** The location of the state of Meghalaya within India.
(Map designed by author and digitally created with the help of Mr Weerachai Sriwai.)

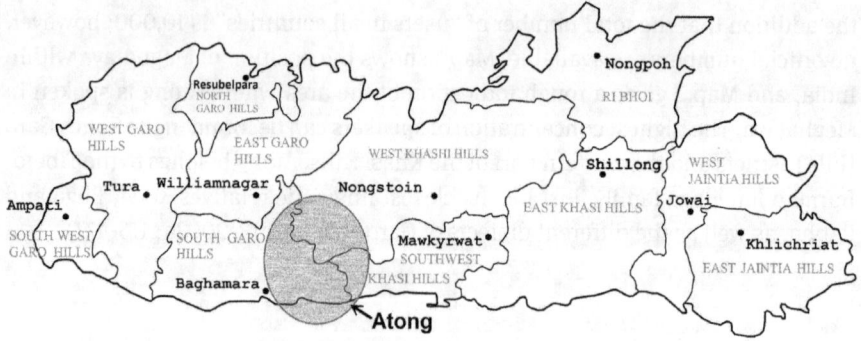

**Map 2** Rough indication of the Atong speaking area in Meghalaya.
(Map designed by author and digitally created with the help of Mr Weerachai Sriwai.)

# A note on Atong orthography

The orthography used in this dictionary was developed by the author, and is described in full in van Breugel 2015a. This spelling system has been in use since the publication of the first edition of the Atong-English Dictionary (van Breugel 2009b) and the Atong story book (van Breugel 2009a, 2015b).

The order of the letters in the Atong alphabet, which is the order in which headwords are sequenced in this dictionary, is presented in Table 1. We can see that each letter has two forms and a name. Each letter is accompanied by an example word in which it is used. Except for the letter *raka*, each letter has a CAPTIAL form and a lower-case form. The two forms of the letter *raka* are in free variation, with the note that the bullet point is more formal, and therefore used in this book, whereas the apostrophe is easier to write when handwriting, or typing on a smartphone or computer.

**Table 1:** The Atong alphabet with example words and English glosses.

| Letters | | Names | Examples | English glosses |
|---|---|---|---|---|
| • | ' | raka | *mym•sa / mym'sa* | 'a fist' |
| A | a | a | *Atong* | 'Atong' |
| B | b | ba | *baju* | 'friend' |
| C | c | cha | *chak* | 'hand, leaf' |
| D | d | da | *dam* | 'place' |
| E | e | e | *era* | 'species of fish' |
| G | g | ga | *gawi* | 'girl' |
| H | h | ha | *ha•ba* | 'rice field' |

**Table 1** (continued)

| Letters | | Names | Examples | English glosses |
|---|---|---|---|---|
| I | i | i | ichi | 'here' |
| J | j | ja | ja•bek | 'curry' |
| K | k | ka | khu•chuk | 'mouth, language' |
| L | l | la | laha | 'resin' |
| M | m | ma | mai | 'rice' |
| N | n | na | net | 'basket' |
| O | o | o | ong ang | 'species of frog' |
| P | p | pa | panchung | 'jackfruit' |
| R | r | ra | rai | 'reed' |
| S | s | sa | symgong | 'species of plant' |
| T | t | ta | tyi | 'water' |
| U | u | u | u•ching ~ ukching | 'leech' |
| W | w | wa | wak | 'pig' |
| Y | y | y | ytykyi | 'like that' |

Note that not all Atong phonemes have their own place or letter in the alphabet. Hence, words starting with the aspirated stops /pʰ, tʰ, kʰ/ are organised under the letters P, T and K respectively. The list of example entry words in Table 2 shows the arrangement of words starting with the aspirated stop /kʰ/, written kh, in the dictionary. The order is: ka, ke, kha, khe, kho, khu, khy, ki, ko, ku, ky.

**Table 2:** Example of the ordering of aspirated stops in the dictionary.

| | | | |
|---|---|---|---|
| kam- | 'to clear the field' | khu•chuk | 'mouth, language' |
| keko | 'species of gecko' | khyw- | 'to drain' |
| kha- | 'salty' | kirin | 'torn (of cloth and paper)' |
| khen• | 'river crab' | ko•rot | 'sugarcane' |
| khi- | 'to count' | kun• | 'a stick' |
| khol | 'skin' | kyn | 'back' |

Table 3 presents an overview of Atong phonemes and the corresponding letters of the Atong alphabet. The phonemes are presented in International Phonetic Alphabet, with the consonants in the left and middle sections of the table, the vowels on the right, and the orthographic ways to present glottalisation at the bottom. The phonemes are grouped together according to their place and manner of articulation. For a full description of Atong phonology, the reader is referred to van Breugel (2014: Chapter 2).

**Table 3:** Atong phonemes (in IPA) and their corresponding letters of the Atong alphabet.

| CONSONANTS | | | | VOWELS | |
|---|---|---|---|---|---|
| Phonemes | Letters | Phonemes | Letters | Phonemes | Letters |
| pʰ | ph | m | m | i | i |
| tʰ | th | n | n | e | e |
| kʰ | kh | ŋ | ng | a | a |
| p | p | r | r | o | o |
| t | t | l | l | u | u |
| k | k | s | s | ə | y |
| b | b | tɕ | ch | ī | ii |
| d | d | dʑ | j | ē | ee |
| g | g | h | h | ō | oo |
| w | w | j | i | ā | aa |
| glottalisation • or ' | | | | | |

The vowel phonemes /ī, ē, ā, ō / are found only in loanwords, and can thus be termed loanvowels. These loanvowels differ from the other vowel because they are never articulated lowered and more centralised in closed syllables, as may happen to the other vowels; and because they can be articulated either long or short in free variation in any environment. The other, non-loanvowel phonemes are only lengthened when extra stress is applied in the syllable in which they occur.

For example, the vowel /i/ is pronounced as the close front unrounded vowel [i] in open syllables, e.g. *ichi* [i.tɕi] 'here', but is usually pronounced as the near close, near front unrouned vowel [ɪ] in closed syllables, e.g. *tin* [tɪn] 'corrugated iron'. In contrast, the loanvowel /iː/ is always pronounced [i], and its length can be freely varied between long and short, e.g. *tiin* [tiːn ~ tin] 'three' as in *tiin baji* 'three o'clock'. The word *tiin* is a loan from Hindi, cf. तीन [t̪iːn] 'three'. Note that not all loanwords in Atong have loanvowels. For more information on loanvowels, the reader is referred to van Breugel (2014: 61–62).

When it occurs after a vowel, the letter i represents a consonant phoneme, namely the glide /j/ [j], as in for example *askui* /askuj/ [askuj] 'star', or *choi•sa* /tɕoj²sa/ [tɕoj²sa] 'a little bit'. In all other cases, the letter i represents the vowel phoneme /i/, e.g. *bisi* /bisi/ [bisi] 'poison', or *piong* /pioŋ/ [pi.oŋ] 'species of bird'. The only exceptions to this rule are loanwords, such as *piktiyr* /piktjər/ [piktjər] 'photo', where the letter i precedes a vowel, yet represents the glide /j/. This word is a loan from English, and it is not unusual for loanwords to have different phonotactics from native words. There is only one Atong word that start with the letter i followed by a vowel, namely the word *ie* 'this, this one'. This word may be pronounced as a single syllable starting with a glide, [je], or with two syllables, [i.e]. However, it has a bound allomorph, or variant form, *i-*, which occurs before phrasal enclitics (see van Breugel 2014: Chapter 17), as in the words *ichi* <i=chi> (this=LOC) 'here' and *iba* <i=ba> (here=ADD) 'this also'.

Note that there are no diphthongs in Atong (see van Breugel 2014:42–43). Diphthongs are defined as "movements from one vowel to another within a single syllable" (Ladefoged 2000: 28). However, the canonical syllable structure of Atong allows for only a single vowel in the nucleus (van Breugel 2014: 37). What may look like diphthongs in the orthography, because of the use of a vowel followed by the letter i, as in the words *kyi•* [kəj$^ʔ$] 'dog', *askhui* [ask$^h$uj] 'star', and *mai* [maj] 'rice', are not phonologically diphthongs, but sequences of a vowel and an off-glide. Analysing these sequences phonemically as vowel plus glide safeguards the canonical (C)V(C) syllable structure of Atong. (van Breugel 2014: 42).

The letter called *raka* is used to indicate glottalisation. Glottalisation is a suprasegmental feature of Atong phonology, operating at the level of the syllable (see van Breugel 2014: 56–60). Phonetically, glottalisation manifests itself as a glottal stop at the end of the syllable. This phenomenon only affects open syllables, and syllables ending in a continuant (which would be written with the letters m, n, ng, w, l or the letter i when it occurs after a vowel). Syllables in Atong can thus either be glottalised, or not glottalised. Van Breugel (2014: 56–58 and 2015a) provide more detailed information about glottalisation.

The presence or absence of glottalisation can change the meaning of a word. It is therefore important to use the *raka* to spell words in which glottalisation occurs. Table 4 is reproduced from van Breugel (2015a: 29) and provides examples of minimal and near minimal pairs of words with and without glottalisation and thus spelled with or without the *raka*.

**Table 4:** Pairs of words with and without *raka* (van Breugel, 2015a: 29).

| WORDS WITH A *RAKA* | | WORDS WITHOUT *RAKA* | |
|---|---|---|---|
| *si•* | 'to sharpen a pointy object' | *si* | 'to peel' |
| *ne•kat* | 'type of bee' | *Nepal* | 'Nepali, Nepalese, Nepal' |
| *cha•* | 'foot, leg' | *cha* | 'tea' |
| *na•* | 'fish' | *na* | 'to hear' |
| *su•* | 'to pound' | *su* | 'to scold' |
| *ri• myla* | 'the penis is small' | *rimyla* | 'slippery' |
| *wal•* | 'fire' | *wal* | 'night' |
| *rong•* | 'stone' | *rong* | 'colour' |
| *man•* | 'to be able, to get' | *man* | 'to crawl' |

**Table 4** (continued)

| WORDS WITH A *RAKA* | | WORDS WITHOUT *RAKA* | |
|---|---|---|---|
| ram• | 'to search' | ram | 'to dry' |
| thyi• | 'blood' | tyi | 'water' |
| taw• | 'chicken, bird' | taw | 'to go up' |
| myng• | 'classifier for persons' | myng | 'classifier for spoken things' |

Glottalisation can be written either with the bullet '•' or the apostrophe after a glottalised syllable. The apostrophe is easy to use when typing on a mobile phone, computer keyboard, or when writing something by hand. The bullet point looks more formal, and is mainly used in printed materials. Both these symbols are called *raka*, which is also the Atong term for glottalisation or glottal stop.

The term *raka* comes from the word *raka* <rak=a> (strong=CUST) 'strong', and was adopted into Atong orthography by the author, because the same word is used with the same meaning in the orthography of Standard Garo.[2] Garo, like Atong, has glottalisation (see Burling 1992: 50), which is written with a so-called raised dot or the apostrophe in the standard orthography. For Atong speakers already literate in Standard Garo, it could be convenient to write glottalisation in a similar way.[3]

Certain interjections, proclauses and ideophones are phonologically different from the rest of the Atong words. In some words belonging to these word classes, the bullet is used to represent a glottal stop, and the combination hm is used to indicate the voiceless bilabial nasal [m̥], as in the interjection *hy•y•y•yw* [hə?ə?ə?əw] 'INTERJECTION OF REBUTTAL', the ideophones •*hm* [?m̥] 'grunt!' and •*hmmmm* [?m̥:]'sigh!', the proclause •*mhm*• [?mhm̥?]'that's right, no'.

---

**2** For a succinct history on Standard Garo orthography, see Burling (2003b: 387 and 1961: 8–9). Standard Garo is defined here as the variation of Garo taught in school in Meghalaya State. Burling (2003b: 387) remarks that "[t]he northeastern dialect on which the written language is based is sometimes called 'A•we'". The Atongs call this standardised form of Garo *Ha•chik*, which is also the general exonym for many people – but not the Atongs and Rugas – who speak one of the mutually intelligible varieties of Garo.

**3** Standard Garo is one of prestige languages in Meghalaya (together with Standard Khasi, and English). Garo and English are the languages of education for Atongs living in Meghalaya, especially or those living in the Garo Hills (see also van Breugel 2014: 17).

# Abbreviations of grammatical categories

| | |
|---|---|
| 1PL | first person plural |
| 1PL/EXCL | plural exclusive |
| 1SG | first person singular |
| 2SG | second person singular |
| 3PL | third person plural |
| 3SG | third person singular |
| AC | attributive clause |
| ACC | accusative |
| ADD | additive |
| ADV | adverbial |
| AFF | affirmative |
| ALT | alternative |
| ATTR | attributive |
| CLF | classifier (in interlinear glosses) |
| Clf. | classifier (indicates which classifier is used with a certain headword) |
| COS | change of state |
| CT | contrastive topic |
| CUST | customary aspect |
| DECL | declarative |
| DISTR | distributive |
| DST | distal demonstrative |
| EMHP | emphatic |
| FACT | factitive |
| FOC | focus |
| GEN | genitive |
| GOAL | goal enclitic |
| IDEO | ideophone |
| IFT | imperious future |
| IMP | imperative |
| IMPEMPH | imperative emphasiser |
| IRR | irrealis |
| LOC | locative |
| MIR | mirative |
| MOB | mobilitative |
| NEG | negative |
| NP | noun phrase |
| OWN | possessive enclitic |
| PL | plural |
| POS | emphatic positive |
| PROG | progressive |
| PROX | proximal demonstrative |
| Q | interrogative |
| QP | quantifier phrase |
| QUOT | quotative |

| RECP | reciprocal |
| SEQ | sequential |
| SPEC | speculative |
| TOP | topic |
| UNC | uncertainty modality |
| WHILE | concomitant action |

# List of abbreviations of types of headword

| | |
|---|---|
| *adj1.* | adjective Type 1 |
| *adj2.* | adjective Type 2 |
| *adv.* | adverb |
| *autoclf.* | auto-classifier |
| *bound.* | bound root |
| *clf.* | classifier (word class) |
| *coll.* | collocation |
| *conj.* | conjunction |
| *dem.* | demonstrative |
| *disccon.* | discourse connective |
| *dtw.* | deictive time word |
| *encl.* | enclitic |
| *evsp.* | event specifier |
| *expr.* | idiomatic expression |
| *ideo.* | ideophone |
| *intens* | intensifier |
| *interj.* | interjection |
| *interr.* | interrogative word |
| *khjyks.* | khathajyksai |
| *n.* | noun |
| *num.* | numeral |
| *postp.* | postposition |
| *ppron.* | personal pronoun |
| *procl.* | proclause |
| *prof.* | indefinite proform |
| *prtcl.* | particle |
| *qtf.* | quantifier |
| *sfx.* | suffix |
| *tw.* | time word (non-deictic) |
| V | verb, used in definitions of event specifiers in the Atong-English dictionary, e.g. -ang 'V away etc.' Instead of the V, a semantically appropriate verb can be inserted, e.g. byt- 'to drive etc.' The result will then be bytang- 'to drive away'. |
| *v.* | verb |
| *vB.* | Primary-B verb (transitive verb that may take goal-marked clausal complements)[4] |
| *vgoal.* | verb that takes a Target argument with the Goal enclitic <=na> |
| *vintr.* | intransitive verb |
| *vØ.* | verb that cannot take any argument, i.e. with zero valency |
| *vpan.* | verb with a prototypically associated noun |

---

4 Primary-B verbs are verbs where "all arguments can be NPs or pronouns but one argument can alternatively be a complement clause" (Dixon 2006: 9). These contrast with Secondary verbs, "whose arguments cannot all be just NPs or pronouns. That is, one argument must be a clause" (Dixon 2006: 9). (See also van Breugel 2014: 464–468.)

| | |
|---|---|
| *vph.* | phasal verb |
| *vs1.* | intransitive verb which can only take one specific argument |
| *vsec.* | Secondary verb (takes goal-marked clausal complements)[4] |
| *vtr.* | transitive verb |

# Labels for the semantic domains of nouns in the Atong-English dictionary

| | |
|---|---|
| ABSTR | abstract noun |
| ACT | human activities, results of or circumstances related to such activities |
| ANIM | animals |
| ART | artefacts including materials and tools used in their production |
| BODY | body parts of humans and animals |
| CORP | diseases or substances produced by the body |
| EMO | emotions and feelings |
| FOOD | food items, ingredients used for food |
| GEO | geographic, geological or natural phenomena |
| KIN | kinship terms* |
| MSRE | measure terms |
| PERS | persons, designations for people or groups of people |
| PLACE | places, landmarks and points used for orientation |
| PLANT | plants and parts of plants |
| QUANT | quantities |
| SHAPE | geometrical and other forms and shapes |
| SUBST | substances (not those used for food or for the production of artefacts) |
| SUPER | ghosts, spirits and other supernatural beings |
| TIME | time expressions |

---

* For each kinship term (except those denoting pairs or groups family members, and are labelled *set*), it is indicated whether it can be directly derelationalised (*drel*), whether it is classificatory (*c*) or descriptive (*d*) whether it can be used referentially (*ref*) or as a term of address (*a*), and whether it can be use reciprocally (*rec*). Note that all address terms can also be used referentially, but not vice versa. For an in-depth description of Atong kinship terms, see van Breugel (2020). Words that are only classificatory when used with their first meaning intended, but descriptive when another meaning is intended, are labelled $c_{(1)}$.

# Symbols

| Symbol | Name | Function |
|---|---|---|
| = | equal sign | enclitic boundary, or indicates that a translation is an enclitic |
| + | plus sign | compound boundary |
| - | hyphen | suffix boundary, or indication that the root is usually followed by another morpheme, or indicates that a translation is a suffix |
| ~ | tilde | separates different forms of a word, i.e. allomorphs |
| § | section sign | is used instead of the word *section* |
| ... | ellipsis | in the Atong-English dictionary, indicates that the constituent words which make up a collocation are attested separately, i.e. used in separate phrases or clauses. |
| [...] | square brackets | in English translations, indicate information that is not stated in the Atong sentences, but either needs to be inferred in the context in which the Atong sentence occurred, or needs to be stated to make the translation coherent or grammatically correct. |

# PART 1: ATONG – ENGLISH DICTIONARY

## What do we see in the Atong-English dictionary?

When we read the Atong-English dictionary, we see that there are words in **bold face**, followed by an abbreviation in *italics*. The words in bold face are called the headwords. A headword and its related text, which explains, translates and exemplifies the use of the headword, is called an entry. Entries start at the margin of the text. Following a headword is an abbreviation in italics, which indicates what type of headword it is, or what word class the headword belongs to. The list of abbreviations can be found in the prologue.

Each word in Atong belongs to a certain word class, and each word class has different grammatical properties. This information is not important if you just want to know the spelling or translation of a word; it is only important if you are interested in the grammar of Atong. Word classes and their grammatical properties are explained in the book *A Grammar of Atong* (van Breugel 2014), which was mentioned above, in the introduction.

When the headword is a noun, indicated by the abbreviation *n.*, we can see that there is another abbreviation in CAPITAL LETTERS following the abbreviation *n.*. This abbreviation in capitals indicates the semantic domain of the noun. Semantic domain has to do with the meaning of the noun. This is particularly useful when there are two nouns with the same spelling but different meanings, for example:

**bytym** *n.* ABSTR. a good smell
**bytym** *n.* CORP/FOOD. fat (of human or animal), grease

The first noun is an abstract noun (ABSTR), which is translated as 'a good smell'; whereas the second noun has to do with the a substance produced by the body of a human or animal (CORP), or with food (FOOD), and can be translated as 'fat (of human or animal)' or 'grease'. This word can be used to refer to fat on the body of a human or animal, and can also be used to refer to fat or grease used in the preparation of food, or as the ingredient of certain dishes.

Some entries in the dictionary have sub-entries. Subentries also have headwords in **bold face**. These sub-entry headwords are related to one or more major entries. Subentries do not start at the margin, but are in line with the indentation of the rest of the text belonging to the headword of the main entry. Consider the entry of the word **si-** 'to starve':

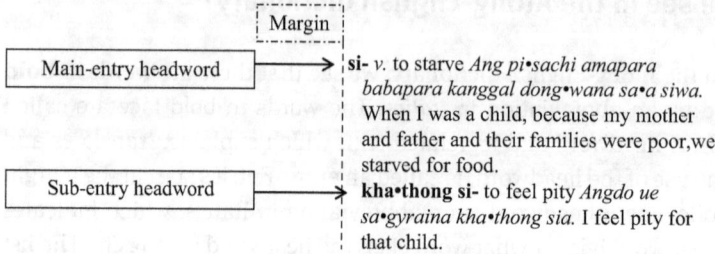

**Figure 1:** Example of a heading with a sub-heading.

The headword **si-** starts at the margin. The entry has a sub-entry, viz. **kha•thong si-** 'to feel pity', which does not start at the margin, but has the same indentation as the rest of the text belonging to the main-entry headword **si-**.

When collocations (see PART 5, §5) are attested in the same phrase, i.e. directly next to one another, they are considered to be *kathajyksai*[5] 'married words'. *Kathajyksai* are presented as headwords in the dictionary, with the abbreviation *khjyks*, as well as with an indication of their word class, as for example:

**gam jym** *khjyks, n.* wealth, riches, fortune

Collocations of which the member words are not attested in juxtaposition, or of which the constituent words are only attested marked separately (by phrasal enclitics) are not given a separate entry in the dictionary, but are presented as sub-entries with the words that form the collocation, as is the case with the collocation *song... nok...*, as we can see in the following example, which is the complete entry of the headword *song*. The same collocation will also occur as a sub-entry with the headword *nok* 'house'. As we can see, the words that make up the collocation are followed by an ellipsis '...', to indicate that they are attested in separate phrases, each with their own phrasal marking, or in separate, but juxtaposed, clauses.

Headwords can be preceded by an equal sign '=' or a hyphen '-', are bound morphemes, which means that they do not occur as separate words in a sen-

---

[5] Note that the word *katha* can be pronounced and thus spelled in different ways, viz. *kata ~ khatha ~ khata ~ katha*. Moreover the term *kathajyksai* is a term which Atong speakers themselves use when talking about collocations. The term is analysed as *katha+jyksai* (word+married.couple).

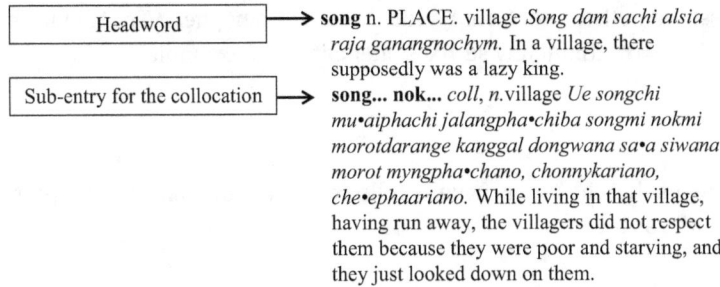

**Figure 2:** Example of a sub-heading for a collocation.

tence, but are rather attached to some other word. For example, the headword =**cha**, which indicates negative polarity on predicates, can be used in *sa•=cha* (eat=NEG) 'don't/doesn't eat' and **-gak,** which is an event specifier meaning 'accidentally' can be used in *sa•-gak=ok* (eat-ACCIDENTALLY=COS) 'has/have accidentally eaten.

Headwords with an equal sign are called enclitics (see van Breugel 2014: 257 and 386 for definitions). Their meaning or function is given in brackets after the abbreviation indicating their headword type. After the brackets, there is a full stop. This configuration separates the grammatical information from the translations and examples. The entry of the imperative enclitic *=bo,* is illustrative.

=**bo** *encl.*(imperative modality). *Sa•bo!* Eat!

If an enclitic has several distinct functions or interpretations, these are then numbered as, as in entry of the locative enclitic.

=**chi**₁ *encl.* (locative). (1) marks Spatial Locations. (2) marks Temporal Locations. (3) marks Temporal Location Clauses. (4) marks Conditional Clauses.

Most of the enclitics cannot be translated with just one word in English. In these cases, when the reader wants to know their exact meaning, it can be found in van Breugel (2014: 257–268, 324–348 and 386–415).

Headwords preceded by a hyphen are suffixes (see van Breugel 2014: 369–375). As with the enclitics mentioned above, the meaning or function of a suffix is given in brackets after the abbreviation indicating their headword type, e.g. the reciprocal suffix *-ruk,* whose dictionary entry looks like this:

**-ruk** *sfx.* (reciprocal). *Na•nange song jan•rukok.* Our villages are very far apart from each other.

Verbs are presented in their root form,[6] and followed by a hyphen.[7] The root form is the form of the verb without any suffixes or enclitics, for example:

**sa•-** *v.* to eat

The hyphen means that the root is usually followed by another meaningful part: a suffix or enclitic; as in:

| | | |
|---|---|---|
| *sa•bo* | <*sa•=bo*> | (eat=IMP) |
| *sa•manok* | <*sa•-man=ok*> | (eat-already=COS) |
| *sa•thokok* | <*sa•-thok=ok*> | (eat-EVERYONE=COS) |
| *sa•chaka* | <*sa•=cha=ka*> | (eat=NEG=IFT) |
| *sa•khuni* | <*sa•=khu=ni*> | (eat=MORE=UNC) |
| *sa•ak* | <*sa•=ak*> | (eat=COS) |
| etc. | | |

In this dictionary, whether a verb is transitive or intransitive is only indicated for very few verbs, the reasons being that many verbs are ambitransitive, and the valency of many other verbs is not know with certainty. For more information, the reader is referred to van Breugel (2014: 351–354).

Some headwords have more than one form or spelling. This can be because different forms are used by different speakers in different villages, or in different grammatical constructions, or simply because different forms can be used without any limitations. Different forms (or allomorphs) or spellings of the same headword are separated by the tilde '~', for example:

**mi•mang ~ me•mang** *n.* SUPER. ghost, spirit of a dead person.
**chang•kui ~ cheng•kui ~ chaw•kyi ~ chaw•ki** *n.* ART. big knife

---

[6] In the light of the lexicographic tradition of Standard Garo, the prestige language of the Garo Tribe, and closely related to Atong, presenting verbal headwords as roots is unusual. In Garo dictionaries and lexica, whether they are presented as headwords, or occur as translations of English headwords, verbs always occur with the bound morpheme <-*a*> at the end (see, for example, Members of the Garo Mission of the American Baptist Missionary Union 1905; Nengminza 1978, 2001; Burling 2004a and b). During my research on Atong in the field, when Atong speakers translated English verbs into Atong during elicitation, they did not always say the verb in the same inflected form. Therefore, and because there is no such category as the infinitive in Atong, I think that it does not make sense to arbitrarily take an inflected form as the entry form for verbs in this dictionary. Hence the root is given as entry form in this dictionary, as was the practice in the previous editions (van Breugel 2009b, 2015c).
[7] Atong verbs do not undergo paradigmatic stem alternation, which means that the form of the root remains the same, regardless of the grammatical environment in which it occurs.

When a word has more than one form, all forms can be found in their own alphabetical place in the dictionary, i.e. they all have their separate entry. This means that there is a separate entry in the dictionary that starts with the headword **me•mang** 'ghost', and another entry, which starts with the alternative form of this word, i.e. **mi•mang**. This way, the different forms can be looked up separately by speakers who usually use just one of them. However, example sentences are only given in one of the entries, so as to save space in the book. The reader is directed to the entry where examples are presented by a notice in brackets, e.g.

=**odo** ~ =**do** *encl.* (topic). (see =*do*)

After the headword and the abbreviations, we find the English translation of the headword. Sometimes, an Atong headword cannot be translated by another word in English. In these cases, a description is given, for example:

**sot** *n.* ANIM. species of very small fly that comes out in the evening and at night and cause itchiness

Sometimes, an Atong headword can be translated by one or more other words in English, but the translation needs some further explanation. For example, the translation of the headword **dam ~ dym** is 'bam!' or 'thud!' followed by the explanation "sound of something heavy hitting the ground":

**dam ~ dym** *ideo.* bam! thud! the sound of something heavy hitting the ground

In most cases, the English translation of a headword is followed by one or more example sentences in Atong to illustrate its use. These Atong examples sentences are written in *italics*. Each example sentence is followed by its English translation in regular font. In example sentences illustrating the use suffixes and enclitics, the suffix or enclitic is underlined to draw attention to it, since it does not occur as a separate word in the sentence, and might otherwise be difficult to find. An example of an enclitic headword with an example sentence is:

=**wa** *encl.* (factitive). *Bisang re•eng__wa__?* Where do you come from?

We can see that after the headword, the abbreviation in *italics* indicates that it is an enclitic. After that, the label of the enclitic is given. In the above example, the label is 'factitive'. Since this is a dictionary and not a grammar book, the labels of enclitics are only explained in the most succinct manner, if at all. Most importantly, example sentences are given to illustrate the use of an enclitic and its translations in all its different grammatical functions. If the reader wants to know the grammatical details, they are referred to *A Grammar of Atong* (van Breugel 2014).[8] In that book, in Appendix 2, when the reader looks up the enclitic headword =**wa**, they will see an indication of the chapter or section in the book where the enclitic is explained in detail, like this:

=**wa** *encl.* (See Chapter 24) factitive enclitic

In very few cases in this dictionary, the scientific name of a plant or animal species is given in the translation of a headword. In these cases, the scientific name is presented in italics, for example:

**palengma** *n.* PLANT. *barebina xariegata*, tree with beautiful white flowers that smell very nice, like magnolia, and are edible

When the translations of an Atong headword are synonymous, only one of the translations will be illustrated with an examples sentence, since the different, synonymous translations are interchangeable in the English sentence. Only when different translations of the same Atong headword are less synonymous do they merit their own examples.

When the translation of an Atong headword is polysemous, its various meanings are indicated with letters between brackets. One such case is the translation *hard* of the Atong headword **rak-**, copied here below. First, the word *rak-* itself is polysemous; hence it has three translations indicated with numbers between brackets, viz. (1) hard, (2) strong, and (3) loud. Second, the translation *hard* is polysemous in English, and three of its uses are recorded in the translations of Atong sentences. Each different meaning of *hard* is indicated with a letter between brackets, viz. (a) (of materials), (b) (with great effort), and (c) (difficult). Each different meaning is followed by one or two example sentences, which have been left out here to save space, but which can be found in the dictionary.

**rak-** *adj1.* (1) hard. (a) (of materials). (b) (with great effort. The translation depends on the context in English). (c) (difficult). (d) (performed with force or vigour. The translation depends on the complement of *rak-* used in the sentence) (2) strong (of natural phenomena). (3) loud.

---

[8] At the time of publication of this volume, *A grammar of Atong* is available in Meghalaya for reading only at the District Library in Baghmara, the University Library of NEHU Tura Campus, the library of Don Bosco new College in Tura, and at the Don Bosco or Sacred Heart Mission in Shallang.

To recapitulate, here are three examples of entries from the dictionary with all the different parts indicated separately.

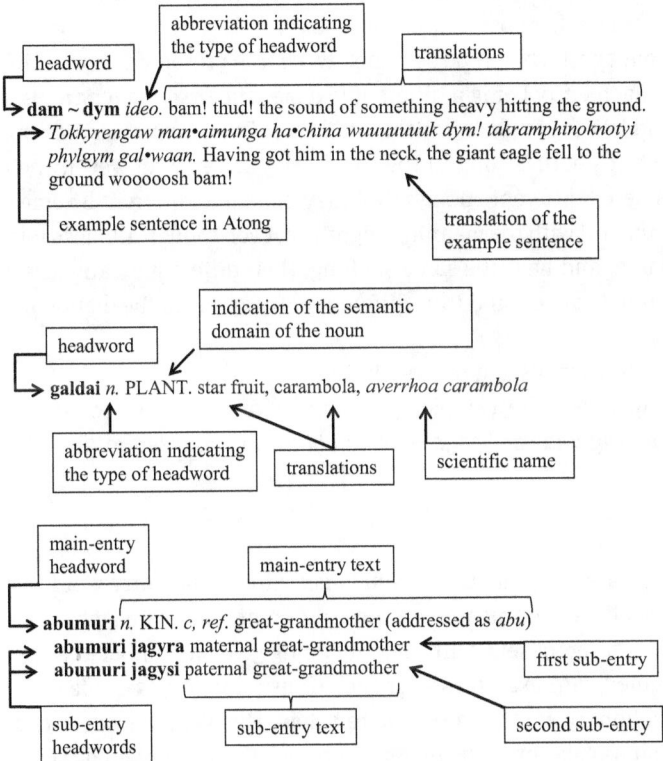

**Figure 3:** Example of dictionary entries with labels for their components.

When the headword is an enclitic or suffix, its function may be indicated before possible translations are given. The enclitic or suffix is underlined in the examples sentences illustrating its use. For example:

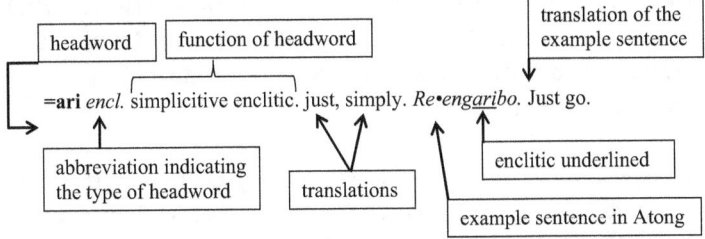

**Figure 4:** Example of an enclitic headword.

Sometimes, two or more entry words sound the same, and are spelled the same, but belong to different word classes, or have very different, unrelated meanings. We can thus say that these entry words belong to different, albeit homophonous or homonymous lexemes. In those cases, the entry words all have a subscript number to keep them apart. As can be seen below, *chat* is presented as four different entry words, each entry being a different lexeme. The first one, **chat-**$_1$, is a Type 1 adjective, and has the English translation *thick (of substances and things), bulky*. **Chat-**$_2$ and **chat-**$_3$ both belong to the word class of verbs, however, their meanings are considered different enough that they belong to different lexemes. Finally, **chat**$_4$ is a numeral with the meaning 'eight'. So, even though all these four words sound the same, and have the same spelling, their different meanings set them apart as different lexemes, and thus different entry words in the dictionary.

**chat-**$_1$ *adj1.* thick (of substances and things), bulky
**chat-**$_2$ *v.* to be fixed together (like a stapled pile of paper or a pile of wood etc.)
**chat-**$_3$ *v.* to promise. *Ang una sot bonga hyn•na chatwa.* I promised to give him fifty rupees.
**chat**$_4$ *num.* eight

In other cases, words are polysemous. This means that a single entry word has two or more semantically related meanings. These different, but related meanings are then presented in different sets, with each set being given a number between brackets, and presented with examples after each translation. An example is the verb *jasa*, whose dictionary entry is copied here below. This verb has two sets of meanings, or English translations. The first set of translations, numbered (1), are synonyms, near synonyms, or very closely related in meaning in English, viz. *to wake up, to get up, to get out of bed*. The second set, numbered (2), consists of only one English verb, namely *to realise*. However, the meanings in both sets (1) and (2) are related, as *realise* is a semantic extension of the notion *wake up*. An example sentence is given directly after each set of translations.

**jasa-** *v.* (1) to wake up, to get up, to get out of bed. *Angdo manapmi jasachawate.* I won't wake up/get up/get out of bed very early in the morning! (2) to realise. *Jasaaimyng chaiwachido biskut ni•okno.* When he realised what was going on, there were no more biscuits.

When nouns in Atong are quantified, i.e. when you say how many of something there are, or how much there is, most of them are used with a classifier (see PART 5, §4). For those nouns for which the classifier is recorded, it is given after the abbreviation Clf. (for *classifier*), followed by a full stop and an example of the quantification of the headword. For example, the classifier used with the noun *tyikhal* 'river', is *chol*. Hence, in the article to the entry word *tyikhal* 'river', the abbreviation Clf. is followed by the classifier *chol*, which is followed by an example of the quantified noun, as we can see in the entry copied below. The order is always Noun – Classifier – Numeral.

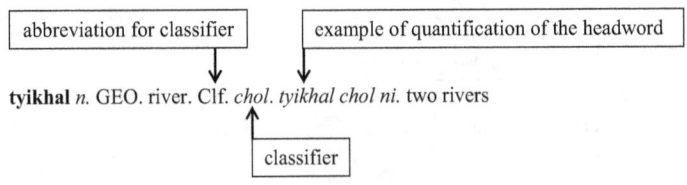

**tyikhal** *n.* GEO. river. Clf. *chol. tyikhal chol ni.* two rivers

**Figure 5:** Example of te use of a classifer in an entry.

In this dictionary, the appropriate classifier is only given for nouns for which they are recorded. Even though the classifiers of many nouns can be predicted based on their denotation, e.g. all nouns denoting animals will use the classifier for animals, *mang*, many nouns in the dictionary are left without information about what classifier they are used with. More fieldwork is needed to establish for each of these nouns what its classifier is.

When a noun is an auto-classifier, it does not need an extra classifier when it is quantified. In those cases, the numeral directly follows the noun. These nouns are labelled *autoclf.* after the indication of their word class. An example of an auto-classifier is the noun $san_1$, of which the entry is copied here below.

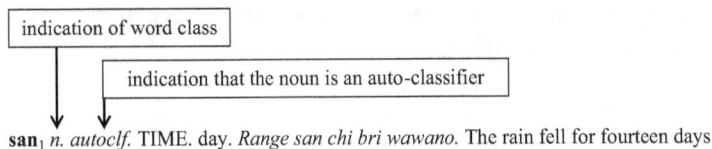

**san₁** *n. autoclf.* TIME. day. *Range san chi bri wawano.* The rain fell for fourteen days.

**Figure 6:** Example of the use of an auto-classifier in an entry.

Loanwords are indicated by mentioning the donor language(s) between brackets.[9] In those cases, the *lesser than sign* '<' will precede the name(s) of the donor language(s). If the word is clearly of Indic[10] origin, but the precise speech variety

---

9 There may be loanwords in this dictionary which the author has failed to recognise as such, due to his limited knowledge of the donor languages.
10 Indic languages are those belonging to the Indic branch of Indo-Iranian subgroup of languages within the Indo-European language family (see Clackson 2007: 6).

could not be established, the word is simply marked as being of Indic origin, like this: < Indic. The following entries, are illustrative.

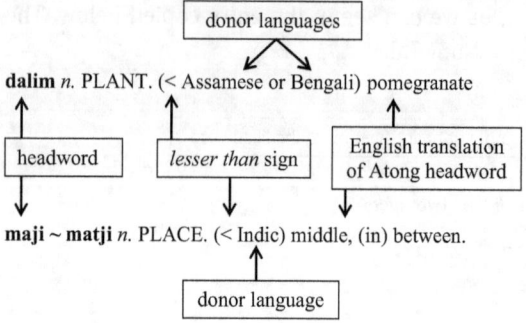

Figure 7: Example of a loanword entry.

Most of the Atong language materials presented in this book were recorded by the author during fieldwork. Some quotes from songs popular during the author's fieldwork are also used as examples in the Atong-English dictionary. Unfortunately, these songs were all published in India at an unknown date on cassette tapes which are long lost. Much to the author's regret, it is therefore impossible to provide complete references to these works. The texts of two of these songs, by Mr Wilseng S Marak and Mr Aristo J. Momin, were written down with the help of my friends, and are published in van Breugel (2019: 471–479). Other quotes come from the author's unpublished field notes. Quotes from these artists are indicated in this dictionary by putting their names in brackets, in small font, after the quoted passage, as in the following entry, where a quote from Mr Wilseng S Marak is used (see van Breugel 2019: 471).

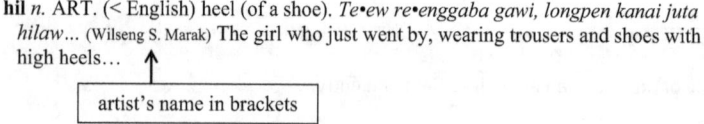

Figure 8: Example of a citation in an entry.

Two final remarks are in order. First, any word or expression in this dictionary containing the enclitic =*gaba*, can also be written with the short form of this enclitic, viz. =*ga*, for example *chepgaba* or *chepga* 'prisoner'. Second, when the reader cannot find a word starting with k, p or t, they are advised to see if it starts with kh, ph or th.

# Atong-English Dictionary

•

For words starting with a *raka*, IPA transcription is provided between square brackets.

•**hm** [ʔm̩m̩] *ideo.* grunt!
•**hmmmm** [ʔm̩ː] *ideo.* sigh! hmmm. *Myng•korokawan monokaimu kynsangdo, utymdo suksuk jywarokno, phylgym pipuknyng•chi. "Ah, •hmmm" noram•takaroknote pipuknyng•chian.* Having swallowed all six of them, after that, they were sleeping comfortably, inside the eagle's stomach. "Aaah, hmmm", they were inadvertently uttering, inside the stomach.
•**mhm**₁ [ʔm̩m̩ʔ] ~ **hm•m** [m̩ʔm] ~ **m•m** [m̩m̩ʔm] ~ •**m** [ʔm] ~ **mm** [mː] *procl.* I disagree. (This proclause is used to disagree with what someone else has said.) *"Mamung dywanchate angdo" nookno. Uchie: "Hm•m ama nang•do tai•nido atongba dywwa.* I haven't added anything", she said. Then: "Yes, you have, mother. Today, you've added something. *"Nang• re•engbo!" "Mm angdo gongcha."* "You go!" "No, I don't want to."
•**mhm•**₂ [ʔm̩m̩ʔ] ~ •**mhmm•** [ʔm̩m̩ʔ] *procl.* That's right. Proclause that is used to agree with a negative proposition. *Ai soke sokchak •mhmm•, soke sokchak.* Oh! I've lost, that's right, I've lost. (Literally: As far as succeeding is concerned, I don't succeed anymore, that's right, as far as succeeding is concerned, I don't succeed anymore.)

## A

**-a**₁ ~ **-ai** ~ **-e** *evsp. V* towards (the deictic centre). *Nang•do hambun isang rai•aphinwachi, jykthangaw bytaibone.* Later, when you come back here, bring your wife, OK.
=**a**₂ *encl.* (customary aspect, indicates that something is/was usually or always the case). *Kan•jotano Ketketa Burae.* Ketketa Bura is skinny. *Phasgaba ha•haw•chenga. Umungsa ha•haw•aimungsa wa•cham tan•a. Umungdo marsja somaidarangchi saw•a.* First, the jungle is cut. Then, having cut the jungle, the old rice stalks are cut. Then, in the month of March, [the dry wood] is burnt. *Ning songsyrekdo, ning Atongdo, dakangdo mamyng thorom ni•wami somaichido waiaw mania.* We heathens, we, the Atongs, in the past, in a time that there were no denominations, worshipped spirits.
**aak aak** *ideo.* caw, caw! the call of the crow
**abek** *n.* ART. long hollow drinking spoon made of a dried fruit with a hole at the top and a hole in the side of the bulge at the bottom used to scoop rice beer (*chyw*) out of the filter (*janti*) which stands in the middle of a large earthen pot (*gora*) filled with fermented rice (*sithi*). The spoon is held by its long slim part and slowly lowered into the liquor in the filter. The liquor seeps into the spoon though the hole in the bulge. Then the spoon is carefully pulled out again, and the liquor is drunk by emptying the spoon in the mouth through the opening in the top of the slim end without touching it with the lips. (see Photo 1, Photo 2 and Photo 3.)
**abi** *n.* KIN. *c, ref, a.* (1) elder sister (2) Mothers-in-law can address each other as *abi* too.
**abong** *n.* PLANT. corn, maize
**abu** *n.* KIN. *c, ref, a, rec*.* (1) grandmother. (2) Also used to address a granddaughter. (3) Also used to address an unrelated elderly woman. *This word is only reciprocal when used as a term of address between grandparent and granddaughter.
**abumuri** *n.* KIN. *c, ref.* great-grandmother (addressed as *abu*)
    **abumuri jagyra** maternal great-grandmother
    **abumuri jagysi** paternal great-grandmother
**abun** *adj2.* (1) next, following (2) neighbouring (3) other, others, other people (4) someone else. Also occurs as quantifier after classifiers.
**achepchep** *n.* ANIM. species of cricket

**achi-** *vintr.* to be born. *Ang Dajongchi achiwa.* I was born in Dajong.

**achok** *adj2* useless.

**achu** *n.* KIN. *c, ref, a, rec\**. (1) grandfather. (2) Also used to address a grandson. (3) Also used to address an unrelated elderly man. *achu ambi* grandparents, ancestors. (4) Also used to talk about or address an elephant when one is in the jungle, because then the word *mongma* is taboo. *This word is only reciprocal when used as a term of address between grandparent and grandson.

**achuambi** *n.* KIN. *set* (1) grandparents (2) ancestors

**achum** *n.* PLANT. species of plant

**achumuri** *n.* KIN. *c, ref.* great-grandfather (addressed as *achu*)

    **achumuri jagyra** maternal great-grandfather

    **achumuri jagysi** paternal great-grandfather

**achuthangmaran** *n.* KIN. *set* a grandfather and his grandchild

**ade** *n.* KIN. *ref, a.* stepmother

**agal** *n.* GEO. forest fire

**aganggi** *n.* ANIM. species of mantis (see Photo 83)

**aganggi gawrai** *n.* ANIM. species of mantis (see Photo 84)

**agos** *n.* TIME. (< English) August

**agre ~ agrai** *adv.* too much. *Umynggymynchi thogigaba morotaw agrai bebe ra•dugana baino.* That's why you should not believe people who lie too much.

**aguk** *n.* ANIM. grasshopper (generic term for the species)

**agychi** *n.* PLANT. species of tree

**agynja** *n.* TIME. (archaic) November

**ah** *interj.* (1) oh! hey! Interjection to get someone's attention. *Uchie: "Ah abu! Ningdo phulis dong•chaba abu kyryina bai, ningsaba" nookno.* Then they said: "Oh, grandma, we are not the police, grandma. Don't be afraid of us". (2) oh? interjection to expresses surprise *Ama! Ah! tai•nido ja•bek thawokte. Na•a phangnado thawai rymcha, atongtykyi tai•ni ja•bek thawoksyi?* Mother! Oh! Today the curry is very tasty, really! You never cook this tasty. Why is the curry so very tasty today?! (3) oh! interjection to express irritation. *"Nang•do ja•bek ga•suaian rymok tai•nido ama" nookno. "Ah! Sa•ariboto jyrymjyrym" nookno.* "You've cooked a great curry today, mother", he said. "Oh just go on and eat quietly!", she said.

**-ai$_1$ ~ -a ~ -e** *evsp.* V towards (the deictic centre). *Nang• jykaw ning songsang byt<u>ai</u>bo.* Bring your wife to our village.

**=ai$_2$ ~ =e** *encl.* (contrastive or new topic). *Magachakdo: "Hai bai•siga biskut sa•khawna" noaidongano. "Hyt man•cha nan g•ba atong budi" nowano pheru<u>e</u>.* The dear said: "Come on, friend! I want to steak the biscuits." What?! No! What are you thinking?!" said the fox.

**=ai$_3$** *encl.* (emphatic positive or strong declarative modality). *Nang•tymdo phylgymaw nukchawakhon<u>ai</u>!* You might not see the eagle at all! *Ooo! Mykgythaldo dong•cha jywmangsama, ma nang• bimang sylw<u>ai</u>.* (Wilseng S. Marak) Ooo! Is it real or is it a dream, but your body is really beautiful!

**ai ai ai** *interj.* interjection to call a pig, Here, piggy piggy!

**aia ~ ai•a** *interj.* oh! wow ! oh no! jeez! hey! Interjection of surprise or grief. *Aia! dam nom•atema.* Jeez! That's cheap! *Aia! Dugaphinok bai•sigaba angaw thogigababa.* Hey! This is too much, friend who has betrayed me!

**aiai ~ aiaiai ~ aiaiaiai** *interj.* ouch! argh! oh no! good grief! Interjection of excitement or shock. *"U•ching ang cha•aw kakaimu thyi jokok" nookno. "Aiaiaiai! sambanggyri akaibo", nookno.* A leech has bitten me on the leg, and blood has come out", he said. "Oh good grief! Pluck some *sambanggyri*", he said.

**aiaw** *interj.* urgh! argh! wow! ouch! damn! huh?! Interjection of surprise, disappointment, anger or pain. *"Ie molaaw ryngbo," noai hyn•okonaro. Myng• sa ryngai chyiokno. "Aiaw! Angdo man•chawa achu! Phekni ido, phekni."* "Smoke this tobacco", he said and gave it to them. One of them tried to smoke it. "Wow! I can't, grandpa! It makes my head spin." *Aia dugaphinok bai•siga, nang•do angaw thogiwado, tyiba karanok neng•ba neng•ok angdo aiaw!* Urgh! This is too much, friend; you have betrayed me, and I am very thirsty and exhausted, damn! *Rakai thetokno.* "*Aiaw chak chotni!*" *nookno.* He pulled hard. "Ouch! You'll rip off my arm!", he said.

**aidia** *n.* ABSTR. (< English) idea, plan

**aiding** *n.* ACT. hopscotch (a children's game)

**=aidonga ~ =aidong ~ =aidok ~ =aronga ~ =arong ~ =arok ~ =edonga ~ =edong ~ =edok ~ =eronga ~ =erong ~ =erok** *encl.* (progressive or durative aspect). (1) progressive interpretation. "*Ata<u>ka</u>rong nono?*" "*Nok ryphi<u>ero</u>k.*" "What are you doing, younger sister?" "I'm plastering the house." (2) durative interpretation. *Angdo tung<u>ai</u>dongate.* I'm hot! *Awangdo tai•samian ichi mu•aidong.* Uncle was just here.

**ain** *n.* ABSTR. custom, law, tradition

**ain niam** *khjyks, n.* ABSTR. laws, customs, traditions, culture

**aina** *n.* ART. (< Indic) mirror. *Aina chaiwachi phalthangaw nuka.* When I look in the mirror, I see myself.

**aiu** *interj.* wow! huh?! hey! oh! oh no! Interjection of surprise or grief. "*Aiu! Ido biphagabamyng syn takwa*" *nowano.* "Hey! I smell a husband's smell", she said.

**aiy** *interj.* interjection of inquiry and surprise: what are you doing?!

**aja** *n.* KIN. elder sister

**ajam-** *v.* to yawn

**ajip** *n.* ART. fan

**ajot** *n.* ACT. children's game played with two groups of unlimited size. Between the two groups sits a person called the 'king'. Children from both groups have to whisper the name of a child from the other group into the king's ear, first a child from one group, then a child from the other group. If two children whisper the same name, the king will call "Ajot!" and the person whose name has been whispered is out. The group that is depleted first loses.

**ajuju** *n.* SUPER. ghost that is a person without legs or which looks like a monkey. It jumps from tree to tree and has a long tongue with which it can hit people, who then melt.

**ak-**$_1$ *v.* to pluck (leaves, fruit etc. not feathers), to pick (flowers)

**=ak**$_2$ **~ =ok ~ =k** *encl.* (change-of-state). (see =*ok*)

**akai** *n.* KIN. *c, ref, a.* (1) aunt: mother's elder sister (2) Also used to address an unrelated married woman older than the speaker.

**akaithangmaran** *n.* KIN. *set* my *akai* (mother's elder sister) and her younger sister's child

**akal** *n.* ACT. famine.

**akan** *n.* ART. wooden rack above the cooking fire used to put meat to dry and to store utensils and baskets (see Photo 15)

**akyrudygyl ~ akrudygyl** *n.* PLANT. pumpkin

**alabok** *n.* ANIM. cattle egret, white heron, *Bubulcus ibis* (species of bird)

**alaga** *adj2/n.* PERS. other/another (person/thing), somebody else, different. *Alaga morotna dymdym damdam hyn•na bai.* Don't just give it to someone else. *Te•ewdo saphawdo wel•ang wel•ang alaga khalaw jalaakno.* Now, the rabbit ran away, zigzag, zigzag, to another hole. *Ytykyisa dyngthang dyngthang songchina hapchina jalthokna ga•akoknowa.* So then, they were compelled to run away to different villages, to different places.

**alamyla** *adv.* (1) a little bit, just a bit, somewhat. *Alamyla neng•takbo.* Rest a bit. *Alamyla na• mang sa mang ni man•awanochym.* He might have gotten a fish or two. (2) ordinary, normal. *Cha•phong gylgabasano, kara khyrynggabararasano, alamylachagabasano.* Strong arms, all tight veins, no ordinary [lads].

**ali** *clf.* classifier for small heaps or piles of things. *narang ali tham.* three piles of oranges

**alsia ~ halsia** *n.* PERS. (< Indic) lazy person

**althu•-** *adj1.* easy

**alu** *n.* PLANT. (< Indic) potato

**alukotar** *n.* ART. (< English) helicopter

**alupren** *n.* ART. (< English) aeroplane

**am ~ ym** *procl.* affirmative, okay. Proclause used to acknowledge another person's statement. *"Ichi taw• banok" nookno. "Ym. Raw•bo" nookno.* "There's a bird trapped here" he said. "Ok, catch it", he said it is said. *"Ang jywchengnine" "Ym."* I will go to bed now." "Okay."

**ama** *n.* KIN. *c, ref, a, rec\** (1) mother (biological or classificatory) (2) Also used to talk about or address a maternal aunt. (3) Also used to address a daughter. \*This word is only reciprocal when used as a term of address between parent and daughter.
  **amapara** *n.* PERS. the people in mother's household.

**amak** *n.* ANIM. macaque, monkey

**aman** *n.* ART. pestle, heavy wooden pole used for flattening rice in an *asam* by pounding. Clf. *goi•. aman goi• ni.* two pestles

**Ambi Chakkhen** *n.* SUPER. When a *kyrydyl* (liana, hanging root) turns into an old lady at night, this lady is called Ambi Chakkhen. The old lady has very long arms and hands with very long nails. She will ask you to scratch her long arm and if you refuse, she will scratch you to death with her long nails. (see Photo 75)

**Ambi Jakbyryt** *n.* SUPER. ghost that is only a hand. It scratches the back of someone walking in the dark.

**ambisuthyk** *n.* ANIM. species of gold-coloured metallic beetle that flips itself back on its feet when it lies on its back (see Photo 85)

**ambithangmaran** *n.* KIN. *set.* a grandmother and her grandchild

**ambret bambret** *n.* ACT. children's game

**ambyrai** *n.* PLANT. species of tree. The small green fruits of this tree look like gooseberries, but are hard and bitter. When you eat them and then drink water, the water tastes sweet. (see Photo 72, Photo 73 and Photo 74)

**amu** *n.* PLANT. species of tree

**amul** *n.* PLANT. species of tree

**=an** *encl.* (focus/identity). *Magachakmi myn•do tyisiwachian miniksuru takjolarianoro.* The deer's fur, when it is whet, just gets flathaired quickly. *Hai, ang ganang. Angan Raja. Angan balthumni.* Come on! I am here. I am the king. I will speak on your behalf.

**anai** *n.* KIN. *c, ref, a.* aunt: father's sister (purely referential term *nai•*)

**anaros ~ kewa** *n.* PLANT. (< Bengali or Assamese) pineapple

**-ang**$_1$ *evsp.* (1) *V* away (from the deictic centre). *Te•ewdo gasamok. rai•na man•anchak. Phetangna dakang walnaka.* Now it has become evening. We can't go anymore. It will certainly be night before we arrive. *Iskyn jan•gaba songsang de•theng jalangok.* He has run away to such a far-away country. *Chabi bichi tanangok?* Where did you put away the key? *Ga•thyngaimuna thyl•angok.* Because he kicked it, it went very far. *Ang ie ha•thapyraaw te•en nygylsang raangaimyng phalni.* I will bring these ashes to the market later, and sell them. (2) *V* without holding back, *V* away, *V* strong(ly), *V* extensively, *V* affluently. *Tai•ni balwa rakai balwaangok.* The wind blew strong today. *Kynsang golphook golphook golphook. golpho kha•wachie walangaidok.* Later on they talked and talked and talked extensively. When they talked, it was becoming night.

**ang**$_2$ *ppron.* I, me, my (first person singular). *Ang nang•aw balni.* I will tell about you. *Na•a angaw tyngkhucha.* You don't know me yet. *Nang• ang noksang mai sa•na rai•abone, hampyi.* You come to my house to eat this evening, ok?

**anga** *ppron.* I (first person singular Agent or Topic. Note that the first person singular pronoun *ang* also occurs as Agent or Topic.) *Sotkat anga baletariok.* I just told a short version. *Dadapara sandijolni anga.* I will quickly search for my elder brothers. *Anga janggiba thyimanok.* I have already died. (Literally: 'As for me, my life has already died.')

**anyng** *n.* KIN. *c*$_{(1)}$, REF, A. (purely referential term *nyng*) (1) aunt: father's sister (2) sister-in-law: husband's elder sister

**aphap** *n.* PLANT. yeast used to make *sithi* 'fermented rice' (see Photo 5)

**aphubawbyl** *n.* PLANT. species of tree
**aphut** *n.* PLANT. species of tree
**apun** *n.* ART. fishing hook
**apunkara** *n.* ART. fishing line
**apunphong** *n.* ART. fishing rod
**aragong** *n.* PERS. a person who is too big for his age
**=ari** *encl.* (simplicitive modality). just, simply. Re•engaribo. Just go. Dong•arini. That's all right. / That's okay. / No worries.
**aro**₁ *conj/disccon/adjective.* (1) and. *pheru aro magachak.* the fox and the deer. *Angna bunduk hyn•etbo, aro angna churiba hyn•etbo.* Give me a gun and give me also a knife. (2) more. *Aro ja•bek hyn•bo.* Give more curry. (3) another. *Ytykyimudo uan, ge•theng, sagaba: "angdo nemkhalancha" nochido aro kamalsang thama chaithiria.* So then, as for that [sick person], if [he] says: "I am not better", [they] will practice divination again at the place of another priest. (4) furthermore, moreover, in addition. *Uchi jalaisa, uchi song mu•aiwa, te•ewchinakhyngkhyng Rongdyng Ha•wai myngjolawa.* Aro rang wagababa *Gandyrungchi* ... That's why this place is still called Badri Rongdyng Ha•wai, it is said. Moreover, some place where the rain falls at Gandyrung ...
**=aro**₂ ~ **=ro** *encl.* (declarative modality). I'm telling you! *Phalthang kanga baguaw awan awan de•thengba ichi dykymchi wenoknoro.* He forgetfully wound his loincloth around his head, I'm telling you! The use of this enclitic in Atong does not always need to be translated into English.
**asa** *n.* PLANT. species of tree
**asakotoi** *n.* PLANT. species of tree
**asalchong•** *n.* ANIM. species of black hairy caterpillar that lives on jackfruit trees
**asalja** *n.* TIME. (archaic) (< Indic) June
**asam** *n.* ART. mortar, big heavy hollowed log in which rice is flattened by pounding it with an *aman*. Clf. *pan. asam pan byryi* four mortars (see Photo 40)
**asel** *n.* PLANT. species of tree
**aset-** ~ **asyt-** *v.* (1) to throw away, to dispose of. *Kereng nokkynchi asetaribo.* Just throw the bones away behind the house. (2) to divorce *Alsia rajae jyk asetai jalangwachie songreangokno.* The lazy king divorced his wives and travelled away.
**asi** ~ **asyi** *n.* KIN. *c, ref, a.* aunt: mother's younger sister (purely referential term *syi*)
**asingja** ~ **asyngja** *n.* TIME. (archaic) (< Indic) September
**aski** ~ **askhui** ~ **askui** *n.* GEO. star
**asok** *n.* ART. (1) type of woven bamboo basket to keep live pigs in to sell at the market (see Photo 31) (2) type of fish trap that is suspended on the edge of a waterfall
**asol**₁ *adv.* really, truly
**-asol**₂ *evsp.* really, truly. *Udo gari rakasolai bytaimu galatwa.* He drove his bike really fast and fell.
**asu** *n.* PLANT/BODY. thorn, fishbone
**asyi** ~ **asi** *n.* KIN. *c, ref, a.* aunt: mother's younger sister (purely referential term *syi*)
**asyngja** ~ **asingja** *n.* TIME. (archaic) (< Indic) September
**asynthalak** *n.* ANIM. species of fish
**asyt-** ~ **aset-** *v.* (see: *aset-*)
**ata** *n.* FOOD. flour
**atak**₁ *n.* PLANT. species of tree
**atak-**₂ *v.* to do what? interrogative verb. *"Aiaw! Angdo chykaidonga." "Atakwa?" "Te•ewmangmangsa tyiruwa na•a."* "Oh! I'm cold!" "What have you done? / What happened?" "I just took a bath, man!" *Kynsangdo atakoknowa•? Jamjolai gopcha amakawe.* What happened later? / What did they do later? They didn't burry the monkey at all.
**atakna** why? *"Na•a atakna jumuaidonga ie ha•thapyraawe?" nowano.* "Why are you collecting these ashes?" they said.
**atakgaba** what kind of? What type of? What sort of? *"Atakgaba raja na•a angna gore lapchagabaaw watetwa" nookno.* What kind of king are you that you send me a good-for-nothing horse?!
**atak... adong...** *coll, v.* what's happening? what's going on? *Bytnaan san sa wal sa phetachawana dong•achawana, unaai songmi nokmi morotdarang "Ataksyrangok, adongsyrangok?"* Because they had been

pulling one day and one night, and did not succeed, then, the villagers said: "What is happening? What's going on?"

**atakai ~ atykyi** *interr.* how? *Ie alsia raja atykyi khengaidok? Atykyian jykaw haldunna man•aidok?* How does this lazy king live? How does he feed his wives?"

**atakna** *interr.* why? *Atakna jalwa bai•sigadyrang?* Why are you running away, friends? *Atakna rai•acha?* Why aren't they coming?

**atha** *n.* MSRE. half

**athamphang** *n.* GEO. the planet Venus, the morning star, the evening star

**athom** *n.* BODY. stomach

**Atong**₁ *n.* ART/PERS. the Atong language, Atong person

**atong**₂ *interr.* what?

**atongba** *prof.* something. *Seino na•a te•ew nang• songsang re•engchido angna atongbaaw ra•bone.* Seino, if you now go to your country, buy me something, OK? Also attested with the accusative enclitic in a different position: *Ge•thengna atongawba hyn•bo.* Give something to him.

**atongtykyi** *interr.* how? how come? why? *Atongtykyi angawe tyngsawnaka?* How will he recognise me? *Atongtykyi tai•ni ja•bek thawoksyi?* Why is the curry so tasty today?!

**atykyi ~ atakai** *interr.* how? "*Ie alsia raja atykyi khengaidok? Atykyian jykaw haldunna man•aidok?*" *noai morotdyrang chanchiphinaidoknoro.* "How does this lazy king live? How does he feed his wives?" thought the people.

**atyw ~ atyyyw** *interj.* interjection of surprise (usually pronounced long). "*Na•nang garu ramgachi phylgym di•etdapai tanangwa*", *nookno.* "*Atyyyw! Ido di•an thawokona, randaido atongtykyi thawarongnaka?!*", *nookno.* "An eagle shat in our dried must ard leaves", she said. "Whaaat?! If this shit is so very tasty, imagine how tasty the meat will be?!" he said.

**=aw ~ =taw** *encl.* (accusative enclitic, indicates definiteness, and various semantic roles). (1) indicates the semantic role of Patient. *Ranustaw nukama nukan-cha?* Have you seen Ranus or not? *Jemi sanchi Dibangkongdangaw matsa kakok.* One day, Dibangkongdang was bitten by a tiger. (2) indicates the semantic role of Material. *Panaw jap kha•aimu kawtawna thymokno.* Having made a defence wall out of wood, they lay in ambush to shoot upward. (3) indicates the semantic role of Road. *Ge•theng ramaw re•engaimyng nukokno.* He followed the road and saw it. (4) marks of the word *cha* 'foot/leg' when used as an instrument. *Samaw cha•aw itykyi tokano.* They trample the grass like this, with their feet. (5) marks the semantic roles of Locations and Sources that are already marked by *=sang. Usangaw rangpynramsangaw ha• haw•ai sa•airongno.* Over there, at Rangpynram they lived of the land. *Hyisangmimaw* from far away.

**awa** *n.* KIN. *c, ref, a.* biological father

**awan**₁ *n.* ART. winnowing basket. *Awanchi mai chawa.* Rice is winnowed in a winnowing basket. (see Photo 27, Photo 41)

**awan-**₂ *vtr.* to forget

**awan awan** absentmindedly, forgetfully. *Phalthang kanga baguaw awan awan de•thengba dykymchi wenoknoro.* He absentmindedly would his loin cloth around his head.

**awang** *n.* KIN. $c_{(1)}$, *ref, a.* (1) uncle: father's younger brother (2) what children call their stepfather

**awyi** *n.* KIN. *c, ref, a.* grandmother (archaic in Badri and Siju)

# B

**=ba**₁ *encl.* (1) indefiniteness. *Nokchi ang dyngdang mu•chiba, sungmaneta anga nang•na.* (Aristo J Momin) Whenever I am sitting at home along, I think of you. (2) addition. *Uchie bythyiba re•engok, machokba re•engok, magachakba re•engok.* Then, the porcupine went and the small deer went and the big deer went. (3) contrastive topic enclitic. "*Angdo usangmi paranggaba*" *noano, biphaba.* "*Na•a bisang rai•awa?*" *nowano, madamba.*

"I am a traveller from over there", said the lad. "Where did you come from?" said the teacher. (4) affirmative modality. *"Nemaidongama bai•siga?" "Nemaidongaba."* "Are you well, my friend?" "I'm well indeed." (5) emphatic modality. *Ue hapchi mu•wachi rangmu chyw ryngsusawanasa Ha•chykkhu•chuksang mykha badri noyi myngwano. Te•ewchina-khyngkhyngba Ha•chyksonggumukdo mykha badri noaria je rangawba.* When they lived at that place, because [they] held the drinking competition with the rain, it is called Mykha Badri in Garo. And indeed until now, all Garo villages just call any rain *mykha badri*.

**ba•-**₁ *v.* to be born, to give birth. *Sa• ba•ak.* The child is born, or She gave birth to a child. *Ang Sijuchi ba•wa.* I was born in Siju. *Angaw ama Sijuchi ba•wa.* My mother gave birth to me in Siju.

**sa• ba•na sa-** to go into labour. *Jyw•gado noksangdo oganangarok nookona, sa• ba•na saarokno.* The mother was at home, pregnant. She was going into labour.

**ba•-**₂ *v.* to carry a child in a cloth on the body. *Dada jojongaw ba•aidonga.* The older brother is carrying his younger brother. (see Photo 123)

**ba•**₃ *interj.* (1) interjection of hesitation. uh..., uhm...., OK then... *"Nang•ba re•engnima?" "Ba• ... Tyngkucha."* "Will you go too?" "Uhm... I don't know yet." (2) interjection to conclude a conversation, OK then (3) interjection to express mutual understanding, OK then. *"Angba re•engni." "Ba•, re•engbo."* "I'll go too." "Ok then, go."

**ba•sek** *n.* ART. cloth in which to carry a baby on the body

**baaa**₁ *ideo.* Moo! the sound a cow makes

**baaa**₂ *interj.* wow! interjection of astonishment. *Biskyn ra•wa, iaw?" "Rong hajal sot chet ra•wa." "Baaa! Damrakate."* "How much did you buy it for?" "I bought it for about ten thousand rupees." "Wow! That's really expensive!"

**baba** *n.* KIN. *c, ref, a, rec\*.* (1) father (biological or classificatory) (2) also used to talk about or address a paternal uncle. (3) also used to address a son. *This word is only reciprocal when used as a term of address between parent and son.

**babaji** *n.* PERS. (< Indic) fortune teller

**babelsi ~ babylsi** *n.* PLACE. kitchen

**babu** *n.* PERS. child or baby (used to address a small child or baby)

**Babyra** *n.* SUPER. supreme god

**bada** *n.* MSRE. a bunch

**badai-** *vtr.* to cross beyond the limit, to pass a certain point. *Changba ge•theng songmi baiaw badaiok.* Somebody crossed the border of his village. *Dolong khagabaaw badaiwachi, ramchi agal saw•gaba ganang.* When you will have passed the hanging bridge, there will be a forest fire along the road.

**badal-** *v.* to unfold *phul badala* flowers unfold. *Ang ie lekhaaw badalaidonga.* I am unfolding this paper.

**badolja** *n.* TIME. (archaic) August

**badym** *n.* PLACE. paddy field, wet rice field (see Photo 23)

**badyng-** *vtr.* to trade, to deal in, to do business in, to peddle. *Dakangmi chasongdo rangdarangaw, rykdarangawsa barudarangawsa badynga.* As for the past era/generation, they traded brass gongs, necklaces and all kinds of ornaments.

**bagan** *n.* PLACE. (< Indic) garden

**bagu** *n.* ART. cloth for man worn around the waist

**bagukhawa** *n.* ART. turban with a knot on the front side of the head

**baguriwa•** *n.* GEO. halo of the moon

**bai** *n.* PLACE. border

**bai•**₁ *n.* KIN. *ref, a.* blood relative, kin, friend
**bai• tyng** *khjyks, n.* blood relative.
**naw... bai•...** *coll, n.* blood relative (see *naw*)

**bai•-**₂ *vintr.* to break. *Balwa rakaimyng wa• bai•ok.* Because of the hard wind the bamboo has broken.

**bai•dam ~ baidam** *n.* PERS. some (people). *Ytykyisa dyngthangdyngthang songchina hapchina jaltokna ga•akoknowa. Bai•dam wa•thaigyrymchi mu•ok, uawdo wa•*

*thaigythym myngok. Bai•dam Rongsa tyikhalmi ha•waichina jalangok.* That's why they were forced to run away to different villages and different places. Some stayed in Wa•thaigyrym; that village is now called Wa•thaigythym. Some stayed in the plains of the river Rongsa.

**bai•khop** *n.* PLANT. species of big broad green and purple bean

**bai•maran chingmaran** *khjyks, n.* KIN. *set.* two distant relatives

**bai•maran** *n.* KIN. *set* two distant relatives.

**bai•sak** *n.* friend.

**bai•sakthangmaran** KIN. *set.* two people who belong to the same *mahari* 'lineage'

**bai•siga ~ bai•sega** *n.* KIN. ref, *a.* kinsman. Also used to address a friend.

**bai•wak-** *v.* GEO. land of broken-off rocks

**baibai** *adj2.* the same. *Ang hanep baibai kha•di khanphinni.* Tomorrow I will wear the same clothes again.

**baidam ~ bai•dam** *n.* (see *bai•dam*)

**baik** *n.* ART. bike

**baisykyl** *n.* ART. bicycle

**baji** *n.* TIME. o'clock. *Tiin baji dong•ok.* It's past three o'clock. Note that hours are counted with numerals borrowed from Hindi (see PART 5, §20).

**baju** *n.* PERS. friend

**bak-**$_1$ *vtr.* to make barren, to weed out all the plants, to scrape with a spade or chopper. *Khudalsang ha• bakwa.* We weeded the land with a chopper.

**bak-**$_2$ *vtr.* to run after someone or something, to chase after. *Kyi• ma•suaw kakna bakaidong.* The dog is chasing after the cow to bite it.

**bakbak ~ bykbyk** *adv.* quickly. *Bakbak rai•a!* Come quickly!

**bakdong** *n.* ABSTR. a forbidden marriage: a marriage between a man and a woman from the same *mahari*, for example M Sangma and M Sangma.

**bakdongmi sa• ~ bakdongmyng sa•** *n.* a child born out of a forbidden marriage.

**baket ~ baltin** *n.* ART. bucket

**baki** *n.* ABSTR. credit *Baki hyn•chawa.* I don't give credit.

**bakrukrak ~ bakrukylak** *adv.* quickly, fast

**baksubipha** *n.* PLANT. species of tree

**bal-** *v.* (< Assamese or Bengali) to speak, to tell, to say, to talk. *Na•a atong khu•chukaw balaidonga?* What language do you speak? "*Nang·mi jorae chang?*" "*Balchawa angdo.*" "Who is your lover?" "I will not tell." *Songgadalawdo Atongsang balchido song pidan dong•achym.* As for *Songgadal*, when you say it in Atong, it should be *song pidan*. *Balphabo, bai•siga.* Keep talking, friend.

**balai hyn•-** to give advice. *Bydyi myng•sa balai hyn•aimyng, baju takphinokno.* After an old man gave advice, they became friends again.

**balwami ~ balwamyng** story telling. *Angmi balwami ichian jametwa. Walnam.* I will finish my story telling now. Good night.

**baletwami ~ baletwamyng** speaking. *Te•ew iaw baletwami somaichi angmi bylsi sotchetsnimangsaai... sotsnisni dong•arikhua.* At the time of speaking, I am only 87... 77 years old.

**bala-** *v.* to hold in the beak. *Daw•kha mang sa pankhambaichi ruti balaai mu•arongnote.* A crow sits in the top of a tree holding a bread in its beak.'

**balaga** *n.* PLACE. outside. *Garu balagachi ramai tanaimuna, ha•basang ha• kamna re•engokno.* Having put the mustard leaves outside to dry, [she] went to the rice field to work.

**balgyto•** *n.* PLANT. orchid

**balmundyri** *n.* GEO. cyclone

**balphak-** *vtr.* to blow something away

**Balphakram** *n.* PLACE. land of the spirits of the dead and national park in the South Garo Hills District

**balpisa** *n.* PLACE. a place to piss

**balsem-** *v.* to talk very long

**balsyruk-** *v.* to whisper

**baltin ~ baket** *n.* ART. bucket

**balwa**$_1$ *n.* GEO/SUBST. wind, air

**balwa-**$_2$ *vs1.* to blow (of the wind)

**bam-**$_1$ *v.* to brood, to sit on an egg. *Taw•kurungchi bamaidong.* The chicken is brooding in her nest.

**bam-**₂ *vgoal.* (1) to bend one's head. *Ha•sang bamai, chaksi jotai, chaksi phai•ai nemen chanchiaidongno.* With his head bent to the ground, fidgeting with his fingers and wringing his hands he was in deep thought. to obey, (2) to obey. *Ie sa•gyrai angnado bamcha.* This child does not obey me. (3) to do willingly. *Amak ge•thengdo rong•pelang sylgabachi kepleplep bamai hyn•takkonoa.* The monkey, as for him, he willingly lay down on his hands and knees with his bum in the air on a flat stone. *"Re•engbo!" "Angdo bamcha!"* "Go!" "I don't want to!" (4) to surrender. *Arong nokmae Englanmi Britis gobormen sason ka•gabana bamchano.* Headman Arong did not surrender to the reign of the British government.
**bamkhup ~ bangkhylok-** *v.* to bend one's head. *Ang (dykymaw) bamkhupaidong pan thotniwana.* I am bending my head. because it will hit the wooden beam.
**ban-**₁ *vintr.* to flow (of rivers). *Symsang tyi Nongal dolongtakai banaidong.* The water of the Symsang river flows under Nongal bridge.
**ban-**₂ *vtr.* to trap, to catch in a trap. *Jaga saakno uchie, taw• pang•ai banokno.* They set traps and then caught many birds.
**bandaw ~ bando** *n.* ART. tree house (see Photo 8)
**Bandija•lang** *n.* SUPER. the tree that lies across the beginning of the road to Balphakram and that acts like a gate for the spirits of the deceased. This tree is also called Dykhija•lang.
**banga** *num.* five
**bangbang** *adj2/v.* empty. *Ang pipuk bangbangaidongkhua.* My stomach is still empty.
**bangbol** *n.* ANIM. species of fish
**Banggal** *n.* PERS. Bengali, Bangladeshi, non-tribal person
**bangganai** *n.* ANIM. species of fish
**banggyri- ~ banggiri-** *v.* earthquake. *Wa•darang bai•rumok banggyriaimu.* All the bamboos are broken because of the earthquake.
**banggyrigaba ~ banggirigaba** *n.* earthquake
**bangka ~ bangkha** *n.* ART. (< Indic) fan
**bangkhylok- ~ bamkhup-** *v.* to bend one's head. *Ang (dykymaw) bangkhylokaidong pan thotniwana.* I am bending my head. because it will hit the wooden beam.
**Banglades** *n.* PLACE. Bangladesh
**bangphak** *n.* ART. vertical posts at the entrance of the bachelors' house between the floor and the horizontal beam above the entrance
**bangsi** *n.* ART. flute
**banthai** *n.* PERS. marriageable boy or man, bachelor
**baphai-** *vtr.* to drop
**baphe** *n.* ANIM. species of large gecko
**bapre ~ baprebap** *interj.* wow! interjection of astonishment.
**bara-** *vtr.* to put in a hole, pan, *wa•sung*, bag etc. *Ang bostachi goi baraairong.* I am putting betel nuts into the bag.
**-barai** *evsp.* always V
**baram-** *adj1.* rough
**baranda** *n.* PLACE. veranda
**barangsi** *n.* PLANT. species of grass
**barat**₁ *n.* ART. lace that pulls the skin of a drum tight
**barat-**₂ *vgoal.* to be ashamed, to be shy. *Nawang na•a! Ang nang•na barataidong.* You idiot! I am ashamed of you. *Ie gawi nang•na barataidong.* This girl is feeling shy towards you.
**barata** *n.* FOOD. (< Indic) paratha, flatbread. *barata phel sa* one paratha
**baratwami ~ baratwamyng** *n.* EMO. shame
**bari** *n.* PLACE. (< Indic) garden
**baru** *n.* ART. traditional ornament
**bas** *n.* ART. (< English) bus
**basak-** *vintr.* to burn and cause a rash to cause irritation or itching. *Thamat ~ thamotba na•jekwa•ba khi•chido basaka.* If you touch the *thamat/thamot* plant and the *na•jek* bamboo they cause irritation. *Thamat basaka. Ta pyi•!* The *thamat* plant causes irritation. Don't touch it!
**basneng•thakgaba** *n.* PLACE. bus stop
**basu** *n.* ANIM. crown feathers of a bird
**bat-**₁ *vtr.* to stick in, to plant by sticking a seed into a hole in the soil. *Kun ha•bykungchi batbo.* Stick the stick in the sand.
**=bat**₂ *encl.* (superlative degree). most, -est. *Uchi mu•tyngabae bylakbatgabae Arong*

*nokma dong•anowa*. The strongest leader was Arong Nokma.

**batdyl** *n.* ART. slingshot

**bathan** *adj2*. lying on one's back. *Bathanai juwbo*. Lie down on your back.

**batkhynyng-** *vtr.* to smash. *Ang botolaw batkhynyngok*. I smashed the bottle.

**baton** *n.* ART. (< English) button

**batphai•-** *vtr.* to throw hard to break something. *Ang botolaw batpai•ok*. I threw the bottle very hard and broke it.

**batpyret-** *vtr.* to smash by throwing something to the ground

**batro ~ u•chingrawi ~ ukchingrawri** *n.* ANIM. species of brown leech that lives in the soil and mud

**baw•-** *vtr.* to dry: to make jerky, to dry vegetables. *Randaiaw baw•aidong*. I am making jerky.

**bawang** *clf.* length of the widely stretched arms and hands

**bawbaw** *ideo.* woof! woof! the sound a dog makes

**bawbyl chambyl** *khjyks, n.* PERS. enemy

**bawbyl** *n.* PERS. enemy

**bawen**$_1$ *n. autoclf.* SHAPE. circle

**bawen-**$_2$ *v.* to move in a circle, to make a circle around something, to encircle, to be rolled up. *Ie nokaw atakna bawenaidong?* Why are you walking around this house? *Dypyw bawenai mu•aidonga*. The snake is lying rolled up. *Pan pang•a, morotaw bawena*. There are many trees, they encircle the man.

**bawili ~ bawyli** *n.* ART. pliers to put logs on the fire

**bawra** *n.* EMO. arrogance

  **bawra tak-** to be arrogant

**beanbebe** *adv.* truly

**bebe** *adv.* truly

  **bebe ra•a** *vtr.* to believe. *Nang• me•mangaw bebera•ama ganag noai?*. Do you believe in the existence of ghosts? *Ang nang•aw bebe ra•cha*. I don't believe you.

**bebylokmai** *n.* PLANT. species of tree

**begyri** *n.* ANIM. cow fly

**bejaw-** *vintr.* to experience the sensation of being tickled. *Nang• angaw thebajawwa,* *ang bejawok*. You tickled me and I feel tickled.

**bek** *n.* ART. bag

**bel•- ~ bil•-** *vtr.* to retract the foreskin from the glans penis. *Nang• ri•aw bel•bo*. Retract your foreskin!

**belcha** *n.* ART. spade (see Photo 26)

**benek-** *v.* to damage

**beng** *n.* PLACE. bank. *Bengmi tangka sarawni angdo*. I will borrow money from the bank.

**bengblok** *n.* ANIM. toad, species of frog

**bera** *n.* ART. a fence

**beraberi ~ dengdyl** *n.* ART. loosely woven bamboo mat used as fence of the balcony or veranda of the house

**beraw-** *adj1.* to contain too much soda (MSG) (said about the taste of food)

**berawri** *n.* species of tree that can be used for firewood and to make posts for houses. The fruits can be used as a spinning top toy

**bering**$_1$ *n.* FOOD. food cooked in a *wa•sung*

**bering**$_2$**-~ bereng-** *vtr.* to cook in a bamboo tube (*wa•sung*) which is sealed with banana leaves and placed in the fire

**beringwa ~ berengwa** *n.* FOOD. food cooked in a *wa•sung*

**betyri** *n.* ART. battery. *Clf. thong. betyri thong• byryi*. four batteries

**bewal**$_1$ *adv.* for some time, a while. *Magachakdo bewal rypaimyng phetaakno*. The deer, having stayed under water for some time, emerged.

**bewal**$_2$ *n.* ABSTR. tradition, habit. *Te•ewtykyi badynggaba dong•cha dakangmi janggi khengwami bewaldyrangdo*. Nowadays, there is no trade in the old traditions.

**bi-**$_1$ **~ bie.** *interr.* (see *bie*)

**-bi**$_2$ *evsp.* intensifier suffix. very, very much, severely. *Banggal myng• sa biskut chyrymbiai paiaidonganote*. A Bengal was carrying a very heavy load of biscuits. *Ketketa Burae duk man•biai pheruna balsakoknowa...* Ketketa Bura, very much struck with grief, answered the fox... *Nang•tymaw mythelbiok*. Thank you very much.

**bi•chamchym** *n.* ART. fragment, bit, small piece

**bi•thyn ~ pi•thyn** *n.* BODY. liver
**bia** *n.* ACT. wedding
   **bia kha•-** to marry, to have a wedding.
   **bia kha•ak** to be married
**biambong** *n.* BODY. biceps
**biawthang** *n.* KIN. *ref, a.* the relationship between a male and the husband of his sister's daughter (*namgaba*)
**biawthangmaran** *n.* KIN. *set.* my wife's elder brother and me together
**biba**₁ *interr.* when? *Na•a bibasa rai•ani?* When exactly will you come?
**biba**₂ *n.* CORP/SUBST. breath, vapour, steam
   **biba jokgaba dam** *expr., n.* a wound
**bibasa** *adv.* wherever
**bibyrokhon ~ bibakoron** *adv.* some day
**bichi** *interr.* where?
**bichiba** *adv.* (1) sometimes. *Bbichiba gisep gisep chiti saietrukarinaka.* We will sometimes write each other letters. (2) somewhere, nowhere. *Ang chabiaw bichiba thagal•ok.* I dropped my key somewhere. *Bichiba ni•wa chabido.* The key is nowhere. (3) never. *Angdo ie biskutaw bichiba ra•cha.* I never buy those biscuits.
**bichiba bichiba** *adv.* sometimes, seldom
**bichylap** *n.* BODY. abdominal membrane
**bida** *n.* a wise person
**bie ~ bi-** *interr.* (1) modifier function. which? *Sam manama. Bie same?* The medicine stinks. Which medicine? (2) pronoun function. which one?, which ones? *Ytykyimu biaw mykchana ytykgarangawe?* So which one(s) am I supposed to fancy, the one/those who do like that? (3) predicative function. where? *Bie nang• jongdyrange? Nang• jonge bie?* Where is your younger brother? Your younger brother, where is he?
**bigaba ~ biga** *interr.* which? what kind of ...? *Bigaaw biskut ra•nima?* Which biscuits shall I buy?
**biji** *n.* ART. injection, injection needle
   **biji su•-** to give an injection. *Clf. phong. biji phong ni.* two injections, two injection needles
**bijyrang-** *vtr.* to hang to dry. *Kha•di bijyrangbo.* Hang the clothes to dry.
**bikha** *clf.* surface of 80 by 80 *pit*

**bil•- ~ bel•-** *vtr.* to retract the foreskin from the glans penis. (see *bel•-*)
**bilding** *n.* ART. (< English) house built with masonry, building (see Photo 12)
**bimang** *n.* BODY/ABSTR. body, appearance
**bimung ~ bimyng** *n.* ABSTR. name. *Angmi bimung Samrat myngwa.* My name is Samrat.
**biins** *n.* PLANT. (< English) species of green bean
**bipha** *n.* PERS. lad, man, male.
**biphagaba ~ biphaga** *n.* KIN. *ref.* husband
**biri** *n.* ART. (< Indic) cigarette
**bisang** *interr.* to where? from where?
**bisangba** *prof.* somewhere.
**bisangmi ~ bisangmyng** *interr.* from where?
**bisi** *n.* SUBST. poison
**biskut** *n.* FOOD. (< English) biscuit. Clf. *kep, phel biskut. kep/phel sa* one biscuit
**biskyn** *interr.* how much? how many?
**bistibal** *n.* TIME. (< Bengali) Thursday
**bisyl** *n.* ART. coin
**bitykyi** *interr.* by which way? *Bitykyi re•engnima? Ie ramtykyima utykyi?* By which way shall we go? By this road or by that one?
**=bo** *encl.*(imperative modality). *Sa•bo!* Eat!
**bo•rang** *n.* ART. tree house (see Photo 8)
**boba** *n.* PERS. crazy man, idiot, fool
**bobi** *n.* PERS. crazy woman, idiot, fool
**bobylawthok** *n.* PERS. fool
**bochi** *n.* KIN. *ref, a.* Only used in the Siju dialect (in the Badri dialect: *ja•chung*) sister-in-law: elder brother's wife
**bochithangmaran ~ buchithangmaran** *n.* KIN. *set.* my wife and her sister or brother together
**boda** *n.* PERS. ignoramus, dumbass
**bodol-** *vtr.* to change. *Nang• jama bodolbo. Manamok!* Change your shirt. It stinks! *Disk bodolbo. Ie jamok.* Change the CD. This one is finished.
**boiom** *n.* ART. a jug. Clf. *thai•. boiom thai• sa.* one jug.
**boisaja** *n.* TIME. (archaic) April
**bok-** *adj1.* white
**bokbok ~ bongbong ~ bong** (1) *v.* to lie, to tell lies. *Ta bongbong! / Ta bong!* Don't lie! (2) *n.* PERS. liar

**bol-** *v.* to burn (like the sensation of a being stung by a stinging nettle); to cause burning, irritation or strong itching. *Thamat bola. Ta pyi•!* The *thamat* plant causes burning. Don't touch it!

**boli** *n.* ART. offering to a spirit. *Songgumukan ue mongmawana wai khurutaisa boli hyn•aisa man•ai sa•thokwano.* Because the whole village prayed and offered to the elephant tusks, they all became very rich.

**bonduk ~ bondyk ~ byndyk** *n.* ART. gun, shotgun

**bonga** *num.* five. this word is only used in the compound numerals. *chi bonga* fifteen and *sot bonga* fifty.

**bongbong₁ ~ bong ~ bokbok** *v.* (1) *v.* to lie, to tell lies. *Ta bongbong! / Ta bong!* Don't lie! (2) *n.* PERS. liar

**-bongbong₂** *evsp. V* too much, *V* more than necessary, *V* in abundance, *V* scandalously much. *Sa•bongbong ryngbongbong.* We eat too much and drink too much.

**bonyng** *n.* KIN. *ref, a, rec.* (1) brother-in-law: the reciprocal relation between a man and his younger sister's husband or a man and his wife's elder brother (2) any man of another clan from the same generation as a male speaker

**borong** *n.* PLANT. cob, part of the fruit where the seeds are set in *abongborong* cob of corn. Jackfruit also has a cob which is called *panchungborong* jackfruit cob

**bosok-** *v.* to itch, to be irritated, to experience the sensation of irritation or itching. *Na•jekwa• khiaimu cha• bosokaidonga.* Having touched the *na•jek* bamboo my leg is itching.

**bosta** *n.* ART. big bag to transport things like rice and betel nut in

**bostu** *n.* ART. thing, things, stuff. Clf. *myng. bostu myng tham.* three things

**bot₁** *n.* ART. a gift

**bot-₂** *vtr.* to court, to woo, to flatter, to give a present to a girl after dating her. *Nang• Turachi nawmyl botwama?* Did you court the girls in Tura?

**botol** *n.* ART. bottle or its volume, bottleful

**breket ~ brekyt** *n.* ART. bracket. *Breketmyng nyng•chi chipgaba katha pang•ai gamchatcha.* The words in brackets are not very important.

**Britis** *n.* PERS. (< English) British.

**bu-** *adj1.* sharp (of pointed things)

**bu•chok-** *vintr.* sharp (of pointy objects)

**bu•chot** *n.* PLANT. mango

**buchithangmaran ~ bochithangmaran** *n.* KIN. *set.* my wife and her sister or brother together

**buchotpan** *n.* PLANT. mango tree

**budok budok** *ideo.* the call of the *daw•budok*

**budu** *n.* PLANT. creeper

**bugyryk** *n.* PLANT. species of vegetable

**bui-** *adj1.* murky, turbid *Tyi buia.* The water is murky/turbid.

**buk-** *vintr.* to grow like a creeper or liana

**bukalang** *adj2.* to have holes in it, with holes in it (of clothes)

**bukylek** *n.* ANIM. species of big grasshopper

**bul-** *vtr.* to dig up, to unearth, to stir. *Noksamsang khudal paiaimyng, tangkaaw bulai Thengthonna hyn•etokno.* Having carried a chopper to the side of the house, he dug up the money and gave it to Thengthon.

**buna** *n.* ANIM. big black and yellow flying insect, possibly a hornet

**burbok ~ bulbok** *n.* PERS. idiot

**burung₁** *n/clf.* a bush, classifier for bushes, or patches or clusters of trees. *Palengma burung bangabanga haw•waan pungphek phingano.* Having cleared an area of five *palengma* trees each, they filled up a granary each.

**burung₂** *bound.* a group. Kinship terms denoting a group of family members form compounds with this word, e.g. *haw•nokholburung* group of fathers-in-law (*haw•nokhol*) and sons-in-law (*kynokhol*). More fieldwork is needed to ascertain the productivity of this morpheme.

**busi** *n.* SUBST. dust that is stirred up by the wind or by human activity

**but₁** *n.* SUPER. spirit that leads you astray

**but-₂** *v.* to squeeze in, to penetrate, to go inside a hole. *Saphawba hang•khalnyng•sang* *butai jalangokno.* The rabbit runs away and squeezes into a hole. *Bandi palyng butangwachi matsa chunggaba gorongokno.* When Bandi penetrated the jungle, he met a big tiger. *Ne•kat wa• hang•khal nyng•sang butangaidonga.* The bees are going into the bamboo hole. *Hang•khalaw butaimuna jalangokno.* Having penetrated the hole, he ran away.

**butang** *n.* PERS. fucker (swearword)

**butbal** *n.* TIME. (< Bengali) Wednesday

**buthu-**$_1$ ~ **buthyw-** ~ **bythyw-** *vintr.* to boil (of water). *Tyi bythywaidok.* The water is boiling.

**buthu-**$_2$ *vtr.* to seal, to close a receptacle by putting something in the opening. *Wa•sung rekchaksang buthuok.* The bamboo tube is sealed with banana leaves.

**butsa** *n.* ANIM. species of big red ant

=**butung** ~ =**butuk** *encl.* (concomitant action). at the time, when, while. *Uchi mu•butung somaichi badri nemen man•ai sa•ano.* At the time they were living there, Badri was very rich. *Ge•theng sa•gyraibutungchiba sansan palyngsang na• punna re•engwa.* When he was a child, he went to the jungle every day to catch fish.

**bychym-** *vtr.* to pull up/out. *Una myng• sagaba sa•banthai myng• sagaba bychymokno, uchiba patangphaariok, dang•angphaariokno.* Then one son pulled the other out [from the water].

**bydyi** *adj2.* old (for persons)

**bydyi badai** *n.* PERS. old couple

**bydyi** *n.* PERS. old man

**byira** *n.* ANIM. cat.

**byira amanthong** *n.* ANIM. jungle cat (the pattern on the skin of this cat is in the shape of an *aman*)

**byirakhem** *n.* ANIM. species of bee

**byisa-** ~ **bysa-** ~ **bywsa-** *v.* to dance

=**byisyk** *encl.* (interrogative modality). how much? how many? *San byisyk mu•ni?* How many days will you stay? *Nang•chi rongbyisyk ganang?* How much money have you got?

**bykbyk** ~ **bakbak** *adv.* quickly. *Bykbyk rai•abo!* Come quickly!

**bykot-** *vtr.* to unsheathe, to take out. *Gal•aimuna kynsangdo phylgymaw uan rykjolaimuna kukuri byk hotaimuna tokkyrengaw tan•thongokno.* After it had fallen to the ground, he ran quickly to the eagle, unsheathed his knife and cut off its neck.

**bykphyl** *adj2.* inside out. *Nang• jama/chola bykphyl.* Your shirt is inside out.

**byl**$_1$ *n.* ABSTR/BODY. muscle, strength. *Ido sa•gyraido hambundo chungwachido ala myla byldo bylnikhon.* In the future that child might really become a bit stronger.

**byl... chak...** *coll, n.* arm/hand. *"Ha•, chamai Bandi, byl neng•chiba chak neng• chiba iaw ryngetphabo" noaimung hyn•aidongano So•reba Bandina.* "Take this, sweetheart Bandi, if your muscles are tired, it your arms/hands are tired, drink this" she said, and gives it, So•re, to Bandi.

**byl... jagydok...** *coll, n.* strength.

**byl chak** *khjyks, n.* strength

**byl-**$_2$ *vtr.* to cut and kill a big animal or person, to slay. *Rangsandi•mai phai rewetangwachian, Raka Motbandaaw byletwa.* When the sun was setting, he slew Strong Motbanda.

**byl•**$_1$ *n. autoclf.* ACT. strike. *Ma•su tan•na byl• sa nangni ge•thengo.* To slaughter the cow, he needs one strike.

**byl•-**$_2$ *vtr.* (1) to roll something into something (2) make a drum and cover it with hide. *Jeen sanchi morot thyiok. Umi sanchi uaw panaw gambiriaw gamsiliaw, gambiri phang sa gamsili phang sa kai•wano. Ytykyisa ge•thengthenge me•aphadyrang chanchiaimu uaw panaw khem byl•okno.* One day, a person died. On that day, those trees, a gambiri and a gamsili were planted. So then, they, the married men, thought and made drums out of the trees.

**bylak-** *adj1.* strong. *Uchi mu•tyngabae bylakbatgabae Arong nokma dong•anoa.* The strongest one who lives there is headman Arong.

**bylbang** *n.* ART. tie beam, horizontal beam that runs over several *manjuri* and forms the base beam of the triangle of the roof (see Photo 11)

**byldyng byldang** *khjyks, adv.* all over the place. *Uchisa matsana makbulna mongmana paichaaimung byldyng byldang jalna ha•bachengok.* Then, not bearing the bears and elephants anymore, they started to run away all over the place.

**bylet** *n.* ART. razor blade

**-bylok** *evsp. V* to a pulp *Berengwa mynwachido su•byloka.* When the stuff cooked in the bamboo cylinder is cooked, it is pounded to pulp.

**bylong-** *vintr.* to too much. *Na•a bylongdugaai thel•nabyi.* 'Don't lie too much. *Bylongok!* It's too much! / Scandalous!

**bylongen** *intens.* very

**bylongok** *interj.* (derogatory) So stupid! Unbelievable! *Bylongokte nang•do angaw taksakchagado.* You are so stupid if you don't help me.

**bylsi** *n. autoclf.* TIME. year. *bylsi thinian.* every year. *Ang kholachi bylsi dong•ok.* I am thirty years old.

**bylu** *adj2.* blue

**byri** *num.* four. this word is only used in the compound numerals *chi byri* fourteen and *sot byri* forty.

**byryi** *num.* four

**byrym** *n.* ART. inside rafter: beam that runs along the ridge board on the bottom of the roof on the inside of the roof and has the *kenchi* 'rafter' as its counterpart; together they form part of the support structure of the roof of a house (see Photo 9, Photo 11)

**byrymbyrym** *adj2.* multi-coloured

**byryp** *adj2.* lying on one's belly. *Byrypai juwbo.* Lie down on your belly.

**bysa- ~ byisa ~ bywsa-** *v.* to dance

**byt-** *v.* (1) to pull. *Odek ang khaw bytai thetok.* The baby pulled my hair and pulled it out. (2) to drive, to ride. *Na•a gari bytna sapama?* Do you know how to drive (a vehicle)? *Na•a baisykyl bytan sapama?* Do you know how to ride a bike? (3) to transport, to lead, to take. *Nang• jykaw isang bytbo!* Take your wife to come here. *Ma•su mang byryiaw bytangaimung sa•akno.* He led the four cows away, and ate them. (4) to guide. *"Bisang rai•khuni?" "Ang nang•aw bytangnaka."* "Where to go from here?" "I will guide you." (5) to draw. *Taw• bytna sapama, na•a?* Do you know how to draw a bird? (6) to shock (electricity). *Waiyr pyi•na bai. Karen bytnaka.* Don't touch the wire. You'll get a shock.

**mai byt-** to harvest rice. *Mai bytwamyngdo pungchina songchina khairata.* We carry the rice harvest down to the granary, to the village.

**bytai-** *vtr.* to lead here, to bring here (by driving)

**bytchirit-** *vtr.* to draw a line

**bytganggang-** *v.* to drive a vehicle over a bumpy road

**bythyi** *n.* ANIM. porcupine

**bythyn** *n.* GEO. shade. *Ichi mu•bo. Ichi bythyn gal•ok.* Sit here. There is shade here. (literally: The shade has fallen here.)

**bythyw-$_1$ ~ buthyw- ~ buthu-** *vintr.* to boil (of water). *Tyi bythywaidok.* The water is boiling.

**bythyw-$_2$** *v.* (1) to block, to be blocked. *Paip bythywok.* The water pipe is blocked. (2) to close the *wa•sung*, or similar containers, with *rai•chak* so that vapour cannot come out. *Wa•sung nemai bythywok.* The *wa•sung* is closed off well.

**bytjekjek- ~ bytjengjeng-** *v.* (1) to give short jerks, to pull jerkily (2) to draw or write scratchily

**bytphin-** *v.* to rewind

**bytphuruk-** *v.* to tear out with the roots. *Ge•theng wa•aw bytphurukwa.* He tore out the bamboo, roots and all.

**bytphyrak-** *v.* to tear apart

**bytruru-** *v.* to drag something

**bytsek-** *vtr.* to abduct, to kidnap, to take away a person, to steal a person. *gawi bytsekgaba* a person who steals somebody else's girlfriend

**bytsorok-** *vtr.* to pull out

**bytwa ~ bytwami ~ bytwamyng** *n.* ACT. harvest

**bytwami** *n.* ACT. tug-of-war

**bytym-$_1$** *adj1.* to smell nice. *Palengma bytyma.* The flower of the *palengma* smells nice.

**bytym**₂ *n.* GEO. a good smell

**bytym**₃ *n.* CORP/FOOD. fat (of human or animal), grease.

**bywsa- ~ byisa- ~ bysa-** *v.* to dance

# C

**cha**₁ *n.* FOOD. tea

**=cha**₂ *encl.* (negative polarity). (1) not. *Udo tyi hungna sapcha.* He does not know how to swim. "*Khasidarangdo noksamchi simen, tota, tin pirinai hama.*" "*Nokkhungchi?*" "*Nokkhungcha! Noksamchi.*" "For the wall of a house, the Khasis mix cement, planks and metal plates." "For the roof?" "Not the roof! For the wall." (2) it's not, it was not, they're not, they were not. "*Kha•rekma ie?*" "*Kha•rekancha, saman.*" "Is this a string-bean ?" "It's not a string-bean, it's weeds." (3) without. *Phywra syw•aimungna garu susetchaai dywetoknoai.* Having pounded rice powder, she added the mustard leaves without washing them, really! (5) not having, because ... not *Tyi ryngchaaimu, kha•ranok.* Not having drunk any water, she was thirsty. / Because she hadn't drunk any water, she was thirsty.

**cha•** *n.* BODY. leg, foot. *cha• kantara* barefoot

**cha•bykung** *n.* BODY. instep

**cha•chok** *n.* BODY. sole of the foot

**cha•choron- ~ cho•choron-** *vintr.* to squat. *Hap sakancha cha•choronai mu•bo.* Because there is not good place to sit, squat!

**cha•dok ~ cha•tok** *n.* BODY. heel

**cha•duk-** *v.* to bump. *Ang nang•sang re•rengwachi, ram manakaimu panchi/panaw cha•dukok.* When I went to your place, because the road was dark, I bumped into a tree.

**cha•dyl** *n.* BODY/PLANT. root, vein

**cha•dylmorong** *n.* PLANT. main root of a tree

**cha•dylsaphek** *n.* PLANT. small root

**cha•gang** *n.* bad rice that is thrown away in the husking process

**cha•godot-** *v.* to stumble. *Gandichi cha•godotwa.* I stumbled over a log.

**cha•gyl** *n.* ART. footstep.

**cha•gywgyw-** *vintr.* to kneel down

**cha•kereng** *n.* BODY. shinbone, shin

**cha•khawak ~ cha•khok** *n.* BODY. hollow side of the knee

**cha•khok ~ cha•khawak** *n.* BODY. hollow side of the knee

**cha•khop ~ ja•khop** *n.* ART. shoe

**cha•kok** *n.* PLACE. hollow between the roots of a tree. *Kynsandgo thik ue napite phepcha•koknyng•sang galatwa.* Later the barber fell exactly into a hollow between the roots of the banyan tree.

**cha•kyw ~ cha•ku** *n.* BODY. (1) knee (2) *clf.* length from the knee to the foot. *Saw•aidongano, thyw•angaidokno, cha•kyw chyigykdarangdo.* He is digging and he is getting deep, about nine knees deep.

**cha•ma** *n.* PLACE. lower side, downstream, bottom

**cha•machi** below, at the lower side of, down from

**cha•man** *n.* ART. footprint

**cha•masang** *n.* PLACE (1) downstream. *cha•masangmi wai* the downstream spirit. *Cha•masangmiaw dinggaraiaw na• chaiokno.Uchiba matdam sa•akno.* He inspected the fish trap(s) downstream for fish. (2) down. *Phepchi pheru ytykyi mu•aidonoaro. Mu•wachie ri•do chu•ret takangokno cha•masang napitsang.* The fox was sitting in the banyan tree like this. While he was sitting there, his penis was hanging down, toward the barber.

**cha•muk** *n.* BODY. medial malleolus

**cha•myn** *n.* BODY. leg hair

**cha•pa** *n.* BODY. sole of the foot

**cha•pakithyk** *n.* BODY. heel

**cha•pathai** *n.* BODY. calf

**cha•pha** *autoclf.* a foot *cha•pha tham.* three feet

**cha•phak** *n.* BODY. groin

**cha•pheret** *n.* CORP. crack in the callous skin of the heel

**cha•phong ~ cha•phung** *n.* BODY. thigh

**cha•phung** *n.* BODY. upper leg

**cha•pungdym** *n.* BODY. hip

**cha•ri** *n.* PLANT. seed for planting, paddy sprouts used for planting

**cha•ri pot-** to plant paddy
**cha•si** *n.* BODY. toe
**cha•sijyw•bydyi** *n.* BODY. big toe
**cha•sitokkyreng** *n.* CORP. eczema between the toes
**cha•syrong-** *v.* to stretch your leg
   **cha•syrongaimu mu•a** *v.* to sit with your legs stretched out
**cha•tok ~ cha•dok** *n.* BODY. heel
**cha•wek₁** *n.* ART. broom to sweep outside the house
**cha•wek₂** *n.* PLANT. chaff
**cha•wekdam** *n.* PLACE. place where the chaff is thrown after winnowing the rice
**chabak-** *vintr.* to fall (of water in a waterfall)
**chabi** *n.* ART. key
**chacha** *adv.* exactly, appropriately, just. *Ue chacha takai kechagaba changbano mykchaaidonga.* Someone might just be fancying someone whom it is not appropriate for this person to marry.
**chachak** *n.* PLANT. tea leaf
**chachakphang ~ chaphang** *n.* PLANT. tea plant
**chachakphang kai•-** to plant a tea plant
**chachek** *n.* ART. tea strainer
**chachura** *n.* BODY. hair on top of the head
**chagak₁** *n.* BODY. palate
**chagak-₂** *v.* to hit, to crash into. *Uchi, pherudo panchi chagakai thyiokno.* Then the fox hit a tree and died.
**chai-** *v* (1) to look at/into. *Aina chaiwachi phalthangaw nuka.* When I look into the mirror, I see myself. (2) to watch. *Angba tibi chaina.* I also want to watch TV. (3) to see. *Angba piktjyr chaina.* I also want to see the photos. (4) to spot. *So•redo Relwakmadareaw khymsawaimyng nokthaichi chairatai balwa ryngai mu•aimyng, khyryk chaisawaidongano.* So•re had married Relwakmadare, and in their small house, they were looking down, enjoying the breeze, and were fully occupied spotting lice. (3) to inspect. *Sathiriaimungna umi chaithirichiba, ba•, matdam sa•akno, aro kynsang ga•samsangphak chaithirichi uawba matdam sa•akno.* Having set out the traps, when he later inspected them again, well, the fish had all been eaten, and later, when he looked again in the evening, an otter had eaten them again.
**changbaaw chaikhu-** to take revenge on someone. *Ang nang•aw chaikhuni.* I will take revenge on you. *Nokha•palchi cha•gyl kyryngaidonga.* Footsteps are making noise outside.
**chaikhaw-** *v.* to spy (on), to peep
**chaira** *n.* ACT. a traditional song
**chairok-** *v.* to attend to, to take care of. *Ge•thenge alu kobi habijabi ytykyi samchakdarangmynggymyn bagan takwano. Ytykyi phangnan ge•thenge uaw chairokjyryngariano, tyi tytjyryngariano.* He made a garden with potatoes, cabbage and all sorts of vegetables. So he took care of them every day and watered them every day.
**chairura-** *vintr.* to look around you
**chairuru-** *v.* to look around
**chaisi-** *vtr.* to hate, to dislike, to be annoyed by something or someone
**chaithawa-** *adj1.* to be beautiful
**chaithum-** *v.* to guard, to watch over
**chaitumgaba** *n.* PERS. watchman
**chak₁** *n.* (1) BODY. arm, hand
   **byl... chak...** *coll, n.* arm/hand (see *byl*)
   **chak chok** *khjyks, n.* hand. *Tykywtokreng chak chok dangchagabachi mai rymetaidongano.* She is cooking rice in a water pot of which the neck is so narrow that you cannot stick your hand in it. (2) *autoclf.* PLANT. leaf. (3) *clf.* classifier for leaves. *rai•chak chak sa.* one leaf to pack food in.
**chak-₂** *v.* to ignite, to light
   **wal• chak-** to make fire, to light a fire, to kindle a fire. *Kumirian wal• chakthiriaimyng te•do na•lam garanawan rymthiriaidongano.* Kumiri rekindled the fire, and cooked the dried fish again.
**chaka** *n.* ART. wheel
**chakchuk** *n.* BODY. elbow
**chakgydok** *n.* BODY. wrist
**chakgytok** *n.* BODY. underarm
**chakkhawan** *n.* BODY. hollow part of the elbow, elbow pit

**chakkhop** *n.* ART. glove
**chakol** *n.* PERS. servant
**chakpha** *n.* BODY. palm of the hand
**chakphakhung** *n.* BODY. back of the hand
**chakphong ~ chakphung** *n.* BODY. arm, upper arm
**chakra** *n.* belongings
**chaksan** *n.* ART. bracelet
**chaksi** *n.* BODY. finger Clf. *goi•*. *chaksi goi•banga*. five fingers
**chaksigysep** *n.* BODY. space in between the fingers
**chaksijotram** *n.* BODY. index finger
**chaksijyw•bydyi** *n.* BODY. thumb
**chaksikhol** *n.* BODY. fingernail
**chaksikhum** *n.* BODY. back of the hand
**chaksirengma** *n.* BODY. little finger
**chaksiweng** *n.* BODY. knuckles
**chaksyrong-** *v.* to stretch one's arm
**chakwak** *clf.* classifier for handfuls. *rong•chakwak chitsa.* eleven handfuls of stones
**chal-**$_1$ *v.* to plant or sow by making a hole in the ground with a stick and putting the seed into the hole. *Ha• khynmanwamungsa maisi khita. Umung abongdarang chala, dachangdarang chala.* Only after collecting the unburnt remains of the jungle from the land, we sow millet. Then we plant maize and we plant sorrel.
**chal-**$_2$ *v.* to support
**chalak** *adj2.* cunning, clever. *Song dam sachi Thengthon mynggaba morot myng• sa ganangno. Ue bylongen chalakno.* In a village lived a man called Thengthon. He was very cunning.
**chalgaba** *n.* ART. a support
**chamai ~ chame** *n.* KIN. *ref, a.* (1) female cross-cousin: mother's brother's daughter or father's sister's daughter (2) the relation of female cousins from intermarriageable families (3) the relation of the parents of a married couple (4) lover, sweetheart
**chamaithangmaran ~ chamethangmaran** *n.* KIN. *set.* a couple of marriageable cross-cousins, a boy and girl who can marry
**chamchia** *n.* PLANT. species of tree
**chame ~ chamai** *n.* KIN/PERS. (see *chamai*)
**chamus** *n.* (< Indic) ART. spoon

**chanchi-** *vB.* to think (about/of). "*Ie alsia raja atykyi khengaidok? Atykyian jykaw haldunna man•aidok?*" *noai morotdyrang chanchiphinaidoknoro.* "How does this lazy king live? How does he feed his wives?" thought the people. *Bandiaw watetna chanchiaidokno.* He thought about sending Bandi. *Jesang ang re•engchiba, man•cha nang•aw awana, chanchia ang nang•awrarasa.* (Aristo J Momin) Wherever I go, I cannot forget you, I think only of you.
**chanchichyp-** *v.* to suppose, imagine. *Morot chanchichypai thik dongokodo, uchian rajaan uaw ajot nosawnaka.* Suppose someone gets it right, then the king will tell him *ajot*.
**chanchok-** *v.* to lean on
**chanchora ~ chanchura** *n.* ANIM. sparrow
**chanet-** *v.* to put on the fire
**chang**$_1$ *interr.* who?
**chang**$_2$ *bound.* multiplied by, times. This morpheme is only used in compound numerals with *khol* 'twenty'. It can be seen as a bound morpheme and written together with *khol* in numerals, viz. *khokchang byryi rong sa* eighty one.
**-chang**$_3$ *evsp. V* suddenly, to stop *V-*ing. *Rong bangamyng chinthai rong ni ra•aimyng songreangte songreangte. Kynsangdo jywchangna nangokno.* Having gotten two melons for five rupees, he journeyed and journeyed. Later, he suddenly had to sleep. *wa•tan•chang-* to cut the bamboo to stop it growing taller
**chang•ai** *n.* GEO. the moon
**chang•khet-** *v.* to be stuck
**chang•khui nagap ~ chang•khui kaldap** *n.* ART. type of big knife with a wavy blade (see Photo 26)
**chang•kui ~ cheng•kui ~ chaw•kyi ~ chaw•ki** *n.* ART. big knife with a curled blade used in the kitchen to prepare food, as well as in the field to cut plants and weeds (see Photo 26)
**chang•kuikatri** *n.* ART. type of big knife with blade that has a rounded hook at the end (see Photo 26)
**changba** *prof.* somebody, someone

**changchon** *n.* BODY. waist

**changgaba** *prof.* (1) someone. *Ang hapchido changba jywaidong.* Someone is sleeping in my spot. (2) whoever. *Changgaba man•ai sa•a changgaba nokdang takga, umi bimyng gumukawan thalai myngaimusa* ...Whoever is rich, whoever is wealthy, having called all their names clearly, ...

**chanpat-** *v.* to build a bamboo bridge

**chanpheng-** *vgoal.* to defend. *Ang ha•songna chanphengni.* I will defend my country.

**-chap₁** *evsp.* V along with, V so as to attach something. *Nokbanthaidyrangaw saw•aimung nokbanthai do•khakhuchi khachapai tangaba mongmawa dora byryi dong•gabaaw ra•ai jalangokno.* Having burnt the bachelors' houses, they took the 20 KG weighing elephant tusks which were tied to the *do•khakhu* and ran away.

**chap-₂** *v.* to stand (be in standing position)

**chapchap** *adv.* packed, close together (as in a crowd)

**chaphang ~ chachakphang** *n.* PLANT. tea plant

**chaphang kai•-** to plant a tea plant

**chara₁** *n.* PLANT. sapling

**chara₂** *n.* KIN. *set.* (1) wife's elder brothers (2) mother's brothers. The *chara* come together when important decisions concerning the *mahari* have to be made.

**charamong** *n.* KIN. (1) wife's eldest brother (2) mother's eldest brother

**charanga** *num.* fifteen

**chasong** *n.* ABSTR. generation, era.

**chat-₁** *adj1.* thick (of substances and things), bulky

**chat-₂** *v.* to be fixed together (like a stapled pile of paper or a pile of wood etc.)

**chat-₃** *v.* to promise. *Ang una sot bonga hyn•na chatwa.* I promised to give him fifty rupees.

**chat₄** *num.* eight. this word is only used in the compound number *chi chat* eighteen

**chatgyk** *num.* eight

**chatom** *clf.* classifier for bagsful. *ra•sunok khatom sa.* one bagful of spring onions

**chaw-₁** *v.* to go by boat, to row. *Rung chawchiba rung bytrongrenga.* When you row the boat, the boat spins.

**chaw•-₂** *v.* (1) to float. *Gorialdo phalhangaw tyichi thyiwatakai chaw•ratai thyiratadoknochym.* The crocodile was floating down the river, pretending to be dead. (2) to drown. *Ie morot tyi hungna sapchaaimu tyi chaw•wa.* Because this person did not know how to swim, he drowned.

**chaw-₃** *v.* to winnow. *Awanchi mai chawa.* Rice is winnowed in a winnowing basket.

**chaw₄** *ideo.* splash! the sound of something plunging into the water. *Magachakdo biskutaw tyisamchi tanaimyng chaw! thorokangokno.* Having put the biscuits by the side of the water, the deer splash! jumped into the water.

**chaw•ki ~ chaw•kyi ~ cheng•kui ~ chang•kui** *n.* ART. big knife with a curled blade used in the kitchen to prepare food, as well as in the field to cut plants and weeds (see Photo 26)

**chawarai-** *v.* to heat meat on a frying pan without salt or water in order to preserve it

**che•e-** *v.* (< Garo) to mock

**che•et** *n.* ANIM. species of green cricket

**chebe** *n.* ANIM. species of bird

**cheen** *n.* ART. (< English) chain, zip fastener, zipper, zip

**chegydek chegydek** *ideo.* call of the *daw•chegydek*

**chek-₁ ~ chyk-** *adj1.* cold. *Aiaw! Angdo chykaidonga.* Jeez, I'm cold! *Tyi cheka.* The water is cold.

**chek₂** *n.* ART. net, fishing net

**chek₃** *num.* ten. *Rong chek hyn•bo.* Give me ten rupees *Mityr chek howwa.* I jumped ten metres.

**-chekchek** *evsp.* V repeatedly. *Bbengblokmyngdo sangumuk pywdyngdyngaimyng mongmamyng mykyranaw syw•chekchekai mu•okno mang ni.* The two of them, the banana-sucking bird and the toad, had been fighting the whole day, and had kept repeatedly picking at the elephant's eyes.

**cheke** *n.* ART. sieve.

**chekjyrym-** *v.* coolish

**chekkyryi** *n.* TIME. cold season

**cheknai** *dtw.* the day after tomorrow

**cheksi** *n.* PLANT. stalk, twig

**chel** *n*. BODY. bosom of a man
**chel•-** *v*. to pry open
**chelbak** *n*. BODY. chest
**chelku** *n*. BODY. rib cage
**chem•-** *v*. (1) to burn up. *Pan wal•chi chemok.* The wood burned up in the fire. (2) to melt away. *Choklet khu•chukchi chem•ok.* The sweet melted away in my mouth.
**-cheng** *evsp*. V first, start to V. *Chang tyrywchengnaka?* Who will take a bath first? *Isolaw sung ra•ai, je kristan donggado Isol phi•ai sa•chenga.* By praising God, anyone who is Christian prays to God and starts eating.
**cheng•-** *adj1*. light, shiny, not heavy
**cheng•khu ~ cheng•khyw** *n*. PLANT. ginger
**cheng•kui ~ chaw•ki ~ chaw•kyi ~ chang•kui** *n*. ART. big knife with a curled blade used in the kitchen to prepare food, as well as in the field to cut plants and weeds (see Photo 26)
**chengchang bengchang** *n*. ACT. noise, racket
**chengcheng** *n*. PLANT. tamarind
**chengchengmachok** *n*. ANIM. spider with a long stomach with yellow stripes which can be fried and eaten. (see Photo 93)
**chenggang-** *v*. upright, erect. *Cho•sa rypaimyng jarawachian myn•an chenggang takariano pherumi myn•do.* He stayed a bit in the water and after a long time, his fur was still upright, the fox's fur.
**chengkhana** *n*. BODY. gills
**chengkhyna ~ chengkana** *n*. BODY. jaw
**chengkhyw ~ cheng•khu** *n*. PLANT. ginger
**chep$_1$- ~ chip- ~ chup- ~ chyp-** *v*. (1) to imprison, to catch, to lock up. *"Khema man•chak" noaimyng koksep wataimyng koksepchi chypangokno.* "You cannot get anymore forgiveness", they said and they made a big bamboo basket and imprisoned him in it. *Breketmyng nyng•chi chipgaba katha pang•ai gamchatcha.* The words in brackets are not very important. (see also *dang•chup-*)
**chep-$_2$ ~ chyp-$_2$** *v*. to close. *Ue Ha•dura waie songchi morot thyinaakodo rong•khalmi nokkhapaw chepchangano. Ue nokkhap chepwachian songgumukmi morotdyrangan naano.* As for that spirit of Ha•dura, when a person had died, the door of the cave would suddenly close. When that door is closed, the people in the village hear it. *Kha•sinai chypangsa dawang takaidonga.* (Gostar R. Sangma) She is slowly closing and opening her eyes.
**-chep$_3$** *evsp*. V alone
**chep-$_4$** *v*. to milk. *Ma•sudut cheparong.* She's milking cow's milk.
**chep-$_5$** *v*. to leak, to deflate. *Maityk tyi chepaidok.* The rice-cooking pot is leaking water. *Robol balwa chapok.* The football has deflated / The football has leaked air.
**chepchap chepchap ~ chepchep chepchap** *ideo*. squeak! the sound a mouse makes. *Muchot chepchep chepchap parawa.* A mouse says "squeak! squeak!" squeak! *Abeknyng•chi muchotsa•gyrai mang byryi chepchap chepchap parawthokaidonga.* Inside the *abek* are four baby mice squeaking squeak! squeak!
**chepgaba** *n*. PERS. prisoner.
**cherym- ~ chyrym-** *adj1*. heavy
**chet-$_1$** *v*. to tear, to tear off (clothes paper etc.)
**chet$_2$** *num*. eight. this number is only used in the compound numeral *sot chet* eighty.
**chetpyrak-** *v*. to tear apart
**chew•khyi** *n*. ART. big knife
**=chi$_1$** *encl*. (locative). (1) marks Spatial Locations. *Song dam sachi alsia raja myng•sa ganangnochym.* In a village, there was a lazy king. (2) marks Temporal Locations. *Wal•chi re•engchawa.* We will not go at night. (3) marks Temporal Location Clauses. *Turasang re•engwachi angna topi ra•bone.* When you go to Tura, buy me a hat. *Rang nemchie ataknakasyi?* Now that the rain has stopped, what the heck shall we do? (4) marks Conditional Clauses. *Balchachido tokni.* If you don't tell it, I'll beat you.
**chi$_2$ ~ chit** *num*. ten. This word is only used in compound numerals. *chit* is only used before *sa*, *chi* before other numerals. *chit sa* eleven, *chi byri* fourteen.

**chiakhol** *n.* PLACE. a well, a source

**-chichi** *evsp.* to *V* into pieces. *Ga•dakchichi-aimuna thypsetthiriokno. Ytykma•chiba uba sa•gyraiba jumu kha•thirithirioknotyi.* They cut him up into pieces and threw him away again. But that child joined together again and again, to our surprise.

**chichot** *n.* PLANT. dud jackfruit; small inedible jackfruit

**chichu-** *v.* to blister

**chichugaba** *n.* CORP. a blister

**Chidymak** *n.* PLACE. Stream on the way to Balphakram. When the spirit of a dead person takes his bath in that river, he forgets everything about his life. This stream is also called *Tyihanggal* or *Tyitykmak*.

**chigi** *n.* PLANT. species of plant

**chigyryng** *n.* ART. traditional snare instrument (see Photo 117)

**-chik ~ -chyk** *evsp. V* as long as you can. *Jaraw jaraw ge•theng sokwa dabatdo sakchykaidongano pheruba.* For a long time, until he did not hold out any longer, he was holding out as long as he could, the fox.

**chikarak-** *v.* to joke

**-chikchak** *evsp. V* in a crowd, swarming. *Bandi nom•khalwachido na•pat syw•chikhakwatykyi chikchak wekwak taksigaaidongano utymba.* When Bandi was tired, the whole crowd beat him like *na•pat* (type of fish) swarming around bait.

**chikchak wekwak** *adv.* swarming around something like fish around bait. (see *-chikchak*)

**chin** *n.* ART/ABSTR. a sign

**chin•thai** *n.* PLANT. melon

**chinara** *n.* PLANT. lemon

**ching** *clf.* classifier for bamboo shoots. *mai•wa ching sa.* one bamboo shoot

**ching•pheng** *adj2.* aslant, slant. *Nok bydyiaimu ching•pengok.* The house, having become old, is aslant.

**chingchongphyrot ~ chomchomphyrot** *n.* PLANT. species of white edible mushroom

**chingchoroi-** *v.* to swing from something

**chini** *n.* FOOD. (< Indic) sugar

**chinik** *n.* CORP. dirt on the body

**chinkak** *n.* PLANT. species of plant

**chip- ~ chep- ~ chup- ~ chyp-** *v.* to imprison, to catch, to lock up. (see *chep-₁*)

**chipchip** *ideo.* chirp! the call of a bird

**chirokhana** *n.* PLACE. (< Indic) zoo

**chisat-** *v.* to vomit, to throw up, to barf

**chisol** *n.* ART. a cross

**chit₁-** *v.* to tear, rip

**chit₂ ~ chi** *num.* ten. this word is only used in compound numerals. *chit* is only used before *sa*, *chi* before other numerals. *chit sa* eleven, *chi byri* fourteen.

**chiti** *n.* ART. (< Indic) letter

**chitthong-** *v.* to tear to shreds

**chiwal-** *v.* to trade, to deal in, to do business in

**cho•chep-** *v.* crumpled. *Dobachi amakba ga•sorotokno. Bonduk baithongaimu kokchengba cho•chepokno nemanchakno.* The monkey slipped and fell in the mud. The gun was broken and the basket was crumpled and not good anymore.

**cho•mot ~ chong•mot₁** *adv.* actually, really, certainly. *"Anga Ketketa Bura dong•cha. Ketketa Bura kanjota, anga mel•a chaibataw"* noaimyng, pheruna Ketketa Bura balwano. Ytykchiba pherue: *"Nang•an cho•mot Ketketa Bura"* nookno. "I'm not Ketketa Bura. Ketketa Bura is thin, I look much fatter", said Ketketa Bura to the fox. But the fox said: "You really are Ketketa Bura".

**-cho•mot ~ -chong•mot₂** *evsp. V* determinedly, *V* certainly, *V* definitely. *San nidyrang dong•phinaidok, nang• noksang rai•anado pa•chong•motchaaidokkhon"* nookno. It has been two days, but maybe he really doesn't dare to come to your house.

**cho•sa ~ choi•sa** *adv.* a little bit

**chogop-** *v.* fully bent but not touching the ground (used only with plants). *Rek chogopok.* The banana tree is bent.

**chogyp-** *v.* to break off and fall down (for branches and big leaves). *Balwana narykhelchak chogypok.* Because of the wind the branch of the coconut tree has broken off and fallen down.

**choi•etja** *n.* TIME. (archaic) (< Indic) March

**choi•sa ~ cho•sa** *adv.* a little bit

**chok-**₁ *clf.* classifier for bunches or small heaps. *ja•ryt chok sa.* one small heap of chillies. *rasunok chok sa.* one bundle of spring onions

**chok-**₂ *v.* to scoop, serve up, dish up, dish out, to comb. *Mai chokbo.* Take some rice.
**khaw chok-** to comb one's hair.

**choka-** *v.* to take apart, to disassemble, to tear off, to cut off. *Pen chokaak.* The pen is disassembled. *Gore mang sa ge•thengmi alu rydymgaaw jamai sa•akno. Una manap chaiwachido gumukan chokarumokno.* A horse had eaten all his potato sprouts.. Therefore, when he looked in the morning, they were all torn off.

**chokchok-**₁ *v.* to drip out. *Una jom•aimu sinthongwachie: "Aiaw!" Noaimu jalangokno. Uchian manapchi chaichido karydylsa thyi•chokchokai mu•aidongno, myn•tyi chokchokai mu•aidongno.* Then, surreptitiously he took his sword and cut her arm in half. "Ouch!" she said and run away. Then, when they looked in the morning, blood was dripping from the hanging root, puss was dripping from it.

**chokchok-**₂ *v.* to sharpen (a pointy object)

**chokchuang-** *v.* to fall head first. *Bandido, phalthang ra• bytgaba daraisang satwyngetwachian chokchuang matsadi•mai chotangsyrangokno.* When Bandi cuts ferociously with the sword which he had brought with him, he cuts off its tail, while the tiger falls head first to the ground.

**chokdeng**₁ *n.* BODY. throat

**chokdeng**₂ *n.* PLACE. the end of a pointy object

**choket-** *v.* to scoop (for solid substances)

**chokhoi** *n.* ART. fishing basket made of bamboo

**choki ~ chuki** *n.* ART. chair

**chokida** *n.* PERS. (< Indic) warden

**choklet** *n.* FOOD. (< English) a sweet, chocolate

**chokrek** *n.* PERS. someone with a touting mouth

**chokset-** *v.* to scoop away

**chol**₁ *clf.* classifier for ways, roads, paths and rivers. *tyikhal chol ni.* two rivers. *ram chol tham.* three roads, paths. *sorok chol byryi.* four roads
**nokchol** *n.* entrance to a house, door

**chol**₂ *n.* ABSTR. idea, plan. Clf. *myng. Na•nangdo myng sa cholawdo taknaka.* We will execute one plan.

**chol**₃ *n.* ACT. (< English) livelihood, way to make a living
**chol chal** *khjyks. n.* livelihood, way to make a living

**chola** *n.* ART. (< Indic) shirt

**cholwat** *n.* PLACE. a space

**chom-** *v.* (1) to stack, to pile up (2) to copulate, to fuck

**chom•** *clf.* classifier for little piles of fruit. *Narang chom• ni hyn•bone.* Give me two little piles of oranges.

**chong** *clf.* classifier for iron nails. *khiil chong ni.* two iron nails

**chong•** *n.* ANIM. insect, bug, lice

**chong•khobok** *n.* CORP. white patches, vitiligo

**-chong•mot ~ –cho•mot** *evsp. V* determinedly, *V* certainly, *V* definitely. (see *-cho•mot*)

**chong•mot~ cho•mot** *adv.* actually, really, certainly. (see *cho•mot*)

**chong•su** *n.* ANIM. caterpillar

**chongchang** *n.* ART. (Siju dialect) bird cage made of bamboo for small birds like *moina*

**chongchyron ~ choncholon-** *v.* to squat. *De•thengdo ramrygynchi di•etna chongchyronwa!* He squatted next to the road to shit!

**chonnyk-** *v.* (1) to look down on *Kynsang phalthangaw chonnykgabaaw naaimyng, alsia rajae jykmyng jalaidokno.* Later, having heard the ones that looked down on him, the lazy king ran away from his wives. (2) to mock, to scorn, to insult. "*Nang•tyme bobamorotkhonte. Bimyng morotsa nang•-tyme?*" *noaimyng chonnykangokno.* You must be crazy people! What kind of people are you?!" he said and scorned them.

**chot-** *v.* to tear (off), to cut. *Sendel chotok.* My sandal is broken. *Aia thetna bai! Ang* *chak chotni!* Ouch! Don't pull! My arm will get torn off! *Gandurian chotkhuchano ue sa•gyraie.* The child had not even had its umbilical cord cut.

**chu•-** *v.* to wrap into something

**chu•ret-** *v.* to hang down. *Phepchi pheru ytykyi mu•aidonoaro. Mu•wachie ri•do chu•ret takangokno cha•masang napitsang.* The fox was sitting in the banyan tree like this. While he was sitting there, his penis was hanging down, toward the barber.

**chu•sok-** *v.* to succeed. *Nang•tyme iawan phalthangthangna hyn•gaawan kamtykyi chu•soketchachido nang•tyme atongtykyi phylgym kawna man•a?* If you cannot succeed in the job that I gave to you, how can you shoot the eagle?

**chuchu** *n.* KIN. *a.* the address term a grandparent uses to their grandson

**chuduk- ~ chyduk-** *v.* to turn upside down, to turn over

**chugup-** *v.* (1) to be on its side. *Rung chugup paitanbo.* Turn the boat on its side. *Chugupai tanwa.* I put it on its side. *Rung chugupok.* The boat is lying on its side. (2) to cover with a lid

**chui** *interj.* interjection to chase away a pig

**chuki ~ choki** *n.* ART. chair

**chula** *n.* PLACE. cooking place (see Photo 18)

**chuli-** *v.* useful. *Tangka poisaba, kamba janggina chulia.* Money and wealth are useful in life.

**chultet-** *v.* to shake off

**chun$_1$** *n.* BODY. trump

**chun$_2$** *n.* FOOD. ground limestone, usually eaten with betel nut

**chung-** *adj1.* (1) big. *Nang•mi nok chungate.* Your house is big. (2) tall. *Ang nang•mi hapsan chunga.* I am as tall as you. (2) to grow. *Ytykyisa, ge•thengtheng cha•wekdamchi panthaiaw kai•wachi, pan chungok.* So then, because they had planted the seeds in the place where the chaff is thrown away, trees had grown.

**chungai rai•cha** *expr.* I don't care.

**chunggalgal-** *v.* to grow up, to become an adult. *"Sa• myng• sa ba•aimung,* *man•dykarok." "Man•dykasola ho•ong. ie Jenkonparaba rai•asyrangchak." "Wel•ang wel•ang chunggalgalwasa ga•nakachym. Jengkonparaba rai•akhuchakhon?"* "After one child has been born, it is difficult." "Difficult indeed, yes. Jengkon and his wife never come anymore." "He will almost certainly be compelled to grow up quickly."

**chungtaw-** *v.* to grow

**chungthai** *n.* BODY. big bosom. *Sam sa mylthai sam sa chungthai.* One of her bosoms is small, one of her bossoms is big.

**chup-$_1$ ~ chep- ~ chip- ~ chyp-** *v.* to imprison, to catch, to lock up. (see *chep-$_1$*)

**chup$_2$~ chyp** *adv.* fully dressed, with all one's clothes on, wearing whatever it is you are wearing. *Ang chyp tyruok.* I took a bath with all my clothes on. *Ang chup re•engariok.* I just went wearing the clothes I was wearing at that time.

**churi** *n.* ART. (< Bengali or Assamese) knife

**churu** *n.* ABSTR. very little food

**chutchut** *ideo.* squeak! the sound a mouse makes

**chuwil chuwal** *khjyks, adv.* spinning. *Thot thyng•thot takwachina dabat sykromaimyng khanetsigaaidongno. Bandi chakwatwamian chuwil chuwal takjolangokno.* He (Bandi) grasped her (So•re) and poured the liquor into her mouth to the last drop. When Bandi let go of her (So•re), her head was spinning.

**chuwyng chuwang** *khjyks, adv.* with a spinning head, dizzily

**chybym** *n.* BODY. forehead

**chyduk- ~ chuduk-** *v.* to turn upside down, to turn over

**chygyl** *n.* ANIM. species of worm that glows in the dark

**chygyp-** *v.* to fall face down on the ground

**chyhyl** *n.* ANIM. species of snail

**=chyi$_1$** *encl.* (conative modality). try to. V *"Na•a sikhal kha•na raiachido, chuchu, ha• ie ang mola hyn•ga ryngan<u>chyi</u>" nookno.* "If you want to hunt, grandson, here, try to smoke this tobacco of mine that I will give you", he said.

**chyi-₂** *v.* to try. *Chaiai chyini gorongnima gorongcha.* Let's try to meet him. (literally: We will try by seeing if we will meet him or not meet him.)

**chyi•-** *adj1.* tired, sleepy

**chyigyk** *num.* ten

**chyk₁- ~ chek-** *adj1.* cold

**-chyk₂ ~ -chik** *evsp.* V as long as you can. (see *-chik*)

**chykhyw** *num.* nine

**=chym** *encl/prtcl.* (irrealis) (1) supposition interpretation. *Song dam sachi alsia raja myng•sa ganangchym.* In a village, there is/was supposedly a lazy king. (2) frustrative interpretation. *Angdo dadaparaaw sandiedongachym.* I am searching in vain for my elder brothers. (3) irresultative interpretation. *Ang tai•sa raja sa lapokchym. Thyiok.* I would have made one hundred [rupees] profit. [But] I lost the game. *Ang ie khata dakangdo tyngchachym. Te•ewdo nemaian tyngok.* I did not know this word before. Now I know it well. *Mura tai•sa ganangchym, te•ew ni•wa.* There was a small stool here a little while ago, but now it's gone. (4) implicative interpretation. *Biskut dyngthangram•achym.* You could have searched for other biscuits. *Ang nang•aw bylongen nukna sykaidongachym. ang phalthang re•engnado sykaidokchym, ytykchiba anga sawamigymyn re•engna man•chaaidonga.* I really want to see you, but something prevents me from doing this. I would have liked to come myself but because I am ill, I am not able to go. *Ha•chyksang balchido sal kolgryksa noai myngnichym.* When you say it in Garo, it would be called *sal kolgryksa. Ytyknaka nogabaaw nuksawaian anga nang•aw peng•wachym.* Angba *re•engnichym.* I would also go. If I had known what was going to happen, I would have prevented you.(5) irrealis particle. *"Jykkymokma, ge•thenge?" "Ho•ong, chym."* "Is he married?" "Yes, supposedly".

**chym•-** *v.* to chew. *Goiaw nemai chym•aimu dakbo.* Chew the betel nut well, then spit it out.

**chymbuk** *n.* ART. magnet

**chympyret-** *v.* to hit with one's fist, to crash head-on

**chyn-** *v.* to offer to the dead. *Ie taw• mama thyigabana chynkhuni.* We will offer this chicken to our dead uncle.

**chyndyk** *n.* ANIM. domestic water buffalo

**chyng•-₁** *adj1.* bright

**chyng•-₂** *v.* to burn. *Ie pan nemai chyng•ni.* This wood will burn well.

**chyng•chet-** *v.* to glitter

**chyngaba** *n.* ART/ACT. offering to a dead person

**chyngmat** *n.* BODY. comb of a rooster

**chyp-₁ ~ chep-** *v.* to close. (see *chep-₂*)

**chyp-₂ ~ chip- ~ chep- ~ chup-** *v.* to imprison, to catch, to lock up. (see *chep-₁*)

**chyp₃ ~ chup** *adv.* fully dressed, with all one's clothes on, wearing whatever it is you are wearing. (see *chup-₂*)

**-chyp₄** *evsp.* V wastefully, V unsuccessfully, V completely, imaginary. *Te•ew una rangsando saniarokno. Sikharba kha•chypanchakno. Ytykthyngai somai jamchypaimuna, jyksang sa•sang waiangokno.* Now the sun was setting. It was too late to hunt. With all this stuff going on, he had wasted his time and he returned to his wife and children.

**chyrym- ~ cherym-** *adj1.* heavy

**chyryt chyryt** *ideo.* squirt, squirt!

**chys** *interj.* tsk! interjection of disapproval. *Chys! Sawthal! Mai sa•na dakang soreachi chaksubo!* Tsk! Don't be so dirty! Wash your hands in the tub before you eat!

**chyw** *n.* FOOD. rice beer, rice wine alcohol, wine, liquor

    **chyw sym•-** *v.* to make. *chyw* by pouring water on the *sithi ~ sythi.*

    **chyw chek-** *v.* to scoop the *chyw* out of the *gora* with an *abek*

**chyw•-₁** *adj1.* high, steep. *ha•kha chyw•a* the mountain slope is steep

**chyw•₂** *n.* PLANT. the new young leaves of a tree

**chywbok** *n.* FOOD. white alcoholic liquid made from fermented rice, white rice beer, white rice wine

**chywgundai** *n.* PLANT. species of plant

**chywgyn**₁ *n.* ACT. the festival of the dead at which the soul of a dead person is sent out of the house to rest in peace. The festival is held around the end of February or the beginning of March. During *chywgyn* people indulge in different activities such as *chyw ryngwa* 'to drink liquor', *khata juw•kynwa* 'to tell stories', *chaira ryngwa* 'to sing songs' and *Wal•jan bytwa* 'to tell the love story about Wal•jan'.

**chywgyn-**₂ *v.* to celebrate the festival of the dead. *Ha•bykungaw morot takaimuba kangkelekaw so•otaimu chigyryngsang dymchyrangsang dakangmi achu ambitykyi dythyichengai takaimu uan me•mang saw•etokno, chywgynokno.* Having made a person out of sand and killed a lizard, making a sacrifice and then burning the ghost, they celebrated chywgyn just like their ancestors.

# D

**da•nang** *interj.* wow!

**da•rat-** *v.* to fall down (for persons)

**daba** *n.* PLANT. coconut

**dabat** *postp.* (this postposition indicates a limit in time, either in the past or future) since, from, until. *Tai•nimyng dabat nang•myngan baju takchaka.* As from today I will not be your friend anymore. *"Ytykchiba na•a angna aro angmyng jykna nang• khengwa dabat ang thyicha dabat angaw mu•ai sa•na hyn•bo" nookno.* "However, you have to keep giving me and my wife food as long as you live, until I die", he said. *Na•a te•ew wen• sa rypbone. Nang• sokwa dabatdo sakchykbo.* Stay under water one more time, okay? You hold out until you can't endure it anymore. **uchina dabat** until then, until that time. *Ang hampyi rai•aphinine, baju. Uchina dabat, ichi mu•bone.* I will come back this evening, my friend. Until then, stay here, all right? **umyng ~ umi dabat** since then, since that time, from that time onward, from then on. *Umi dabatsa iawe Dabat myngwanoro.* Since then, it is called Dabat, I'm telling you.

**dabia** *n.* a demand. (< Indic) *Na•a atongaw dabia angmyngaw ang nang•na hyn•arinaka.* Whatever your demand of me is, I'll give it to you.

**dabogos** *n.* ART. skewer

**dachang** *n.* PLANT. roselle, *Hibiscus sabdariffa* (see Photo 48)

**dada** *n.* KIN. *c, ref, a.* (1) elder brother. (2) Also used to speak about or address a related older male relative of your own generation: cousin. (3) Also used to address an unrelated man older than the speaker.

**dada… phaw•jong~phawjong…** *coll.* elder brother. *Re•engphinaribo dada, jokangphinaribo phaw•jong.* Go back, elder brother; run back, elder brother.

**dagi** *n.* CORP. scar

**dai-** *v.* to be bigger, greater.

**daiaiok** over, finished

**dai•-** *v.* to wash away (as in a landslide). *Rang wawana, ha• nom•aimu ha•byri dai•ok.* Because of the rain the ground had become soft and therefore the mountain washed away. *ha• dai•ok* there has been a landslide

**daijol-** *v.* to overstay

**daikhalaisa** *expr., conj.* moreover. *Ytykyian joton kha•chiba, kha•chiba man•anchakno, daikhalaisa gore kha•phaksang bykphylangokno.* So then, when he tried, he couldn't do it; moreover, he rolled around and ended up on the horse's underside.

**dainingrum** *n.* PLACE. (< English) dining room

**dairamphin-** *v.* to work overtime

**dairukruk-** *v.* become more and more, to increase

**dak**₁ *n.* CORP. freckle

**dak**₂ *n.* CORP. phlegm from the lungs

**dak-**₃ *v.* to spit out. *Goiaw nemai chym•aimu dakbo.* Chew the betel nut well, then spit it out.

**dakal ~ takal** *n.* SUPER. witch, demon

**dakan-** *v.* to dress someone else

**dakang₁** *adv.* previously. *Gandrung songchamdo dakang mynggaba Songma Songgni Khychu Badri nogabaan.* The old village of Gandrung is the previously so called Songma Songgni Khychu Badri.

**dakang₂** *dtw.* past, in the past, before, earlier. *Dakangdo, mamung khem ni•wachido dymchyrangsangsa chywgyn ryngwano.* In the past, when there were no drums, they celebrated the festival of the dead only with the *dymchyrang*. *Gam man•ni udo uan, tangka poisa. Uan gam mynga, dakangmi chasongdo. Te•ewsa kepasyti noai myngaidonga. Chasongna kri gam myngariaro, tangka poisa.* He will obtain wealth, money. Earlier generations called that "wealth". Now they call it "capacity". According to my generation this money is called "wealth".

**dakang₃** *postp.*(takes goal-marked complement) ago, before. *Bylsi sana dakang jyk khymok ge•thengdo.* He got married one year ago. *Sa•na dakang chaksua.* Before eating, we wash our hands. *De•theng angna dakang re•engwa.* He left before me.

**dakanggaba** *n.* TIME (1) first. *Dakanggabado jineralmitingchengni. Umungsa song gumuk thom•aimung ha•ba ha•ryn ha•rynaw sowalni.* First they will start with a general meeting. Then the whole village comes together and they will divide the *ha•ba* plot by plot. (2) the first. *Uchi thymaimyng, dakanggaba bobaan dirichengokno.* Then, having lain in ambush, the first crazy person got hold of (the horse's tail) first. (3) the first time. *Dakanggaba Turachi mu•wachi Mobbinaw gorongwa.* The first time I stayed in Tura I met Mobbin.

**daket tak-** *v.* to rob. *Sa•khawchiba dak takchiba patok tancha.* When someone steals and robs, he is not put in jail.

**dakham** *n.* ART. very small wooden stool consisting of one rectangular wooden board to sit on and two small rectangular wood blocks attached underneath as supports (see Photo 37)

**dakmanda** *n.* ART. long women's dress tied around the waist, skirt (see Photo 122)

**dala₁** *n.* ART. bamboo mat made of *wa•tyng* for drying rice, chillies or other vegetables and leafy greens in the sun, also called *damplak* (see Photo 41 and Photo 42)

**dala₂** *n.* PLANT. branch of a tree not directly attached to the trunk, young plant *dala sa* one branch *dala phek sa* one branch

**dalchini** *n.* PLANT. species of tree

**daldi** *n.* PERS. beloved person, love, darling

**dalibibi** *n.* ART. doll. Clf. goi.• *dalibibi goi• sa* one doll

**dalim** *n.* PLANT. (< Assamese or Bengali) pomegranate

**dalni tatdepgaba** *n.* ART. ramp of a door

**dam₁ ~ dym** *ideo.* bam! thud! the sound of something heavy hitting the ground. *Tokkyrengaw man•aimungna ha•china wuuuuuuuk, dym! takramphinoknotyi phylgym gal•waan.* Having got him in the neck, the giant eagle fell to the ground wooooosh, bam!

**dam₂** *bound.* PLACE. place. *jaboldam* rubbish heap *cha•wekdam* place where the chaff is thrown

**dam₃** *clf.* classifier for villages *Song dam sachi alsia raja myng• sa ganangchym.* In a certain village there was supposedly a lazy king.

**dam₄** *n.* ABSTR. price. *Ie ma•sugari dame biskyn?* What is the price of this bullock cart?

**dam₅** *n.* ART. bamboo mat

**-dam₆** *evsp.* V truly, really. *Khurutna sapgaba morotawsa songgumukchiba songchi pang•ramaria. Udo pang•aido sap<u>dam</u>cha. Myng• sa myng• ni ytykyi sapa.* There are just so many people in the village who know how to perform an incantation. Many of them really don't have the skill. Only one or know how to do it.

**dama** *n.* ART. drum of the Mandai people, a bit smaller than the Atong *khem*. The *dama* is used during the Garo festival of Wanggala. *dama tam•-* to beat a drum

**-damdam** *evsp.* V in different places, V one after the other, V continuously

**damdyl** *n.* ART. bamboo mat that is used as the side of a house. Clf. *khap / khaw•* /

*jyw•. wa•da damdyl saw•a.* a mat is made from *wa•da. damdyl wat-* to wicker a mat (see Photo 9 and Photo 10)

**damplak** *n.* ART. bamboo mat, also called *dala,* made of *wa•tyng* for drying rice, chillies or other vegetables and leafy greens in the sun (see Photo 41 and Photo 42)

**damrak-** *adj1.* expensive

**damthol** *n.* ART. a rolled up mat

**dan-** *v.* (1) to spread out, to lay out (mats etc.). *Na•aw khan•tongai danwa.* She laid the fish down and cut it in pieces. *Palongchi kombol danbo.* (2) spread a blanket over the bed. (when preparing it to go to sleep). *Kombol palongchi danbo.* Spread the blanket out over the bed.

**dandan-** *v.* to be pressed with one's back against something, to lean against something. *Dandanai mu•bo.* Sit with your back against the wall (or any other supporting object). *Panchi dandanaidonga.* He's leaning against a tree.

**dang•-** *v.* (1) to enter, to go/come in. *Ytykyisa ue Arong nokma thyiwamisa saepe bondyk paiaimu sipaidyrang dang•na man•okno. Sipaidyrang dang•wachie kan•tyra guli nyi kawphetphetai rai•aaknokhon. Uchian songchi dang•ok.* That's why, after head man Arong's death, the gun-carrying sahibs were able to come in. When the sahibs entered, they might only have fired without bullets. Then they entered the village. *Noksang byk dang•jolai jalangoknoai.* He quickly ran into the house. (2) to visit. *"Ma• baba, atykyimu walawa?" nookno amakaw, amakmi sa•dyrange. "Ue nang• awangpara nokchi dang•phakawa na•a" noatakokno.* "But daddy, why are you so late? It is already night", the monkey's children said. "I visited your uncle" he said. (3) to set (of the sun) *Re•enwachian rangsan dang•aimu walokno.* When he left, after the sun had set, it was night. (4) *vph.* to enter into a mental state, to start. *Anga nang•aw nukjyryngaria uchian anga nang•aw nukjyryngwachian nang•na kha•galwa dang•ok.* I just saw you every day, then, when I saw you every day, I started loving you.

**dang•chup-** *v.* to lock up. *Gawigamuba olrukanchakno. Mama manithanggamuba olrukanchakno. Ytyken jyrym barataimu dang• chupai mu•arokno.* He didn't talk with his wife anymore. He didn't talk with his father- and mother-in-law anymore. He just locked himself up in the house like that, quietly and ashamed.

**dangkhym-** ~ **dangthym-** *v.* to collapse (of a road or bridge), to go into a hole

**danyl** *n.* ART. shield

**dap-₁** ~ **dep-** *v.* (1) to be on top. *Jemi sanchi rong• rymrym dapetaimung Warma sep nogaba jagysimi chak bai•thongokno.* One day, a rock rolled on top of [it, and] broke so-called Warma sahib's hand/arm. *Mungmae nokchol chypgatykyi hang• khalawan dapoknoro.* The elephant stood on top of the hole like a door blocking an entrance. (2) to put on top (3) to press (4) keep together by force, pinch together (5) to pinch. *Rukwakchakaw khen• depok.* The toad's paw got pinched by a crab. (6) to stack.

**-dap₂** ~ **-dep** *evsp.* (1) *V* on top. *Na•nang garu ramtananggachi phylgym di•etdapai tanangwa.* An eagle shat on the mustard leaves that we had left outside to dry. (2) *V* more, *V* and add, also, as well, in addition. *Ue myng• tham myng• byryigaba pi•dapokno.* He asked for three or four more children. (3) to crush. *Sambanggyri akaiokno, tokdepdepaimu pha•ato.* He plucked some sambanggyri it is said, crushed it, and applied it.

**dapet** *adv.* insipid, not tasty. *Ja•bek dapet dapet takaidong.* The curry is not tasty.

**darai** *n.* ART. sword (see Photo 121)

**darang₁** ~ **dyrang** *n.* PERS. people, anyone, everyone

**darang₂** *qtf.* all. *Darang matan re•engok, jamok.* All the animals went, all of them.

=**darang₃**~ =**dyrang** *encl.* (1) plural number, used to emphasise that there are many or that there is a lot of something. *Ytykyimu sa•darangba pang•anoa.* So then, there were many children. *Maiawdo pang•ai sa•cha phorenmi morotdyrangdo.* They don't eat a lot of rice, all those foreign people. *Ang ie maidyrangaw sa•chawa.* I will not eat all that rice. (2) indicator of an approximate amount or approximate time. *Imi kilomityr kolgykdyrangtykyi rang ni•wa.* About twenty kilometre from here, there is no rain. *Angdo maja 13 tarikchi chaphang phang 99 ang nok rygynchi kai•ok. Te•ew phang 150-darang sakkhunichym.* A few days ago, on the 13th, I planted 99 tea plants near my house. Now I might fit another 150 or so more. *Umungdo marsja somaidarang-chi saw•a.* So then, around the time of the month of March, we burn the land.

**darangba** *prof.* anybody, anyone, nobody, no one. *Ang songchi darangba Atong khu•chuk olna man•cha.* In my country there is nobody speak Atong with.

**dareng** *n.* PLACE. edge

**dari-** *v.* to commit adultery

**daw-** *v.* (1) to peel. *Khopphylak dawarok.* She's pealing the skin off a fruit. *Narykel dawaidong.* He's peeling an orange. *Taw•ti dawbo.* Peel the egg. (2) to open. *Kha•sinai chypangsa dawang takaidonga.* (Gostar R. Sangma) She is slowly closing and opening her eyes. *Nokkhap dawbo!* Open the door!

**daw•-** *bound.* ANIM. bird. This is the bound form of the word *taw•* 'chicken, bird', which appears in names of birds and compounds with the root 'bird' in them.

**daw•blok** *n.* ANIM. bulbul bird

**daw•budok** *n.* ANIM. lineated barbet, *Psilopogon lineatus* (species of bird)

**daw•chegydek** *n.* ANIM. species of bird

**daw•gamdot** *n.* ANIM. eagle

**daw•gep** *n.* ANIM. duck

**daw•kha** *n.* ANIM. black crow

**daw•kharasun** *n.* PLANT. species of onion, crow onion,

**daw•kruha•sym** *n.* ANIM. common emerald dove, Asian emerald dove, grey-capped emerald dove *Chalcophaps indica* (species of pigeon)

**daw•kumai** *n.* PLANT. species of tree

**daw•kyru ~ daw•kru** *n.* ANIM. pigeon

**daw•mai•tak** *n.* PLANT. species of tree

**daw•nok** *n.* (Badri dialect) bamboo bird cage especially for small birds like myna

**daw•phaw ~ do•pho** *n.* ANIM. owl

**daw•phylgym** *n.* ANIM. eagle

**daw•pynchyrep** *n.* ANIM common tailorbird, *Orthotomus sutorius* (see Photo 112)

**daw•reng** *n.* ANIM. hawk, kite or falcon (the author does not know which one of these it is)

**daw•rigi ~ daw•rygi ~ daw•rugoi** *n.* ART. traditional headband ornamented with chicken feathers (see Photo 115)

**daw•rugu** *n.* ANIM. species of bird of prey

**daw•sik** *n.* ANIM. parrot

**dawel-** *v.* circular

**de** *inter* okay then, well

**de•et- ~ di•it- ~ di•et-** *v.* to shit, to do number two, to defecate. *Udo de•etna re•engwa.* He went for a shit.

**de•etset- ~ di•itset- ~ di•etset** *v.* to pick one's nose. *Nakhung de•etsetaronga.* He is picking his nose.

**de•theng ~ ge•theng** *ppron.* he/she, third person singular pronoun, usually referring to animate beings

**de•thengtheng ~ ge•thengtheng** *ppron.* they, third person plural pronoun, usually referring to animate beings

**dekdek-** *v.* to shiver, to tremble. *Dekdekai thyiok.* He died shivering.

**dekhep ~ dykhep** *vtr.* to make someone cry

**dekoresyn** *n.* ART. (< English) decoration

**del- ~ dyl-** *v.* to sting (of a bee etc.)

**delang ~ dylang** *n.* ART. little house for the spirit of a dead person built close to the house where the dead person is burnt to keep his remains and ashes. The spirit of the deceased will live in this little house until it is burnt in the ceremony called *me•mang saw•eta* about one year after his death and the spirit will go to *Balphakram.* (see Photo 126 and Photo 127)

**dem•-** *v.* to fold

**demdong-** *adj1.* weak, soft

**dempharai** *n.* ART. lengthwise cut long bamboo strip used in the construction of a house

**deng-** *v.* to untie. *Ang chakaw dengbo.* Untie my hands.

**dengdyl ~ beraberi** *n.* ART. an open whickered type of bamboo mat used as fence of the balcony or veranda of a house (see Photo 9)

**dengga** *n.* PLANT. species of small leafy green

**denggu** *n.* ACT. crime, extortion, naughtiness. *Ang denggu takni na•a. Na•a biskutaw paiai jalbone.* I'll do the extorting, oh, you. You carry the biscuits away, okay?

**denjyr** *n.* (< English) danger

**dep-₁ ~ dap-** *v.* (1) to be on top. (2) to put on top (3) to press (4) keep together by force, pinch together (5) to pinch. (6) to stack. (see *dap-₁*)

**-dep₂ ~ -dap** *evsp.* (1) *V* on top. (2) *V* more, *V* and add, also, as well, in addition. (3) to crush. (see *-dap₂*)

**di•** *n.* CORP. shit

**di•but** *n.* ANIM. dung beetle

**di•chongkhamai** *n.* BODY. cloaca

**di•chongkhanthyi** *n.* BODY. pygostyle

**di•chyrak-** *v.* to have diarrhoea

**di•it- ~ de•et- ~ di•et-** *v.* to shit, to do number two, to defecate. *Udo di•itna re•engwa.* He went for a shit.

**di•itset- ~ di•etset ~ de•etset-** *v.* to pick one's nose. *Nakhung di•itsetaronga.* He is picking his nose.

**di•khal** *n.* BODY/PLACE. bottom, arse, anus.

**di•khaldgisep ~ di•khalgesep** *n.* BODY. arse crack

**di•kyntyk** *n.* PLACE. toilet

**di•mai** *n.* BODY. tail

**di•phathai** *n.* BODY. buttock

**di•pyru-** *v.* to have diarrhoea

**di•pyryw-** *v.* to shit one's pants

**di•sep** *n.* BODY. arse crack

**di•sepra** *n.* BODY. arse crack

**di•thap** *n.* ART. diaper. *Jyw•gaba sa•garaiaw di•thap pha•etaidonga.* The mother is putting a diaper on the child.

**di•thap** *n.* MSRE. half (of a volume). *Gylas di•thapan phingancha.* The glass is not half full *Gylas di•thaptharaan.* only half a glass.

**di•thom** *n.* BODY. gizzard, ventriculus, gastric mill, gigerium

**digi** *n.* GEO/PLACE. well, ditch

**dikirin-** *vtr.* to tear (clothes, paper etc.)

**diksyneri** *n.* ART. (< English) dictionary

**dil** *n.* CORP. body smell, body odour. *Mongmadil manama.* The body smell of an elephant stinks.

**dile** *n.* ACT. (< English) delay

**din** *n.* PLACE. bedroom

**dinggarai** *n.* ART. fish trap. *Morot myng•sa manapmi sirimynmyn Dabatwarisang re•engaimungna dinggarai saakno.* A man went to Dabat Wari very early in the morning and set a fish trap. (see Photo 32)

**diphing-** *vtr.* to fill. *Gylaschi tyi diphingbo.* Fill the glass with water. *Gylas phingok, diphingna man•chaka.* The glass is full; you cannot fill it anymore.

**diphu₁** *n.* CORP. a fart

**diphu-₂** *v.* to fart

**dipot** *n.* ART. teapot. Clf. *thai•. dipot thai• ni.* two teapots

**diprin ~ dipyrin** *n.* PLANT. species of vegetable

**diri- ~ dyri-** *v.* to hold (onto). *Ytykyisa ge•theng gore di•maichi diriwano.* So then, he held on to the horse's tail. *Ytykchiba uaw nukaimu: "Aiaw! ie dakang amapara ha•ba wylwachi byira joketgabachym" noaimu, rykangaimyng di•maichi diriaidonganote.* However, having seen it: "Hey! this is probably be the cat, that escaped from my mother's rice field", he said, and chased it, and held it by its tail, I'm telling you!

**dirikhap-** *v.* to catch

**diritat-** *v.* to hold firmly

**disembyl** *n.* TIME. (< English) December

**disko** *n.* PLACE. (< English) disco

**distrik** *n.* PLACE. (< English) district

**disu-** *v.* to piss, to pass urine, to do number one, to urinate

**disudap-** *v.* to piss on top of

**disutyi** *n.* CORP. piss, urine

**disutyitup** *n.* BODY. urine bladder

**=do ~ =odo** *encl.* (topic) (1) Topic marker. *Ning songsyrekdo, ning Atongdo dakangdo mamyng thorom ni•wami somaichido waiaw mania.* We heathens, we the Atongs, in the past, in times when there was no religion, we worshipped spirits. *Manap chaiwachido gumukan chokarumokno.* When he looked in the morning, the potatoes were all torn. (2) marker of the condition (protasis) in Conditional Clauses. *Turasang re•engchido angna tupi ra•bone.* If you go to Tura, buy me a hat. *Hap pidan ramna man•okodo jytnaka?* After you have found a new place, you'll immediately move?

**do•de** *n.* ANIM. peacock

**do•dokhichong** *n.* PLANT. species of plant

**do•jenjok** *n.* GEO. Big Dipper (star sign), Ursa Major (constellation)

**do•khakhu** *n.* ART. carved, ornamented and colourfully painted king post of the bachelors' house above the entrance in between the tie beam (*bylbang*) and the peak of the roof (see Photo 14)

**do•pho~ daw•phaw** *n.* ANIM. owl

**doba** *n.* SUBST. mud

**doi•- ~ joi•-** *v.* to drag, to catch (by dragging a net through the water), to hold, to grasp, to scoop into a receptacle

**dojanggre** *n.* ART. a ladder with only one axis to which the runs are attached

**dok-**$_1$ *v.* (1) to take off (clothes) (2) to unplug (3) to take apart, to disassemble (4) to unblock

**dok-**$_2$ *v.* to weave

**dok**$_3$ *num.* six. this word is only used in the compound numerals. *chi dok* sixteen and *sot dok* sixty.

**dokhan** *n.* PLACE. (< Indic) shop

**dokra** *n.* ART. bag

**doksylok-** *v.* to be detached

**dol**$_1$ *n.* CORP. wrinkles

**dol**$_2$ *n. autoclf.* MSRE. group. *dol ni.* two groups

**dol•romrom-** *v.* to roll up

**dolan** *n.* PLACE. big building, big house

**dolong** *n.* ART. bridge

**dolrorom-** *vtr.* to roll up. *lekhaaw dolrorombo.* Roll up the paper.

**dong•-**$_1$ ~ **dong-** *v.* (1) to be. *Ue hape Chigachak te•ew Kol India kolani hapan dong•wachymno.* That place Chigachak is now supposedly the Coal India Colony place. (2) to be enough, to be sufficient, to be OK. *"Mai sa•khunima?" "Dong•ok."* "Will you eat more rice?" "It's enough." *Mamung dong•cha.* It doesn't matter. (3) to be convenient (4) to have passed, to be past. *Bylsi chykhywdyrang dong•phinokno.* Nine years have passed. *No baji dong•ok.* It's past nine o'clock.

**dong•arini ~ dongarini** *expr.* That's all right. That's okay. No worries, That will do.

**dong•taw-** *v.* to be enough, to be sufficient, to suffice. *Aia! dong•tawanchakte ang tangkado!* Oh no! I don't have enough money!

**dong•-**$_2$ *v.* to arrive. *Rai•akno rai•akno, nokthangchina dong•okno.* He went and went, and arrived at his own house.

**phet... dong•...** *coll, v.* to arrive, to reach (see *phet-*)

**dong•wa** *n.* ACT. event

**dongang-** *v.* to arrive

**dora** *clf.* weight of 5 kg. *Nokbanthaidyrangaw saw•aimung nokbanthai do•khakhuchi khachapai tangaba mongmawa dora byryi dong•gabaaw ra•ai jalangokno.* Having burnt the bachelors' houses, they ran away taking the four *dora* (20 kg) weighing elephant tusks which were tied to the king post of the bachelors' house.

**dorai** *n.* PLANT. lady's finger, okra ~ okro ~ ochro (see Photo 53)

**dorma ~ dolma** *n.* ART. salary. *Jahasmi kepten man•ok, dolma chungai sa•ak.* He became the captain of a ship, and got a big salary.

**dosi** *n.* ABSTR. (< Indic) blame. *"Acha, na•a angmyng goreaw dosi hyn•ok" nowano rangramyng rajado.* "So, you blame my horse", said the king of the sky.

**dot** *clf.* classifier for cylindrical objects like candles and bananas and logs (but not for batteries). *wa• dot sa.* one culm of bamboo. *kendel dot sa.* one candle. *pan dot sa.* one log

**drakha** *n.* PLANT. (< Assamese or Bengali) grape

**dram** *n.* ART. drum, barrel

**-duga₁** *sfx.* (excessive degree). *V* too much, too *V*, very much. *Ta syng•dugasi.* Don't ask too much! *Ie taw•do damrakdugaa.* These chicken are too expensive. *Waiphinwami gesepchi baratdugaaimu.* Upon his return, he felt very much ashamed.

**duga-₂** *v.* to be too much. *Aia! Dugaphinok bai•sigaba angaw thogigababa.* Oh! it's too much, friend who has betrayed me."

**duk** *n.* ABSTR. (< Indic) sorrow, sadness. *Itykgaba chasongchi mamungba duk ni•wa.* In that era, there was no sorrow.

**duk man•-** to be sad. *Ketketa Burae duk man•biai pheruna balsakoknowa* .... *Ketketa Bura* was very sad, and answered the fox ...

**duk sak-** to suffer. *Aiaw! Biskynba bylsi nidyrang dong•phinai duk sakwachido de•thengna mamyng tangka poisa, de•thengmi duk sakwana, wak rakhiganaba, tangka poisa hyn•chano.* Alas! How many? About two years had passed in which he suffered; he wasn't given any money, and for his suffering, for guarding the pigs, he wasn't given any money.

**duk dong•-** to be grieved. *Morottykyi nukanchakno, sa•na man•chaaimu, ryngna man•chaaimu, duk dong•aimu, nalbas sa•akno, ue sa•gyraie.* He didn't look human anymore, not having gotten any food, not having gotten any drink, being grieved and nervous, the child.

**dukhup- ~ dykhyp-** *vtr.* to dress someone, to put clothes on someone else

**dukung** *n.* PLANT. species of plant

**dukung-** *vtr.* to dam, to make circular a wall of stones in the water in the river to trap fish. *Bai•sigathangmaran tyi dukungokno. Na•do ramramanchakno.* The friends dammed the water. There was an abnormal quantity of fish.

**dum-** *v.* to gather, to swarm. *Hajambutungchi umyng khu•chuksang sotamai dumna dang•thokokno.* When he was yawning a swarm of flies entered his mouth.

**duma** *n.* PERS. crowd

**duma-** *v.* to gather (of people)

**dumut-** *adj1.* moulded

**dumuta** *n.* PLANT. species of edible mushroom

**dung-** *v.* to put something into something

**dung•-** *v.* (1) to climb. *Amakdo wel•ang wel•ang pankhambaisang dung•khatai jalangokno. Pherudo pan dung•na man•cha.* The monkey quickly ran away, climbing to the top of a tree. The fox cannot climb trees. (2) to mount or ride a horse. *Bildo te•awba gore dung•na sapchanotyi.* Bil does not know how to mount/ride a horse, to our surprise.

**dupliket** *n.* ART. (< English) a fake

**durrrmeme** *ideo.* eeeeee! the sound of a bleating goat The pronunciation of this English word in Atong orthography would be e•e•e•e•e•e•, or [ɛʔɛʔɛʔɛʔɛʔ] in the symbols of the International Phonetic Alphabet.

**durymytdyl** *n.* PLANT. species of liana

**dygri** ABSTR. (< English) degree

**Bechylyrdygri** Bachelor's degree

**Masteldygri** Master's degree

**dykdyk** *adv.* for a short while, quickly

**-dykdyk** *evsp.* about to *V. Ransan songdykdykangaidok.* The sun is about to set.

**dykha** *n.* FOOD. wine drunk during the *chywgyn* festival.

**dykhep- ~ dekhep-** *vtr.* to make someone cry

**Dykhija•lang** *n.* SUPER. the tree that lies across the beginning of the road to Balphakram and that acts like a gate for the spirits of the deceased. This tree is also called Bandija•lang.

**dykhyp- ~ dukhup-** *vtr.* to dress someone, to put clothes on someone else

**dyksyl ~ tyksyl** *n.* ART. pan for cooking rice (see Photo 43)

**dykyl** *n.* PERS. (1) Khasi person (pejorative) (2) cannibal

**dykym** *n.* BODY/PLACE. head, upside, upper side, top

**dykym sa-** to have malaria

**dykymphak** *n.* PLACE. side where the head is, space above the head. *Dokra dykymphakchi syithaiwa.* The bag hangs above your head.

**dykyret- ~ dykyryi-** *vtr.* to threaten

**dykyryng-** *v.* to make noise on purpose
**dyl-₁ ~ del-** *v.* to sting (of a bee etc.)
**dyl₂** *n.* PLANT. root, vine
**dyl-₃** *v.* to lead. *Songmongaw dylgabae Dilbangkongdang Umangchalmang mu•tynwano.* The leaders of Songmong were Dibangkongdang and Umangchalmang.
**dylang ~ delang** *n.* ART. little house for the spirit of a dead person built close to the house where the dead person is burnt to keep his remains and ashes. The spirit of the deceased will live in this little house until it is burnt in the ceremony called *me•mang saw•eta* about one year after his death and the spirit will go to *Balphakram*.
**dylgaba** *n.* PERS. leader.
**dym₁ ~ dam** *ideo.* bam! thud! the sound of something heavy hitting the ground. (see *dam₁*)
**dym-₂** *v.* to grow (of plants), to sprout
**dymbrubru** *adj2.* shiny
**dymbyl** *n.* PLANT. species of tree or its leaf which can be dried and smoked like tobacco
**dymbyra dymbyra** *adv.* scattered about
**dymchyrang** *n.* ART. type of traditional snare instrument played by plucking (see Photo 116)
**dymdam₁** *adj2.* naked
**dymdam₂** *adv.* gratuitously, simply
**dymdym damdam** *khjyks, adv.* carelessly, just, anyway. *Alaga morotna dymdym damdam hyn•na bai.* Don't just give it to someone else.
**dyng-** *v.* to fight
**dyngdai-** *v.* to dangle
**dyngdang** *adj2.* alone. *Biphagaba thyiokno. Kynsangdo gawigabado dyngdanganokno.* The husband died. Then the wife was alone.
**-dyngdyng** *evsp. V* persistently. *Taw•reksyrupmyng bengblokmyngdo sangumuk pywdyngdyngaimyng mongmamyng mykyranaw syw•chekchekai mu•okno mang ni.* The banana-sucking bird and the toad persistently flew and fought all day, and kept pounding the elephant's eyes repeatedly, the two of them.

**dynggyni** *n.* ANIM. species of fish
**dyngthang** *adj2.* different
**dyngthangmancha** *adv.* especially
**dypyleng-** *v.* to flatten, to make flat. *Gari bengbylokaw dypylengok, ytykyimu bengbyloke pylengok.* The car flattened the toad, so the toad was flat.
**dypyw** *n.* ANIM. snake
**dypywha•saw** *n.* ANIM. species of small snake
**dypywkaram** *n.* ANIM. species of black snake
**dypywkheng** *n.* ANIM. species of green snake (possibly a species of pit viper, see Photo 110)
**dypywnokma** *n.* ANIM. anaconda
**dypywpoda** *n.* ANIM. cobra
**dyra-** *v.* to rape
**dyrang₁ ~ darang** *n.* PERS. people, anyone
**=dyrang₂ ~ =darang** *encl.* (1) plural number. many, a lot of. (2) an approximate amount or approximate time. (see *=darang₃*)
**dyri- ~ diri-** *v.* to hold (onto). (see *diri*)
**dythyi-** *vtr.* to kill (only used for animals)
**dytyi** *n.* KIN. *c, ref, a.* uncle: father's elder brother
**dytyithangmaran** *n.* KIN. *set.* my *dytyi* (father's elder brother) and his younger brother's child
**dyw-** *v.* to add. *Na•sawmung alumung thiksa na•sawmungsa soda dywaimyng rymai sa•a.* 'We cook and eat it with fermented fish and potatoes, with fermented fish and added soda.

# E

**-e₁ ~ -ai ~ -a** *evsp. V* towards the deictic centre. (see -a ~ -ai ~ -e)
**=e₂ ~ =ai** *encl.* (contrastive/new topic). *Magachakdo: "Hai bai•siga biskut sa•khawna" noaidongano. "Hyt man•cha nang•ba atong budi" nowano pherue.* The dear said: "Come on, friend! I want to steak the biscuits." What?! No! What are you thinking?!" said the fox. *Rang nemchie ataknakasyi?* Now that the rain has stopped, what the heck shall we do?

**echaluk** *n.* ANIM. snail
**edisyn** *n.* ART. (< English) edition
**edres** *n.* PLACE. (< English) address
**egyro** *n.* PLANT. species of tree
**ek-** *v.* to separate. *Pheruna hyn•cha sa•wana amak, pherudo jalokno. Baju ekokno. Kynsangdo amakdo dyngdanganok.* Because the monkey gave nothing to the fox, the fox ran away. The friends separated. Later the monkey was alone.
**elmoni** *n.* SUBST. aluminium
**elong** *n.* ANIM. species of fish
**-eng** *sfx.* (makes a numeral into a fracture). -(e)th. *kolgyk<u>eng</u>* one twentieth. *Kolgyk<u>eng</u> chyigyk<u>eng</u> jesykyn ganang phalchido sotbongaba chynaria, kamalnado.* A twentieth, a tenth, however much the revenue of the sale, even fifty percentage, you just offer it to the priest
**engkal ~ ingkal** *n.* ART. handkerchief
**engsyri** *n.* ART. bamboo strip that runs on top of the bamboo floor of a house and has *wa•rap* as its counterpart underneath the floor to keep the bamboo strips that make up the floor in place
**epril** *n.* TIME. (< English) April
**epyl** *n.* PLANT. (< English) apple
**era** *n.* ANIM. species of fish (see Photo 103)
**eskrup** *n.* ART. hinge
**-et** *sfx.* (causative). On transitive verbs this suffix indicates that the action is manipulated, more intense, or the suffix emphasises that the Patient argument is affected

# G

**ga•-**$_1$ *adj1.* good, nice (as a character trait or a property of things). *morot ga•gaba* a good person. *"Na•ange bichi sa•naka, biskute?" "Acha na•nange iawe hyiawchi tyi ga•gabachi tyrywai sa•na bai•siga" noaidongano.* "Where shall we eat then, the biscuits?" "Well, we'll eat them over there, at that nice river while we take a bath", he said. *Dymchyrangsangsa chywgyn ryngaisa na•nangdo ga•chawa.* Celebrating *chywgyn* only with string instruments is not good enough for us.
**ga•-**$_2$ *v.* to trample, to trod
**mai ga•-** to thresh rice
**ga•ak-** *vsec.* to be compelled to, to be forced to. *Mongma wa ni•wamian man•ai sa•chak, khanggal dong•ok. Ytykyimu hapsan nukhung rajasa mu•chido man•ai sa•na neng•ok. Ytykyisa dyngthangdyngthang songchina hapchina jalthokna ga•akok.* Because the elephant tusks were gone, they the people of Badri were not rich anymore, they became poor. So then, if they would stay together in the hundred houses, they would run out of wealth/food. Therefore they were all compelled to run away to different villages and places.
**ga•dak-** *vtr.* to cut up, to cut into pieces, to cut up, *Phylgym chungga•awdo ga•dakaimu ra•akno, kokchenggumuk.* Having cut up the big eagle, they took it with them, a whole *kokcheng* full.
**ga•dap-** *v.* to step on
**ga•dukduk-** *v.* to prod with one's legs or feet. *Gore jalna rakbebeokno. Kha•sinkhalai jalkhalna noaimyng ga•dukdukchiba rakkhalai rakkhalai jalariokno.* The horse ran really quick. Having told it to run slower, whenever he prodded it with his legs, it just ran faster and faster.
**ga•jonong-** *v.* to trample on, to crush, destroy. *Uchi rukpekba: "Hai angba. Ang ha•bilchi nok takai mu•gabaaw phangnan mongmae ga•jononga."* Then the frog said: "Come on, me too. The elephant always crushes the earthen house in which I live."
**ga•jyret-** *v.* to crush with one's foot
**ga•khat-** *v.* to climb. *Amakdo pan ga•khatna man•a.* Monkeys can climb trees.
**ga•kynyng-** *v.* to trample on, to crush, to destroy. *Uchi rukpekba: "Hai angba. Ang ha•bilchi nok takai mu•gabaaw phangnan mongmae ga•kynynga."* Then the frog said: "Come on, me too. The elephant always crushes the earthen house in which I live."
**ga•phak-** *v.* to hit with one's foot while walking

**ga•phynek-** *v.* to stamp to death

**ga•pyret-** *v.* to stamp to death, to crush with one's foot

**ga•pyryw-** *v.* to stamp through something, to pierce by stamping. *Thik thak saphaw butangga rong•khalawan hai•ba mongmaba ga•pyrywman•oknote.* The elephant stamped exactly through the hole where the rabbit had squeezed into.

**ga•reret-** *v.* to tread on, to step on something

**ga•ryngreng-** *v.* to kick

**ga•sokhok-** *v.* to stumble

**ga•sokhok aksokhok** *khjyks, adv.* stumbling

**ga•su-** *adj1.* splendid, cool, terrific

**ga•sylek-** *v.* to sprain one's foot

**ga•syrot-** *v.* to slip and fall. *Wetsa re•engrawrawwachian dobachi amakba ga•sorotaimu bunduk bai•thongsyrangokno.* Once when they went a little further, the monkey slipped and fell in the mud and the gun broke in pieces.

**ga•tha ~ gatha** *n.* PERS. idiot, fool, crazy person (masculine)

**ga•thi ~ gathi** *n.* PERS. idiot, fool, crazy person (feminine)

**ga•thymbylong** *v.* to make a hole in a road or bridge by stamping. *Ge•theng dolongchi ga•thymbylongok.* He made a hole in the bridge by stamping on it.

**ga•thyng-** *v.* to kick

**=gaba ~ =ga** *encl.* (1) attributive. (a) on clauses. *Thawgaba symgaba phangnan sa•rongchagaba jilami bostudyrangaw raai hyn•aimung kha•sin kha•sin gumukawan palyngchi jalgabadyrangaw jykthangthangaw jumuphynetaaknowa.* Having brought and given tasty, sweet things from the district which are usually never eaten, they were slowly able to get their husbands back, who had run away into the jungle. *Nok ang mu•gaba gurumok.* The house in which I lived has collapsed. *O chame, angmi nang•na kha•galgabaaw nang•mi kha•tongchi dang•etna man•phanima?* O sweetheart, will you be able to insert also my love for you into your heart? (b) on numerals. *Byryigaba magachakdo jalan-gaimyng kawna man•ancha.* The fourth deer ran away, so we could not shoot it. *Bigaaw biskut ra•nima?* Which biscuits shall I buy? (c) on the relative time word *dakang. Dakanggaba turachi mu•wachi mobinaw gorongwa.* The first time I stayed in Tura, I met Mobbin. *Ning songchigabadarangdo nemthokaidonga.* We, all the villagers, are all well. (c) on the bound interrogative formative. *Bigaaw biskut ra•nima?* Which biscuits shall I buy? (2) derelational. *Morot sa•banthaigabaaw kynchi baaimu daw•reng kawwano.* A man carrying his son on his back shot the eagle. (3) relational (not productive). Found in only a few words, namely *nokgaba* landlady, landlord, *gawigaba* wife, *biphagaba* husband.

**gada** *n.* ANIM. donkey

**gadang** *n.* ART. shelf

**gadang gadang** *adv.* step by step

**gadaw-** *v.* to lift one's chin up

**gajol** *n.* PLANT. species of red carrot

**-gak** *evsp.* V accidentally. *Chaiparangai tokaimyng, biphagaba nakhungaw tokgakmanaimyng thyisyrangokno.* While she was beating him, she was looking away, and she accidentally hit her husband's nose, and he died on the spot.

**gakgu-** *v.* to nod one's head

**gakji** *n.* PLANT. lemon

**gal$_1$** *n.* CORP. scar

**gal$_2$** *n.* EMO. pride, arrogance. *Phalthang khu•chuk dumgaba sotmaiaw hongkhotna man•chaaimyng thygabaaw gal takokno.* He took pride in the flies which had gathered in his own mouth, and had died not being able to come out.

**rasong... gal...** *coll, n.* praise and pride (see *rasong*)

**gal•-** *vintr.* to fall down. *Kynsangdo rai•wachie napitdo mongma matsana nekarawrawna kyrethyngaimyng phepmyng gal•syrangokno napitdo.* Later, when (the animals) were coming, he feared the tigers, the elephants, the ones that were continuously coming closer, so much, he fell out of the banyan tree, the barber.

**gal•ruru-** v. (1) scatter all over the place. *Bostu gal•ruruaimu rum serabera takok.* Because things are scattered all over the place, the room is a mess. (2) to fall through something. *Dokra pyrywaimu tangka bisyl gal•ruruok.* Because my pocket had a hole in it, the coins fell through it.

**galat-** v. to fall. *Tyikhal patwachi rong•rimylaimu ga•sokhokaimuna, kokcheng galatokno, saphawba galatokno.* When they were crossing the river, because the stones were slippery, the *kokcheng* fell and the rabbit fell too.

**galcha-** v. to boast. *Bakbak rasong taknado thapthap galchanado man•chawa.* It's not easy to quickly get praise, to get pride.

**galdai** n. PLANT. star fruit, carambola, *averrhoa carambola*

**galjak ~ kaljak** n. ANIM. catfish (see Photo 100)

**galon** n. ART/MSRE. jerry can

**gam ~ kam** n. ABSTR. work, wealth, riches, matters, activities. *Kam kha•na harataidong angdo.* I'm lazy. / I don't want to work. *Kam ni•wa.* Worthless. *haw•ai kamai sa•gaba* someone who works hard and struggles to survive. *Gam man•ni udo uan, tangka poisa. Uan gam mynga, dakangmi chasongdo. Te•ewsa kepasyti noai myngaidonga. Chasongna kri gam myngariaro, tangka poisa.* He will obtain wealth, money. Earlier generations called that "*gam*". Now they call it "capacity". According to my generation this money is called "*gam*".

**gam jym** *khjyks,* n. ART. wealth, riches, fortune. *Tangka poisa ha•gylsak gam jymaw khaiaimu, de•thenge parangangokno.* Having collected money and riches, he wandered away.

**gambiri** n. PLANT. species of tree of which traditional drums called *khem* were made

**gamchat-** v. valuable, important. *Breketmyng nyng•chi chipgaba katha pang•ai gamchatcha.* The words in brackets are not very important.

**gamchatga(ba)** n. ABSTR. value

**gamsa** n. ART.(< Indic) a cloth

**gamsili** n. PLANT. species of tree of which traditional drums called *khem* were made

**gan•chang** n. ART. rack above the *tyinok* for plates and other kitchen utensils; rack under the *akan*, where meat is put to dry above the fire (see Photo 15)

**gan•theng ~ ga•theng** n. PLANT. the stem of a leaf or fruit

**gan•thong** n. ART. stick, handle (of knife etc.), stump (of a tree). Clf. *thong. gan•thong thong• ni.* two sticks

**ganang** v. locative/existential verb, to exist, to be, there is/are, to have, to live. *Nang sa•gyrai ganangma?* Do you have children? *Tanka ni•chiba ganangchiba ang nang•aw nemnuka.* Whether you have money or not, I like you. *Ie songchi nok kolachitsa ganang.* There are thirty one houses in this village. *Song dam sachi alsia raja myng• sa ganangchym.* In a village supposedly lived a lazy king.

**gandai** n. ART. big beam that forms the base of the house and rests on the *rong•thai* (see Photo 11)

**gandalak** n. ANIM. species of frog which says *gagagagaga*

**gandi** n. PLANT. a log

**gandurian** n. BODY. umbilical cord

**gandyrui** n. BODY. bellybutton, navel. Clf. *goi•. gandyrui goi• korok* six bellybuttons

**gang-** v. to be erect, to have an erection, to have a hard on. *Nang• ri• gangama?* Do you have an erection?

**-ganggang** *evsp.* bouncily, bumpily, going up and down in a bouncing way. *Ram thymbylong, ytykyimyng gari bytganggangwa.* The road is full of potholes, so the car was bouncing.

**ganggawa** n. ANIM. mosquito

**gangma** n. CORP. pimple

**gangphu-** v. to swell, to blow up (like a chapatti on the fire)

**gangthai** n. BODY. fin (of fish)

**ganthai ~ ganthi** n. ANIM. cicada

**gantheng** n. PLANT. stalk

**ganthirengreng** n. ANIM. species of cicada that makes a very loud and high pitched sound

**gapsan ~ hapsan** *adj2.* (1) the same. *Ie hapsan nok.* This is the same house. (2) as ...as .... *Ang nang•mi/nang•myng hapsan chunga.* I am as big as you. (3) together. *Na•ang gapsan sa•nine.* We'll eat together, OK?

**garamak** *n.* SUBST. soot

**garan** *n.* FOOD. jerky

**gari** *n.* ART. (< Indic) vehicle, car

**Garo** *n.* PERS/ART. Garo (person or language)

**garu** *n.* PLANT. mustard (see Photo 58)

**garuthai** *n.* PLANT. species of grain (see Photo 51)

**gasam**₁ *tw.* afternoon, evening, later part of the day. *Gasam tin bajichi re•engni.* We will leave this afternoon at three o'clock. *myia gasam* yesterday evening/afternoon. *Tai•ni gasam re•engphinni.* I will go back this evening/afternoon.

**gasam**-₂ *vØ.* to be evening. *Gasamok.* It has become evening. *Gasamnaka.* It will soon be night.

**gasam gasam** *adv.* sometimes, seldom

**gasamphang ~ gasamphak** *tw.* afternoon, evening, later part of the day. *Ga•samsangphak chaithirichiba na•awba matdam sa•akno.* When he looked again in the evening, an otter had eaten the fish again.

**gat**-₁ *v.* to dig

**gat**-₂ *v.* to put in/on, to load into/onto. *Phagongmachi sa• gataimyng tyinyng•sang dang•angokno.* Having put the child on his shoulders he entered into the water.

**-gat**₃ *evsp.* V up onto, to start V-ing. *Chengwami achu ambido khemaw rangaw ra•gatnaan...* The first ancestors wanted to collect drums... *Atongbatykyi ga•sokok aksokok hungthamakaimuna saphawba ha•khungchina mangatokno.* Somehow, the rabbit stumbled and crawled up onto the river bank.

**gatdap-** *v.* to stack, to put on top

**gatha ~ ga•tha** *n.* PERS. fool, crazy person (masculine)

**gathi ~ ga•thi** *n.* PERS. fool, crazy person (feminine)

**gawak** *n.* ABSTR. disease

**gawang ~ guwang** *n.* ANIM. spider

**gawanghu•raw** *n.* ANIM. species of spider (see Photo 94)

**gawangsyryng** *n.* BODY. spider web

**gawasu ~ gawsu***n.* BODY. rib. Clf. *tyn. gawasu tyn tham.* three ribs

**gawi** *n.* PERS. female, marriageable girl, girl (unmarried)

**gawigaba ~ gawiga** *n.* KIN. *ref.* wife

**gawsu ~ gawasu** *n.* BODY. rib

**ge•theng ~ de•theng** *ppron.* he/she, his/her, third person singular pronoun, usually referring to animate beings

**ge•thengtheng ~ de•thengtheng** *ppron.* they, their, third person plural pronoun, usually referring to animate beings

**gebeng** *n.* ABSTR. width, breadth

**geer** *n.* ART. (< English) gear (of a vehicle)

**gegydek** *n.* PLANT. species of plant

**gekgek** *ideo.* the call of the hornbill

**geng** *clf.* classifier for long vegetables. *rasunok gengsa.* one spring onion

**genji** *n.* ART. tank top. Clf. *khung. genji khung ni.* two tank tops

**gepgep** *ideo.* quack quack! the call of a duck

**ges ~ kes** *n.* ART. strut: beam that runs, as the long side of a rectangular triangle, between the *manjuri* and the *gandai* in the structure of a house (see Photo 9)

**gesep ~ gysep ~ gisep** *n.* PLACE. space, interval. (see *gysep*)

**getphul ~ lekhaphul** *n.* bougainvillea (see Photo 81)

**giching ~ gyching** See *gyching*.

**ginggang** *adj2.* having, with

**gisep ~ gysep ~ gesep** *n.* PLACE. space, interval. (see *gysep*)

**gisep gisep ~ gysep gysep** *adv.* from time to time. *Na•nage song jan•rukok. Umi gymyn bichiba gisep gisep chiti saietrukarinaka.* Our countries are very far from each other. Therefore we will sometimes write each other letters from time to time.

**git** *n.* ART. (< Indic) a song, music where there is singing

**githyng ~ gythyng ~ githing ~ gi•thyng** *adj2.* unripe, uncooked, raw

**gobormen ~ golmen** *n.* PERS. government

**godot-** *v.* to bump. *Cha• rong•chi godotwa*. I bumped my foot on a stone

**gogak** *n.* ANIM. beetle (see Photo 88)

**gogat-** *v.* to carry on the shoulders

**gogylek** *n.* ANIM. cock, rooster, cockerel

**gogyrek** *n.* PERS. a baby with its neck bent sideways while it is being carried on the back

**goi** *n.* PLANT. betel nut, areca nut, *Areca catechu*

**goi•** *clf.* non-specific classifier

**goichara** *n.* PLANT. a betel nut sapling, a young betel nut tree

**goilapan** *n.* FOOD. betel nut and paan/pan

**goira** *n.* GEO. thunder. *Goira kawa*. The thunder roars. *Goira byl• tan•ok*. Lightening has struck.

**Goira** *n.* SUPER. the god of thunder. *Goira kawa*. The god of thunder shoots. / The thunder roars. *Goira byl• tan•ok*. The god of thunder has struck. / Lightening has struck.

**gol ~ gool** *n.* ACT. goal. *Ge•theng gol sa•ak*. He scored a goal (in football).

**golap** *n.* PLANT. (< Indic) rose

**golmal ~ gormal** *n.* ACT. a fight, a quarrel, chaos

**golmen ~ gobormen** *n.* PERS. (< English) government

**golpho₁** *n.* ART. story
  **golpho kha•-** (1) to talk extensively. *Kynsang golphook golphook. Golpho kha•wachie walangaidok*. Then, they talked and talked. While they were talking, night was falling. *Tai•sa anga wen•sado uaw golpho kha•ak*. I have just talked about that. (2) to gossip. *Boba myng• sagabachi myng• abun bobachi te•ewdo bobarara myng• ni golpho kha•rukokno*. The first crazy person to the other crazy person, now, among crazy persons, the two of them gossiped to each other.

**golpho-₂** *v.* to talk extensively

**gom₁** *n.* PLANT. (< Indic) wheat, pasta

**gom-₂** *v.* to bend

**gomagundai** *n.* PLANT. species of thick banana

**gomga ~ gomgaba** *n.* ANIM. leech

**gompara ~ gompyra** *n.* ANIM. species of large, poisonous, black ant

**gomynda** *n.* PLANT. pumpkin (see Photo 50)

**gomynthyri** *n.* PLANT. species of vegetable

**gondu** *n.* ANIM. rhinoceros, rhino

**gong- ~ gong•-** *v.* to agree, to be willing. "*Tanka rong raja ni phalni.*" "*•hmm• gongcha anga.*" "I'll sell it for two hundred rupees." "No, I don't agree." "*Ha• ambi ang chakaw khenetkhu*" nowano. "*Gon•gchak*" nowano. "Hey grandchild, scratch my arm some more!" she said. "I don't want to anymore", she said.

**gongchit** *n.* ANIM. stag beetle

**gongdang** *adj2.* bent. *Pan gongdang takgabachi ne• nangwanote*. There was a bees' nest hanging from a bent tree branch.

**gonggong-** *v.* to bend over

**gop-** *v.* to bury, to hide, to cover up. *Morot thyigabaaw hanep gopnaka*. Tomorrow they will bury the dead person. *Nang• baba noksamchi tangka gopgaba ganangno*. Under your father's house lies buried money. *Thengthon morot tangka bisyl pang•ai khaigabaaw nukokno. Ytykyimyng hap dam sachi syruk syruk gopaidongano*. Thengthon sees a man who was carrying a lot of coins. So then, he is secretly hiding them.

**gopgylang** *adj2.* hollow

**gopjyrujyru-** *v.* to look down with one's head bent down

**gopram** *n.* PLACE. grave

**gora** *n.* ART. large earthen pot in which rice liquor (*chyw*) is made (see Photo 4)

**gorai ~ gore** *n.* ANIM. horse

**gore dung•-** to ride a horse

**gorial** *n.* ANIM. (< Indic) crocodile

**gorong-** *v.* to meet. *Ram thong• sachi bydyi gorongokno*. Halfway, they met an old man.

**-gorop** *evsp.* V with a whole group, V together. *Uchie jyrym tymaimyng raw•okno, jinmatykyi pyi•goropokno*. Then, the people of the village, having quietly lain in ambush, caught Thengthon, they grasped him all together.

**gorothop** *n.* PLANT. species of small leafy green

**gorweng ~ wengwang** *n.* ANIM. species of cicada that makes the noise of a screaming baby or a woman being murdered

**grem** *clf.* (< English) gram, gr.

**gremyr** *n.* ART/ABSTR. (< English) grammar

**guchung** *n.* ART. ladder (see Photo 9)

**guduk-** *v.* to wiggle, to budge, to move slightly, to be unstable, to wobble, to move unstably. *Rong• gudukaimu galatok.* Because the stone moved, I fell. *Na•lam gudukwachie te•ewdo tyi thangpytpytaimyng jyksaiaiawan Nawengawmu Kumiribaawma•khamoknowa.* When the *na•lam* (species of fish) wiggled, water splashed on the married couple Naweng and Kumiri and burned them. **guduk tak-** almost verb. *"Aia! Udo magachakdo khorate" noaimyng rykoknowa. Tharapna guduk takwachiba tarakai jalariano magachake.* "Hey, this deer is lame!" he said and chased after it. When he almost caught up with the deer, it run away fast, the deer.

**gugyreng** *n.* ANIM. species of grasshopper (see Photo 86)

**gugyrengsa•** *n.* ANIM. species of bright-green grasshopper

**gukchepchep** *n.* ANIM. species of grasshopper (see Photo 87)

**gukmadym** *n.* ANIM. grasshopper (see Photo 89)

**gul- ~ jul-** *v.* to walk through the jungle with difficulty.

**gulgul galgal** *ideo.* growl! growling noise that the stomach makes. *Pipuk gulgulgalgal takaidonga.* My stomach is growling.

**gumi** *n.* KIN. *d, ref, a.* brother-in-law: (1) elder sister's husband (2) husband's elder brother

**gumithangmaran** *n.* KIN. *set.* my husband and my younger sister or younger brother together

=**gumuk** *qtf.* all, whole, everybody, everyone, everything. *Songgumukan ue mongmawana wai khurutaisa boli hyn•aisa man•ai sa•thokwanochym.* The whole village had supposedly become rich because they made offerings to the elephant tusks. *Ytykyisa dyngthangmancha ge•theng walchi walgumuk thymai chaisyrangwano.* So then, at night, the whole night, he especially lay in ambush and watched intensely.

**gumuk gamak** *khjyks, qtf.* altogether, in total. *Gumuk gamak angna ma•su mang raja sa hyn•etwa angnado.* Altogether, they gave me a hundred cows.

**gumukan** *expr.* (1) everybody, everyone. *Dakangmi pichammi kamdyrangdo, gum ukan songsyrek dong•butungchido bylongen han•senga.* As for how things were in the past, when everybody practiced animism, we were very happy. (2) all (of them). *Na•dyrangdo uaw rukpek bisi ryngaimu gumukan thyithokoknowa.* The fish, having drunk the frog's poison, all died. *Thawgaba symgaba phangnan sa•rongchagaba jilami bostudyrangaw raai hyn•aimung kha•sin kha•sin gumukawan palyngchi jalgabadyrangaw jykthangthangaw jumuphynetaaknowa.* Having brought and given [them] tasty, sweet things of the region which are not usually eaten, [they] recollected all their husbands, who had run away to the jungle

**gumuksangan ~ gumuksang** *prof.* everywhere. *Gumuksangan ukching ganang.* There are leeches everywhere. *Ge•theng gumuksang re•engok.* He went everywhere.

**gun montyro man•ga(ba)** *n.* PERS. person who can control the spirits

**gunda** *n.* PERS. criminal, brawler, fighter

**gungsynung** *n.* PLANT. species of tree

**guri** *n.* GEO. mist, fog. *Guri thupa.* The fog is thick.

**guruchup-** *v.* to be shrouded in clouds. *Waimong nukcha, guruchupok.* Waimong mountain is not visible, it is shrouded in clouds.

**gurum- ~ gyrum-** *v.* to collapse, to break off and fall down. *Banggyriaimu nok gurumok.* Because of the earthquake the house has collapsed. *Narykhelchak gurumok.* The leaf of the coconut tree has broken off and fallen down.

**gusu-** *v.* to cough

**gusum-** *v.* spoiled, off (only used with meals). *Mai ja•bek gusumok.* The rice and curry are spoiled.

**gusylak** *n.* CORP. abrasion. *ang randai gusylakok* I got an abrasion.

**guthini ~ guthyni** *n.* ART. (1) spear (2) bamboo spear which is part of an elephant trap (3) walking stick

**guthum ~ gythym ~ gythum** *n.* ART. village

**guwang ~ gawang** *n.* ANIM. spider

**gyching ~ gycheng ~ giching**$_1$ *n.* PLACE. vicinity, side

**gychingchi ~ gychengchi ~ gichingchi** next to, near, close to. *Jyksaian phong•gychengchian mu•aidonga.* The married couple are sitting near the cooking place.

**gyching ~ giching**$_2$ *adj2.* aslant, slant, diagonal, at an angle, inclined

**gyching ~ giching**$_3$ *n.* ABSTR. angle, inclination

**gychingching mu•**-to be tilted, to make an angle

**gyl-**$_1$ *adj1.* strong. *Usang, songga Manggagremi banthaidarangba rai•aaithokaidongano, Rakarelwakmadare, Gyrynggyrang, Saljapang, Aragundi, Motbanda, Asyngduraparaba gumukan rai•athokaidongano. Chakphong gylgabasano, kara khyrynggabararasano, alamylachagabasano.* They are all coming to there, the young man from the strange village of Manggare: Rakarelwakmadare, Gyrynggyrang, Saljapang, Aragundi, Motbanda, Asyngduraparaba, they are all coming. They are men with strong arms and tight veins all over, they are not ordinary men.

**gyl-**$_2$ *v.* to collect, to gather

**gylarong ~ siwi** *n.* PLANT. species of sea bean or its pod (see Photo 69)

**gylas ~ gilas** *n.* ART. (< English) glass or its volume, glassful. *cha gylas ni* two glasses of tea. Clf. *goi•. Gylas goi• tham bai•ok ge•thene.* He has broken three glasses.

**gylgyl-** *v.* to roam

**gylja** *n.* PLACE. church

**gyljanok** *n.* PLACE. church

**gymyn** *postp.* (takes genitive-marked complement) (1) because (of), that's why, reason, cause. *Ue gam pang•wami gymyn kam pang•wami gymyn ge•thengtheng mykbyrukokno.* Because of this wealth and these riches, they had become jealous of one another. *Umi gymynsa ie hapawe Badri Rongdyng Ha•wai noyi te•chinakhyngkhyng myngwano.* That's precisely why this place is still called Badri Rongdyng Ha•wai up till now. (2) with, for. *Ge•thenge alu kobi habijabi ytykyi samchakdarangmynggymyn bagan takwano.* He had made a garden with potatoes, cabbages and all sorts of vegetables. (3) about. *Uan jorami gymyn cho•sa golpho ka•etwa.* I have told a bit about that love match.

**gyngjangjang** *n.* neck

**gyp** *ideo.* thunk!, tap!, bam! a hitting sound. *Ue uawdo kunsang gyp satetok.* He hit him bam! with a stick.

**gyrum- ~ gurum-** *v.* to collapse, to break off and fall down. (see *gurum*)

**gyrym** *n.* PLANT. bush, patch

**gyryp-** *v.* to cover

**gyryw-** *v.* to shake (an object that you can pick up, a non-fixed object)

**gysep ~ gisep ~ gesep** *n.* PLACE/TIME. (1) space, interval. *chaksigisep* the space between one's fingers (2) between. *Manap walgysepchi ningdo kam rakai takok.* Between morning and night, we worked hard. (3)(after an noun phrase marked by the genitive enclitic =*mi ~ =myng*) in the meantime, meanwhile. *U<u>mi</u> <u>gesep</u>chian de•thengdo ha•gylsakaw chol takanchychiba jamok.* Meanwhile, he had tried all [sorts of] livelihoods, and had spent it all. (4) )(after an noun phrase marked by the genitive enclitic =*mi ~ =myng*) upon. *Waiphinwa<u>mi</u> <u>gesep</u>chi baratdugaaimu...* Upon his arrival he felt very much ashamed.

**gysep gysep ~ gisep gisep** *adv.* from time to time. *Na•nage song jan•rukok. Umi gymyn bichiba gisep gisep chiti saietrukarinaka.* Our countries are very far from each other. Therefore we will sometimes write each other letters from time to time.

**gythym ~ guthum ~ gythum** *n.* PLACE. village

**gythyng ~ githyng ~ githing ~ gi•thyng** *adj2.* unripe, uncooked, raw

# H

**ha** *interj.* (1) hey! interjection to get someone's attention. *Ha! achudyrang, nangtyme bisang?" nooknote.* Hey grandsons, where are you going? (2) huh? interjection to express surprise. *Uchi ue sa•gyraiba: "Achu!" nookno, "Achu!". "Ha? Atong?"* Then, the child said: "Grandpa!", "Gran dpa!". "Huh? What?"

**ha•** *n.* SUBST. soil, earth, land

**ha• haw•-** *v.* to cut/clear the land to make a *ha•ba*. *Phasgaba ha•haw•chenga. Umungdo marsja somaidarangchi saw•a.* First, the land is cleared. Then, around the month of March, the land is burnt.

**ha• haw•ai sa•-** *expr.* to live from agriculture, to live from tilling the soil. *Ue gawile ha• haw•ai sa•arongnotyi, usangaw Rangpynramsangaw ha• haw•ai sa•airongno.* That woman lived from agriculture and she tilled the soil there at Rangpynram.

**ha• kam-** *v.* to clear the field, to cut the jungle to make a field, to tear out or cut weeds

**ha• kham-** *vpan.* to burn the land (land that is prepared for agriculture). *Umungdo marsja somaidarangchi saw•a. Saw•aisa ha•khamchido khynna nangcha. Ha•khamchachido ha• khyn-chenga. Ha• khynman•wamungsa maisi khita.* Then, around the month of March, the land is burnt. If it is burnt and everything burns, it is not necessary to collect any remaining cinders. If not everything is burnt, the remaining cinders are collected first. Only after collecting the cinders, millet is sown. (see Photo 20)

**ha• khyn-** *vpan.* to collect the remaining cinders after burning the field to make it ready for agriculture. (see *ha• kham*)

**ha•** *procl.* Take this. Take this from me. Here, take this. Here. Here you go. *Ha•, chabi.* Take this, the key. "Chama?" "Ho•ong" "Ha•. Ryngbo." "Tea?" "Yes" "Here you go. Drink."

**ha•ba** *n.* PLACE. rice field, swidden, slash-and-burn field, dry rice and vegetable field on the slope of a hill made by cutting away and burning the jungle. Clf. *tym*. *ha•ba tym ni.* two slash-and-burn fields (see Photo 21 and Photo 25)

**ha•bacheng-** *vB.* to start, to begin. *Uchisa matsana makbulna mongmana paichaaimung byldyng byldang jalna ha•bachengok.* Then, not bearing the tigers and elephants anymore, they started to run all over the place.

**ha•bachenggaba** *n.* ABSTR. (1) beginning (2) Genesis

**ha•banok** *n.* PLACE. rice field house (see Photo 21)

**ha•bong** *n.* ANIM. species of ant (see Photo 91)

**ha•bykung** *n.* SUBST. sand

**ha•byreng** *n.* PLACE. old *ha•ba* (see Photo 22)

**ha•byri** *n.* GEO. hill, mountain. Clf. *thut*. *ha•byri thut tham.* three hills/mountains

**ha•chak** *n.* ART. wages

**ha•chang ha•chang** *adv.* one after the other. *Ha•chang ha•chang phorenmi morot phetaaidonga.* Foreigners are arriving one after the other.

**ha•chepchep** *n.* ANIM. grasshopper

**Ha•chyk** *n.* PERS/ART. Garo (person or language)

**ha•dawak** *n.* PLACE. lower side of a hill, low ground

**ha•gun** *n.* PLACE. old plot of land in a *ha•ba*. *Bai•damdo haw•angman•gaba ha•gun sa•angman•gaba ha•rynthangthangaw kanga.* Some people occupy their old already cut plot, their own parcel which is already used completely.

**ha•gylsak ~ ha•gyrsak** *interj.* interjection of astonishment. my word! Jeez! unbelievable! *Myla ha•gylsak sa•gyraido!* My word! That child is small.

**ha•gylsak ~ ha•gyrsak** *n.* ABSTR/PLACE. (1) the world, the earth. *Mekalaia ha•gelsakgumukchi wabatsyranggaba.* Meghalaya is the rainiest place on earth. (2) everything, all, nothing. *Mamungba ha•gylsak ni•wa.* There is absolutely nothing.

**ha•jagyra** *n.* ACT. the first weeding of the *ha•ba*. *Mai kai•manwamungsa ha•jagara kama.* Having planted the rice, we weed the land for the first time.

**ha•kha**₁ *n.* PLACE. hillslope, a steep slope, upper side of a hill, high ground. *Ha•khasang tawangbo.* Climb to a higher part of the hill.

**ha•kha-**₂ *adj1.* very tight *Ian ha•khaai kha•bo.* Tie this very tightly.

**ha•khong** *n.* PLACE. valley

**ha•khung** *n.* GEO. river bank

**ha•khyng** *n.* PLACE. land that belongs to a rich person or *nokma*.

**ha•kym** *n.* a bond between two families, where the woman marries a man who belongs to her father's mahari. If a woman marries a man from a mahari other than her father's, her husband has to undergo a process called *jyw• ra•a* as part of an affidavit into the wife's father's mahari to legalise his bond as member of the wife's father's mahari.

**ha•mai** *n.* ANIM. species of white earthworm

**ha•mang** *n.* SUBST. soil, earth, clay

**ha•mangkyrang** *n.* ANIM. scorpion

**ha•mangrong** *adj2.* brown

**ha•mat** *adv.* troublesome

**ha•pal** *n.* PLACE. (1) field. *Bydyi myng•sa ha•pal khamaidongano, ma•susang ha•pal waiaidongano.* An old man was working in a field, he was ploughing the field with his cows. (2) outside *Nokha•palchi mu•ni.* We'll sit outside the house.

**ha•rongrong** *n.* PLACE. lower side of a hill, low ground. *Ha•rongrongsang wylangbo.* Go down to a lower part of the hill.

**ha•ryn** *n.* PLACE. plot of land, parcel. *Songgumuk thom•aimyng ha•ba ha•ryn ha•rynaw sowalni.* The whole village gathers and will divide the *ha•ba* parcel by parcel.

**ha•saw** *n.* ANIM. species of black snake with red neck and head

**ha•sel** *adv.* for no reason, uselessly, troublesome

**ha•sel... ha•mat...** *coll, adv.* troublesome. *Gythym phakthangthangsangmi morotdarangan khata man•aimyng nemchaka. Ha•sel dong•ok, ha•mat dong•ok.* The people from three neighbourhoods heard the news, and turned bad. There was trouble.

**ha•song** *n.* PLACE. village and surrounding lands, area that can comprise several *gythym ~ gythum ~ guthum* and the surrounding lands, country

**ha•thapyra** *n.* SUBST. ashes

**ha•thywkong** *n.* GEO. a puddle

**ha•tykylok** *n.* GEO. a puddle

**ha•wai** *n.* PLACE. plain area

**habijabi** *adj.* all sorts of

**hachi-** *v.* to sneeze

**hai** *procl.* Let's go! Come on! *Hai! Rai•naka!* Come on! Let's go! *"Rai•nakama?" "Hai!"* "Shall we go now?" "Let's go!"

**hai•e ~ hai•-** *interj.* (1) pause filling interjection. Let me see...; uh...; erm... ; whatchamacallit? *Utykyi dongwano ie Wiliamnagarmi hai•e Do•renggo Wa•dachongmi histyri.* That's how it is, Williamnagar's... uh... Do•renggo Wa•dachong's history. (2) this thing, such and such happened, be like..., whatever. *Uchie karydyl chunggaba hai•wano: "Ha• ambi ang chakaw khenetkhu" nowano.* Then the hanging root was like: "Hey grandchild, scratch my arm!" she said. *"Phylgymsa hai•wa na•a ue. Garu ramgabachi na•nang garu ramtananggachi di•etdapai tangwa" nookno.* "The giant eagle did this, oh you! In the dried mustard, in our dried mustard, he left a big shit", she said. *Ytykyi phetaaimungna hai•okno, nokchina janggalan: "Bie nang•jongdyrange? Nang• jonge bie?" nookno janggalchi syng•okno.* So when they had arrived, it was like..., at home, all of them: "Where is your younger brother? Your younger brother, where is he?" she said. *Uchiansega hai•okno, pheru nuksegaakno sa•wamiaw.* Then in turn, this thing happened, the fox spotted some food. *Rymai sa•wachie, amakdo pan ga•khatna man•ano. Khu•sumdo hai•okno. Khu•sumdo ga•khatna man•chano.* Having cooked and eaten it, the monkey is able to climb into a tree. As for the turtle, he did whatever. The turtle cannot climb trees.

**haida** *procl.* I don't know.

**hajal ~ hajar** *bound. num.* (< Indic) thousand. Despite being written as separate orthographic words, this numeral is phonologically bound to the following multiplier, e.g. *hajal sa* [hadʑal'sa] 'one thousand'.

**hajam-** *v.* to yawn

**hajira** *n.* ART. (< Indic) daily wages

**hal-** *v.* to feed (give food to someone)
  **mu•thai hal-** *v.* to breastfeed

**hala kha•-** *v.* to wake someone up, to disturb someone

**haldun-** *v.* to feed, to provide for. *"Ie alsia raja atykyi khengaidok? Atykyian jykaw haldunna man•aidok?" noai morotdyrang chanchiphinaidoknoro.* "How does this lazy king live? How does he feed his wives?" thought the people.

**halsia ~ alsia** *n.* PERS. (< Indic) lazy person

**ham-** *v.* to build, to construct

**hama ~ nokhama** *n.* PLACE. under, underneath, below, space between the floor or the base of something and the ground. *Taw•sa•grai nokhamaaw jalphakangaidonga.* The chicks are running under the house (from one side to the other). *Ang tankabek palonghamachi.* My wallet is under the bed.

**hambun** *dtw.* later (but not today), in the future

**hampyi** *dtw.* in the late afternoon, in the evening

**han•cheng** *n.* SUBST. sand

**han•dykmai** *n.* CORP. jungle fever
  **han•dykmai sa-** to be ill with jungle fever

**han•dyng** *n.* ART. beam that forms part of the base structure of a house, perpendicular to the *gandai* (see Photo 9 and Photo 10)

**han•saw-** *v.* to enjoy

**han•seng-** *v.* be happy, be joyful, to enjoy. *Angdo Garo Hillschi bylongen hans•senga.* I very much enjoy being in the Garo Hills. *Na•do amak di sa•aimu hansengthokaidoknoa.* After the fish had eaten the monkey's shit, they were all very happy.

**han•tung-** *adj1.* to be a dangerous place. *Mungma pang•wachi palyng han•tunga.* When there are a lot of elephants the jungle is a dangerous place.

**han•tung-** *v.* to feel secure, to feel safe. *Morot pang•ai rai•bo, hanthungkhala.* Go with lots of people, you will feel safer.

**hanep** *n.* tomorrow. *Hanep san saanok.* Tomorrow there will be one day left.

**hang-** *v.* to warm one's hands by the fire, to dry meat or other things near the fire

**hang•khal** *n.* GEO. cave, hole. Clf. *khal*. *hang•khal khal chatgyk* eight caves

**hanggal** *n.* SUBST. charcoal, cinder

**hangkyn** *n.* ANIM. species of termite

**hangkyn raja** *n.* ANIM. species of termite (see Photo 92)

**hanthi-** *v.* (1) to divide. *Songgumuk thom•aimyng ha•ba ha•ryn ha•rynaw hanthini.* The whole village gathers and will divide the *ha•ba* parcel by parcel. (2) to share. *Je ha•ryn ni•gababado uan hanthirukai haw•a.* As for those whoever does not have a plot, those mutually share and clear the land. *Angdo dyngthangmancha nang• kha•galchido nang•mi gamaw angna hathiphabo.* If you love me especially, share your wealth with me / divide your wealth for me.

**hap₁** *n.* GEO. place. *Umigymynsa ie hapawe Badri Rongdyng Ha•wai noanowa, aro rangawba mykha badri myngwanowa.* That's why this place is called Badri Rongdyng Ha•wai, and the rain is also called *mykha Badri*.

**hap₂** *n.* QUANT. half

**haphu-** *v.* to blow

**hapjyt-** *v.* to move house

**happen** *n.* ART. (< English) short pants, shorts

**hapsan ~ gapsan** *adj2.* (1) the same. *"Ie bek dyngthanma?" "Yhy•, hapsan."* "Is this bag different?" "No, it's the same." (2) as … as. *Ang nang•mi/nang•myng hapsan chunga.* I am as big as you. (3) together. *Na•nang gapsan sa•nine.* We'll eat together, OK?

**harat-** *v.* lazy, reluctant,. *Kam kha•na harataidong.* I'm lazy.

**hat-** *v.* to copulate, to fuck. *Teraka krismassomaichi ue gawiaw babylsichi nokwengchi hatok angdo.* Last year at Christmas I fucked that girl on the floor in the kitchen.

**haw•₁** *n.* KIN. *drel, c, ref.* uncle: mother's brother (addressed as *mama*)

**haw•₋₂** *v.* to clear/cut the jungle to make a rice field **haw•ai kamai sa•-** *expr.* to work hard to survive. *Ue songmi morot haw•ai kamai sa•gaba gumukan.* The people of that village are all people who work hard and struggle to survive.

**haw•maran** *n.* KIN. *set.* my *haw•* (mother's elder or younger brother) and his (elder or younger) sister's unmarried child

**haw•nokhol** *n.* KIN. *ref.* father-in-law, deceased father-in-law's heir (addressed as *mama*)

**haw•nokholburung** *n.* KIN. *set.* a group of fathers-in-law (*haw•nokhol*) and sons-in-law (*kynokhol*)

**hawchi** *dem.* over there, yonder

**hawe ~ haw-** *dem.* remote demonstrative. that over there, those ones over there. *Angdo hawe nokchi mu•aidong.* I am staying in that house way over there. *"Bisang re•engoktyi, Lawroie?" "Hawsang, nygylrygynsang re•engok."* "Where has Lawroi gone?" "He has gone way over there, to behind the market."

**hawtyi** *adv.* for some time. *Thorokaimyng hawtyi rypokno magachake.* Having jumped down, he stayed in the water for some time, the deer.

**hen•** *n.* ART. knot

**heng•-** *v.* widely spaced, sparse

**henraiting** *n.* ART. handwriting

**het-** *v.* to clean an orifice or hole, to blow one's nose. *Nakhal hetbo.* Clean your ears!

**hetmastel** *n.* PERS. (< English) headmaster

**hetmadam** *n.* PERS. (< English) headmistress

**hijra** *n.* PERS. (< Indic) gay person, homosexual

**hil** *n.* ART. (< English) heel (of a shoe). *Te•ew re•enggaba gawi, longpen kanai juta hilaw...* (Wilseng S. Marak) The girl who just went by, wearing trousers and shoes with high heels...

**hira** *n.* SUBST/ART. (< Indic) diamond

**histyri** *n.* ABSTR. history

**hit-** *v.* to command, to order. *Usangphak nang• re•engbo, usangphak nang• re•engbo". Ytykyi hitrumokno.* "You go that way, you go that way." He commanded them like that. *San sachi umyng gawigaba uaw kam kha•khalna hitoknowa.* One day, his wife ordered him to do some work.

**hm•m ~ •mhm• ~ m•m ~ mm ~ •m** *procl.* I disagree. (see *•mhm•₁*)

**ho-** *v.* to jump

**ho•ong** *procl.* I agree, That's right.

**hochorokchorok-** *v.* to jump like a deer

**hogol** *n.* snoring

**hogol ra•-** *v.* to snore. *Juwchenwachi nang• hogol ra•wa.* For the first part that you were asleep you snored

**hojokjok-** *v.* to jump up and down

**hok-** *v.* to call loudly

**holdiasop** *n.* CORP. jaundice, yellow fever

**Holen** *n.* PERS/PLACE. (< English) Holland, Dutch *Holenmorot* Dutchman, Dutchwoman

**hongkhot-** *v.* (1) to come out, to exit *Tam•ai chaichie te•do byirakhem hongkhotruruaimu kaksyrangokno pheruawdo.* When he tried hitting it, the bees all came out and bit the fox all over. (2) to ejaculate, to cum.

**hopat-** *v.* to jump like taking a step, i.e. with one's legs apart, not with both legs together

**hospytyl** *n.* PLACE. (< English) hospital

**hot-** *v.* to extract, to draw a knife

**hu•raw** *n.* ANIM. gibbon, *Hylobates hoolock*

**huhu** *interj.* Hello? Is someone there?

**huk-** *v.* to sweep together

**huksetgaba ~ huksetga** *n.* ART. dustpan

**hung-** *v.* to swim. *Kha•sinsin gorialdo hungtawai takokno.* The crocodile swam very slowly upstream.

**tyi hung-** *vpan.* to swim. *Na•a tyi hungna sapama?* Do you know how to swim?

**hup-** *v.* to suck

**husyring** *n.* ANIM. rabbit

**hy• ~ hy•y ~ yhy•** *procl.* I don't agree.

**hy•y•y•yw** *interj.* interjection of emphatic admonishment, vehemently indicating that the interlocutor is wrong and should be careful about what he/she is saying

**hyiawchi** *dem.* over there, yonder

**hyiawe ~ hyiaw-** *dem.* (very remote demonstrative) that way over there, that one way over there, those ones way over there. "Bichi nang• noke?" "Ang nokdo hyawe, ha•byrikhambaichi." "Where is your house?" "My house is way over there, on the top of the hill." *Hyiawe Rongsumyng ha•banokan.* Those ones way over there are the rice-field houses of Rongsu.

**hyits ~ hys ~ hyis ~ tyis ~ tys ~ yis** *interj.* (interjection that expresses disapproval or indignation) Hey! Ugh! What?! What the...?! Tsk-tsk! *"Hai bai•siga, biskut sa•khawna." "Hyt man•cha nang•ba atong budi".* "Come on, friend, I want to steal those biscuits." "What?! No! What kind of idea is that?!"

**hyn•-** *v.* to give. (1) When the recipient is marked with the goal enclitic =*na* the object given was meant to become the possession of that recipient. *Tangkaaw Samratna hyn•bo.* Give the money to Samrat. *Angna tangka ratja banga hyn•etbo.* Give me five hundred rupees. (2) When the recipient is marked by the locative enclitic =*chi*, the recipient is a temporary keeper of the object given, i.e. the locative-marked recipient will give the object to someone else later. *Tangkaaw Samratchi hyn•bo.* Give the money to Samrat (so that he can give it to someone else). This locative construction in Atong is functionally equivalent to the Thai construction using the word ฝาก in: ผมฝากเงินให้เขาไปให้คุณ I give him the money to give to you. (3) to wish. *Nang•ba happy new year hyn•etaidong.* We wish you a happy new year too.

**hynggek** *n.* ANIM. hornbill

**hys** *interj.* (see *hyits*)

**hyt** *interj.* (1) interjection of irritation or anger (2) interjection to chase an animal away

**hyw** *interj.* interjection of admonishment, indicating that the interlocutor is wrong and should be careful about what he/she is saying *Alsia rajae: "Ie gari biskyn?" Phalgabae: "Hyw! na•a ra•nae syng•e syng•chagaba, taknae syng•e syng•chagaba."* The lazy king: "How much is this cart?" The salesman: "Hold on! You are not going to buy anything. You're just asking. You're just pretending, just asking."

## I

**ie ~ i-** *dem.* proximal demonstrative. (1) this, these. *Ie baik biskyn?* How much is this bicycle? *Ie komyla biskyn?* How much are these oranges? *Ian Badrimyng oltho.* This is the meaning of Bari. (2) this one, these ones. *"Ha•, lekha khung tham." "Ma•, ytykchido, iaw nangchawa. Iaw khung niawsa poraini."* "Here, three books." "Very well, in that case, I don't need this one. I will read only these two." (3) he, she, it. *Ido tyngchawa.* He doesn't know.

**ileksyn** *n.* ACT. election

**ilektrisiti** *n.* ART. electricity

**inchi** *n.* clf. (< English) the width of the upper joint of the thumb, i.e. the joint under the nail

**India** *n.* PLACE. India

**inggeech** *n.* ACT. engagement. *Inggeech kha•ak.* I am engaged to be married.

**Inggylan** *n.* PLACE. (< English) England

**Inggylis ~ Inglis** *n.* (< English) English (language or person)

**ingkal ~ engkal** *n.* ART. handkerchief

**insuren** *n.* ART.(< English) insurance

**iskyn ~ isykyn** *dem.* (1) indicating the extent of a property. *"Biskyn chunga, na•an?" "Iskyn chunga."* How big was it, the fish? "This big." (the size is also indicated by stretching out the arms.) (2) indicating emphasis on a certain property. *Na•a myltengteng. Ang isykyn madam kam kha•phinok.* You are still too small. I am already working as a *teacher* (emphasis on *teacher*).

**Isol ~ Isor** *n.* ABSTR. (< Hindi) God

**istyr** *n.* TIME Easter. (< English) *Bichi ister sa•nima, na•a?* Where will you celebrate Easter?

**isykyn ~ iskyn**$_1$ *dem.* so much, so many, such, this much, this many. *Anga iskyn gamaw* hanthietok. iskyn jan•gaba songsang de•theng jalangok. I have divided so much of [my] wealth. He has run away to such a faraway country. Kynsangdo bean bebe iskyn san iskyn somai thik kha•aknoaro. Later, truly, they agreed to meet on such and such a day and at such and such time.

**isykyn ~ iskyn**$_2$ adv. this. Biskyn chunga?" "Iskyn chunga." "How big?" "This big."

**itha ~ ita** n. ART. brick. itha thut sa one brick

**itihas** n. ART. history. Ian Badrimi itihas khata machongdo jamok. The story of Badri, the words of the first mother/clan, is finished.

**itykgaba** n. such. Itykgaba chasongchi mamungba duk ni•wa. In such eras/generations, there was no sorrow.

**itykyi** adv. like this. Angdo itykyi balaimu tanarinaka. Having spoken like this, I'll just stop.

**itym** ppron. (third person plural personal pronoun) they, them, their

# J

**ja**$_1$ n. autoclf. TIME. month, moon. Ja phetok. The moon has risen.

**ja**$_2$ interj. interjection to chase a cow away

**ja•bek** n. FOOD. curry (anything that is eaten with rice)

**ja•chung** n. KIN. d, ref, a. Badri dialect: sister-in-law: (1) wife's elder sister (2) elder brother's wife. Siju dialect: wife's elder sister

**ja•chungthanmaran** n. KIN. set. my husband and my elder sister together

**ja•ga** n. ART. a trap. Ja•ga saakno uchie, taw•pang•ai banokno. They set traps and then caught many birds. ja•ga sagaba someone who sets traps, an enemy

**ja•garaw ~ ja•gyraw** n. species of tree of which the big leaves are used as rai•chak to pack food in

**ja•garu** n. PLANT. species of vegetable

**ja•jol ~ ja•gol** n. PERS. person with long legs

**ja•khop ~ cha•khop** n. ART. shoe

**ja•naw** n. KIN. ref, a. (1) elder sister. (2) Also used to address an older female cousin or a woman older than the speaker. (The word abi is more respectful as a term of address for both referents.)

**ja•nawburung** n. KIN. set. a group of sisters

**ja•nawmaran** n. KIN. set. two sisters

**ja•phang** n. PLACE. foot of a tree.

**ja•ryt** n. PLANT. chilli pepper (see Photo 56)

**ja•rytbok** n. PLANT. species of plant

**jabol** n. SUBST. garbage

**jaboldam** n. ART. garbage heap

**jabyra** n. PERS. fool, crazy person

**jada** n. PERS. stupid person, idiot

**jadu** n. ACT. (< Indic) magic

**jagat-** v. to experience that one's soul leaves one's body and temporarily enters an animal

**jagybeng ~ jagebeng** n. ART. beams that form the supporting structure of the floor of a house and on top of which the bamboo floor can be attached (see Photo 11)

**jagydok** n. BODY/ABSTR. biceps, strength. **byl... jagydok...** coll, n., strength (see byl) Ido sa•gyraido hambundo chungwachido alamyla byldo bylnikhon, jagydokdo jagydoknikhon. In the future that child might really become a bit stronger, it might really get strength.

**jagyra** n. PLACE. (1) right, right hand, right hand side (2) maternal. Only used with the words abumuri and achumuri. abumuri jagyra maternal great grandmother' achumuri jagyra maternal great-grandfather

**jagyryng** n. ACT. shadow cast by a person

**jagysi** n. PLACE. (1) left, left hand, left hand side (2) paternal. Only used with the words abumuri and achumuri. abumuri jagysi paternal great grandmother achumuri jagysi paternal great-grandfather

**jahas** n. ART. (< Indic) ship

**jai-** v. to scold someone
**naw... jai...** coll, v. to scold (see naw-)

**jai•-** v. to oppose, to refuse. Unasa Ketketa Burae pheruna jai•sakna chol man•cha cho•motaimyng... Because Ketketa Bura could really not come up with an idea to oppose the fox, ...

**jajong** n. GEO. moon

**jajyreng**$_1$ n. EMO/ABSTR. confusion; danger

**jajyreng-**₂ *vgoal.* to worry. *Ang nang•na jajyrenga.* I am worried about you.

**jakhal-** *v.* to use. *"Aiaw! byldo bylte jagydokdo jagydokte!" noaimu matsaba bylgumukaw jakhalai chultetsrangaidongano.* "Wow! He's very strong!" [it] said, and it used all its strength to shake Bandi off. *Garu noaiba jakhalanchak; Mandai noaiba jakhalanchak.* The names Garu and Mandai are not used anymore.

**jakhalthaw-** *adj1.* very useful

**jakhep** *clf.* as much as is contained in the palm of the hand when clenched, the quantity contained in the closed palm

**jaksan** *n.* BODY. wrist

**jaksithem** *n.* ART. ring

**jaksyl** *n.* ART. bracelet

**jakun** *n.* ACT. the second weeding of the *ha•ba. Mai kai•manwamungsa ha•jagara kama. Ha•jagara kamaisa kamaimung kynsange jakun kama.* Having planted the rice, we weed the land for the first time. Having cleared the weeds for the first time, we will clear them for a second time.

**jal-** *v.* to run (away). *Iskyn jan•gaba songsang de•theng jalangok.* He ran away to such a far country. *Magachakdo lengla, lengla jalaidonganowa.* The deer is running like a cripple, it is said. **jal... pyw...**, *v.* to flee.

**jaljeng** *n.* ART. cupboard

**jalphakang-** *v.* to run from one side to the other

**jam-** *vph.* (1) to finish, to deplete, to use up. *Sa•wa jamkhucha.* I have not finished eating yet. *"Rang nemchengama na•nang chyw jamchenga" noai rangmu chyw ryngsusai chyichie, range san chi byri wawano.* "Will our liquor finish first or will the rain stop first?" they said and while they were trying to compete with the rain in drinking, the rain fell for fourteen days. *Ang amaparami nokchi randai sa•na jamcha.* At my mother's house, there is always meat to eat. (2) to complete, to be complete, be total. *Awa nakhal thawbia, awa nakhal thawbia noaimyng, sa•dyrangai nakhalgumukan jamai khan•aimu sa•akno.* Having said that father's ear is very tasty, the children completely cut off his ears and ate them. (3) to gather. *Raja songnae matkhakhetdo jamok.* In order to appoint a king, all the animals had gathered. (4) to succeed. *Beanbebe Bandiba re•engbebeai chaiaichie, manjuri cha•dyl sukphinaidokno. Bytai chaichie, jamchaaidokno.* Truly, Bandi went, and when looked, the king post had taken root again. When they tried to pull, they didn't succeed. (5) to defeat. *Chunggaba mungmaawan aidiasang jamanowa.* The big elephant was defeated with a plan.

**jamangwa** *n.* end. *ja jamangwasang* at the end of the month

**jama** *n.* ART. (< Indic) shirt

**jamang-** *v.* to set (of the sun). *Rangsan jamangaidok.* The sun is setting.

**jambu** *n.* PLANT. species of tree

**jamkhamwa** *n.* ABSTR. the last one

**jamura** *n.* PLANT. pomelo

**jan•-** *adj1.* far. *Ha•ba jan•rukwaan nukruketchawa.* The rice fields are very far apart from each other, you will not see each other.

**jang-** *adj1.* quick

**jang•jot** *adj2.* biconcave, curved on both sides like the inner surface of a sphere, narrow in the middle. *Cha•e dabakun tykyi ympong jang•jot takarioknotyi.* As for his legs, they looked like coconuts on sticks: bulgy in some parts and very thin in others.

**janggal** *n.* QUANT. everybody, everything, all, all of them, all of it. *Sa•gyrai mylgbami dadadarangawdo janggalawan monokokno. Phylgym chunggaba monokrumokno myng• korokawan.* As for the brothers of the small child, they were devoured. The big eagle had devoured them all, the six of them. *Mai ja•bek pang•ai hyn•aimu, ge•thengdo janggalan sa•ak.* Having been given a lot of food, he ate all of it.

**janggi** *n.* ACT. (< Indic) life.

**janggi khenwa** *khjyks, n.* ACT. life. *Ang janggi khengwagumuk Atong khu•chuk balwa.* I have spoken Atong all my life.

**janggi thyi-** *vpan.* to die. *De•thengmi janggi thyimanok.* He has already died.

**janira** *n.* ART. mirror

**janti** *n.* ART. filter for rice beer (*chyw*). Woven cylindrical filter made of reed that stands in the jug (*gora*) to form a permeable membrane between the fermented rice on the outside of the filter and the alcoholic water inside the filter. Water is poured onto the fermented rice and the alcoholic liquid is collected inside the filter and scooped out with an *abek*. (see Photo 2)

**januari** *n.* TIME. (< English) January

**jap**₁ *n.* ART. trap to drive away enemies. A pile of rocks is stacked on a hill behind a plank. The plank is tied to a tree. When the enemy comes, the rocks are released, roll down and crush the enemy.

**jap-**₂ *v.* to pile up

**japang** *n.* PLANT. tree trunk

**japrukruk** *adv.* one on top of the other, in a pile

**jarambong** *n.* GEO. full moon

**jarang-** *vintr.* to shine

**jaraw-** *adj1.* (for) a long time

**jari-** *v.* to be startled. *"Oi!" nooknoaro uchie, jariaimu mongma: "Atong?" noai jariaimu su•kherekaimu wa khaw•sa bai•okno.* "Oi!" he said, and then, because it was startled, the elephant said "What?", and because it was startled, it crashed, and broke a tusk.

**jaria** *n.* ART. (< Indic) influence, pathway

**jasa-** *v.* (1) to wake up, to get up, to get out of bed. *Angdo manapmi jasachawate.* I won't wake up/get up/get out of bed very early in the morning! (2) to open one's eyes. *Jasaaimyng chaiwachido biskut ni•okno.* Having opened his eyes, when he looked, the biscuits were gone. (3) to realise. *Jasaaimyng chaiwachido biskut ni•okno.* When he realised what was going on, there were no more biscuits.

**jaseng•-** *v.* to shine

**jasyri** *n.* ACT. the experience that one sees oneself in a different place while one is asleep as if one's soul leaves one's body

**jat** *n.* PERS. (< Indic) tribe, race

**jatha** *n.* ART. a spear

**jatram** *n.* PLANT. species of medicinal plant

**jaw•-** *v.* to fry

**je** *prof.* indefinite proform, any, whichever, whatever, anyone, a certain, one. *Je ja•bek rymai hyn•gabado sa•aribo, mythelaribo.* Whatever curry they prepare for you and give you, just eat it and appreciate it. *Jemi sanchi Dibangkongdangaw matsa kakok.* On a certain day, Dibangkongdang got bitten by a tiger. *Je kristan donggado Isol phi•ai sa•chenga.* Anyone who is Christian prays to God and starts eating. *Jeen sanchi morot thyiokno.* One day, somebody died.

**jechiba** *prof.* anywhere, wherever.

**jekhai** *adv.* (1) for example, for instance (2) as. *Rongkhaisang jalanggaba Thometsangrepha Rangkhaimadophae nukkhung sot dokmyng, jekhai Atongsang balchido nokkhung rum• thaman ganangno.* Thometsangrepha Rangkhaimadopha, who ran away to Rongkhai, had sixty houses, as you would say in Atong: *nokkhung rum• tham.*

**jekjek-** *v.* to shake from side to side

**jel-** *v.* to increase, to multiply, numerous. *Mahari jela.* The family is big.

**jenethene** *adv.* somehow. *Jenetne rajamyng noksang phetangokno.* Somehow they reached the king's house.

**jeng** *n.* PLANT. species of plant from which brooms are made

**jero** *num.* (< English) zero

**jesangba** *prof.* somewhere. *Ge•theng jesangba re•engok.* He has gone somewhere.

**jesykyn** *prof.* however much, however many. *Jesykyn nang•chi ganang chynaribo, kamalna.* However much you have, just offer it to the priest.

**jetja** *n.* TIME. (archaic) (< Indic) May

**jetykyiba** *prof.* somehow. "*Nang•tyme iawan phalthangthangna hyn•gaawan kamtykyi chu•soketchachido nang•tyme atongtykyi man•a uawe.*" "*Acha achu, ninga jetykyiba takanchyini aro uaw kawna re•engarini ningdo achu.*" "If you yourselves did not succeed at the task I gave you, how can you do that?" "Okay, grandpa, we'll try somehow, and we'll just go and shoot it, we will, grandpa."

**jila** *n.* PLACE. region

**jineralmiting-** *v.* ACT. (< English) to hold a general meeting. *Dakanggabado jineral mitingchengni. Umungsa songgumuk thom•aimung ha•ba ha•ryn ha•rynaw sowalni.* First they will start with a general meeting. Then the whole village comes together and they will divide the *ha•ba* plot by plot.

**jingjong-** *adj1.* wiggly, unstable

**jingkha ~ jingka ~ sawel ~ sawyl** *n.* PLANT. species of luffa or loofah vegetable (see Photo 51)

**jingonget-** *v.* to shake

**jinma** *n.* QUANT. group, herd. *Bajudyranggumukan jinmami palyngsang sikhal kha•na re•engwa.* All the friends went to the jungle in a group to hunt. *mongmajinma* a herd of elephants

**Jisu** *n.* PERS. Jesus

**jit- ~ jyt-** *v.* to move. *Rong•awan jitna jamchano.* He could not move the rock. *Hap pidan rama man•okodo jytnaka?* When you have found a new place, you'll immediately move?

**jitymryma-** *vtr.* to roll something (for the purpose of transporting it)

**joba** *n.* PLANT. Chinese rose (see Photo 77)

**jobaphul** PLANT. the flower of the Chinese rose (see Photo 77)

**joi•- ~ doi•-** *v.* to drag, to catch (by dragging a net through the water), to hold, to grasp, to scoop into a receptacle

**jojong** *n.* KIN. *c, ref, a.* (1) younger brother (2) Also used to talk about or address a related younger male of one's own generation: cousin (3) Also used to address a young male unrelated person younger than the speaker.

**jok-** *v.* (1) to escape. *Matsado jenethene jokaimyng jalangoknoro.* Having somehow escaped, the tiger ran away. (3) to avoid. *Isol Babyra jora nokgaba tanetman•gabaawdo una jokna man•chawa.* We cannot avoid what God [and] Babyra, the keepers of matches in love have already planned for us. (3) to go free. *Na•a ie sastiaw rakna man•chido jokangni.* If you can endure this punishment, you will go free. (4) to leak out. *Robolmi balwa jokok.* Air has leaked out of the football. (5) to come out. *U•ching kakaimu thyi jokok.* Because he got bitten by a leech, blood came out. (6) to finish. *"Sala! mylteng te•euan jora chaina, roalan jokkhucha!"* "Damn! You are still too small now to see your lover; you have not even finished primary school yet!" (7) to jump because something startled you.

**re•eng... jok...** *coll, v.* to go back (see *re•eng-*)

**jokal** *n.* ART/PERS. (< English) comic strip, cartoon, anime; a character from one of these categories

**-jokjok** *evsp. V* up and down. *Ne•kat kakaimyng, ge•theng hojokjokwa.* Because he got stung by a bee, he jumped up and down.

**jokruru-** *v.* to flow into. *Tyikhalmyng tyi badymha•china jokruruaaidok.* The water flows from the river into the rice field.

**jokset-** *v.* to drain

**jol**$_1$ *n.* GEO. region, area

**jol-**$_2$ *v.* to roll up

**-jol**$_3$ *evsp.* (< Hindi) (1) *V* quickly. *Magachakmi myn•do tyisiwachian miniksuru takjolarianoro.* The deer's fur, when it is wet, just goes flat very quickly, I'm telling you. (2) as soon as. *Ba•jolwamian khu•chuk olna sapjolariokno.* It could just speak as soon as it was born. (3) *V* accidentally. *Kynsangdo thik ue napite phep cha•koknyng•sang galatwae dang•jolangwano.* After that, OK, the barber accidentally fell into a hole under the banyan tree. (4) intensifier suffix. *Jamjolai gopcha amakawe.* The monkey was not buried at all. *Ie cha•ba nemjolanchak.* There is something very wrong with this leg/foot.

**joljol**$_1$ *adv.* quickly. *Gore jalna rakchido joljol jalangarinaka.* If the horse is very fast, [I] will just quickly run away.

**-joljol**$_2$ *evsp. V* quickly. *Dykhimi balgabatykyi kha•sin kha•dymai re•engcha; jaljoljolangaidongano.* Like Dykhi had said, he didn't go slowly; he ran quickly.

**jolpi** *n.* ART. bamboo fish trap

**jom•-** *v.* to sneak, to sneak up on somebody

**jomphol** *n.* ART. hoist, crow bar, pry bar

**jong** *n.* KIN. *drel, d, ref, a.* (1) younger brother. (2) Also used to address a younger male cousin or (3) an unrelated man younger than the speaker.

**jonggi** *n.* ART. type of fish trap that has an opening on the op with long spikes pointing inwards

**jongkhu•ri** *n.* youngest brother. "*Dykhimi jongku•riawba bylawmangmangdo chaianchyini*" *noaimyng, ramchi jywpengsawaidongano.* "Dykhi's youngest brother is all strength, I will test him", (the tiger) said and it was sleeping and blocking the way while waiting.

**jongsyri** *n.* KIN. *d, ref, a.* (1) brother-in-law: spouse's younger brother (2) female's younger sister's husband (A male's younger sister's husband is *bonyng*.)

**jonja** *n.* PERS. twin

**jonong- ~ jorong-** *v.* to dissolve. *Chini tyichi jorongok.* The sugar has dissolved in the water.

**jora**₁ *n.* PERS. (< Hindi) partner, love (person), match in love

**jora**₂ *clf.* (< Hindi) classifier for things that occur in pairs. *sendel jora sa* one pair of sandals *mykren jora sa* one pair of eyes

**jorong- ~ jonong-** *v.* to dissolve. *Chini tyichi jorongok.* The sugar has dissolved in the water.

**jot-** *v.* (1) to point. *Jong, na•a re•engaribo, chaksi jotetgaba thongthong re•engaribo.* Brother, you just go. Just go straight in the direction of the finger with which I point. (2) to fidget. *Waiphinwami gesepchi baratdugaaimu [...]. Ha•sang bamai, chaksi jotai, chaksi phai•ai nemen chanchiaidongno.* When he returned, he was very much ashamed. With his head bent to the ground, fidgeting with his fingers and wringing his hands he was in deep thought. (3) to prod

**jotthat-** *v.* to prod

**jotkhyngkhyng-** *v.* to mash

**joton** *n.* ACT. attempt, try

**joton kha•- to** try, to make an attempt *Hanep rai•na jotong kha•ni.* I will try to go tomorrow.

**jotpyryw-** *v.* to pierce

**juk-** *v.* to wink

**jul-₁ ~ gul-** *v.* to walk through the jungle with difficulty.

**jul-₂** *v.* to jack up, to lift up, to dig up. *Rukwakdo pan mylgaba paiatakaimu rong•aw julokno.* The toad took a small stick and jacked up the stone.

**jul**₃ *n.* PLACE. mine, coalmine

**julai** *n.* TIME. (< English) July

**jumang ~ jywmang** *n.* ACT. dream. (see *jywmang*)

**jumu-** *v.* (1) to collect. *Thawgaba symgaba phangnan sa•rongchagaba jilami bostudyrangaw raai hyn•aimung kha•sin kha•sin gumukawan palyngchi jalgabadyrangaw jykthangthangaw jumuphynetaaknowa.* Having brought and given [them] tasty, sweet things of the region which are not usually eaten, [they] recollected all their husbands, who had run away to the jungle. *Thengthondo ge•theng nokaw songmyng morot saw•gabamiaw ha•thapyra hanggalaw jumusawaidonganoro.* Thengthon was collecting the ashes and charcoal of this house, which had been burned down by the villagers! (2) to join. *Uawbathiri ga•dakchichiaimuna thypsetthiriokno. Ytykma•chiba uba sa•gyraiba jumu kha•thirithirioknotyi.* They again cut him into pieces and disposed of him. But it, the little child, joined back together yet again!

**jun** *n.* TIME. (< English) June

**jut-** *v.* to encourage. *Ge•theng angaw sa•khawkhalna jutwa.* He encouraged me to steal.

**juta** *n.* ART. (< Indic) shoe. *Te•ew re•enggaba gawi, longpen kanai juta hilaw...* (Wilseng S. Marak) The girl who just went by wearing trousers and shoes with high heels...

**jyk** *n.* KIN. *ref.* spouse

**jykjak-** *v.* to be noisy, to make noise

**jykmong ~ jykmongma** *n.* PERS. first wife of a man who has two wives

**jyknyi** *n.* PERS. widow, widower
**jykrat-** *v.* to accuse of adultery
**jykri ~ jykyryi** *n.* PERS. widow, widower
**jyksai** *n.* KIN. *set.* a married couple, husband and wife
**jyktyi** *n.* PERS. second wife of a man who is already married
**jykyryi ~ jykri** *n.* PERS. widow, widower
**jym** *n.* riches, wealth
**jyngjang** *adj2.* dense
**jyryk-** *adj1.* to have a nutty taste
**jyrym ~ jyryp** *adv.* quietly, silently. *Uchie jyrym thymaimyng raw•okno.* At that moment, having lain quietly in ambush, they caught him.
**-jyryng** *evsp. V* daily, *V* repeatedly, *V* all the time. *Ge•thenge alu kobi habijabi ytykyi samchakdarangmynggymyn bagan takwano. Ytykyi phangnan ge•thenge uaw chairokjy-ryngariano, tyi tytjyryngariano.* He made a garden with potatoes, cabbage and all sorts of vegetables. So he took care of them every day and watered them every day.
**jyryng jyryng** *adv.* always
**jyryngnam** *adv.* always
**jyryp ~ jyrym** *adv.* quietly, silently. *jyryp mu•bo* sit quietly
**jyrypet-** *v.* to shut somebody up, to make someone be quiet
**jyt- ~ jit-** *v.* to move. *Rong•awan jitna jamchano.* He could not move the rock.
**jyw-** *v.* (1) to lie (down) (both the movement and the position). *Matsae, ramchi jywpengsawaidongano.* The tiger lay, blocking the road. (2) to sleep. *Ning jywsukbutungdyrangaw atakna hala kha•wa?!* Why did you wake us up while we were comfortably asleep?!
**jyw- siri- khjyks.** *v.* to sleep. *Thengthondo noksangmyng jywai siriaimyng hongkhotokno.* Then, Thengthon came out of his house, having slept.
**jyw•**₁ *clf.* classifier for strips of bamboo and woven bamboo mats. *wa•tyng jyw• sa.* one bamboo strip. *damdyl jyw• ni.* two bamboo mats
**jyw•**₂ *n.* ART. a flattened bamboo used to make mats. *jyw• sa.* one flattened bamboo

**jyw•**₃ *n.* KIN. *c, ref.* biological mother
**jyw•para** *n.* PERS. mother's household, mother's family
**jyw• wa** *n.* PERS. parents
**jyw•burung** *n.* KIN. *set.* a group of mothers and daughters
**jyw•bydyi** *n.* PERS. old woman, woman with children
**jyw•gaba** *n.* PERS. someone else's biological mother
**jyw•mong ~ jyw•morong** *n.* KIN. *ref.* eldest of a group of sisters
**jyw•mrong solkari ~ jyw•myrong sorkar** *n.* PERS. central government
**jyw•para** *n.* PERS. mother's house, mother's household
**jyw•ri ~ jyw•ryi** *n.* PERS. child who lost its mother
**jywbythyn-** *v.* to lay on one's belly
**jywdap-** *v.* to lie on top of something
**jywgebeng-** *v.* to lie on one's side
**jywkapkap-** *v.* to lie on one's belly
**jywkarang-** *v.* to lie on one's back
**jywmang ~ jumang** *n.* ACT. dream. *Atong jywmang nukwa?* What did you dream about? *Taija walchi jywwachi jywmangsang banggirigaba nukwa.* Last night at night when I was sleeping, I saw an earthquake in my dream.

# K

**=k ~ =ak ~ =ok** *encl.* (change of state).(see *=ok*)
**=ka ~ =naka** *encl.* (imperious/certain-future modality). *Cha rynaimyng, mai sa•aimyng, rai•naka.* After drinking tea and eating rice, we will go. *"Angaw thogiwado! Tai•nido tanchomotchaka anga ge•thengaw kakai sa•naka" nookno pherue.* "You have betrayed me! Today, I will certainly not spare him. I will certainly bite and eat him.
**ka•- ~ kha•-** *adj1.* bitter (taste)
**ka•dymbai** *n.* BODY. chin
**ka•myn•** *n.* BODY. beard
**ka•ran- ~ kha•ran-** *v.* to be thirsty (for something). *Tyiba ka•ranok bai•siga angdo.* I am thirsty for water, my friend.

**ka•syrak-** *v.* to drink from the bottle (i.e. without using a glass). *Tyi ka•syrakai ryngbo.* Drink the water from the bottle!

**ka•wak- ~ kha•wak-** *v.* to open one's mouth (widely), to say aaah, to have one's mouth open. *Dada, choi•sa ka•wakbone.* Elder brother, open your mouth a little. *Khu•thipna harataimyng ka•wakai mu•gabamyng manamthynggabana sotmai dumaai pang•phinokno.* Because he was so lazy that he did not want to close his mouth, and because he sat there with his mouth open, which stank so much, flies swarmed into it.

**kabal ~ kabar** *n.* ART. (< English) cover

**kabin** *n.* ANIM. species of big black ant

**kachi tyimuk nukgaba** *n.* PERS. person who can see into the future

**kai•-** *v.* to plant (of things that you stick into the ground, like paddy and tree sprouts). *Ha• khynmanwamungsa maisi khita. Umung abongdarang chala, dachangdartang chala. Ytykyimungsa chalmanwa machotwamungsa mai kai•chenga.* Only after collecting the unburnt remains of the jungle from the land, we sow millet. Then we pant maize and we plant *dachang*. Then, only after we finish planting these do we plant rice. *Cha•chakphang kai•aimu, balwa sakwa.* After planting tea trees, we enjoyed the wind.

**kaithuk** *n.* ANIM. flea

**kak$_1$** *ideo.* slap! the sound of something hitting or slapping

**kak$_2$** *n.* ART. lid. *potolkak* lid of a bottle

**kak-$_3$** *v.* to bite. *Jemi sanchi Dibangkongdangaw matsa kakok.* One day, Dibangkongdang got bitten by a tiger. *"Ang nang•aw kakai sa•ni"* nowano pherue. I will devour you, he said, the fox.

**kak-$_4$** *v.* to close with a lid

**kakdep-** *v.* to bite on something

**kakhirok** *n.* ANIM. head lice, pubic lice, crabs

**kakkhap-** *v.* to bite down on, to firmly hold between the teeth or in the beak

**kakmyn•** *n.* BODY. antenna (of insect), feeler

**kakpyret-** *v.* to crush by biting

**kal** *n.* ART. horn (traditional instrument, see Photo 119)

**kal•tek ~ kal•thek** *n.* ANIM. species of big red ant

**kala** *n.* PERS. deaf person, deaf man, deaf woman

**kalai** *n.* ART. loincloth

**kaljak ~ galjak** *n.* ANIM. catfish

**kalthek** *n.* ANIM. species of big red ant

**kaltyk** *n.* PERS. person who never washes himself/herself

**kam$_1$ ~ gam** *n.* ACT/ABSTR. work, wealth, riches, matters, activities. *Kam kha•na harataidong angdo.* I'm lazy. / I don't want to work. *Kam ni•wa.* Worthless. *haw•ai kamai sa•gaba* someone who works hard and struggles to survive. *Gam man•ni udo uan, tangka poisa. Uan gam mynga, dakangmi chasongdo. Te•ewsa kepasyti noai myngaidonga. Chasongna kri gam myngariaro, tangka poisa.* He will obtain wealth, money. Earlier generations called that *"gam"*. Now they call it "capacity". According to my generation this money is called *"gam"*.

**kam-$_2$** *v.* to clear the field, to cut the jungle to make a field, to tear out weeds

**kam-$_3$** *v.* to suffer a penalty. *Sa•khawgabaaw jurimana kamna nangni.* Thieves have to suffer a penalty.

**kamal** *n.* PERS. priest

**kan-** *v.* to wear. *Te•ew re•enggaba gawi, longpen kanai juta hilaw...* (Wilseng S. Marak). The girl who just went by, wearing trousers and shoes with high heels...

**kan•$_1$** *n.* BODY. body (of a human)

**kan•-$_2$** *v.* to last. *Bigaba nygylmi ra•wa ie? Tyngen kan•okte ido. Angba ytykgaba botolaw ra•nichymte.* From which market did you buy this? It lasts very long. I should buy one such a bottle too.

**kan•jot-** *adj1.* slim, skinny, thin (of a person). *"Ang ytykram•phinai kan•jota na•a atykyi kakai sa•na?" noai balwano Ketketa Burado pheruna.* "I am so thin, why eat me?" said Ketketa Bura, to the fox.

**kan•ol-** *v.* to stretch

**kan•peng** *n.* BODY. side of the body
**kana** *n.* PERS. blind person, blind man, blind woman
**kana** *v.* blind. *Taw•reksyrupmyng bengblokmyngdo sangumuk pywdyngdyngaimyng mongmamyng mykyranaw syw•chekchekai mu•okno mang• ni. Rangsan tiin char bajichian mongmae kanaaknowa.* The bananasucking bird and the toad persistently flew and fought all day, and kept pounding the elephant's eyes repeatedly, the two of them. The sun [points] three or four o'clock, [and] the elephant has become blind.
**kana theka** *khjyks, n.* ACT/FOOD. (< Indic) (1) have food and drink (liquor), food and drink (liquor) (2) have a drink (of liquor), liquor
**kanggal** *n.* PERS. poor person, pauper
**kangguru** *n.* ANIM. (< English) Kangaroo
**kangkang** *n.* ANIM. species of edible frog, green with black spots, which lives in caves and the hollows of stones at the side of a river
**kangkylek** *n.* ANIM. species of lizard with red neck, said to drink human blood (see Photo 111)
**kantara** *n.* PLACE. emptiness
**kanting-** *v.* to tear spontaneously
**kap-** *v.* to catch, to close
**kapangsi** *n.* ANIM. a clam
**kapkap-** *v.* to lie flat on one's belly. *Rukpeke rong•phelang sylgabachi kapkapai hyn•oknowa.* The frog presented himself on a beautiful flat stone lying flat on his belly.
**kapkung** *n.* ANIM. snail
**kar-** *v.* to peel off. *Abong karai sa•a angdo* I eat the corn while peeling off the seeds with my hand.
**kara** *n.* ART/BODY. (1) rope (2) vein
**karan** *n.* PLANT. seed, kernel, fruit stone
**karang** *n.* BODY. wing
**karat ~ ka•rat** *n.* ANIM. squirrel
**karaw** *n.* ABSTR. (1) debt, obligation. *Nang• aw ang karaw balni nang• angmi bostu sa•khawchido.* I'll tell you what your debt will be if you steal my things. *Gawi bytsyrukai jalangchido gawimi maharidarang karaw balni.* If you secretly run away with a girl, her family will tell you your obligations. (2) trouble. *Nang•do uaw takchido karaw man•nine.* If you do that, you'll be in trouble.
**karen** *n.* ART. electricity. *San thamok karen ni•wa.* It has been three days and/that there is no electricity.
**karydyl ~ kyrydyl** *n.* PLANT. species of liana (hanging root). It is believed that at night, when you fall asleep under a *karydyl*, it will change into *Abu Chakkhen*. (see Photo 75)
**kata ~ khata ~ katha$_1$ ~ khatha** *n.* ART. (< Assamese or Bengali) word.
**kata ~ khata ~ katha ~ khatha jyw•khynwa** *expr.* to tell long epic stories during the festival of *chywgyn*, one story usually takes one night or longer to tell.
**kata ~ khata ~ katha ~khathajyksai** *expr.* collocation
**kata ~ khata ~ khata ~ katha ra•-** *vtr.* to obey, to listen to (to heed someone's advice). *Jonggabae babamyng kata ra•chano.* The younger brother did not obey his father.
**katha$_2$** *n.* ART. shallow bamboo basket
**katija** *n.* TIME. (archaic) (< Indic) October
**katua ~ khatua** *n.* (< Hindi) ANIM. turtle, tortoise
**kaw$_1$** *n.* PLANT. type of fruit
**kaw-$_2$** *v.* to shoot
**kaw•warai** *n.* BODY. gill
**kawbipha** *n.* PLANT. species of tree
**kawrawraw** *adv.* easily, effortlessly, without effort. *"Ama, angdo mai sa•naka." "Ym, kawrawraw ga•wa, te•en sa•bo."* "Mom, I want to eat rice now." "Yes, that will be easy, you'll eat in a bit."
**ke- ~ khe-** *v.* appropriate. *Ue chacha takai kechagaba changbano mykchaaidonga.* Some one might be fancying someone whom it is not appropriate for this person to marry. *Nang• mykchagaba biphae khecha.* The boy you fancy is not suitable (for you to marry).
**ke•ret** *n.* CORP. gall
**kebyl** *n.* ART. (< English) cable

**keji** *clf.* (< English) kilogram, kg
**kek-**₁ *adj1.* blunt (of pointed things)
**kek-**₂ *n.* FOOD. (< English) cake
**kek-**₃ *v.* to chop wood
**kek-**₄ *v.* to grow
**keko** *n.* ANIM. species of large brown tokay gecko with narrow white stripes on its back and white-and-brown ringed tail and brown eyes (see Photo 109)
**kel-** *v.* to hide behind or in something. *Mykhang baketchi kelaidong.* She's hiding her face in a bucket.
**kelki ~ khelki** *n.* ART. (< Hindi) window
**kemyra** *n.* ART. (< English) camera
**kenchi** *n.* ART. outside rafter: big beam that runs from the ridge board to the bottom of the roof on the outside of the roof and that has *byrym* 'inside rafter' as its counterpart; together they form part of the support structure of the roof of a house (see Photo 11)
**kendyl ~ kendel** *n.* ART. (< English) candle *kendyl dot sa* one candle
**kensi ~ kesi** *n.* ART. (< Indic) scissors
**kep**₁ *clf.* classifier for small flat things. *biskut kep sa.* one biscuit
**kep**₂ *n.* PLACE. cave (from English *cave*)
**kepleplep ~ kepreprep** *adv.* to b on one's hands and knees with one's bum in the air. *Amak ge•thengdo rong• pelang sylgabachi kepleplep bamai hyn•takkonoa.* The monkey, as for him, he willingly lay down on his hands and knees with his bum in the air on a flat stone.
**kereng ~ kyreng** *n.* BODY. bone
**kes ~ ges** *n.* ART. strut: beam that runs, as the long side of a rectangular triangle, between the *manjuri* and the *gandai* in the structure of a house (see Photo 9, Photo 11)
**keset ~ kheset** *n.* ART. (< English) cassette, tape
**kesi ~ kensi** *n.* ART. (< Indic) scissors
**ket-** *adj1.* tight. *Jama keta.* The shirt is tight.
**kethylik** *n.* ABSTR. Catholic
**kewa ~ anaros** *n.* pineapple
**kewal ~ khewal** *n.* ART. a peddle, oar. *khewal phong sa* one oar
**kha-**₁ *adj1.* salty
**kha**₂ *ideo.* caw! the call of a black crow
**kha**₃ *interj.* interjection to threaten somebody and to warn that you might fight, war cry, Beware! Beware of ...! This interjection precedes a clan name and in some cases the proper name of a mythical person in a story. *Kha Marak!* Beware Marak! This interjection can also be used before the clan name of the person who says it as a way of self-support, i.e. 'Beware of me!' *"Kha Bandi Goira!" noangthiriaidonga Bandiba.* "Beware of Bandi and the god of thunder!" Bandi is saying again.
**kha-**₄ *v.* to tie. *Nokbanthai do•khakhuchi khachapai tangaba mongmawa dora byryi don•gabaaw rai•ai jalangokno.* They took the elephant tusks weighing twenty kilos which were kept tied to the *do•khakhu* of the bachelors' house and ran away. *"Angawdo gorechi cha•aw nemen khabone" nookno.* "As for me, tie my legs well to the horse", he said it is said.
**kha•-**₁ **~ ka•-** *adj.* bitter (taste)
**kha•**₂ *n.* ABSTR. fighting spirit. *Ge•thengdo kha•rara taka, angba kha• ganang, ge•thengnado kyrecha, takrukarini.* He has fighting spirit, but I also have fighting spirit, I am not afraid of him and will just fight with him.
**kha•-**₃ *v.* to do, to work. *Kam kha•aidonga ang.* I'm working. *Angna phone kha•etboto!* Call me (on the phone)! *Sikharba kha•chypanchakno.* They didn't waste their efforts on hunting anymore.
**kha•-**₄ *v.* to pour liquid into a jug
**kha•at-** *v.* to work with, to handle. *Ge•theng koila kha•ata.* He works with/handles coal.
**kha•dang-** *vgoal.* to care for with great love. *Ama thyiaimu akai sa•gyraina kha•danga.* After the mother died, her elder sister took care of the child with great love. *Nang•mi amana kha•dangai symsakbo.* Take care of your mother.
**kha•di** *n.* ART. clothes. *Raityngchi kha•dian phingsawok.* The clothes line is full of clothes.

**kha•dong-** *v.* (1) to hope, to be hopeful (2) to be courageous. *Te•ewe ningan kyryiphina. Alsia rajado kha•dongaria. Thoroksyrangok una, ningdo jalgabaak.* Now it is us who are afraid of him. The lazy king is just courageous. He jumped out [of the banyan tree] and so we became the ones who ran away.

**kha•gal-** *vgoal.* to love. *Ang nang•na kha•gala.* I love you.

**kha•pak-** *vgoal.* to miss. *Ang songna kha•paka.* I miss my village.

**kha•pet-** *v.* to be angry. *Ang bajuaw kha•petaidong.* I am angry with my friend.

**kha•phak** *n.* BODY. chest

**kha•ran- ~ ka•ran-** *v.* to be thirsty. *Tyiba ka•ranok bai•siga angdo. Ang tyi cho•sa ryngna.* I am thirsty, my friend. I want to drink a little water.

**kha•rek** *n.* PLANT. yardlong bean, also known as the long-podded cowpea, asparagus bean, snake bean, or Chinese long bean. The subspecies name is *sesquipedalis*. (see Photo 56 and Photo 57)

**kha•rekrek-** *v.* to vomit, to barf, to chunder

**kha•rongthai** *n.* BODY. chicken heart

**kha•rongthai** *n.* BODY. kidney

**kha•si-** *v.* to shun, to not like and ignore. *Ge•thengthengrara gorongrukokno gorongaimyngdo te•do, kha•sirukarokno.* They met each other and having met, they did not like each other and ignored each other.

**kha•sin ~ khasin** *adj2.* slow

**kha•sin kadym** *adj2. khjys* slow. *Uaw badaiangwachian bean bebe darairaragabasang dolong khagabachina phetangoknowa, Bad ido. Dykhimi balgabatykyi kha•sin kadymai re•engcha. Jaljoljolangaidongano.* When he crosses beyond that point, truly Bandi arrives at a bridge made entirely out of swords. As Dykhi had said, he does not go slowly. He is running quickly.

**kha•sop** *n.* BODY. lung

**kha•sym-** *adj1.* bitter-sweet

**kha•thol** *n.* BODY. wattle (of a chicken)

**kha•thong** *n.* BODY. heart. *O chame, angmi nang•na kha•galgabaaw nang•mi kha•thongchi dang•etna man•phanima?* O sweetheart, will you be able to insert also my love for you into your heart?

**kha•wa** *n.* PERS. lover

**kha•wak- ~ ka•wak-** *v.* to open one's mouth, to open one's mouth widely, to say aaah, to have one's mouth open. (see *ka•wak-*)

**kha•wak khu•wak** *adv.* with open mouth. "*Me•mangma morotma ie sa•gyraido?*" *noaimu kha•wak khu•wak chaisawthokaidongano.* "Is that child a ghost or a man?" they said and all were surely watching him with open mouth.

**khabak₁** *clf.* as much as the arms can encompass, an armful

**khabak-₂** *v.* to embrace, to grab firmly as in an embrace.

**khachol** *n.* ANIM. (< Lyngam) species of fish

**khadok** *n.* ANIM. (< Lyngam) species of fish

**khagynyk** *n.* ANIM. species of fish

**khai-** *v.* to carry on one's back with a strap tied around the head (like a basket that is carried on the back but that has a strap that is put around the head). *Ge•thengthengdo na• khynaimyng bai•sigathangmaran rukpekba tang sa, amakba tang sa• khaiaknowa.* As for them, having collected the fish, the frog and the monkey carried one basket each. (see Photo 35)

**khaithyi-** *v.* to hang oneself to death. *Ie nokchi morot phalthangaw khaithyiwa.* In this house somebody has hanged herself.

**khak** *interj.* woosh! someone running away fast. *Banggale biskutaw tanaimyng: "Aia! Udo magachakdo khorate" noaimyng rykoknowa. Te•do tharapna guduk takwachi, khak! ytykyi jalwano magachake.* The Bengal put down the biscuits and said: "Hey! That deer is lame." When he almost caught up with it, whoosh! it sped off like that and ran away.

**khakhet** *qtf.* all, everything, everybody

**khakhudyl** *n.* PLANT. species of plant

**khakrok** *interj.* woosh! someone running away very fast. "*Ian jan•naka*" *nowachi te•do magachakdo, te•do khakrok jalangokno, ramgonggai takaimyng.* "This is far enough", the deer says, and whoosh! runs off in another direction.

**khal**₁ *clf.* classifier for orifices, holes and caves. *hang•khal khal ni.* two caves. *nakhungkhal ni.* two nostrils

**khal**₂ *n. autoclf.* PLACE. hole.

**=khal**₃ *encl.* (1) comparative degree. more, -er as in bigg*er*, larg*er* and green*er*. *Gore jalna rakbebeokno. Kha•sinkhalai jalkhalna noaimyng ga•dukdukchiba rakkhalai rakkhalai jalariokno.* The horse ran really quickly. Having told it to run slower, whenever he prodded it with his legs, it just ran faster and faster. *Ang nang•na daiai chungkhala.* I am bigger than you. *Ning kholdo pinak. Doba nuketcha. Nang khole pibok. Nangchi nuketkhala.* Our skin is dark. The mud is not visible, Your skin is white. It is more visible on you. (2) superlative degree. most, -est as in bigg*est*, larg*est*, green*est*. *Mamyngawan nangchawa raja na•a angna nang•myng gore jalna rakkhalgabaaw hyn•etaribo" nookno.* "I don't need anything, o king, you just give your fastest running horse", he said.

**-khal•** *evsp.* event specifying suffix used on a non-finite verb in certain types of causative imperative clauses. *Ang nonoaw nokchi mu•khal•na balbo.* Tell my younger sister to wait at home. *Ang nonoaw noksang re•eng(cha)khal•na pengbo.* Prevent my younger sister to go home.

**khalbong** *n.* PERS. person who eats scandalously much, a glutton

**khali** *adv.* (< Indic) only, exclusively. *Phorenna daiai man•ai sa•gaba ni•wa. Mai sa•waba nalcha, khali alu biskut, gom.* There are no richer people than white foreigners. They don't eat a lot of rice, only potatoes, biscuits and pasta.

**khaljong** *n.* ANIM. species of fish

**khalpak** *n.* ART. strap of a basket (see Photo 27, Photo 28)

**khalput** *n.* PERS. dirty person

**khalthyng-** *v.* to be nauseating

**khaltyi** *n.* SUBST. soda

**-kham**₁ *evsp.* to V for a long time

**kham-**₂ *vintr.* (1) to burn, to scorch. *Rangsan khama.* The sun burns. / It's hot. *Saw•aisa ha•khamchido khynna nangcha. Ha•khamchachido ha• khynchenga.* By burning it, if the soil is scorched (completely burnt), you don't need to collect [unburnt pieces of forest material], if the soil is not scorched (completely burnt), you need to collect [unburnt pieces of forest material] first. (2) to scald. *Nawengawmu Kumiriawba khamoknowa, tyi.* Naweng and Kumiri, got scalded, by the water. (3) to be on fire, to burn. *Nok khamaidong.* The house is on fire. **ha• kham-** *vpan.* (see *ha•*)

**khambai** *n.* PLACE. top, upstream

**khambaisang** *n.* PLACE upstream

**khambykthai** *n.* PLANT. species of edible tuber

**khamphung** *n.* PLANT. species of edible tuber

**khampyryw-** *v.* to have a hole in something as the result of burning, to burn through

**khamthymbylong-** *v.* to have a hole in a road or bridge as the result of burning, to burn a hole into something, to get damaged by fire *Dolong wal•sang khamthymbylongok.* The fire burned a hole in the bridge. / The bridge was damaged by the fire.

**khamynkhap** *n.* ANIM. (< Lyngam) species of fish (see Photo 101)

**khan**₁ *clf.* classifier for objects like logboats. *rung khan ni.* two boats

**-khan**₂ *evsp.* V in addition, also V. *Na•a angna messagedarang watwachi Inglissang saietbone, ang baju darangba tyngkhan.* When you write messages to me, write in English, so that my friends will also know. *Nang• photodarangaw Facebookchi pang•ai upload kha•bota ytykaimung share kha•bo ang bajudarang chaikhanna.* Upload many pictures on Facebook so then I can share them, my friends can also want look at them.

**khan**₃ *n.* PLANT. cassava, tapioca

**khan-**₄ *v.* to suckle

**khan•-** *v.* to slaughter, to chop, to mince, to cut

**khan•chot-** *v.* to cut. *Nang• ang khaw khan•chotbo.* You, cut my hair.

**khan•peret- ~ khan•pyret** *v.* to split, to cut open

**khan•phyt-** *v.* to cut a solid object in half lengthwise

**khan•pyrak-** *v.* to cut a hollow object in half lengthwise

**khan•thong•-** *v.* to cut in half

**khan•tongthong•-** *v.* to cut up in pieces

**khana** *n.* (< Indic) PLACE. port, harbour, station

**khang-**₁ *v.* to occupy. *Bai•damdo haw•angman•gaba ha•gun sa•angman•gaba ha•rynthangthangaw khanga.* Some people occupy their own parcel which is already cleared and used up completely.

**khang-**₂ *v.* to solidify

**khanmynchyw** *n.* PLANT. Species of tapioca (see Photo 60)

**khansynen** *n.* PLANT. species of edible tuber red on the outside and white on the inside

**khansyrui** *n.* ANIM. earthworm (see Photo 98)

**khanta ~ khantha** *n. autoclf.* TIME. hour. *Imi Dajongchina cha• aw re•engchido khantha sa nangni.* From here to Dajong, if you go on foot, takes one hour.

**khap**₁ *clf.* classifier for flat pieces of hard materials. *tota khap sa.* one plank. *tin kahp sa.* one sheet of corrugated iron. *damdyl khap sa.* one *damdyl. so•rekhap khap ni.* two pieces of mica.

**khap**₂ *n.* ART. (< English) cup, teacup or its volume, cupful. Clf. *goi•. thai•. Khap goi•/thai• tham hyn•bo.* Give three teacups. *Cha khap tham hyn•bo.* Give three cups of tea.

**khap**₃ *n.* SUBST. flat piece of hard material. *so•rekhap khap sa.* a piece of mica.

**khap•-**₄ *v.* to be cooked without *mai•tyi ~ maiti. jabek khapgaba* curry without *mai•tyi ~ mai•ti*

**khapeng-** *v.* to hinder

**kharok** *n.* ANIM. species of very small fish

**Khasi**₁ *n.* PERS. Khasi

**khasi-**₂ *v.* to castrate, to remove the testicles

**khasin ~ kha•sin** *adj2.* slow

**khasot** *clf.* classifier for bundles *ra•sun khasot sa.* a bundle of onions

**khat**₁ *interj.* interjection to chase a dog away

**khat**₂ *n.* CORP. ringworm (a fungal infection)

**khat-**₃ *v.* to slaughter. *Biana wak khatna raw•aidonga.* We are catching a pig to slaughter for the wedding.

**khata ~ khatha ~ kata ~ katha** *n.* ART. (< Assamese or Bengali) word.

  **kata ~ khata ~katha ~ khatha jyw•khynwa** to tell long epic stories during the festival of *chywgyn*, one story usually takes one night or longer to tell.

  **kata ~ khata ~ katha ~ khathajyksai** collocation

  **khata ~ khatha ~ kata ~ katha ra•-** *vtr.* to obey, to listen to (to heed someone's advice). *Jonggabae babamyng khata ra•chano.* The younger brother did not obey his father.

**khatchi** *n.* ART. sickle

**khatdep-** *v.* to wrap, to wrap up, to fold

**khatua ~ katua** *n.* ANIM. turtle, tortoise

**-khaw**₁ *evsp. V* secretly, *V* surreptitiously. *Arong nokma chaikhawwachi Arong nok mami mukhangaw khiemu thyiokno.* When headman Arong looked surreptitiously, having hit Arong's face, he died.

**khaw**₂ *n.* BODY. hair (of the head)

**khaw•-**₁ *clf.* classifier for teeth, planks, sheets of corrugated iron for roofs and flattened bamboos used to make mats (*jyw•*) when they are in a mat. *damdyl khaw•. sa* one *jyw•* of a *damdyl, wa khaw• ni.* two teeth, two tusks (of elephant). *tota khaw• tham.* tree planks. *tin khaw• byryi.* four sheets of corrugated iron

**khaw•-**₂ *v.* to catch water in the palms of one's hands. *Paipmi tyi ge•theng khaw•aimu ryngok.* He caught the water from the pipe in his hands and drank it.

**khawakwak-** *v.* to vomit, to barf, to chunder

**khawcha•ryng** *n.* BODY. sideburn

**khawchi ~ khawkhai ~ khawkhi** *n.* BODY. grey hair

**khawchyryng** *n.* BODY. scalpel hair

**khawdam-** *v.* to put down

**khawkham** *n.* ART. pillow

**khawkhirok** *n.* CORP. dandruff

**khawkhuthuk** *n.* ART. cloth for men worn around the head

**khawphyng** *n.* ART. turban

**khawra** *n.* CORP. a hair that has fallen out of the head

**khawratcha** *n.* PLANT. species of tree

**khawsuk** *n.* PLACE. source of a river

**khe- ~ ke-** *v.* to be proper, appropriate, suitable. *Nang• mykchagaba biphae kecha.* The boy you fancy is not suitable (for you to marry).

**khel** *n.* ACT. (< English) care

**khele-** *v.* (< Indic) to play
  **rophyl... khele...** *coll, v.* to joke around, to play around (see *rophyl-*)

**khelegaba** *n.* ACT. game. *khelegaba myng sa* one game

**-khelek** *evsp.* V for fun

**khelhi ~ kelki** *n.* ART. window

**khem** *n.* ART. big drum (traditional instrument) played during *chywgyn* (see Photo 118)

**khema** *n.* ACT. forgiveness. *Nang•tym angaw wet sado khema kha•khubo.* Please forgive me one more time.

**khen-** *v.* to scratch. *Machok kan• panchi khenaronga.* A deer is scratching his body against a tree.

**khen•** *n.* ANIM. river crab (see Photo 103)

**khen•jasyri** *n.* ANIM. a river crab that is walking on the road

**khen•khorong** *n.* BODY. crab's pincher, crab's claw

**kheng-** *v.* to be alive

**khengchek** *adj2.* green, blue

**khengkhang** *adj2.* eternal

**khengsyryk** *adj2.* dark green

**khengwa** *n.* ACT. life

**khep-$_1$ ~ khup ~ khyp-** *v.* to close, to cover, to spread out, to put on clothes

**-khep$_2$** *evsp.* V firmly. *Raw•khepbo!* Hold it firmly!

**khep-$_3$** *v.* to cry

**khep-$_4$** *v.* to pinch, to cut with scissors. *Rong•khalchi khonokaimu khen• chak khepok.* When I felt under the stone, a river crab pinched my hand.

**khep-$_5$** *v.* to weave a bamboo mat (*damdyl*)

**khereng-** *v.* to resist, to struggle, to make a great effort. *"Ta wat! Ta wat!" noaimyng bawen sene bawenaimyng thopangaidongano, tungangaidongano. Bandi kherengwachido taw•sa•gyrai watwa watwatykyi dymbyra dymbyra takangaidongano.* "Don't let him get away! Don't let him get away!" they said, and they ganged up on him in seven circles. When Bandi resisted, the crowd broke up and they scattered like chicks.

**khet$_1$** *v.* to be stuck

**khet-$_2$** *v.* to tie the cloth in which you carry a baby (*ba•sek*)

**khewal ~ kewal** *n.* ART. a peddle, oar. *khewal phong sa* one oar

**khi-$_1$ ~ khi•-** *v.* to hit (a target), to touch. *Arong nokma chaikhawwachi, Arong nokmamyng mykhangaw khiaimyng thyiokno.* When headman Arong peeked, headman Arong's face was hit, and he died.

**khi-$_2$** *v.* to count

**khiil** *n.* ART. nail (made from iron). *khiil chong sa* one iron nail, bamboo strip that is used to tie beams together

**khingcheng** *adj2.* aslant, slant

**khirip ~ kyiryp** *n.* PLANT. species of edible plant of which the leaves are mashed and dried and then cooked to a pulp

**khit-** *v.* to sprinkle, to sow seeds. *Ha• khynmanwamungsa maisi khita.* Only after collecting the unburnt jungle material from the land, we sow millet.

**khok-** *v.* to remove (skin, bark, peel, dress etc.)

**khokalang** *n.* PERS. bold person

**khol$_1$** *n.* BODY. skin (of human, animal or), hide (of animal), scale (of fish). *ma•sukhol* cow-hide, cow-skin

**khol$_2$** *num.* twenty

**kholachi ~ kholechyi** *num.* thirty

**khol chang ...** *num.* twenty times ..., used only in compound numerals. *khol chang byryi rong ni* eighty two. *khol chang byryi* literally means 'twenty times four'.

**khole** *num.* twenty. this word is only used in compound numerals.

**kholechyi ~ kholachi** *num.* thirty

**kholgyk ~ kholgryk** *num.* twenty. the variant *kholgryk* is a loan from Garo.

**kholjisop** *n.* CORP. infection of the inner ear, labyrinthitis

**kholnang** *n.* rice used for seeding, unhusked rice. **kholnang khit-** to sow rice

**kholthyrai-** *v.* to peel, to shed skin, to come off (of skin). *Rangsan khamaimu ang nakhung kholthyraiok.* After the sun burnt it, the skin on my nose came off.

**khom•-** *v.* to sit with one's head in one's lap and one's legs pulled up

**khomchuk-** *v.* to bend over

**=khon** *encl/prtcl.* (speculative modality or status). (1) encl. *Rangsan rangbyrymaidonga, waini<u>khon</u>.* The sun is blocked by clouds, it might rain. (2) prtcl. "*Nang• ama nygylsang re•engwama?*" "*Ho•ong, <u>khon</u>.*" "Did your mother go to the market?" Yes, maybe."

**khonchi** *n.* CORP. leprosy. *khonchi man•ok ~ khonchi sa•ak* to have leprosy

**khong•-** *v.* to bark

**khonok-** *v.* to search by feeling. *Ichi rong• khalchi khen• ganangthel mang sa mang nido, ganangthelnaba ganang. Hai, nang• usang khonokbo ang isang khonoknaka.* Here in these holes under the stones there are crabs for sure. Let's go, you feel and search over there and I will feel and search over here.

**khophynga** *n.* ART. cloth for women worn on the head with a knot at the back of the head

**khopja ~ kopja** *n.* ART. hinge

**khoppalak ~ khoppylak** *n.* PLANT/BODY. (1) the husky skin of an onion, garlic, corn etc. (2) skin or peel of fruit (3) eggshell

**khorat ~ khorot ~ korot** *n.* ART. (< Indic) a saw

**khori** *n.* ART. watch

**khorop-** *v.* to break (only used for bamboo). *wa• khoropok* the bamboo is broken.

**khorot** *n.* ART. a saw (see Photo 26)

**khoryndachong** *n.* ANIM. silkworm

**khosylak** *n.* CORP. abrasion

**=khu** *encl.* (1) incompletive status. *Angdo sawamigymyn te•ewrawrawdo re•engna man•<u>khu</u>chaaidonga.* I cannot yet go school because of my illness. *Sa•akma sa•<u>khu</u>cha?* Have you eaten already or not yet? *Sa•aidong<u>khu</u>a.* I'm still eating. (2) polite imperative. *Baba, angna tangka ratja ni hyn•<u>khu</u>.* Dad, please give me two hundred rupees.

**khrukhru** *ideo.* roo-koo! the call of a pigeon

**khu•bisi-** *v.* to hate, to dislike

**khu•cheng-** *v.* to bite one's teeth firmly together, to grit one's teeth. *Phalthang sokwa dabatdo tyinyng•chi rong•chi pyi•aimyng wa khu•chengphinai sakchykaidokno.* Until he could not hold out any longer, he sat under water as long as he could bear it, holding on to a stone and gritting his teeth.

**khu•chi-** *v.* to dislike. *Mongma ranai sa•na ang khu•chia.* I don't like to eat elephant meat.

**khu•chit** *n.* CORP. harelip

**khu•chuk** *n.* BODY/ART. mouth, language

**khu•chul** *n.* BODY. lip

**khu•eng** *n.* PERS. person with a crooked or slant mouth

**khu•gri** *n.* PERS. someone who does not talk much, a quiet person

**khu•hamgaba** *n.* ART. grammar

**khu•jylok** *n.* PERS. someone with an open mouth

**khu•ma** *n.* PERS. dumb person, someone who cannot speak

**khu•mong-** *v.* to conspire. "*Ramchi hampyi na•nangdo watchaka ge•thengawdo, sala! Ge•thengaw watkhuna so•otthelarinaka*" *noai khu•mongangokno.* "This evening we will seize him on the road, the bastard! We will kill him after all to banish him once more", they freely conspired.

**khu•rang** *n.* CORP. voice

**khu•rasak-** *v.* to promise. *Rongdyng maharimu Jaksongram matsanokphandaimi matsamu takrukaisa Rongdyng maharidyrange dokra Ha•beng khung sa ri•pan patsai sympak khung sa hyn•naka noyi khu•rasakokno.* Because the Rongdyng family fought with the tigers from the bachelors' house of Jaksongram, the Rongdyng families promised to give one Ha•beng bag, one woman's dress and one sleeping mat.

**khu•sak-** *v.* to answer, to reply, to respond

**khu•sep** *n.* BODY. corner of the mouth
**khu•sum ~ ku•sum** *n.* ANIM. tortoise
**khu•sylip ~ khu•sylyp-** *v.* to whistle
**khu•symang** *n.* BODY. facial hair, beard, moustache
**khu•thikhu•thyraiga(ba)** *n.* mumbling
**khu•thoro-** *v.* to call out
**khu•thym-** *v.* to kiss
**khu•tip-** *v.* to close one's mouth. *Hongkhotphinna man•chaaimyng khu•thipwachie sotmai mang sene man•symokno.* Not being able to come out again, when he closed his mouth, he swallowed the seven flies.
**khu•tyi ~ khu•ti** *n.* CORP. spit, spittle, saliva. *khu•tyi thandapai bala* to speak with a lot of spittle
**khu•tyisot-** *v.* to spit
**khuchia** *n.* ANIM. species of fish
**khuchina** *n.* ANIM. eel
**khuchylep** *n.* PERS. blabbermouth, someone who cannot keep secrets and talks a lot
**khudal ~ kudal ~ wa•khu** *n.* ART. hoe, chopper (see Photo 26)
**khugyri ~ koksi** *n.* ART. small basket made of bamboo and reed. The word *khugyri* is used in Badri; *koksi* is used in Sijyw. (see Photo 27)
**khul** *n.* ART. pillow stuffing
**khuli-$_1$ ~ kuli-** *v.* (< Indic) to open
**khuli$_2$** *n.* FOOD. opium
**khung$_1$** *clf.* classifier for flat things (and photos even when displayed on a computer screen). *tangka khung sa.* one banknote. *piktiyr khung sa.* one photograph, picture
**khung$_2$** *n.* BODY. shell of a crab, tortoise etc., carapace
**khup- ~ khep- ~ khyp-** *v.* to close, to cover, to put on clothes, to spread out
**khuru** *clf.* length from the tip of the thumb to the tip of the middle finger when one puts one's hand down on the table on these points (Old English: span)
**khurung$_1$** *adj2.* wanting to lay an egg
**khurung$_2$** *n.* ART. a chicken's nest
**khurut-** *v.* to perform an incantation to determine which spirit makes someone ill. Prototypically associated noun: *wai*
**khurutgaba** *n.* PERS. someone who performs an incantation to determine which spirit makes someone ill.
**khusi dong•- ~ dong-** *v.* to be happy. *Ge•thengtheng khusi dong•thamakaimyng gore di•maichi phalthang chak diriga sangwalaimyng watokno.* They were so excessively happy that they forgot their own hands which were holding the horse's tail and they let go.
**khutai** *n.* top (of a house)
**khyi-** *adj1.* sour
**khym-** *v.* to marry. *Ue alsia rajae jykba myng•ni khymanoro.* That lazy king married two wives.
**khymbal** *n.* PLANT. species of tree
**khymgaba** *n.* PERS. spouse
**khymkhalymphong** *n.* PLANT. species of tree
**khyn-** *v.* to pick up, to gather, to collect. *Tyipaichi sukyrung khynok ningdo.* We have gathered river snails in the Tyipai river. *Ha•khamchachido ha• khynchenga, ha•khynmanwamungsa maisi khita.* If the land is not burnt, we collect the unburnt jungle material from the land first, only after collecting it, we sow millet.
=**khyngkhyng** *adv.* (1) still. *Ue ha•byriawe te•ewchinakhyngkhyng Atong khu•chuksang Matsa Chaw•kyi Asetram myngwano.* Up till now, we still call that mountain *Matsa Chaw•kyi Asetram* in the Atong language. (2) despite, even though. *Angdo te•ewrawrawdo isykynmi duknakhyngkhyngdo nemai mu•phaaidonga.* I'm doing well, despite my suffering.
**khyntyri** *n.* ART. (see Photo 11) (1) the wooden beam running over the top of the roof of a house, comparable to a ridge board. (2) purlin (other speakers call purlins *wa•khaw ~ wa•khaw*)
**-khynyng** *evsp.* to V into pieces. *Ang randaiaw da•dakkhynyngok.* I have cut the meat into small pieces.
**khyp-$_1$ ~ khup- ~ khep-** *v.* to close, to cover, to put on clothes, to spread out
**khyp-$_2$** *v.* to take a bite

**khyryithang ~ khyrythang** *n.* KIN. *c, ref, a.* nephew: male's sister's son or female's brother's son

**khyryithangsyw•** *n.* KIN. *d, ref.* great-nephew: the son of my husband's sister's daughter or the son of my brother's wife's daughter

**khyryk** *n.* ANIM. louse (plural: lice)

**khyw-** *v.* to drain, to shake out fluid

**kilip ~ kylip** *n.* ART. (< English) clip

**kilomytyr** *clf.* (< English) kilometre

**kingreng kingcheng** *adv.* like a chicken without a head, like crazy. *Thot thyng•thot takwachina dabat sykromaimyng khanetsi-gaaidongno. Bandi chakwatwamian chuwil chuwal takjolangokno. Taw• tokai watetwatykyi usang kingreng usang kingcheng takjoletarioknoti.* He (Bandi) grasped her (So•re) and poured the liquor into her mouth to the last drop. When Bandi let go of her, her head was spinning. Like letting go of a beheaded chicken she ran around like a chicken without a head.

**kirin**₁ *adj2.* torn (of cloth and paper)

**kirin-**₂ *vintr.* to be torn *Longpen kirinok.* The trousers are torn.

**kitap** *n.* ART. (< Arabic) book

**klas** *n.* PLACE. (< English) class. *Ando klas wan mangmangsaan dong•phaarikhuwa.* I have only gotten as far as class one.

**ko•rot** *n.* PLANT. sugarcane

**kobi** *n.* PLANT. cabbage

**koila** *n.* SUBST. (< Indic) coal

**koilagari** *n.* ART. (< Indic) coal truck

**kok** *n.* ART. basket

**kokalang** *n.* PERS. a bold person

**kokbal** *n.* ART. enormous basket made of bamboo and used to store rice and vegetables in the kitchen or granary (see Photo 30)

**kokcheng** *n.* ART. type of basket made of bamboo carried on the back with a strap around the head, smaller than a *koktang* (see Photo 27)

**kokdam ~ koktang** *n.* ART. big basket made of bamboo used to carry goods and worn on the back with a strap from the head. *Kokdam* is said in Badri, *koktang* in Siju (see Photo 27)

**koke** *n.* ANIM. species of large gecko that lives in trees

**kokpylak** *n.* PLANT. chaff

**koksep** *n.* big woven bamboo cage used to keep chicken in when they are sold on the market (see Photo 29)

**koksi ~ khugyri** *n.* ART. small basket made of bamboo and reed. In Sijyw *koksi* is used; *khugyri* is used in Badri. (see Photo 27)

**koksi kongdang** *adv.* in a disorderly way

**koktang ~ kokdam** *n.* ART. big basket made of bamboo used to carry goods and worn on the back with a strap from the head. *Kokdam* is said in Badri, *koktang* in Siju. (see Photo 27)

**kokkylek ~ toktokkylek** *n.* PLANT. species of plant (see Photo 65)

**Kol India** *n.* PLACE. Coal India

**kol** *n.* ACT. telephone call. *Kol khabo!* Call her/him/them!

**kolachi ~ kolechyi** *num.* thirty

**kolachita** *n.* PLANT. bitter gourd, *Momordica charantia* (see Photo 56)

**kolani** *n.* PLACE. (< English) colony

**kom-** *v.* to feel like a loser. *Angdo barataimyng komok.* Me, being ashamed, I felt like a loser.

**kombol** *n.* ART. blanket

**komok-** *v.* to feel insulted

**kompiutyr ~ komputer** *n.* ART. (< English) computer

**kompyl ~ kongpyl** *adj2.* bent. *kun• kompyl* a bent stick / the stick is bent

**komyla** *n.* PLANT. orange

**kon•** *adj1.* winding

**kong•-** *v.* to have rabies

**kongken naken** *adv.* zigzag, winding

**konglarong** *adj2.* orange

**kongpyl ~ kompyl** *adj2.* bent *kun• kompyl* a bent stick / the stick is bent

**kongtoksi** *n.* PERS. used in the expression *halsia kongtoksi* lazy person (pejorative)

**kontrektyr** *n.* PERS. contractor

**kopi** *n.* FOOD. (< English) coffee

**kopja ~ khopja** *n.* ART. a hinge

**koplak** *n.* BODY. egg shell
**korea** *n.* ART. big metal wok (see Photo 43)
**korok** *num.* six
**korong ~ kyrong** *n.* BODY. horn (of an animal)
**koros** *n.* ABSTR. expenses
**korot ~ khorot ~ khorat)** *n.* ART. (< Indic) a saw
**kri** *postp.* in accordance with, according to. *Chasongna kri gam myngariaro, tangka poisa.* According to my generation this money is called "wealth". *Morotna kri, hapna kri.* Different people and different place have different customs.
**krimichong** *n.* ANIM. parasitic worm that lives in the bowels
**krismas** *n.* TIME. (< English) Christmas
**kristan ~ kristen** *n.* PERS. (< English) Christian
**kror** *num.* (< Hindi) ten million, a crore
**krrrrr** *ideo.* the sound of someone smoking viciously. *Ryngokno ryngokno ryngokno, krrrr jamoknotyi.* He smoked and smoked and smoked, krrrrr, he finished it!
**krymkraw ~ kyrymkyraw** *adv.* together, in unison, in co-operation
**ku•sum ~ khu•sum** *n.* ANIM. tortoise
**kudal ~ khudal ~ wa•khu** *n.* ART. hoe, chopper
**kukuri** *n.* ART. dagger, type of knife with a blade with an obtuse angle used to survive in the jungle
**kulal ~ kular ~ kural ~ kurar** *n.* ART. (< Indic) axe (see Photo 264)
**kuli- ~ khuli-** *v.* (< Indic) to open
**kulthuk ~ kyltyk ~ kyltuk** *n.* PLANT. (see *kyltyk*) (see Photo 80)
**kuma** *n.* PLANT. species of tree
**kun-** *adj1.* curly
**kun•₁** *clf.* classifier for stick-like things. *nokwek kun• banga.* five brooms
**kun•₂** *n.* ART. a stick
**kung-** *vintr.* to be dammed up, to be enclosed by a dam or circle of stones
**kunremrem-** *v.* to curl, to be curly. *Ge•thengmi khawdo khunremrema.* His hair is curly.
**kural ~ kular ~ kulal ~ kurar** *n.* ART. (< Indic) axe

**kuti** *n.* ART. (< Indic) cow shed
**kutuuukutukutuk** *interj.* interjection to call a dog
**kyi•** *n.* ANIM. dog
**kyi•wa** *n.* BODY. canine teeth
**kyimang** *n.* ANIM. fruit fly
**kyiryp ~ khirip** *n.* PLANT. species of edible plant of which the leaves are mashed and dried and then cooked to a pulp
**kyisym-** *adj1.* salty
**kykgul** *n.* BODY. eyelid
**kykulwil** *n.* PERS. person who has an ear infection and has lost his balance
**kyl-** *v.* to hide, to avoid
**kylchap** *n.* PLANT. cotton
**kylip ~ kilip** *n.* ART. (< English) clip
**kyltyk ~ kulthuk ~ kyltuk** *n.* PLANT. species of tree which, in the dry season, brings forth bunches of bright red and yellow hairy beans with black seeds. When you eat the seeds, you get sleepy. (see Photo 80)
**kymbal** *n.* PLANT. species of tree
**kymkha** *n.* PLANT. species of berry
**kymkhalongphong** *n.* PLANT. species of plant
**kyn** *n.* (1) BODY. back. *Amakgawigaba biphagabaaw kynaw thup, thup tokaidonganowa.* The monkey's wife is hitting her husband on his back, tap! tap! it is said. *Ang nang•aw kyn kyn symni.* I will follow you closely. (2) PLACE. behind. *Kereng nokkynchi asetaribo.* Just throw the bones away behind the house.
**kynsang.** (1) behind. *Ge•thengdo nokkhapmyng kynsang mu•aidong.* She's behind the door. (2) after, then, later. *Phetangaidokno, dong•aidoknotyi, sane san chidok wal chidok re•engwamyng kynsangsa.* He has arrived, the has reached his destination, to our surprise, after going for sixteen days and sixteen nights. *Sagaltyisamchi poreok poreok poreok poreok. Kynsangdo chunggaba kam man•ok.* At the seaside, he studied hard. Later/Then, he got a well-paid job.
**umi ~ umyng kynsang** thereafter. *Rangmu chyw ryngsusaai chyichie, range san chibri wawano. Umi kynsang san khole*

san sa chyw ryngkhuanowa. "While competing in drinking with the rain, the rain fell for fourteen days. Thereafter they still drank liquor for twenty one days.

**kynbyret** *adv.* backwards, reverse motion. *Kynbyret rai•bo.* Move back!

**kyndam** *n.* PLACE. land behind a village

**kyngjung-** *v.* to turn one's back to someone. *Angsang kynjungbo.* Turn your back to me.

**kynjung-** *v.* to make an about turn

**kynkom-** *adj1.* old and crooked

**kynkongbang** *n.* SUPER. woman ghost without back

**kynkyreng** *n.* BODY. spine

**kynokhol** *n.* KIN. *d, ref, a.* son-in-law: deceased testator's son-in-law, or husband of a household's heiress

**kynpha-** *v.* to be last, to be late. *Ie sa•gyraido kynphaai phetdangok.* This child arrived last. *Somai kynphaak.* It was already late.

**kynphak-** *v.* to sleep in. *Rang wawamigymyn manapchi kynphakwa.* Because of the rain I slept in this morning.

**kyp-** *v.* to fit tightly, to fit and close off

**kyreng ~ kereng** *n.* BODY. bone

**kyrewami ~ kyriwami** *n.* ABSTR. danger

**kyrong ~ korong** *n.* BODY. horn (of an animal)

**kyrurua** *vintr.* to roll (by itself)

**kyrydyl ~ karydyl** *n.* PLANT. species of liana that looks like a long arm with elbow joints as it hangs between trees. It changes to *ambi chakkhen* at night, an old lady with long arms and hands with long nails. She will ask you to scratch her long arms, and when you refuse, she will scratch you to death with her long nail. (see Photo 75)

**kyryi**$_1$ *n.* EMO. fear

**kyryi-**$_2$ *vgoal.* to be afraid of *Ang mongmana kyryia.* I am afraid of elephants.

**kyryiwa** *n.* EMO fear. *Phina re•enggabadyrangba kyryiwa ganang.* The ones that came to ask were afraid.

**kyryk**$_1$ *ideo.* someone running, running and jumping like a deer. *Magachakdo hawtyi kyryk kyryk kyryk re•engaimu ...* The deer went a little away out of sight ...

**kyryk-**$_2$ *v.* clear, transparent

**-kyrym** *evsp.* to *V* in a group. *Taw• dang• kyrymangok, taw•nokchi.* The chickens have all entered their coop.

**kyrymkyraw ~ krymkraw** *adv.* united, together, in cooperation

**kyryng-**$_1$ *adj1.* tight. *Chakphong gylgabasano, kara khyrynggabararasano, alamylachagabasano.* They are men with strong arms and tight veins all over, they are no ordinary men.

**kyryng-**$_2$ *v.* to make noise, to make a sound, to sound. *Nokha•palchi cha•gyl kyryngaidonga.* Footsteps are making noise outside. *Amake wak tokokno. "Thup!" nowachie "Wek!" noai kyryngokno. "Ma•, atong kyryngaidong?"* The monkey beat the pig. When there was a "Slap!" it said "Squeal!" it is said. When he said "Huh? What is making that sound? he said.

**kyrynggaba** *n.* ART. sound, noise

**kyryngwa** *n.* ART. sound, noise

**kyryw** *n.* ART. thin strip of bamboo used to make rope, bamboo rope

**kyrywkeng** *n.* ANIM. parasitic worm that lives in the flesh of animals and humans. It is believed that when a *kyrywkeng* is crossing the road, and a pregnant woman steps over it, she will have a miscarriage.

**kyw** *procl.* I am here! (answer to a search call)

# L

**laha** *n.* SUBST. resin

**lain** *clf.* (< English) classifier for a collection of objects lined up on shelves

**laisak** *n.* PLANT. cabbage

**laisen** *n.* ART. (< English) licence

**lait** *n.* GEO/ART. (< English) light

**laityr** *n.* (< English) lighter

**lak** *num.* (< Hindi) a hundred thousand, a lakh

**laklak-** *v.* to prod in an orifice or hole for pleasure, to nag

**-langlang** *evsp.* intensifier suffix. very. *Rykangwachi te•do magachakdo thamat gyrymchi jywsawaidonganote, sak<u>langlang</u>.* When [he] chased [him], now, the deer was sleeping next to a bush of stinging nettles, I'm telling you, very red.

**lap**₁ *n.* ABSTR. profit, interest, gain, value. *Ang tai•sa rajasa lapokchym, thyiok.* I just made a hundred rupees profit but I lost the game.

**lap-**₂ *v.* to gain, to make profit, to be profitable. *Ha•gylsakaw chol takai chyichiba nang•aw khymaido angdo mamyng lap ni•okte.* When you try all sorts of small jobs, it will not benefit me to be married to you. *Asetaribo. Lap ni•wa.* Just throw it away. It's worthless.

**lapchagaba** *n.* PERS. a good-for-nothing. *"Atakgaba raja na•a, angna gore lapchagabaaw watetwa" nookno.* What kind of king are you [that] [you] send me a good-for-nothing horse?

**lapan** *n.* PLANT. pan/paan leaf, betel leaf

**las** *n.* ABSTR. (< English) the last one

**lasgaba** *n.* ABSTR. (< English) the last one, last

**lathia** *n.* ANIM. species of fish

**law** *n.* PLANT. cucumber-like vegetable (see Photo 49)

**lechu** *n.* PLANT. lychee

**lekadaw•reng** *n.* ART. kite

**lekat-** *v.* to waste time. *Na•a mai syw•khalna balwachymte. Na•a te•ewchinaba lekataidongkhua.* You said that you would pound some more rice. Until now, you have been wasting time and you are still wasting time.

**lekha** *n.* ART. (< Indic) book, paper

**lekhaphul ~ getphul** *n.* PLANT. bougainvillea (see Photo 81)

**leklek-** *v.* to prod in an orifice or hole

**leng** *n.* CORP. bronchitis. *leng man•ok* to have bronchitis

**lengla** *adv.* crippled. *"Ang khora taknane, na•a paiai jalbone bai•sigane" noaidonganote. Ytykyimyng magachakdo lengla, lengla jalaidonganowa.* "I will pretend to be lame, you carry away the biscuits, ok, friend?" he is saying. So then, the deer is running as if it is crippled.

**lepstik** *n.* ART. (< English) lipstick. *Khu•chul pisak nongwa lepstik.* On her red lips, she put lipstick

**letrin** *n.* PLACE. (< English) toilet

**lityr** *clf.* (< English) litre

**lolal** *n.* ART. (< English) road roller

**longpen** *n.* ART. (< English) a pair of pants, trousers, long pants, long trousers. *Te•ew re•enggaba gawi, longpen kanai juta hilaw...* (Wilseng S. Marak). The girl who just went by wearing trousers and shoes with high heels... Clf. *khung. longpen khung sa* a pair of pants.

**loskor** *n.* (< Indic) PERS. highest rank in the system of customary law of the Garos, judge

**lukchok ~ rukchok** *n.* ANIM. species of frog

**lukchokchok** *n.* ANIM. species of small gecko that creeps up the walls of houses at night

**lukpekpek ~ rukpek** *n.* ANIM. species of frog

**lukwak ~ rukwak** *n.* ANIM. toad (see Photo 104)

**lyp** *adv.* gollop, eat all at once. Used in the expression *lyp takai sa•-* to gollop

# M

**ma-**₁ *v.* to lose. *Sanglas palyngchi maak angdo.* I lost my sunglasses in the jungle.

**=ma**₂ *encl/prtcl.* (interrogative modality). (1) *encl. Nang•ba re•engnima?* Will you be going too? (2) *prtcl. Nang•ba re•engni ... ma?* Will you be going too?

**ma**₃ *conj.* or *Ie ma ue?* This one or that one?

**ma•** *interj.* okay, okay then, (very) well then. (Nepale:) *"Ytykchido ang re•engsigama nang•mi phal?"* (Thengthone:) *"Ma• ytykchido dongarini, ang chakdyrangaw dengbo" nooknoro.* (The Nepali:) "But shall I go instead of you?" (Thengthon:) "Very well then, in that case, it will be most convenient, untie my hands", he said.

**ma•am-** *v.* to moan

**ma•ang-** *vintr.* to set (of the sun). *Rangsan ma•angaidok.* The sun is setting.

**ma•chot-** *v.* to be an orphan

**m•m ~ hm•m ~ mm ~ •mhm•** *procl.* I disagree. (see •mhm•₁)

**ma•rek** *n.* ANIM. when this bird sings, you know that someone will visit the village

**ma•su** *n.* ANIM. cow

**ma•subipha** *n.* ANIM. bull

**ma•subolot** *n.* ANIM. ox

**ma•suchawkhol** *n.* star sign: the Big Dipper
**ma•sugari** *n.* ART. bullock cart
**ma•sutan•dam** *n.* PLACE. place where cows are slaughtered
**machak-** *v.* to be vengeful. *Ang nang•aw machakaidonga* I am vengeful towards you.
**machok** *n.* ANIM. species of large deer, maybe sambar deer
**machong** *n.* KIN. woman-founder of a clan
**machot-** *vph.* to finish, to stop. *Chalmanwa machotwamyngsa mai kai•chenga.* Only after finishing the sowing, we begin to plant the rice. *Mongmaaw chaina sa•wa machotok.* I stopped eating to look at the elephant.
**madam** *n.* PERS. female teacher
**madong** *n.* PERS. somebody who has married a person from the same mahari as themselves.
**magachak** *n.* ANIM. barking deer
**magal** *n.* ANIM. species of fish
**magana-** *v.* to lose. *Byk galatokno rutido. Uawdosega pherusa kakkhypaimu sa•akno. Ytykyimuna kynsangdo daw•khado maganaak.* The bread fell down. The fox then grabbed it with its mouth, and ate it. And so the crow had lost its bread.
**magyna** *adv.* in vain
**mahari** *n.* PERS. relatives (of the same clan), family (of the same clan), clan. Among the Garos, there are five maharis: Sangma, Marak, Momyn (Standard-Garo spelling: Momin), Sira and Areng
**mai**₁ *n.* PLANT/FOOD. rice
   **mai ga•-** *v.* to thresh the rice with one's feet so as to separate the grains from the ores
   **mai pylakgaba** *n.* PLANT flowering rice (see Photo 19)
   **mai sa•-** *vpan.* to eat. *Mai sa•akma?* Have you eaten already?
**mai**₂ ~ **mei** *n.* TIME. (< English) May
**mai•byram** *n.* CORP. cracks in the skin of the cheek
**mai•cheng** *n.* PLANT. mezenga, edible shrub with scented (nice smelling) leaves and soft thorns, *Zanthoxylum oxyphyllum*
**mai•do** *n.* PLANT. species of creeper
**mai•in** *n.* PLANT. species of plant

**mai•kerep** *n.* PLANT. species of plant
**mai•raja** *n.* TIME. (archaic) December
**mai•tyi ~ mai•ti** *n.* FOOD. juice from *ja•bek*, curry juice, the watery liquid or broth that is the result of cooking *ja•bek*
**mai•wa** *n.* PLANT. bamboo shoot *mai•wa ching sa* one bamboo shoot
**mai•wakhyi** *n.* FOOD. fermented bamboo shoots
**mai•wek** *n.* ANIM. species of bird that is believed to call every time somebody comes to visit the village
**maichek~ maichyk** *n.* FOOD. cold rice
**maidan** *n.* PLANT. new rice (just harvested). *Mai mynokodo maidan syla thoka.* When the rice is ripe, we celebrate the new rice festival.
   **maidan syla thoka** *expr.* to celebrate the after-harvest festival
**maidu** *n.* PLANT. species of plant
**maigasam** *n.* FOOD. meal eaten in the later part of the day or evening, dinner
**maiguru** *n.* PLANT. species of rice from which beer is made for the *chywgyn* festival (see Photo 54)
**maijyk ~ maimijyk** *n.* ANIM. dragonfly
**maijyreng** *n.* FOOD. leftovers of cooked rice dried in the sun used to feed the pigs
**maikap** *n.* PLANT. hay
**maikhol** *n.* PLANT. the skin of a grain of rice
**maikholnang** *n.* PLANT. unpeeled rice (see Photo 41)
**maikhoppylak ~ maikhoppalak** *n.* PLANT. stalks left over after threshing rice
**maikhyt** *n.* FOOD. burned rice
**maikung** *n.* ACT. second rice harvest (in November)
**maimanap** *n.* FOOD. meal eaten in the morning, breakfast
**maimijyk ~ maijyk** *n.* ANIM. dragonfly
**mainyl** *n.* PLANT/FOOD. sticky rice
**mainym** *clf.* length from the elbow to the tip of the fist
**maip ~ mep** *n.* ART. (< English) map
**maipal** *n.* PLANT flowering rice, rice flower
**maipalak ~ maipylak** *n.* PLANT. stalks left over after threshing rice
**maiphang** *n.* PLANT. paddy

**mairong** *n.* PLANT. husked rice, uncooked rice

**mairongkholnang ~ mairongkhol** *n.* PLANT. unhusked rice

**mairugu ~ meringgu ~ meringgaw ~ merenggaw** *n.* PLANT. mushroom (edible)

**maisan** *n.* FOOD. meal eaten in the middle of the day, lunch

**maisen** *n.* FOOD. sticky rice in a banana leaf

**maisi** *n.* PLANT. millet

**maityk** *n.* ART. pot for cooking rice

**maja** *dtw.* the day before yesterday, some time ago, a few days ago, in the recent past. *Angdo maja 13 tarikchi chaphang phang 99 ang nok rygynchi kai•ok.* A few days ago, on the 13th, I planted 99 tea plants near my house.

**maji ~ matji** *n.* PLACE. (< Indic) middle, (in) between. *Ytykyi dolnitakai sualmanwamyng kynsang, rajatakai matjichi myng sa mu•ni.* After making two groups like that, one person will stand in the middle, like the king. *Angdo Sandishmyng Bittermyng matjichi mu•aidonga.* I am sitting between Bitter and Sandish. *Ang gumukmyng matjichi mu•aidonga.* I'm sitting in the middle of everybody.

**makbul** *n.* ANIM. bear

**makja** *n.* TIME (archaic) (< Indic) January

**mal-** *v.* familiar, easy to deal with

**mama** *n.* KIN. *c, ref, a.* (purely referential term *haw•*) (1) uncle: mother's brother (2) Also used to address my father-in-law. (3) Also used to address an unrelated man older than the speaker in a respectful way.

**mamung ~ mamyng** *n.* ABSTR. (1) nothing. *"Atong dywwa ama?"* nookno. *"Hy? Ni•wate baba, anga mamungawan dywanchate."* "What did you add, mother?" he said. "Huh? Nothing, my son; I have added nothing, I'm telling you!" (2) no. *Ning songsyrekdo, ning Atongdo dakangdo mamyng thorom ni•wami somaichido waiaw mania.* We heathens, we the Atongs, in the past, in a time when there was no religion, we worshiped spirits.

**mamylet** *n.* FOOD. omelette

**man**$_1$ *n.* ABSTR. respect **man ra•-** *v.* to respect. *Ge•thengaw man ra•na nanga.* You have to respect him.

**man**$_2$ *n.* PLANT. species of plant which looks like *ring*, but which is inedible and has much bigger stems and leaves, growing more than two metres tall (see Photo 64)

**man-**$_3$ *v.* to crawl, to creep. *Khyryk khawchi manaidonga.* Lice are crawling in my hair. *Atongbatykyi ga•sokok aksokok hungthamakaimuna saphawba ha•khungchina mangatokno.* Somehow the rabbit, stumbling and barely swimming, crawled onto the river bank.

**-man**$_4$**•~ -man•** *evsp.* already V-ed, resultative suffix. *Sa•manok ~ sa•man•ok* angdo. I've already eaten.

**man•-** *vB.*(1) to be able, can. *Sa•gyrai rai•na man•a.* The child can walk. *"Ie angmi mola dolai hyn•gabaaw; nang•tym iaw ryngna man•chido nang•tym uaw phylgymaw man•ni"* nookno. "This is my tobacco, which I will divide and give you; if you can smoke it, you'll get the eagle", he said. *Umigymynchi anga ytykgaba janggi khengnamangmangba man•chawamigymyn tai•ni ang jyk asetwamigymyn anga ytykyi parangna nangok"* nowano. Therefore, because I was not able to get even a simple life, today, because my wife has divorced me, I have to wander like this", he said. (2) to be possible. *"Na•a bisang rai•wachym?" "Angdo sagal nalsasang rai•wachym." "O! manchaka. Tiin baji dong•wachido man•chak."* "Were would you be going?" "I plan to go to the other side of the sea." "Oh! That's not possible anymore. After three o'clock, it's no longer possible." (3) to be allowed to. *Darangba iawdo khaw•ai ryngna man•cha.* Nobody is allowed to drink from it. / Nobody can drink from it. (4) to get, to obtain. *Kynsangdo chunggaba kam man•ok sagal tyisamchi.* Later, he got a very good job at the seaside. *Kawbutungchi thik thokyrengaw man•okno.* When he shot the giant eagle he got it exactly in the neck.

**man•ai sa•-** (1) to eat in great amounts. *Uchian magachakdo biskutaw man•ai* man•ai man•ai man•ai man•ai sa•aidokno. Then the deer ate the biscuits in great amounts. (2) to be rich. *Morot bilding chunggabachi mu•gaba man•ai sa•ak.* The man who lives in that big house is rich.

**man•dapami ~ mandapwami** *n.* ABSTR. profit, interest, gain

**man•dyk-** *adj1.* physically difficult, complicated, troublesome, cumbersome. *Angdo gylgylrongchawanasa te•ew nokchi rang waaimu kam kha•na haratok, man•dykok kam kha•na.* As for me, precisely because I usually roam around, now that I'm at home and it's raining, I've become lazy; it's cumbersome to work.

**man•symrukruk-** *v.* to inherit

**manak-** *adj1.* dark (due to te absence of light)

**manam-**₁ *adj1.* to stink, to smell bad. *Di• manama.* Shit stinks.

**manam**₂ *n.* GEO. bad smell, stench

**manap**₁ *tw.* morning. *Wal thywwamigymyn ang manapchi jywchepwa.* Because I went to sleep late at night, I slept alone in the morning.

**manap-**₂ *vØ.* to be morning. *Manapok.* It has become morning. *Manapnaka.* It will soon be morning.

**manapmi** *adv.* very early in the morning. *Kynsangdo manapmi sirimynmyn re•engaimungna Dabatwarisang dinggarai saakno.* Then, having gone to Dabatwari very early in the morning at the break of dawn, he put up his fish traps.

**mandal** *n.* PLANT. *erithrina superosastricta*, species of tree with thorns and very red flowers which blossom in the late part of the dry season

**mang** *clf.* classifier for animals, knives and tools. *bythyi mang sa.* one porcupine. *chaw•kyi ~ chang•kui mang sa.* one big knife

**mangcha** *n.* SUPER. a ghost that is a corps risen from the dead, zombie

**manggisi** *n.* PERS. corps, dead body. *Thymanggami manggisiaw synai tanok.* We have laid the dead person's corps on the bed.

**manggywak** *n.* ANIM. millipede (see Photo 97)

**mangka** *n.* ANIM. species of fish

**mangkhrang ~ mangkhram ~ mangkyrang ~ mankyrang** *n.* ANIM. scorpion

**mangkung** *n.* ANIM. cricket (generic name for the species)

**mangkyrang ~ mankyrang ~ mangkhrang ~ mangkhram** *n.* ANIM. scorpion

=**mangmang**₁ *encl.* only, just, exclusively, even This enclitic indicates exclusivity and mirative modality simultaneously, i.e. an emotion of annoyance, anger, apology or disappointment, resentment or disbelief on the part of the speaker. *Te•ew-mangmangsa tyiruwa na•a.* I only just now took a bath. *"Angna mamyngawan nangchawa. Ytykchiba na•a angna myng samangmangaw takbo" nookno.* "I don't need anything. But you have to promise me only one thing", he said.

-**mangmang**₂ *evsp.* only, just, exclusively, continuously, as best you can, barely, shoddily. This suffix also indicates that the action was done poorly or negligently, or not fully as it was supposed to be done. *Ue morote hang•khal saw•ai tangka bisylaw gopanchanoro. Sam tanaimyng alamyla ytykyi samaw gopmangmangariwanoro.* That man did not bury the coins in a hole dug in the ground. He just put some grass over them, like that; he shoddily buried them only in grass.

**mangneng-** *v.* to whine, to quarrel. *Ytykyimu isangdo jykdo sa•do, sa•do jongdo mai okhiaimu mangnengaidokno mangnengaidokno. Babado biba rai•naka?" noaidokno.* So then at home, because they got hungry, his wife and children are whining and whining. "When is dad coming back?" they are saying.

**mangnengruk-** *v.* to quarrel

**mangsong-** *v.* to plan

**mani**₁ *n.* KIN. *c, ref, a.* (1) aunt: father's sister. (2) Also used to address one's mother-in-law.

**mani-**₂ *v.* to worship, to pay respect to someone. *Ning songsyrekdo ning atongdo dakangdo mamyng thoromaw ni•wami somaichido waiaw mania.* We heathens, we the Atong, in the past, in times when there was no religion, we worshipped spirits.

**manjuri ~ manjyri** *n.* ART. supporting post for a house or other such structure, king stud (see Photo 9 and Photo 10)

**mankyrang ~ mangkyrang** *n.* ANIM. scorpion

**manram-** *v.* to crawl

**mansylang ~ manthylang** *n.* BODY. spleen

**mantaw** *n.* PLANT. species of gourd

**mantawbylati** *n.* PLANT. tomato

**mantawthai** *n.* PLANT. type of vegetable

**=maran** *encl.* (reciprocal, indicating a reciprocal relationship between two or more persons). *Bai•sigathangmaran tyi dukungokno.* The friends dammed up the water.

**marirang** *n.* PLANT. species of plant

**mars** *n.* TIME. (< English) March

**marudyl** *n.* SUPER. species of liana that turns into a ghost that makes a loud noise of rummaging through the jungle, breaking branches and trees. But when you look in the morning, there are no broken branches and trees. The noise is so loud and frightening, that when you are within a hundred metres of this ghost, you cannot sleep. When you cut a *marudyl*, a blood-like, red liquid comes out.

**mastel** *n.* PERS. (< English) male teacher

**mat₁** *n.* ANIM. animal

**mat₂** *n.* CORP. a wound

**mat-₃** *v.* (1) sharp. *Kukuri matanchak.* The dagger isn't sharp anymore. (2) to cut. *Chaw•kyi ang cha•aw matwa.* The knife cut my hand. *Cha• matok* my hand is cut. (4) to wound. *Thalasang satetok. Chybym thyi• hongkhotokno. Sarai matok.* She hit him with a plate. Blood came out of [his] forehead. She had wounded him.

**matburung ~ matpalyng** *n.* ANIM. wild animal. *Matburung* is a loan from Garo.

**matchirit-** *v.* to scratch, to be scratched, to have an ulcer on your skin

**matdam** *n.* ANIM. otter

**matdi** *n.* ANIM. wild water buffalo

**matgba** *n.* BODY. a wound, something which is sharp or which cuts/can cut.

**mathai** *n.* ANIM. bachelor elephant, solitary male elephant

**matji ~ maji** *n.* PLACE. (< Indic) middle, (in) between. *Ytykyi dolnitakai sualmanwamyng kynsang, rajatakai matjichi myng sa mu•ni.* After making two groups like that, one person will stand in the middle, like the king. *Angdo Sandishmyng Bittermyng matjichi mu•aidonga.* I am sitting between Bitter and Sandish. *Ang gumukmyng matjichi mu•aidonga.* I'm sitting in the middle of everybody.

**matpalyng ~ matburung** *n.* ANIM. wild animal. *Matburung* is a loan from Garo.

**matrong** *n.* ANIM. jungle goat

**matsa** *n.* ANIM. tiger

**matsadu** *n.* SUPER. creature which is human during the day and becomes a tiger at night

**matsamykhang** *n.* PLANT. species of plant. The leaves of this plant can be dried, and then used to make *aphap*.

**maw** *n.* PLANT. species of tree

**mawkhol** *n.* PLANT. bark (of a tree)

**mawsa ~ mosa** *n.* KIN. *c, ref, a, rec.* (1) male cross-cousin: father's sister's son or mother's brother's son (2) the relation of male cousins from intermarriageable families (3) a male friend belonging to an intermarriageable family

**mawsathangmaran ~ mosathangmaran** *n.* KIN. *set.* two boys of different *maharis* (lineages), for example Marak and Sangma, in two possible relationships depending on the gender of the speaker. Male speaker: my elder sister's son and my son. Female speaker: my elder brother's son and my son.

**me•ama** *n.* PERS. married woman

**me•apha** *n.* PERS. married man

**me•mang ~ mi•mang** *n.* SUPER. ghost, spirit of a dead person

**me•mang saw•et-** ceremony performed a year after someone's death. The spirit of the dead person then leaves the house and goes to *Balphakram*.

me•mangdanggai *n.* PLANT. species of tree

me•mangguchung₁ *n.* PLANT. species of liana, woody vine that grows in the jungle as winding branch with an undulating pattern twirling itself around other trees for support. The name of this plant translates as 'ghost ladder' (see Photo 66)

Me•mangguchung₂ *n.* PLACE. the way up over the hill before the spirit of the deceased reaches *Chidymak* on their way to Balphakram

me•mangkereng *n.* ANIM. stick insect, walking stick, an animal from the order of *phasmadotea* (see Photo 95)

me•mangkereng *n.* SUPER. skeleton ghost

me•mangkoksi ~ mi•mangkoksi *n.* PLANT. pitcher plant

me•mangkyi ~ mi•mangkyi *n.* ANIM. species of small frog that says *pekpekpekpek*

Me•mangmaisansa•ram *n.* PLACE. the place on the way to Balphakram where the spirits of the dead eat their lunch. There is a place on a hill in the Badi area that bears this name.

me•mangsawdet *n.* CORP. wart

me•mesi *n.* ANIM. flying insect

meee *ideo.* meh-eh-eh! naa! the sound a goat makes

megalaia *n.* PLANT. species of creeper of which the sap supposedly helps in the healing process of wounds

Megam *n.* PERS. Megam, Lyngam

mei ~ mai *n.* TIME. May

mejakbal *n.* ANIM. alligator

mejistret *n.* PERS. magistrate

mejoryti *n.* ABSTR. majority

Mekalaia *n.* PLACE. Meghalaya

mel•- *adj1.* fat (of person)

melanggaw *n.* ANIM. poisonous red or black ant

memaboro *n.* PLANT. species of nice smelling rice

menpart *n.* PERS. most important or most salient person

mep ~ maip *n.* ART. map

meringgu ~ meringgaw ~ merenggaw ~ mairugu *n.* PLANT. mushroom (edible)

mes *n.* ANIM. sheep

mew *ideo.* meow! the sound of a cat

=mi ~ =myng *encl.* (genitive). (1) marks the semantic role of Possessor *Bai•dam rongsa tyikhalmi ha•waichina jalangok.* Some ran away to the plains of the river Rongsa. *Taw•reksyrup mang sa, ge•thengmyng thupaw phangnan mongma phai•ai sa•rongwana, mongma mathaiaw thapna re•engaidonganowa.* Because the house of a banana-sucking bird was always eaten up by an elephant, he went to beat up the bachelor elephant. (2) marks the semantic role of Source. *Mai bytwamungdo, maiaw ha•basangmyng pungchina songchina khairata.* After harvesting, the rice is carried down from the rice field to the granaries in the village. (3) marks the Standard of comparison in equative constructions *Ang nang•mi hapsan chunga.* I am as tall as you. (4) marks nominalisation after the factitive enclitic =wa. *Angmi balwami ichian jametwa.* My story ends here.

mi•mang ~ me•mang *n.* SUPER. ghost, spirit of a dead person.

  mi•mang saw•et- ceremony performed a year after someone's death. The spirit of the dead person then leaves the house and goes to *Balphakram.*

mi•manggambyrai *n.* PLANT. species of plant

mi•manggrai ~ mi•manggyrai *n.* PLANT. species of plant

mi•mangkoksi ~ me•mangkoksi *n.* PLANT. pitcher plant

mi•mangkyi ~ me•mangkyi *n.* ANIM. species of small frog that says *pekpek pekpek*

mili- *v.* (1) to assemble, to meet, to come together. *Song sulsang rai•na dakang na•nang ichi milini.* We will meet here before going to the neighbouring village. (2) to find. *"Tupi pidan ra•wama, nygylchi?" "•mhm•, milicha."* "Did you buy a new cap at the market?" "No, I didn't find one." (2) appropriate. *Ie sendel milicha, skulsang re•engna.* These sandals are not appropriate for going to school.

**milimityr** *clf.* millimetre

**mimi-** *v.* to laugh, to laugh at someone. *Ie biphae gawigumukaw mimia.* This guy is laughing at all the girls.

**mimikakak-** *v.* to shake with laughter

**mimiwami ~ mimiwamyng** *n.* ACT. a joke

**miniksuru-** *v.* to be flat-haired (of animals). *Magachakmi myn•do tyisiwachian miniksuru takjolarianoro.* When the deer's fur is wet, it just quickly gets flat-haired.

**minit ~ minyt** *n. autoclf.* TIME. minute. *Usang rai•na minit kholachidarang nangni.* To go there, you'll need about thirty minutes.

**mirang** *n.* BOCY. neck feathers of chicken

**mistyri** *n.* PERS. mason, house builder and painter, vehicle repair man

**misyn** *n.* PLACE. mission

**miting-** *v.* to hold a meeting. *Bai•sigathangmaran myng• tham mitingaidoknowa.* The friends are holding a meeting.

**mityr** *clf.* metre

**mm ~ m•m ~ •mhm• ~ hm•m** *procl.* I disagree. (see •*mhm•*₁)

**mmmm mmmm** *ideo.* Screeeech! the call of an eagle.

**=mo** *encl/prtcl.* (confirmative tag). (1) *encl.* "*Tyt! di•phuram•ama?*" *nookno.* "*ho•ong manamaidongmo*" "Hey! Did you just accidentally fart?" he, said. "Yes, it stinks, doesn't it?" (2) *prtcl. Tan•manokona thyiok udo, mo.* Because they had cut him up, he died, that one, isn't.

**mobail** *n.* ART. mobile phone

**mobil** *n.* SUBST. motor oil, engine oil

**mochok** *n.* PLANT. sapling

**moharas** *n.* PERS. Majesty

**moila** *n.* SUBST. dirt, filth

**moina** *n.* ANIM. common myna (a species of bird) *Acridotheres tristis* (see Photo 113)

**mojekjek-** *v.* to shake (a fixed object)

**mojet- ~ mojot-** *v.* to suck

**mok** *n.* PLANT. species of plant

**mon** *clf.* unit of 40 kg *mon sa* one unit of 40 kg (possibly from English *maund*, which in British India was 37.3242 kg.)

**monchara asu** *n.* PLANT. species of tree

**mondoli** *n.* ABSTR. the Church, congregation, church community. *Ang mondolina kham kha•ni* I am going to do work some for the Church.

**monggolbal** *n.* TIME. (< Bengali) Tuesday

**mongma ~ mungma** *n.* ANIM. elephant

**mongmabipha ~ mungmabipha** *n.* bull

**mongmachong•** *n.* ANIM. caterpillar

**mongmachong•su** *n.* ANIM. giant caterpillar

**mongmamathai ~ mungmamathai** *n.* ANIM. bachelor elephant

**mongmawa• ~ mungmawa•** *n.* BODY. elephant tusk

**mongnal** *n.* PLANT. lotus

**mongreng ~ mongyreng** *n.* ART. very big knife on a long pole (see Photo 26)

**monok-** *v.* to swallow, to devour. *Goilapan chym•aimu monokbo.* After chewing the betel nut and paan, swallow it. *Sa•gyrai mylgabami dadadarangawdo janggalawan monokokno. Phylgym chunggaba monokrumokno myng• korokawan.* As for the brothers of the small child, they were devoured. The big eagle had devoured them all, the six of them.

**montyri** *n.* PERS. minister

**morot**₁ *n.* PERS. (< Hindi) Clf. *myng•*. (1) person, people, human, human being *Morot myng• sa ganangno. Uba jyw•taraanokno, wa• ni•okno.* There was a person. She was a single mother, there was no more father. "*Angdo morot myng• seneaw wetsachi so•otna man•gaba*" *nookno.* "I am [someone] who can kill seven people in one blow", he said. (2) someone. *Te•ewba song dam sachi Thengthon mynggaba morot myng• sa ganangno.* Now, in a village, there was someone called Thengthon. (3) man. *Ge•thengdo morot wa• sa•agaba.* He is a tough guy. (Literally 'He is a man who eats bamboo.')

**morot-**₂ *v.* to grate

**mosa ~ mawsa** *n.* KIN. *c, ref, a, rec.* (1) male cross-cousin: father's sister's son or mother's brother's son (2) the relation of male cousins from intermarriageable families

(3) a male friend belonging to an intermarriageable family

**mosathangmaran ~ mawsathangmaran** *n.* KIN. *set.* two boys of different *maharis* (lineages), for example Marak and Sangma, in two possible relationships depending on the gender of the speaker. Male speaker: my elder sister's son and my son. Female speaker: my elder brother's son and my son.

**mot-** *v.* to shake a fixed object. *Manjuri cha•dyl dymphinaidokno. Motchaaidongano; gudukchaaidongano.* The king post had grown roots again. They were shaking it, but it did not budge.

**motorajip** *n.* ART. electric fan

**mrimri ~ rimirimi** *adv.* when squinting. *Hyiawe morot re•enggabawdo jan•dugaaimung, mrimri nuketariokte.* That person going way over there is too far, I can only see him when I squint. *Mykren rimirimi takaidong.* My eyes are almost closed because I'm so tired.

**=mu ~ =myng₄ ~ =mung ~ =mungna ~ =muna** *encl.* (sequential) and, after. (see *=myng₄*)

**mu•-** *v.* (1) to stay, to sit (be in sitting position), to sit down, to be at, to live somewhere. *Phylgymdo nukanchano. Atongba sa•ai mu•arongno.* The giant eagle did not see him. He was sitting and eating something. *Dakangdo Dawa maharisa ichi mu•wanokhone.* Perhaps in the past the Dawa family lived here. *Chokichi mu•bo.* Sit down on the chair. *Dakanggaba Turachi mu•wachi Mobbinaw gorongwa.* The first time I stayed in Tura I met Mobbin. (2) to keep *V*-ing, durative verb. *Bengblokmyngdo sangumuk pywdyngdyngaimyng mongmamyng mykyranaw syw•chekchekai mu•okno mang ni.* The two of them, the banana-sucking bird and the toad, had been fighting the whole day, and had kept repeatedly picking at the elephant's eyes.

**mu•chonchyron- ~ mu•choncholon-** *v.* to squat. *Gari hapalchi mu•chonchyroai chaibo.* Look under the car squatting. (see Photo 43)

**mu•dap-** *v.* to sit on something

**mu•khuchok** *n.* BODY. nipple

**mu•peng-** *v.* to sit and block someone's view

**mu•pyret-** *v.* to crush by sitting on something

**mu•rong-** *v.* to shit on

**mu•si-** *v.* to stay somewhere uncomfortably

**mu•symbylek- ~ mu•symblek-** *v.* to sit on the floor (with one's bum touching the floor)

**mu•ten-** *v.* to look after, to watch, to keep company

**mu•thai** *n.* BODY. breast (of woman), bosom
**mu•thai hal-** to breastfeed

**muchi** *n.* ANIM. species of fish (see Photo 100)

**muchot** *n.* ANIM. mouse, rat. *Abeknyng•chi muchotsa•gyrai mang byryi chepchap chepchap parawthokaidonga.* Inside the *abek* are four baby mice squeaking eek eek.

**mudu** *n.* PLANT. papaya

**muja** *n.* ART. sock

**muk-** *v.* to smoke (to produce smoke, like a fire does)

**mukta** *n.* ART. perl

**mukthai** *n.* PLACE. asperity, protrusion

**mula** *n.* PLANT. white radish

**muluwa•** *n.* PLANT. species of bamboo

**=mung ~ =mungna ~ =myng₄ ~ =muna** *encl.* (sequential) and, after. (see *=myng₄*)

**mungma ~ mongma** *n.* ANIM. elephant

**mungmabipha ~ mongmabipha** *n.* ANIM. bull, male elephant

**mungmamathai ~ mongmamathai** *n.* ANIM. bachelor elephant

**mungmawa• ~ mongmawa•** *n.* BODY. elephant tusk

**muni** *n.* ACT. a magic spell. *Ma• pynwasama muni ma• ang mykrenaw, ma• nang•chi gan ang atongba jadu.* (Wilseng S. Marak) Whether my eyes are covered by a magic spell, you have something magical.

**mura** *n.* ART. stool (to sit on) (see Photo 38)

**muri** *n.* FOOD. popped rice

**musuri** *n.* ART. mosquito net

**mychym-** *v.* to smile at someone. *Gawi angaw mychymaidok.* A girl is smiling at me.

**myia** *dtw.* yesterday

**myk₁** *clf.* forearm length: the length from the elbow to the tip of the middle finger; ell, cubit

**myk-₂** *v.* to tell lies

**mykasyrep** *n.* PLANT. species of plant

**mykbu-** *v.* to be jealous, to be envious

**mykbyryw- ~ mykbryw- ~ mykbyru-** *v.* (1) to have itchy eyes (2) to be jealous

**mykbyruk- ~ mykburuk-** *v.* to be jealous of one another. *Ue gam pang•wamigymyn kam pang•wamigymyn ge•thengtheng mykburukokno.* Because of all the wealth and riches, they had become jealous of one another.

**mykcha-** *v.* to like somebody, to fancy somebody

**mykchagaba** *n.* PERS. sweetheart, girl or boy that you fancy

**mykchel-** *v.* to shine in the eyes. *Rangsan angaw mykchelaidong.* The sun is shining in my eyes.

**mykchep-** *v.* to look down upon, to despise, to scorn, to underestimate. *Atakna nang•do angaw mykchepa?* Why do you despise me?

**mykdaw ~ mykdo** *n.* CORP. night blindness

**mykgythal** *n.* ABSTR. reality. *Mykgythaldo dong•cha, jywmangsa.* It's not reality, it's just a dream.

**mykha badri** *n.* GEO. long period of incessant heavy rainfall.

**mykhal-** *v.* to be older than someone. *Ge•theng ang mykhalgaba.* He is my elder.

**mykhang₁** *n.* (1) BODY. face. (2) PLACE. front, in front of. *Kyi• nokmykhangchi mu•aidonga.* The dog is sitting in front of the house.

**mylhang₃** *n.* TIME. future. *Nang•ba mykhangsangba ytykyi takna bai.* Don't you ever do that in the future.

**mykhang-₂** *v.* to face. *Isang mykhangbo.* Face this way. *Mykhangrukbo.* Face each other. *Ang ge•thengsang mykhangaidong.* I'm facing him.

**mykjyw-** *v.* to doze off

**mykkep** *n.* BODY. temple

**mykkhi** *n.* CORP. slime from the eyes

**mykpeng-** *v.* CORP. cross-eyed. *Ie morot mykpengok.* This person is cross-eyed.

**mykpeng mykpeng** *n.* PERS. name to call a cross-eyed person

**mykphylyp-** *v.* to blink with one's eyes

**mykrak-** *v.* to hold a wake (often used with the incorporated noun *wal* 'night'). *Wa•gaba thyigabana sa•dyrangba jyw•gabamyng wal mykrakaidonga.* The children and the mother are holding a wake for the dead father.

**mykraket-** *v.* to warn. *Ang nang•aw mykraketarong.* I'm warning you.

**mykren ~ mykyren** *n.* BODY. eye

**mykren wa•thok songphin-** to gaze in amazement. *Bandiba manjuriaw kawraw bytjasaaimu, phalthang phagongmachi paiai rai•aaidonganote, Bandiba. Bandi paianggabaaw mykren wa•thok songphinai Gryngrang chaisymaidongano.* Bandi pulled the king post out with ease, and came carrying it on his shoulder, I'm telling you, it was Bandi. Gryngrang gazed at the carrying Bandi in amazement. *mykren ronronok.* My eyes are almost closed because I am so tired *mykren rimirimi ~ mrimri takaidong.* My eyes are almost closed because I'm so tired.

**mykren tan•-** to wink

**mykren nuk-** *vpan.* to see. *Nang• walchi mykren nukama?* Can you see at night?

**myksep** *n.* BODY. corner of the eye

**myksolkhare** *n.* BODY. ring finger

**myksong-** *v.* (1) to plan, to intend *Ang kymna myksongarongchym ytykchiba man•ni ma man•chabai kymna.* I intend to get married but maybe I will not be able to. (2) to decide *Baba na•a ha•gylsakchi angaw watai bisang alagachi nokhor hitgabatykyi na•a parangna myksongaidong?* Son, why did you decide to leave me and wander and work as a slave in someone else's house? (3) to mean. *Nokma asolaw ha•kynggore nokgaaw myksongwa dong•cha.* The word *nokma*, really, does not mean the owner of all the land.

**myksu-** *v.* to wash one's face. *Angdo phangnan jasa•aimyng myksua.* I always wash my face after getting up.

**myksul** *n.* next. *Myksulmu jahas chawpatmanwachi gorongaiok madamaw.* He sure met her when he crossed over with the next ship, his teacher.

**myksymyl ~ myksmyl** *n.* BODY. eyebrow

**myksyram ~ myksram** *n.* BODY. eyelash

**mykthoram** *n.* BODY. middle finger

**myktoksi** *n.* PLANT. plant with beautiful white flowers with a yellow heart that look like big jasmine flowers

**myktyi** *n.* CORP. tear. Clf. *thothak*. *myktyi thothak ni*. two tears/teardrops

**myktyiwatram** *n.* BODY. side of the hand under the index finger

**mykyren ~ mykren** *n.* BODY. eye (for examples, also of the sub-entries, see *mykren*)

**mykyren wa•thok songruk-** to look with wide open eyes

**mykyren tan•-** to wink

**mykyren ronronok** to close one's eyes because you are tired

**mykyren rimirimi ~ mrimri takaidong.** my eyes are closing because I'm so tired.

**myl-** *adj1.* small, little. *Rukwakdo pan mylgaba paiai takaimu rong•aw julokno.* The toad, carrying a small piece of wood, levered a stone. *Ie nok myla.* This house is small. *Sala! mylteng te•ewan jora chaina, roalan jokkhucha!* Damn! You're too little right now to see your lover; you haven't even finished primary school!

**mylthai** *n.* BODY. small bosom. *Rong sa mylthai, rong sa chungthai.* One big bosom, one small bosom. *Phak sa mylthai, phak sa chungthai.* On one side a big bosom on the other a small bosom.

**mym•**$_1$ *n. autoclf.* BODY. a fist

**mym•**$_2$ *clf.* classifier for fists and things that are like a fist

**mym•-**$_3$ *v.* to be like a fist

**myn-** *vintr.* ripe, cooked, ready. *Mai mynokodo maidan syla thoka.* When the rice is ripe, we celebrate the new rice festival. *Ie panchung mynkhucha.* This jackfruit is not yet ripe. *Ja•bek mynok.* The curry is ready.

**myn•** *n.* BODY. body hair (of humans), fur (of animals)

**myn•dyluk** *n.* PERS. person without body hair (this word is used jokingly)

**myn•sym-** *v.* to be hairy with small hairs

**myn•symok** *n.* BODY. a small body hair

**myn•tyi** *n.* SUBST/CORP. puss; resin; latex of jackfruit, thick fluid of various fruits. *Myn•tyi gumukan takapa.* All thick fluids are sticky. / All puss/resin/latex is sticky.

**myndyni** *adv.* a certain way to wicker a bamboo mat (*damdyl*). *myndyni khepa* to wicker the *damdyl* in a *myndyni* way

**myng**$_1$ *clf.* classifier for spoken things, games and for the words *chol* and *bostu*. *golpho myng sa.* one story. *khata myng ni.* two words. *chol myng sa.* a plan. *bostu myng tham.* three things

**myng-**$_2$ *v.* to call someone/somebody a name. *Wiliamnagalaw symsanggre noai mynga.* Williamnagar used to be called Symsanggre. *Angmi bimung Braiton myngwa.* My name is Braiton. *Angmyng amaaw Goje M Sangma myngwa.* My mother is called Goje M Sangma.

=**myng**$_3$ ~ =**mi** *encl.* (genitive). (see =*mi*)

=**myng**$_4$ ~ =**mung** ~ =**mu** ~ =**mungna** ~ =**muna** *encl.* (sequential) and, after. *Babami tangka ratja banga piaimung jalangok.* He asked five hundred rupees from his dad and ran away. *Manapmian mai ja•bek rymaimungna re•engariok.* After cooking rice in the morning, she just left.

=**myng**$_5$ ~ =**mung** ~ =**mu** *encl.* (comitative) and, with. *Te•ewe rukpekmyng amakmyng bai•siga kha•wano.* Now, the frog and the monkey are friends. *Na•sawmung alumung soda dywaimyng rymai sa•a.* We cook and eat it with fermented fish, potatoes and soda. *Jahasnamu, cha ryngnamu, bagajinamu raja ni tangka jamok.* The two hundred rupees were all spent on the ship, on drinking tea and on the fortune-teller.

**myng•** *clf.* classifier for humans. *myng• thamkhua.* there are still six persons left

**mynga-** *v.* to call upon someone or something. *Mani myngwaan, hapawan dyngthangdyngthang myngaa, thokthok myngaa. Ie cha•masangmi wai khurutchido ue hyisangmiaw Banggladesmi thyl•*

*Kongosmi jaria ha•gyrsakgumukawan myngani.* As for what we call the worshipping, different places are called upon, they are called upon according to the division. When he summons the downstream spirit, the priest will call upon the influence of all those faraway places up till Bangladesh and the influence of Kongos, all of them.

**myngkhelek-** *v.* to call somebody by a nickname. *Ang nang•aw Matsumoto myngkheleka.* I call you by the nickname Matsumoto.

**myngkheleka** *n.* ART. nickname

**myngnang-** *v.* suitable

**myrumyru** *adv.* not clearly

**myryng myryng** *adv.* barely. *Hyiawchi ha•banokba ganang, myryng myryng nuketaria.* There is also a rice field house, it is barely visible.

**mysepai chai-** *v.* to look/watch with one eye

**myt-** *v.* to extinguish, to be extinguished, to be out. *Wal• mytok.* The fire is out.

**myte** *n.* SUPER. deity, god

**mythel-** *adj1.* to appreciate, to thank, to be thankful, to be grateful. *Nang• taksagaaw mythela.* I appreciate your help. *Anga nang•tymaw mythelbiok.* I thank you very much.

# N

**na-**$_1$ *v.* to hear. *Kyrynggaaw naakno.* She heard the noise.

**na**$_2$ *interj.* nah. Expresses disapproval. *Kyn sang phalthangaw chonnykgabaaw naaimyng, alsia rajae: "Na! anga ytykyi cholie cholisemchaaidok" noaimyng ....* Having heard those who scorned him, the lazy king says: "Nah! I won't succeed".

=**na**$_3$ ~ =**ona** *encl.* (1) marker of a Beneficiary. *Hanep anga nang•na golpho balni.* Tomorrow, I will tell you a story. (2) marker of Recipient. *Na•a angna tangka hyn•chama?* Won't you give me any money? (3) marker of the Standard of Comparison in comparative clauses. *Ang nang•na chungkhala.* I am taller than you. *Khaw kan•wana daiaido na•nange ichi chaiai mu•waan ga•sukhalnaka.* It is better that we sit here and watch than that you cut hair. (4) marker of a Spatial or Temporal Limit. *Dada chungkhalgado, nokchina phetangokno.* The elder brother reached home. *Nygyltyi ni tyi thamchina hongkhotanchakno ue.* He didn't come out for two or three weeks. (5) marker of a Destination. *Bisangnasa nang•tyme?* Where exactly are you going to? (6) marker of an Emotor. *Ang nang•na kha•gala.* I love you. *Machana makbulna mongmana paichaaimung byldyng byldang jalna ha•bachengok.* Not bearing the tigers, bears and elephants anymore, they stared to run away all over the place. (7) marker of the complement of the postposition *dakang*. *Nang•tymmi nanggabaaw nang•tymmi pi•aidongabaaw, nang•na dakangan phetangok, nang•na dakangan udo re•engsawok.* [The curse] which you needed, which you were asking for, had arrived before you and it has certainly left before you. (8) marker of a Desiderative Clause *Bisang rai•na bai•siga?* Where do you intend to go, friend? *Hai bai•siga, biskut sa•khawna.* Come on, friend, I want to steal the biscuits. *"Cha•masangba chaiok! Khambaisangawba chaiok. Cha•masangmi chaichiba matdam sa•ak, khambaisangmi chaichiba matdam sa•ak. Biaw chaikhuna? Angna ni•ok!" nookno.* "I looked downstream, I looked upstream! Whenever I looked downstream, the otters had eaten it. Whenever I looked upstream, the otters had eaten it. Where else am I supposed to look? I have nothing else!" he said. (9) marker of a Purpose Clause. *Ning ue phylgymaw kawna re•engnane.* We are going to shoot that eagle, okay? *Biana wak khatna raw•aidonga.* We are searching a pig to slaughter for the wedding. *Bandiaw watetna chanchiaidokno.* He was thinking about sending Bandi. (10) Marker of a Complement Clause of Primary-B and Secondary Verbs. *Ang phalthangan re•engnado sykaidokchym.* I would like to go myself, but I can't. (11) marker of Complement

Clauses of intransitive verbs. *Mamung tangka ni•wa aro sa•na ryngnaba ni•wa.* I have no money and I have nothing to eat or drink. (12) marker of a Reason Clause. *Balphakram ha•byrigumukokona rai•sotna man•cha.* Because Balphakram is all hills, you cannot go there directly.

**na•** *n.* ANIM. fish
  **na• pun-** to fish, to catch fish
**na•chan** *n.* ANIM. firefly
**na•cheng** *n.* ANIM. river shrimp or river prawn
**na•garang** *n.* ANIM. species of electric fish
**na•gungphel** *n.* ANIM. species of fish
**na•jek** *n.* ANIM. species of fish
**na•jekwa•** *n.* PLANT species of bamboo that causes irritation when you touch it.
**na•kha** *n.* ANIM. species of fish
**na•lam** *n.* ANIM. species of blue, purple river fish that tastes particularly good when prepared in a bamboo tube (see *bering-*) (see Photo 103)
**na•lamsusyrakdyl** *n.* PLANT. species of liana
**na•langtaupal** *n.* ANIM. species of fish
**na•luk** *n.* ANIM. tadpole (see Photo 105)
**na•matsa** *n.* ANIM. species of fish
**na•nang** *ppron.* we, first person plural inclusive
**na•nyl** *n.* ANIM. electric eel
**na•pat** *n.* ANIM. species of fish
**na•phok** *n.* ANIM. species of fish
**na•rong** *n.* ANIM. species of fish (see Photo 103)
**na•ru** *n.* ANIM. fish poison
**na•rym** *n.* ANIM. species of fish
**na•rymkhu** *n.* ANIM. species of fish (see Photo 102)
**na•sak** *n.* ANIM. species of red fish
**na•saw** *n.* FOOD. fermented fish
**na•wachak** *n.* ANIM. species of fish
**na•wak** *n.* ANIM. species of fish
**nabak** *n.* ART. knot
**nadanggap** *n.* ANIM. flat blood sucking parasite on humans and animals
**nadanggorot** *n.* BODY. oesophagus, food pipe, gullet
**nadekaram** *n.* BODY. earlobe
**nagok** *adj2.* deaf

**nai•** *n.* KIN. *drel, c, ref.* aunt: father's sister. (addressed as *anyng, anai,* or *mani*)
**nai•maran** *n.* KIN. *set.* my *nai•* (father's elder or younger sister) and her (elder or younger) brother's unmarried child
**nai•nokhol** *n.* KIN. *d, ref.* mother-in-law (addressed as *mani* or *anai*)
**nai•nokholburung** *n.* KIN. *set.* a group of mothers-in-law and daughters-in-law
**nai•nokholthangmaran** *n.* KIN. *set.* my *nai•* (father's elder or younger sister) and her (elder or younger) brother's married child
**naija** *dtw.* next year, at some time in the far future
**nak-** *adj1.* black
**=naka ~ =ka** *encl.* (imperious/certain-future modality). *Cha rynaimyng, mai sa•aimyng, rai•naka.* After drinking tea and eating rice, we will go. *"Aia dugaphinok bai•siga! Angaw thogiwado. Tai•nido tanchomotchaka anga ge•thengaw kakai sa•naka"nookno pherue.* "It's too much, friend! You have betrayed me. Today, I will certainly not spare him. I will certainly devour him.
**nakamai** *n.* ART. small basket to sow rice from
**nakhal** *n.* BODY. ear. *nakhal sam sa* one ear
  **nakhal ruru-** to have an ear infection
  **nakhal na-** *vpan.* to hear. *De•thengmi balgabaaw nakhal ta na!* Don't listen to the things he says!
**nakhalcha•dan** *n.* BODY. part of the head behind the ear
**nakhalthek** *n.* ART earring
**nakhong** *n.* BODY. backside of the ear
**nakhung** *n.* BODY/CORP. nose, snot (liquid), mucus (from the nose). *nakhung goi• sa* one nose
**nakhung ra•taw-** *v.* to snort
**nakhungdi•** *n.* CORP. hard piece of snot
**nakhungkhal** *n.* BODY. nostril. Clf. *khal. nakhungkhal ni.* two nostrils
**nakhungmyn•** *n.* BODY. nose hair
**nakhungthek** *n.* ART. nose piercing
**nal-** *v.* to gorge, to stuff one's face
**nalbas** *adv.* (< English) nervous. *Nalbas sa•akno, ue sa•gyraie.* The child was nervous.

**nalsasang** *n.* PLACE. the other side
**namakai ~ nakamai** *n.* ART. small basket used to sow rice out of
**nambal ~ nombol** *n.* ABSTR. (< English) number
**namchyk** *n.* KIN. *d, ref, a.* niece: (1) female's brother's daughter (2) male's sister's daughter (Denotes the same relation as *namgaba*.)
**namchyksyw•** *n.* KIN. *d, ref.* grand-niece or great-niece: (1) the daughter of my husband's sister's daughter (2) the daughter of my brother's wife's daughter
**namgaba** *n.* KIN. *d, ref.* niece: (1) male's sister's daughter (2) female's brother's daughter (addressed as *namchyk*)
**namnokhol** *n.* KIN. *d, ref, a.* daughter-in-law
**namnokholburung** *n.* KIN. *set.* group of daughters-in-law
-**nang**$_1$ *evsp. V* in a beautiful or nice way
**nang-**$_2$ *v.* to bear fruit
**nang-**$_3$ *v.* to hang (down from). *Pan gongdang takgabachi ne• nangwanote. Ne• nanggaba okumwachi jywsawthiriokno.* Bees were hanging from a bent tree. Under the hanging bees lay the deer, fast asleep again.
**nang-**$_4$ *v.* (1) to need, to have to, must. *Nang•na atakgaba syldaraidarangaw nangni?* What kind of sword would you need? Then, you need/have to prepare some liquor for the priest. *Ido pherudo ang sa•awdo poresemetchagaba kakai sa•na nangok ido" noaimung chanchichypai gorialba.* This fox, who certainly did not teach my children, must have devoured them, the crocodile thought. *Umi chywba sym•ai rymna nangni, ue kamalna.* (2) to have to call someone by a certain term. *Nang•do ge•thengaw mani nanga.* You have to call her *mani*.
**nang•** *ppron.* you (singular), second person singular
**nang•tym** *ppron.* you, second person plural
**nangchomot-** *adj1.* important
**nanggandai**$_1$ *adv.* naked
**nanggandai**$_2$ *n.* PERS. naked person
**nanggodolong** *n.* PERS. naked person

**nangthaigaba** *n.* CORP. swelling, abscess
-**nap** *evsp. V* with all one's heart. *Anga babasangan jalnapaimu anga songthangsangan waiphinnaka.* I will run back to my father with all my heart, and I will return to my own village.
**napit** *n.* PERS. hairdresser
**narang** *n.* PLANT. orange
**narot** *n.* PLANT. species of edible tuber
**narykel ~ narykhel** *n.* PLANT. coconut
**narykeltyi** *n.* PLANT. coconut water
**nasengkhet** *n.* ANIM. tick
**nasi-** *v.* to be annoying to listen to, to hurt one's ears. *Kyi• parawchido ang nasia.* It's annoying to hear the dog bark. *Kyi•parawga nasia.* The barking dogs make my ears hurt.
**nat-** *v.* to scrub, to scour, to clean by scrubbing, to remove by scrubbing. *Wa natbo.* Brush your teeth. *Nang• nonoe tyigatchi nataidok.* Your younger sister is washing the dishes at the water place. (see Photo43)
**nathek** *n.* ART. earring
**natheng** *n.* BODY. cheek and cheekbone, side of the head
**nathyra** *n.* ANIM. tick
**natym-** *v.* to listen (to). *Golpho balna pang•cha natymthokbo.* I want to tell a story, everybody listen.
**naw**$_1$ *n.* KIN. *drel, d, ref, a.* (1) younger sister. (2) Also used to address a younger female cousin or (3) an unrelated woman younger than the speaker.
**naw... bai•...** *coll, n.* younger sister. *Waiphinwami gesepchi baratdugaaimu, wa•na jyw•na baratai, bai•na tyngna baratai, nawna bai•na baratai, de•thengdo dang•anaan chaithylaisa mu•arongno, gopjyrujyrutykyi.* Upon his return, he felt very much ashamed, he shied away from his father and mother, he shied away from his blood relatives; upon entering, he sat only looking away into the distance, his head bent.
**naw-**$_2$ *v.* to scold. *Tura re•engni, dongchachido angaw baba nawni.* I will go to Tura, otherwise father will scold me.
**naw... jai...** *coll, v.* to scold, to quarrel.

**nawang** *n.* PERS. retard, half-brain, fool, stupid, confused person

**nawchak** *n.* ANIM. species of fish

**nawmyl** *n.* PERS. marriageable girl

**nawsyri** *n.* KIN. *d, ref, a.* sister-in-law: (1) spouse's younger sister (2) younger brother's wife (Elder brother's wife is *ja•chung* or *bochi*.)

**=ne** *encl/prtcl.* (affirmation-seeking tag) (1) encl. *Uchi Nepaldo: "Na•a ang ma•su mang rajasaaw tynangsegabone" nookno. "Ym" noaimyng Thengthonba tynangokno.* Then the Nepali said: "You lead my hundred cows away, ok?" "Yes," he said and Thengthon led them away. (2) *prtcl. Nemai re•engbo bai•siga, ne.* Go carefully, my friend, ok?

**ne•** *n.* ANIM. bee (possibly also wasp and/or hornet, see also *ne•kat*)

**ne•balang** *n.* ANIM. species of mantis (see Photo 84)

**ne•kat** *n.* ANIM. species of bee, or general name for a type of insect (bees, wasps, hornets) (see Photo 108). More research is necessary to find out which species called *ne•* are bees, which are wasps and which are hornets.

**ne•katthup** *n.* ART. hive, bee's nest (see Photo 107)

**ne•wal** *n.* ANIM. species of bee

**Nedyran** *n.* PLACE. (< Dutch) The Netherlands, Dutch

**Nedyranmorot** PERS. (< Dutch and Hindi) Dutchman, Dutchmen

**nek-** *adj1.* close, near. *Rame tyi nekokno.* The road is very close to the water. Literally: As for the road, the water is very near.

**neksem-** *adj1.* to be very near

**nem-₁** *adj1.* (1) good. *Umi golpho nemate.* His stories are good. *Nang• rai•awa nema.* It was good of you to come. *Nemarini.* It's ok. / It's all right. (2) to stop (of rain). *Rang nemchengama na•nang chyw jamchenga?* Will the rain stop first, or will we run out of liquor first?
    **nemai** *adv.* well. *Ang ge•thengaw nemai tynga.* I know her well.

**nemcha** *v.* bad, wrong. *Ie madam nemchate. Angaw tokwa.* This teacher is bad. She hit me. *Ie cha•ba nemjolanchak, uching kakaimu thyi jokok" nookno.* There is something wrong with his leg; leeches have bitten it and blood has leaked out", he said.

**nem-₂** *v.* to get better, to heal. *Ytykyimu, khurutaimu chanchichypai nemok.* So then, having performed the incantation, [the patient] has supposedly healed/gotten better.

**nemkhal-** *v.* to get better, to improve. *Wai khurutaimu, sa•ai ryngaimu, nemkhalchiba nemkhalchachiba ue morotnado dykdyksa chaisakni.* Having performed the incantation, having eaten and drunk, he will observe the patient for a short while, to see if he has improved or not.

**nemen** *intens.* very. *"Te•ew wen sa rypa nang•do nemen sylnaka" noaidongano pherue.* "If you go into the water once more you will certainly be very beautiful" said the fox.

**nemgyni** *n.* ABSTR. advantage, good fortune, good luck

**nemnuk-** *v.* to like. *Songmyng nokmyng morotdarang ge•thengaw bylongen nemnukano.* The villagers liked him a lot.

**neng•-₁** *adj1.* (1) tired (after making an effort) *Sangumuk taw•reksyrupmyng bengblokmynge myng• ni thopaimyng neng•ba neng•okno.* The whole day, banana-sucking bird and the toad, the two of them, had ganged up on it, and it had become tired. (2) difficult. *Hapsan nokkhung raja sa mu•chido man•ai sa•na neng•oknowa.* If [they] would stay together in the hundred houses, [it] would be difficult to get rich.

**neng•-₂** *vsec.* to lack, to fail to. *Ha• chamai Bandi, byl neng•chiba chak neng•chiba iaw ryngetphabo!* Take this sweetheart Bandi, when you lack strength, when your hands are tired, drink this! *Wak rakhiwami gesepchian de•thenge maimynawan man•ai sa•na neng•okno.* When he was herding pigs, he lacked proper food.

**neng•thak- ~ ning•thak-** *v.* to rest, to take a rest, stop for a while

**Nepal** *n.* PERS/ART/PLACE. Nepali (person and language), Nepal (country)

**nesynyl haiwe** *n.* PLACE. national highway

**net**₁ *n.* ART. basket worn on the waist to put in the harvested rice (see Photo 27 and Photo 28)

**net-**₂ *v.* to shine. *Changba bydyi myng• sa khen• raw•arongnote wal• netaimu.* Some old person was catching river crabs, shining a light (made by fire).

**netwak** *n.* network

**ni**₁ *num. bound.* two

**=ni**₂ *encl.* (uncertainty modality). *Morot so•otgabaaw gobormen so•otsigani.* Murderers will be killed by the government. *Atongsang balchido san kolgyksa noai myngnichym.* If you speak Atong, you would say san kolkgik sa.

**=ni**₃ **~ =nyi** *encl.* (< Hindi) (privative, indicates a referent that has been left out). without. *Chininyi·cha takbo.* Make tea without sugar. (see also *=ri*)

**ni•- ~ nyi•-** *v.*(< Hindi) (negative locative/existential verb) to not exist, not have, there isn't, there aren't, there wasn't, there weren't etc. *Songma Songgni mynggaba songba ni•ok.* The so-called village of Songma Songgyni does not exist anymore. *Mamung tangka ni•wa aro sa•na ryngnaba ni•wa* I don't have any money, and I don't have anything to eat or drink. *Ang sa• ni•wa.* I don't have children. *Te•ewe, Balphakramchi mongma nyi•wa.* There are no elephants in Balphakram now. Dakangdo, Garo Hillschi Banggal ni•wa. *In the past, there were no Bengals in the Garo Hills. Ning songsyrekdo ning atongdo dakangdo mamyng thoromaw ni•wami somaichido waiaw mania.* We pagans, we the Atong, in the past, in times when there was no religion, we worshipped spirits. *Somaido nyi•ok* There's no time left.

**ni•et-** *v.* to switch off, to turn off. *Lait ni•etbo.* Switch off the light!

**ni•wa** *expr.* it's nothing, never you mind. "*Atong khaiwa?*" "*Ni•wa na•a.*" "What are you carrying?" "It's nothing." "*Bisang re•engni?*" "*Ni•wa.*" "Where are you going?" "Never you mind." *Ni•wa na•a!* It's none of your business!

**niam** *n.* ACT. custom, law, tradition

**ning** *ppron.* we, us, first person plural exclusive

**nisan-** *v.* to aim. *Bunduk ra•aimuna matsaaw kawna noaimuna nisanaroknotyi.* Having taken the gun, he said he wanted to shoot the tiger, he aimed, to our surprise.

**nisi-** *v.* to poison. *Sanarai sa•chido ang nang•aw nisina man•chaka.* If you eat a centipede, I will certainly not be able to poison you.

**no-**₁ *v.* (1) to say. "*Dadapara sandijolni anga*" *noaimu re•engarokno.* I'll quickly search my elder brothers", he said, and went on his way. *De•thengdo "ang re•engchawa" noai balwa.* He said that he would not go. *Mai sa•akma nobo* Say: "*mai sa•akma?*". (2) to call. *Mama nobo.* Call him *mama*.

**=no**₂ **~ =nowa** *encl.* (hearsay evidential predicate enclitic, termed quotative, that indicates that the information in the clause was not witnessed by the speaker, but that the speaker has the information from hearsay). *Song dam sachi morot myng• sa man•ai sa•bigyba ganangnochym.* In a village was supposedly a very rich man, it is said. *Rong•khasang galaimyng thyiokno.* He fell down a cliff and died, they said. "*Man•okma bai•siga, biskute?*" *noaidonganowa.* "Did you get the biscuits, friend?" he was saying, it is said.

**nobembyl ~ nobembol** *n.* TIME. (< English) November

**noga** *n.* ART. tree-house (see Photo 8)

**noga ~ nogaba** *expr.* so-called. *Dakangmi somaido ning pi•sa mylbutungchido nokma nogado man•ai sa•gasa.* In the past, in our childhood, when we were small, so-called *nokma* were only rich persons

**nogek**₁ *n.* ART. broom

**nogek**₂ *n.* PLANT. the cut off part of the plant of which brooms are made (see Photo 34)

**nok** *n. autoclf.* PLACE. house. *nok tham.* three houses (see Photo 7)

**song... nok...** *coll, n.* village (see *song*)

**nokbanthai** *n.* PLACE. bachelors' house. Before Christianity each village had a Bachelors' house for every clan that lived in the village. In this house lived young, unmarried men. They would practice fighting, hunting, singing, storytelling and all kinds of things that young men would have to learn before getting married. Women and members of other clans were not allowed to enter the bachelors' house. (see Photo 14)

**nokchama** *n.* KIN. *d, ref.* the relationship of the parents of a married couple

**nokchina ~ nokna** *n.* KIN. *ref.* (Siju dialect, *nokrom* in Badri dialect) the heiress of a household or her husband.

**nokchol** *n.* ART. door, entrance

**nokdang** *n.* PERS. the family that live together in one house

**nokgaba** *n.* PERS. landlady, landlord, house owner, God

**nokha•pal** *n.* PLACE. outside, outside the house

**nokhama**$_1$ **~ hama** *n.* PLACE. under, underneath, below, space between the floor or the base of something and the ground. *Ang tankabek palongnokhamachi.* My wallet is under the bed. *Taw•sa•grai nokhamaaw jalphakangaidonga.* The chicks are running under the house (from one side to the other).

**nokhama**$_2$ *n.* ART. a supporting structure

**nokhap** *n.* PLACE. a level piece of land on which a house is built

**nokhol ~ nokhor** *n.* PERS. slave

**nokkhap** *n.* ART. door

**nokkhung ~ nukkhung** *n. autoclf.* ART. roof (see Photo 9)

**nokma** *n.* PERS. village headman, rich man, respected man. *Dakangmi somaido ning sa•gyrai mylbutungchido nokma nogado man•ai sa•gasa, gam pang•gasa nokma mynga.* As for the past, when we were small children, a so called *nokma* was a wealthy person, only someone with a lot of wealth was called *nokma*.

**nokna ~ nokchina** *n.* KIN. the heiress of a family or her husband

**nokphandai** *n.* PLACE. bachelors' house

**nokphin-** *v.* to return home. *Nokphinniba utymdo.* They will return home.

**nokrom** *n.* KIN. *ref.* (Badri dialect, *nokna ~ nokchina* in Siju dialect) the heiress of a household or her husband. *Nang•do ie nokchian mu•na nangni. Nang•an ie nokmi nokrom.* You have to stay in this house. You are the heiress.

**noksam** *n.* ART. (1) wall of a house. *Noksamchi simen, tota, tin pirinai hama.* They build the walls of their houses with a mix of cement, planks and corrugated iron. (2) the piece of land a house is built on. *Nang• baba noksamchi tangka gopgaba ganangno.* Under your father's house lies buried money.

**noksuk** *n.* PLACE. the side of an object that faces the wall

**noktapa** *n.* ANIM. species of small gecko that creeps up the walls of houses at night

**nokthai** *n.* ART. a small house separate from mother's house, small house next to the main house

**nokwa• ha•chak** *khjyks, n.* bamboo used to build a house.

**nokwek**$_1$ *n.* ART. broom (made from *nokwek*$_2$). Clf. *kun. nokwek kun tham.* three brooms

**nokwek**$_2$ *n.* PLANT. the cut off part of the plant of which brooms are made (see Photo 34)

**nokweng** *n.* ART. floor

**nol** *n.* ART. fence, fenced enclosure
 **nol kha•-** to make a fence

**nom•-**$_1$ *adj1.* soft, weak, easy. *Dam nom•a.* It's cheap. *Tam•ai chaichie te•do byirakhem hongkhotruruaimu kaksyrangokno pheruawdo. Nom•angaidokno udo.* When he tried to hit it, the bees all came out and bit the fox all over. The fox became weak.

**nom•-**$_2$ *v.* to loosen. *Gethengdo majuriaw bytai nom•ok.* He loosened the king post by pulling it.

**nombok-** *v.* unconscious, tired after eating a lot. *Tai•sa nombokok.* A little while ago he was unconscious.
 **nombok thyibok** *khjyks, adj2.* unconscious, almost dead. *Uchian ne• thopai kakokno pheruawdo. Nombok thyibok.* At

that moment, the bees swarm and sting him, the fox. He is unconscious.

**nombol ~ nambal** *n.* ABSTR. (< English) number

**nong-** *v.* to apply, to put (on the skin or body, like a cream or medicine), to smear, to spread, to crush and smear. out *Khu•chul pisak nongwa lepstik* She has put lipstick on her red lips. *Ja•ryt chamussang nongaidong.* I'm crushing and smearing out the chillies with a spoon. *sam nong-* to put medicine (on a wound or on the skin).

**nono** *n.* KIN. *c, ref, a.* (1) younger sister. (2) Also used to talk about or address a related younger female of one's generation: cousin or (3) to address a young unrelated female person younger than the speaker.

**norok** *n.* PLACE. (< Hindi) hell

**nosto dong•-** *v.* damaged, defective

**nuk-** *v.* (1) to see. *Uchie phylgym chunggabaaw nukokno.* Then he saw a very big eagle. (2) to look (like), to resemble. *Ie kyi• matsatykyi nuka.* This dog looks like a tiger. *Sa•banthaigaba noaian tynganchakno. Kan•jotokno. Morottykyi nukanchakno.* He did not recognise his so called son. He was very skinny. He did not look human anymore. (3) to find. *Nang• Atonggawiaw sylai nukama?* Do you find Atong girls pretty? (4) to know, to see something coming. *Ytyknaka nogabaaw nuksawaian anga nang•aw peng•wachym.* If I had known what was going to happen, I would have prevented you.

**nukcham-** *v.* to predict, to see into the future

**nukhu** *n.* PLACE. courtyard

**nukkhung ~ nokkhung** *n.* ART. roof (see Photo 9)

**nygyl** *n.* PLACE. market

**nygyltyi** *n.* autoclf. TIME. week. *Nygyltyi rai•agadyrangchi rai•ani.* He will come sometime next week. When this word is quantified two or more times in a row, after the first time, the word *tyi* is used instead of the word *nygyltyi* for each iteration, e.g. *nygyltyi sa tyi ni.* a week or two

**nygyltyityi.** every week.

**nyi•$_1$ ~ ni•-** *v.* (< Hindi) to not exist, not have, there isn't, there aren't, there wasn't, there weren't etc. (see *ni•-*)

**=nyi$_2$ ~ =ni** *encl.* (< Hindi) (privative, indicates a referent that has been left out). without. *Chini<u>nyi</u>·cha takbo.* Make tea without sugar. (see also *=ri*)

**nyng** *n.* KIN. *d, ref.* (1) aunt: father's sister (2) sister-in-law: husband's elder sister (addressed as *anyng*)

**nyng•** *n.* PLACE. inside. *Noknyng•chi* inside the house

**nyng•thyw-** *adj1.* thorough. *Haida nyng•thywai balchido pang•bia somai nangnagaba ganang, sotkat anga baletariok.* I don't know, if I would tell it too thoroughly, would there be enough time; so I just told a shorter version.

# O

**O** *interj.* (attention seeking) hey! o...! *"O jojong! Rai•bo.* Hey, younger brother! Come here.

**ooo** *interj.*(acknowledgement) oh. *Bisang re•engwa, nangmi bajue?" "Nygylsang." "Ooo."* "Where did your friend go?" "To the market." "Oh."

**obosta** *n.* ACT. event. *Te•ew sansachi ge•thenge ytykgaba obosta dong•wano.* One day, such an event happened to him.

**odek** *n.* PERS. baby

**=odo ~ =do** *encl.* (topic). (see *=do*)

**ogynang- ~ oknak- ~ oknang- ~ okgynang-** *v.* to be pregnant

**ogynanggaba ~ okgynanggaba** *n.* CORP. pregnancy

**oi** *interj.* oi! hey! interjection to draw someone's attention. *Oi, na•a atakaidonga?* Hey you! What are you doing?

**oikor** *n.* ART. alphabet, letter

**oja** *n.* PERS. medicine man, traditional herbal doctor

**=ok$_1$ ~ =ak ~ =k** *encl.* (change of state) (1) change-of-state interpretation. *Nang• ie tupi bimi ra•<u>ak</u>?* Where did you get that cap? *Bisang re•eng<u>ok</u>?* Where did he go? *Churi matanch<u>ak</u>.* The knife is not sharp anymore. (2) intensifier interpretation (on Type 1 adjectives). *Ja•bekan thawok!* The curry is very tasty! *Ie lekha chatok.* This book is very thick. / This book has become very thick.

**ok**₂ *n.* EMO. hunger

**okgynanggaba ~ ogynanggaba** *n.* CORP. pregnancy

**okha-** *v.* to be full after eating. *okhaakma, bai•siga?* Are you full, friend?

**okhi-** *v.* to be hungry. *Mai okhiedok angdo.* I'm hungry. Alternative spelling: *Mai okhiaidok angdo.*

**okhuchak** *n.* CORP. stomach pain

**okhynyng-** *v.* to break a round hollow object in half (crosswise)

**okma** *n.* BODY. the front of the body, belly, underside. *Nawgabaaw ja•nawgaba okma chi ba•aidok.* The elder sister is carrying her younger sister on the front of her body.

**okmyng-** *v.* to starve

**oknak- ~ ogynang- ~ oknang-** *v.* to be pregnant

**oksephang** *n.* CORP. pain in the lower abdomen

**oktobyl** *n.* TIME. October

**oktyk** *n.* PLACE. bottom of ravine or cliff

**ol-** *v.* to speak, talk. *Uba sa•gyraiba sengjolaaknoai. Ba•jolwamian khu•chuk olna sapjolariokno.* The child had quickly become intelligent, really! It could just speak as soon as it was born. *Aia ta oldugasi!* Hey! Don't talk too much!

**oltho ~ ortho** *n.* ABSTR. meaning

**=ona ~ =na** *encl.* (see =*na*₃)

**ong ang** *n.* ANIM. big edible frog that makes the sound *ong ang.* The old spelling was *ong•ang.*

**ong** *n.* ANIM. wasp

**opis ~ ophis** *n.* PLACE. office

**opiser ~ ophiser** *n.* PERS. officer

**ortho ~ oltho** *n.* ABSTR. meaning

**Ostrelia** *n.* PLACE. (< English) Australia

**ostro** *n.* ART. weapon

**oto** *n.* ART. auto rickshaw

**otorewain** *n.* ACT. (< English) auto rewind

**P**

**pa•-**₁ *adj1.* low, plain, flat, thin (of things)

**pa•-**₂ *v.* to perch. *Sympak chunggabachi phylgym pa•ai mu•sawaidongano.* The eagle is perching in a *sympak* tree.

**pagawa** *n.* PLANT. the white, spongy inside of a banana tree

**pai-**₁ *v.* to carry by hand. *Banggal myng•sa biskut chyrymbiai paiaidonganote.* A Bengal is carrying a heavy load of biscuits.

**pai-**₂ *vgoal.* to support, to tolerate, to bear. *Ang khol rangsanna paicha.* My skin does not tolerate the sun. *Uchisa matsana makbulna mongmana paichaaimung byldyng byldang jalna ha•bachengok.* Then, not bearing the bears and elephants anymore, they started to run away all over the place.

**pai•ra ~ phai•ra** *n.* ART. type of basket (see Photo 28)

**paila** *n.* ART. scale (for weighing)

**paip** *n.* ART. (< English) water pipe (see Photo 43)

**paitaw-** *v.* to lift up

**pakara ~ pakyra** *n.* PLANT. stalk of a fruit, cord for *kukuri*

**pakrai ~ pakri** *n.* ART. horizontal beam that runs perpendicular to the *bylbang* under the roof to form the base of the roof of a house (see Photo 9, Photo 10, Photo 11)

**pakyl** *n.* ART. centre strap of a sandal

**pal**₁ *n.* PLANT. flower. *Pal man•ok.* The flower is blossoming. *pal mochoka* a flower bud

**pal-**₂ *v.* to bloom. *Pan palaidonga.* The tree is in bloom. *Balgyto• palaidonga.* The orchid is in bloom

**palak** *n.* ART. bamboo spoon: piece of bamboo split in half and used to stir (see Photo 36)

**palengma** *n.* PLANT. *barebina-xariegata*, tree with beautiful white flowers that smell very nice, like magnolia, and are edible

**palong** *n.* ART. bed

**palyng** *n.* PLACE. jungle

**pan**₁ *n.* PLANT. tree, firewood. *pan phan sa* one tree *pan dot sa* one log

  **pan wa• khjyks,** *n.* PLANT. plants, vegetation, plants and trees.

**pan... wa•...** *coll, n.* plants, vegetation, plants and trees. *Bandi balaidongano: "Panaw wa•aw khi•wama phalthangawan khi•wama?* Bandi said: "Did you hit a tree, or did you hit yourself?"

**pan**₂ *clf.* classifier for apparatus, appliances, mechanical and electrical things or gadgets, cars, bikes, bicycles, mortars and umbrellas. *gari pan sa.* one car. *redio pan sa.* one radio. *satha pan sa.* one umbrella. *thep pan sa.* one tape. *tibi pan sa.* one TV. *asam pan tham.* three mortars

**pan•pyrak-** *v.* to cut breadthwise

**panachol** *n.* PLANT. mushroom (not edible)

**panbai** *n.* PLANT. firewood

**panchak** *n.* PLANT. leaf

**panchan** *n.* PLANT. species of plant

**panchengrong** *n.* PLANT. species of tree plant

**panchoka** *n.* PLANT. small log

**panchong** *n.* PLANT. tree trunk

**panchung** *n.* PLANT. jackfruit

**panchungchong•su** *n.* ANIM. species of black hairy caterpillar that lives on jackfruit trees (see Photo 99)

**panchyksi** *n.* PLANT. twig *panchyksi goi•sa* one twig

**panchyreng** *n.* ART. non-supporting horizontal beam that forms part of the structure of the side of a house and to which the *damdyl* can be attached (see Photo 11)

**pandala** *n.* PLANT. twig

**pandawsik** *n.* PLANT. species of tree

**pang•-** *adj1.* a lot, many, much. *Aia bai•siga, nang•chido sa• pang•ate!* Wow, friend, you have a lot of children! *pang•a bylsidarangmi kynsangang* many years later. **pang•aiba** (1) a lot as well, also a lot. *"Tyyyk!"* achudo: *"Na• man•wa" nogabana pang•aiba tanangaidonganowa.* "Soooo", at his grandfather's, who says "you got fish", he also leaves a lot behind. (2) ever. *Nang•mi janggi khengwagumuk kha•galwamu na•a khakhetaw na•a takchido nang•mi janggigumukchi pang•aiba na•a kanggal dong•chawa.* When you love your whole life and are honest, you won't ever be poor.

**pang•cha** not many, not a lot of, a little, a little bit, a few. *Sa•aknoai magachakdo man•ai man•thing man•ai man•thing biskutaw biskut pang•chaanokno.* The deer was eating the biscuits in great amounts as fast as possible, and there weren't many left. *Anga Durakhalmigymyn pang•cha balna sykaidonga.* I want to tell a little about Durakhal.

**pan wa•** *khjyks, n.* PLANT. plants, vegetation, plants and trees.

**pang•wami** *n.* ABSTR. quantity, abundance

**pangkol** *n.* PLANT. guava

**pangkollipa** *n.* PLANT. species of tree

**pangkywal** *n.* PLANT. guava

**pangyrym** *n.* PLANT. jungle thicket

**panju** *n.* PLANT. firewood

**panjyl** *n.* PLANT. species of tree

**pankhol** *n.* PLANT. bark (of a tree)

**panmaikung** *n.* PLANT. species of tree

**panmang** *n.* PLANT. species of tree

**panmatha** *n.* PLANT. species of tree

**pannok** *n.* wood shed, woodstore, wood stock house (see Photo 11)

**panphek** *n.* PLANT. sapling, young tree

**panrasun** *n.* PLANT. species of tree

**pansok** *n.* PLANT. terminal bud, end of a branch where the tree grows

**panthai** *n.* PLANT. type of fruit

**panthjong** *n.* PLANT. species of tree

**panthong** *n.* ART. wooden stick

**pantiki** *n.* PLANT. wood chip

**pape** *n.* ANIM. species of big, brown gecko

**papol** *n.* FOOD. (< Indic) pasta

**papret-** *v.* to throw to death

**para** *n.* GEO. river junction

**=para** *encl.* (associative plural). X and company, X and those associated with him/her. *Nang•dadaparado usang phylgym chunggaaw kawna re•engwanote.* Your elder brothers went that way to shoot a giant eagle. *amapara* mother and those in her household.

**parang**₁ *n.* PLANT/ART. reed, thatch (see Photo 7)

**parang-**₂ *v.* to wander, to go astray

**-parang**₃ *evsp. V* without destination, *V* without goal, *V* aimlessly, *V* absentmindedly, *V* inattentively. *Chaiparangai tokaimyng, biphagaba nakhungaw tokgakmanaimyng thyisyrangokno.* While she was beating him, she was looking away, and she accidentally hit her husband's nose, and he died on the spot. *Gumukan jalparangok.* Everybody wandered off.

**parap-** ~ **pyrap** *adj1.* to be (too) salty

**paraw-** *v.* to call (of animal), to shout (of animal and human)

**parawchyrik-** *v.* to shout loudly. *Morot sorokchi khepai parawchyrikaidong.* The man was shouting loudly on the road because he was angry.

**pargunja** *n.* TIME. (archaic) (< Indic) February

**pat-**$_1$ *v.* to cross *Sikhar kha•na re•engokno re•engokno re•engokno. Tyikhal goi•saaw patna nangokno.* They went hunting, they went and went. Then they had to cross a river.

**-pat**$_2$ *evsp.* V across. *Te•edo jahastaw chawpatangokno.* Now he sailed the boat away to the other side.

**patal** ~ **phatal** ~ **phathal** ~ **pathal** *n.* GEO. (< Indic) stone

**patyl** *n.* ART. slingshot

**pawai** *n.* ART. bowl to serve curry in or its volume, bowlful, classifier for curries. *"Atong ja•bek sa•ak?" "Alu na•saw pawai sa, taw• khirip pawai sa."* "What curry did you eat?" "Potatoes with fermented fish and chicken with *khirip.*"

**pawdyr** *n.* SUBST. (< English) powder, baby powder

**peel** ~ **pheel dong•-** ~ **dong-** *v.* (< English) to fail. *Ge•theng lekha nemai poreancha, ytykyimu poreka peel dong•ok.* He did not study the book well, so then he failed his exam.

**peket** *clf.* (< English) classifier for packets. *Sigyret peketsa ganangkhuama?* Do you still have a packet of cigarettes?

**peking** ~ **pheking** *n.* ART. (< English) luggage, packing

**pekpek** *ideo.* croak! ribbit! the call of a frog

**pel-** *v.* to copulate, to fuck

**pelang** ~ **peleng** ~ **pylang** ~ **pyl•eng** *adj2.* flat

**peleng-** ~ **pel•eng-** *v.* to deflate

**peleng** ~ **pelang** ~ **pylang** ~ **pyl•eng** *adj2.* flat

**pen** *n.* ART. (< English) pen

**peng** ~ **peng•-**$_1$ *v.* to prevent, to hinder, to obstruct. *Ytyknaka nogabaaw nuksawaian anga nang•aw peng•wachym.* If I had known what was going to happen, I would have prevented you. *Ang mykhangchi mu•pengna bai. Ang nukcha.* Don't obstructively stand in front of me. I can't see. *Nang• lekhaaw komputyrchi tanpengna bai.* Don't obstructively put your book on the computer.

**peng•-**$_2$ *v.* to curse. *Takgaba Rywgabasang Phatigaba Raronggabasang phalthang peng•ai tananggabaaw ra•phinkha•na dengetkhalna.* The supreme god wanted to lift the curse that he himself had put on the village.

**penta** *n.* PLANT. species of plant

**pepylok** *n.* ANIM. species of bird

**pereng-** *adj1.* straight

**peret-** ~ **pheret-** *v.* to split, to crack, to burst, to explode

**pering tongtong** *adv.* straight

**pering-** *v.* straight

**-pha** *evsp.* (1) V also, V in addition, V along with, V together. *Na•a abundyrangtykyi kam kha•ai chaiphabota!* You! Try to do some more work, like other people! *Na•tyme goi•byisyk man•phawa ie bylsie?* How many [baskets full of rice] did you get altogether this year? (2) please. *Amukawae, cha•masang chaichengphabo!* Father of Amuka, please look down below first! (3) intensifier suffix. *Saphawdo patna man•phachano.* The rabbit could not possibly cross.

**pha•-** *v.* to dare. *Noksang rai•naba pha•phinchaaidok.* I really don't dare to go home.

**pha•at-** ~ **pha•et-** *v.* to apply, to put on, to put on a wound, to apply to a wound. *Sambanggyri akaiokno, tokdepdepaimu pha•atokno.* He plucked *sambanggyri*, crushed it and put it on the wound. *Jyw•gaba sa•garaiaw di•thap pha•etaidonga.* The mother is putting a diaper on the child.

**pha•lak** *n.* ART. piece of old cloth used to clean things

**pha•lap** *n.* PERS. whore, prostitute

**phadyr** *n.* PERS. (< English) Father (priest)

**phagongma ~ phagungma** *n.* BODY. shoulder

**phai•-** *vtr.* (1) to break. *Taw•reksyrup mang sa, ge•thengmyng thupaw phangnan mongma phai•ai sa•rongwa.* A banana-sucking bird's nest always got broken and eaten by an elephant. (2) to translate. *Atongsang phai•bo* Translate it into Atong.

**chaksi phai•-** to wring one's hands. *Ha•sang bamai, chaksi jotai, chaksi phai•ai nemen chanchiaidongno.* With his head bent to the ground, fidgeting with his fingers and wringing his hands he was in deep thought.

**phai•ra ~ pai•ra** *n.* ART. type of basket (see Photo 28)

**phai•thong-** *v.* to break a solid object in half (crosswise)

**phaikana ~ paikhana ~ phaikhana** *n.* PLACE. toilet

**phaithawa ~ phaithopa** *n.* BODY. cheek

**phak$_1$** *n.* SHAPE. (1) side (2) half (which is the result of a longitudinal section, i.e. a cut along the length of something).

**phak$_2$** *clf.* classifier for halves of objects cut lengthwise

**-phak$_3$** *evsp.* (1) *V* lengthwise. *Wa•aw tan•phakbo.* Cut the bamboo lengthwise. (2) *V* vertically, *V* upright. *Khi•okno udo, parang goi•sado. Aro goi•tham songphakokno. Goi•tham songphakgaawba uawba kawthiriokno.* He hit it, the culm of reed. The stuck three more into the ground. He shot all three, which she had put up. (3) *V* off, *V* to get rid of someone or something. *Bengblokmyngdo sangumuk pywdyngdyngaimyng mongmamyng mykyranaw syw•chekchekai mu•okno mang ni. Ytykyi tokphakchiba man•chakno. Ytykyi satphakchiba man•chaknowa.* So then, the two of them, the banana-sucking bird and the toad, had been fighting the whole day, and had kept repeatedly picking at the elephant's eyes. He could not beat them off anymore. He could not fight them off anymore. (4) *V* around. *Dypyw ang chakaw wenangphakwa.* The snake would itself around my arm. (4) indicates that the action or part of the action takes place at a side or the sides of something. *Rai•phakangok na•a angmi rygynaw.* (Wilseng S Marak) I'm telling you, she passed behind me. *Taw•sa•gyrai nokhamaaw jalphakangaidonga.* The chicks are running under the house (i.e. they go under at one side and come out from underneath on the other side). (5) *V* for a little while (6) *V* together in one blow. *Angtykyi wetsachi morot myng• sene so•otphakna man•gabaaw atykyi hitramna?* Why are you trying to command a person like me, who can kill seven persons at once?

**phak-$_4$** *v.* to throw out, to empty, to gush out

**phakdemel** *n.* PLANT. species of plant

**phakphaklak-** *v.* to spill

**phakset-** *v.* to throw away (for solid substances and things)

**phakthangthang** *n.* PLACE. side by side. *Nok thai• ni phakthangthangsang* Two houses standing side by side.

**phakwal** *n.* BODY. armpit

**phakwil phakwal ~ phakwyl phakwal** *coll., adv.* side by side

**phakweng-** *v.* to row *wa•rok phakwengaidong* rowing a bamboo raft (see Photo 124)

**phal$_1$** *n.* ABSTR. (1) share, shift of work (2) instead of. *Ang re•engsigama nang•mi phal?* Shall I go instead of you?

**phal-$_2$** *v.* to sell. *Ang ie narykhel te•en nygylsang raangaimyng phalni.* I will bring these coconuts to the market and sell them later.

**phalong ~ phalwang** *n.* PLANT. species of plant

**phalthang** *ppron.* self

**phalthangthang** *ppron.* selves

**phalwang ~ phylwang** *n.* PLANT. species of tree

**phan** *clf.* classifier for trees; classifier for food packed in bundles. *rai•chakpan phan sa.* one tree

**phang** *clf.* classifier for grass, trees and flowers. *narang phang sa.* one orange tree. *narang rong sa.* one orange

**phangnan** *adv.* (1) always. *Phangnan rukpek mu•gabachido tyi ganang.* There is always water where there are frogs. (2) never. *Thawgaba symgaba phangnan sa•rongchagaba jilami bostudyrangaw raai hyn•aimung khasin khasin gumukawan palyngchi jalgabadyrangaw jykthangthangaw jumuphynaakno.* Having brought and given tasty and sweet things from the district, which are usually never eaten, they slowly recollected all their husbands who had run away into the jungle.

**phangphyl** *adj2.* upside down. *phangphylok* to be turned over, to be upside down

**phanthai** *n.* PLANT. type of sour fruit

**phari** *n.* (< Indic) a wound

**phas** *n.* ABSTR. (< English) the first one

**phasgaba** *n.* ABSTR. (< English) first, the first one. *Phasgaba ha•haw•chenga. Umungsa ha• haw•aimungsa wa•cham tan•a.* First we clear the jungle. Then, having cleared the jungle, we cut the old rice stalks. *Angdo phasgaba.* I am the first.

**phat**$_1$ *clf.* classifier for clothes. *ri•pan phat sa.* one loincloth

**phat-**$_2$ *v.* to chuck away, to throw out

**phathi-** *v.* to bless, to bestow upon

**phatsai** *n.* ART. woman's dress

**phaw•jong ~ phawjong** *n.* KIN. drel, c, ref, a. (1) elder brother. (2) Also used to address an older male cousin or (3) a man older than the speaker.
    **dada... phaw•jong~phawjong...** *coll, n.* elder brother (see *dada*)

**phaw•jongmaran ~ phawjongmaran** *n.* KIN. set. two elder brothers

**phe-** *v.* to disembowel, to gut (fish),

**phe•ep ~ phep** *n.* PLANT. banyan tree

**phe•epmisi ~ phepmisi** *n.* PLANT. species of tree

**phe•phong** *n.* BODY. floater organs of a fish

**phebaw** *n.* PERS. person with a swollen cheek or swollen tonsils

**phebuari** *n.* TIME. (< English) February

**pheel dong- ~ pheel dong•- ~ peel dong- ~ peel dong•-** *v.* (< English) to fail. *Ge•theng lekha nemai poreancha, ytykyimu poreka peel dong•ok.* He did not study the book well, so then he failed his exam.

**phek**$_1$- *clf.* classifier for smaller branches of trees. *dala phek sa.* one branch

**phek-**$_2$ *v.* drunk. *Ge•thengdo phekok.* He is drunk. / He was drunk.

=**phek**$_3$ *qtf. encl.* (distributive). *Palengma burung banga banga haw•waan pungphek phingano.* They cut five bushes of *palengma* (the tree *Barebinia xariegata*) each and one rice stock house each was filled. (This means that the soil was so fertile that they could cultivate a lot of rice on a relatively small piece of land.) *Wa•tyng tyngphekna ma•su mangphek hyn•wa.* For every bamboo strip, [they] gave [me] one cow.

=**phekphek** (reduplicated form of the distributive enclitic) More fieldwork research is needed to find out if there is a difference in meaning between the simple and reduplicated forms of the distributive enclitic. *Myng•phekphekan bunduk ra•angrumokno.* Each to them took a gun.

**pheking ~ peking** *n.* ART. (< English) luggage, packing

**phekphek-** *v.* flipping and turning (like fish do on dry land or in a dammed-up fishing place). *Na•rongdo phekphekramphinaidoknowa.* The fish are flipping and turning a lot.

**phel** *clf.* classifier for flat, baked things and coins. *biskut phel sa.* one biscuit. *tangka phel sa.* one coin

**pheng•chang-** *v.* to hold something in front of something else

**phep ~ phe•ep** *n.* PLANT. banyan tree

**pheret**$_1$- ~ **peret-** *v.* to split, to crack, to burst, to explode

**pheret**$_2$ *n.* CORP. crack in the skin

**pheru** *n.* ANIM. fox

**-phet**$_1$ *evsp.* V detrimentally, V scandalously much. *Man•gabaaw sa•phet ryngphet.* The rich eat and drink scandalously much.

**phet-**$_2$ *v.* (1) to arrive (at), to reach. *Ue raja nygylchina phetokno.* The king arrived at/reached the market. *Jenetne rajamyng noksang phetangokno.* He somehow reached the house

of the king. (2) to come out of the water, to emerge. *Bewal rypaimyng phetaakno.* Having been in the water for some time, he emerged.(3) to rise (of the sun and the moon) *Ja phetok.* The moon has risen.
**phet... dong•...** *coll, v.* (1) to arrive, to reach. (2) to succeed. *Bytnaan san sa wal sa phetachawana dong•achawana...* Because they had been pulling for one day and one night, and did not succeed, ...

**phet-**₃ *v.* to swell up

**-phetphet** *evsp. V* repeatedly and intensely, *V* like your life depends on it. *Sipaidyrang dang•wachie kan•tyra gulinyi kawphetphetai rai•aaknokhon.* When the soldiers had entered the village, they started repeatedly firing blanks. *Dykhimi balgabaaw sung ra•aisa jalphetphetangaidongano.* Remembering what Dykhi has said, he ran like his life depended on it.

**phi-** *v.* to invite. *Beanbebe montyridyrngba Bilaw phina takyi hongkhotangthokokno.* The ministers truly all went out to invite Bil.

**philm ~ philym ~ philim** *n.* ART. (< English) film, movie

**-phin ~ -phyn** *evsp.* (1) directional event specifier. *V* backward, *V* back, *V* again. *Hai! Noksang rai•phinnaka.* Come on! Let's go back home. *Ytykyi taw•phinbo.* Go back up like that. (2) intensifier suffix. *V* fully, obviously *V, V* fully, *V* totally, *V* completely, over-*V*, etc. *Ge•theng nang•na ytykphinai kha•galano.* She loves you so much, she says. *Phylgym gungami tokkyrengaw man•aimungna ha•china wuuuuuuuk dym! takram•phinoknotyi phylgym gal•waan.* Having hit the giant eagle's neck, [it] [fell] right down to the ground "woooshshsh ... boom!" to [our] surprise.

**phing-** *v.* full. *Gylaschi tyi phingok.* The glass is full of water. *Gylas phingok, diphingna man•chaka.* The glass is full; you cannot fill it anymore. *Nang•mi kha•thong bangbang dong•chido ang phingetni.* If your heart is empty, I will fill it.

**phingpyryt- ~ phingpurut-** *v.* to overflow

**pho•ot ~ phot** *n.* ANIM. mythical black amphibian like a salamander

**phok-** *v.* to lift up, to uproot, to swell

**phone ~ phoon** *n.* ART/ACT. (< English) telephone, telephone call. *Angna phone kha•etboto!* Call me (on the phone)!

**phong**₁ *clf.* classifier for cylindrical objects and for long sharp or pointy objects

**phong**₂ *ideo.* brap! blarp! the sound of someone farting

**phong**₃ *n.* ART. wooden handle of big knives, axes and spears

**phong•** *n.* ART. fire place for cooking

**phong•khal** *n.* ART. stones to put a cooking pot on

**phong•thu** *n.* ART/PLACE. stones to put a cooking pot on, fire place for cooking

**phoren** *n.* PERS/PLACE. (< English) (1) white foreigner (2) country of white foreigners

**phot ~ pho•ot** *n.* ANIM. mythical black amphibian like a salamander that eats people.

**phowa** *n.* the span between the tip of the thumb and the tip of the index finger when spread out and placed on a surface

**phryngphrang askui** *n.* GEO. the morning star

**phu•chul** *n.* ANIM. monitor lizard (see Photo 14, Photo 106)

**phuk- ~ puk-** *v.* to be stuck. *Tokkyrengchi asu phukaidonga.* There is a fishbone stuck in my throat.

**phulis ~ pulis** *n.* PERS. (< English) police

**phulkobi** *n.* PLANT. cauliflower

**phuruk-** *v.* (1) to become uprooted. *Bildo kyryiaimyng pandyrangchi pyi•chiba panba baiariokno, wa•chi pyi•chiba wa•ba phurukariokno.* As for Bil, because he was afraid, when he held on to the trees, the trees would break, when he held on to bamboo, the bamboo would just become uprooted. (2) to break (for plants and trees)

**phuset-** *v.* to spit out

**phuthi** *n.* ANIM. species of fish

**phutsul** *n.* ANIM. species of water monitor (lizard) that supposedly can eat humans, also called water dragon. *Kha Dawa* nochachido phutsul ra•arianoro. If you don't say "Kha Dawa" the phutsul will get you.

**phyl•-** vØ. to transform, to change into. *Imi wa• juw• wak phyl•wa.* Her father and mother have changed into pigs.

**phylgym** *n.* ANIM. eagle

**phyltawtaw** *adv.* jerkingly (over a rough road)

**phylwang ~ phalwang** *n.* PLANT. species of tree

**phylyp-** *v.* to blink (with one's eyes). *Mykrenmi phylyp chaiwaan.* (Gostar R. Sangma) I looked at her with blinking eyes.

**-phyn₁ ~ -phin** *evsp.* V backward, V back, V again, over-V, V overtime, V fully, obviously V, V fully, V totally, V completely

**phyn-₂ ~ pyn-** *v.* to cover. *Ang kombol phynaidonga.* A blanket is covering me. / I am lying under a blanket.

**phyryw₁** *adj2.* to have a hole in it

**phyryw-₂** *v.* hollow

**phyt-** *v.* to slice

**phywra** *n.* FOOD. rice powder

**pi•-** *v.* (1) to ask, to request. *Pherudo: "Te•ewba nang• sa•aw atana nokchi tana?" noaimung, ue myng• tham myng• byryigaba pi•dapokno.* The fox said: "Now why would you keep your children at home?" and he asked for three or four more. (2) to beg. *Pherudo: "Haito mosa, na•a angna hyn•chakama?" noaimu ytykyi chaitawai pi•ai mu•arokno.* The fox said: "Come on, friend, won't you give anything to me?", and sat like that looking up and begging. (3) to pray. *Isolaw sung ra•a.* I pray to God.

**pi•sa** *n.* TIME. childhood

**pi•thyn ~ bi•thyn** *n.* BODY. liver

**pi•ti ~ pi•tyi** *n.* FOOD. rice beer (gold coloured)

**pi•tyng** *n.* ART. thread, necklace

**pibok** *adj2.* white, unripe, very light green

**picham** *adj2.* old (of things)

**pidan** *adj2.* new

**pidio** *n.* ART. (< English) video

**pido** *n.* ACT. game played with small stones. The game can be played by just one person or a small group of people. The player has to throw first one, than two, than three etc.

stones up into the air, make difficult hand gestures, and catch the stone again, sometimes, via first juggling them on the back of their hand. A player's turn is over when they fail to catch the stones.

**pijyw** *n.* PLANT. rice seeds for sowing, newly harvested rice, unhusked rice that is thrown away when cleaning a portion of rice before cooking it

**pikheng** *adj2.* alive

**piktiyr** *n.* ART. (< English) picture, photo

**pinak** *adj2.* black

**ping-** *v.* to block the way

**piong** *n.* ANIM. species of bird

**pipuk** *n.* BODY. belly, intestines, bowels, stomach

**pirin-** *v.* to mix

**piryt** *n.* BODY. gall bladder

**pisak** *adj2.* red, blond

**piseri** *n.* (< English) fishery

**piit** *clf.* the length of two fists and two thumbs when one joins them

**po•tolong** *n.* PERS. person with a naked chest

**plak** *n.* ART. (< English) plug

**pok-** *v.* to swell

**pokotia** *n.* PERS. freeloader, sponger, person who takes advantage of the kindness of others

**poop** *n.* FOOD. (< Indic) triangular pastry eaten with tea

**porai- ~ pore-** *v.* (< Indic) to read; to study

**poram-** *v.* to fly over

**pore- ~ porai-** *v.* (< Indic) to read; to study

**porika** *n.* ACT. (< Indic) exam, examination

**pot-** *v.* to plant by sticking a sprout in the mud. **cha•ri pot-** to plant paddy

**powa ~ pywa** *n.* QUANT. a bowl of rice

**puk- ~ phuk-** *vintr.* to be stuck. *asu pukok* the fishbone is stuck

**puksuk** *n.* BODY. waist, side of the body

**puktyng** *n.* BODY. small intestine

**pulis ~ phulis** *n.* PERS. (< English) police

**pun-** *v.* to catch with a fishing rod and fishing hook
   **na• pun-** to fish, to catch fish

**pung** *n.* ART. granary, rice stock house. *Mai bytwamyngdo pungchina songchina khairata.*

We carry the rice harvest down to the rice stock house, to the village. (see Photo 13)

**purun** *n.* ANIM. goat

**pusipusi ~ puspus** *interj.* (< English) interjection to call a cat. Here, kitty kitty!

**pyi•-** *v.* (1) to touch. *Kopiuterskrin pyi•na bai.* Don't touch the computer screen. (2) to grasp, to grab, to hold onto. *Uchie songmyng morotmyng jyrym thymaimyng Thengthonaw raw•okno pyi•goropokno.* Then the people of the village, having quietly lain in ambush, caught Thengthon, they grasped him all together. *Pherue jaraw jaraw rong•chi pyi•thataimyng rypaidokno.* The fox is staying submerged, holding on to a rock with all his force.

**pyi•khap-** *v.* to catch with one's hands

**pyi•khep-** *v.* to hold firmly

**pyi•khyrep-** *v.* to crush with one's hand

**pyi•ram-** *v.* to feel for, to search by feeling

**pyi•ru-** *v.* to collapse

**pyi•thyng-** *v.* to hold on to, to grass

**pyjyw-** *v.* to sow seeds by scattering them

**-pyl** *evsp. V* rapidly

**pylang- ~ peleng ~ pelang** *adj2.* flat. *Gari bengbylokaw depylengok, ytykyimu bengbyloke pylengok.* The car flattened the toad so the toad was flat. *rong• pelang* a flat stone

**pyn- ~ phyn-** *v.* to cover. *Ooo mykgythaldo dong•cha jywmangsama, ma nang•bimang sylwaai! Ma pynwasama muni ang mykrenaw, ma nang•chi ganang atongba jadu.* (Wilseng S. Marak) Ooo, it's not real, it's only a dream, but your body is really beautiful! Are my eyes covered by a spell, or do you have something magic?

**pyn•-₁** *adj1.* dense, thick

**pyn•-₂** *v.* to pack, to wrap up, to pack in a banana leaf, to cook in a banana leaf. *Manapmian mai ja•bek rymaimungna, mai ja•bek mynmanaimungna rymai sa•aimungna, maisangumuk pyn•aimungna, hai•aw garu balagachi ramai tanangokno.* It was early in the morning, when she cooked rice and curry, and when the rice and curry were ready, and she had eaten, and packed lunch, she put some mustard leaves out to dry in the sun.

**pyndap-** *v.* to cover

**-pyrak** *evsp. V* and cut, *V* to pieces *Atakna ie chola chetpyrakok?* Why is this shirt ripped to pieces?

**pyrap-** *adj1.* to be too salty

**pyryi-** *v.* mature

**-pyryt** *evsp.* over-*V*. *Tyi glaschi phingpyrytok.* The glass is too full of water.

**pyryw- ~ pyru-** *v.* to pierce, to make a hole in something

**pyt-** *v.* to wrap neatly as a present

**pyw-** *v.* (1) to fly. *Daw•khado, pywramangarokno.* The crow just flew away. (2) to flee. *Tankaaw sa•khawaimyng, pulis ge•thengaw sandiaimyng, song sulsang pywangok.* After stealing the money, because the police were searching for him, he fled to the neighbouring village. (3) to jump. *Atongbatykyi ang kha•petaimyng goremyng pywratna nangokodo goreba ni•ni nang•tymba ni•ni.* If I'll somehow get angry, and if I need to jump off my horse, the horse won't survive, and you won't survive.

**jal... pyw...** *coll, v.* to flee (see *jal-*)

**pywa ~ powa** *n.* QUANT. a bowl of rice

**pywgak-** *v.* to crash (in flight)

**pywtaw-** *v.* to jump over something. *Muraaw ang pywtawa.* I jump over the small stool.

# R

**ra-** *v.* (1) to bring. *Ytykyimu nukwachie phalthangmi gawigaba cha raaknoro.* So then, when he looked, his wife had brought tea. (2) to take to. *Ang ie bostuaw te•en nygylsang raangaimyng phalni.* I will take these things to the market later, and sell them.

**ra•-** *v.* (1) to get, to buy. *Wai chunggaba dong•chido purun ra•a.* If it is a big spirit, you get a goat. *Kha Dawa nochachido phu•chul ra•ariano.* When you don't say "Kha Dawa!", the monitor lizard will just get you. *"Nang• ie tupi bimi ra•ak?" "Turami ra•ak."* "Where did you buy that cap?" "I bought it in Tura." (2) to take

(from). Ge•thengdo uaw thymai chaiaimyng ue morot re•engman•wachi, uaw tangkaaw ra•akno. Having hidden and watched him, when that person left, [he] took that money.
**bebe ra•-** vtr. to believe. *Ang nang•aw bebe ra•cha.* I don't believe you.
**hapsan ra•-** *Gumukan hapsan ra•na nanga.* We have to consider everybody as being the same.
**hogol ra•-** v. to snore. *Jywchengwachi nang• hogol ra•wa.* When you were asleep first, you snored.
**katha ra•-** vtr. to listen to, to heed someone's advice. *Wa•mi jyw•mi balgabaaw katha ra•chagabae anga ytykgachina dong•ok.* Because I did not listen to the words of my parents, I have become like this.
**ra•ai sa•-** vtr. to marry off, to marry someone to someone else. *Mamathanggaba sa•mynchykgana khyrethangaw ra•ai sa•naka.* Mother's brother will marry his daughter to her cousin.
**sung ra•-** v. (1) to remember (2) to praise (when used in relation to *Isol* 'God')
**ra•ang-** v. to take (away), to bring. *Thengthondo tangkaaw ra•angokno songsang.* Thengthon took the money home. *Mai okhiaimyng sa•na ra•anggaba maisanaw sa•na re•engbutungchi sa•butungchi ...* They were hungry, and while those who left to eat went away to eat the lunch they had brought with them, ...
**ra•gat-** v. to collect. *Chengwami achuambido khemaw rangaw ra•gatnaan.* Our first ancestors wanted to collect drums.
**ra•rung-** v. to revoke, to take back.
**ra•sak-** v. (1) to welcome. *Tyinyng•sangba nawmyl sylsylgabasa ra•saksawa.* Under water, only beautiful girls welcomed me. (2) to accept, to receive. *Angdo myng•sa agrai ra•sakchawa.* I will not accept/receive more than one person.
**ra•sek-** v. to snatch
**rabak ~ rabak rabak** adv. quickly, fast
**rabal** n. PLANT. (< English) rubber tree
**Rabuga** n. SUPER. god who created the world according to ancient religion

**rai** n. PLANT. reed
**rai•-** v. (1) to go. *Hai, rai•naka* Come on, let's go. (2) to come. *Alsia rajado phepchi synthibutungchi te•ewe napit myng• sa rai•phaaknoro.* While the lazy king was lamenting in the banyan tree, a barber came along.
**rai•a-** v. to come. *Phorenmi morot rai•adonga, phorensangmi rai•aidonga.* Foreign people are coming, they come from foreign countries. *"Angdo hanep nang•sang re•engni." "Rai•abo."* "I will go to your place tomorrow." "Do come."
**rai•byt-** v. to carry around
**rai•chak** n. ART. big leaf used to pack food (see Photo 45, Photo 46, Photo 47)
**rai•ganggang-** v. to go/drive/ride over things on a bumpy road. *Rong•aw rai•ganggangwa.* I bumped over a stone while going.
**rai•phak-** v. to go through; to hit with one's elbow while walking. *Amakdo songjinma rai•phakangaidoknowa.* The monkey is going through the whole village.
**rai•ram-** v. a motion like shit coming out of the body
**rai•sotwa** n. ACT. a shortcut
**rai•wil-** v. to walk around something. *Kynsangdo matsado morotsyn man•aimyng rai•wilokno alsiado. Rai•wilwilokno.* Later, having caught the smell of a human, the tiger walked around the Lazy King. He went round and round.
**raidi** n. PLANT. turmeric
**raima** n. PLANT. cane
**raithai** n. PLANT. tree with thorns on its stem
**raityng$_1$** n. PLANT. rattan
**raityng$_2$** n. ART. clothes line, washing line. Clf. tyng. *raityng tyng tham.* three washing lines. *Raityngchi kha•dian phingsawok.* The clothes line is full of clothes.
**raja-$_1$ ~ ratja-** bound. num. hundred. *raja sa ~ ratja sa* one hundred. Despite being written as separate orthographic words, this numeral is phonologically bound to the following multiplier, e.g. *raja sa* [radʐa'sa ~ rad:ʐa'sa] 'one hundred'.
**raja$_2$** n. PERS. king

**rajami khu•symang ~ rajamyng khu•symang** *n.* PLANT. species of light green creeper, which overgrows trees (see Photo 71)

**rak-** *adj1.* (1) hard. (a) of materials, the opposite of soft. *Biskutdo rakokte. Sa•na man•anchak.* These biscuits are hard. You can't eat them anymore. (b) with great effort. (The translation depends on the context in English). *Rakai thetokno.* He pulled hard. *Montyridyrangba kyryiaimyng cha•aw thik dongaian rakai kha•akno.* The ministers, because they were afraid, tied his legs tightly, exactly as he had asked. (c) difficult. *Atong balna raka.* It's hard to speak Atong. (d) performed with force or vigour. The translation depends on the complement of *raka* used in the sentence. *Na•a angna nang•myng gore jalna rakkhalgabaaw hyn•etaribo.* Just give me your fastest running horse. (2) strong (of natural phenomena). *Tai•ni balwa rakai balwaangok.* The wind blew strong today. / There was a strong wind today. (3) loud. *Jan•rukaimu rakai olrukokno.* Because they were far away from each other, they spoke loud.

**raka** *n.* ART. the first letter of the Atong alphabet, glottal stop, glottalisation

**rakhi-** (1) *vgoal.* to protect from, to guard against. *Ning ha•bachi mongmana amakna mai sa•niwana rakhiaronga.* We are protecting our rice field against elephants and monkeys, so that we will eat rice. (2) *v.* to guard *"De, na•a ichi mu•sigabone bai•sigane, biskut rakhibone" nookno.* Okay, now you stay here friend, Okay, guard the biscuits, okay?", he said. (3) *v.* to look after. *Myng• sa morot man•ai sa•gabachi wak rakhina ga•akoknoaro.* He was forced to look after the pigs of a rich person.

**rakhigaba** *n.* PERS. guard, caretaker

**ram**$_1$ *bound. n.* PLACE. place. With this meaning, the word is only found in names of places, e.g. *Balphakram* 'the place where the wind blows'.

**ram**$_2$ *n.* PLACE. road, way, path. Clf. *chol. ram chol tham.* three roads, paths, ways. *Ram watbo!* Get out of the way!

**ram-** $_3$ *v.* to dry in the sun, to put in the sun to dry. *Garu balagachi ramai tanaimuna, ha•basang ha• kamna re•engokno.* After she had put the mustard leaves outside to dry in the sun, she went to work in the rice field.

**ram rai•-** *vpan.* to go. *Ytykyimu ie ha•byritykyi ram rai•ano.* So they went over this mountain.

**ram•-**$_1$ *v.* (1) to search. *Bai•sigathanggaba pheruaw ram•aimyng nukaiokno.* Having searched his friend the fox, he found him. (2) to try. *Mongma ytykyi ha•kha wylna ram•butungchi thik thak saphawba tharapaioknoro.* When the elephant tried to go down the hill side, the rabbit caught up with it.

**-ram•**$_2$ ~ **-ram**$_4$ *evsp. V* inadvertently, *V* unintentionally, *V* by accident. *Phulistykyi nukramphinokno bunduk paigana.* They inadvertently looked like the police, because they were carrying guns.

**ramga** *n.* PLACE. the side of an object that faces away from the wall

**ramram**$_1$ *adj2.* normal, ordinary. *Ramram rangawdo mykha badri myngcha.* Normal rain is not called *mykha badri. Ie ramram dong•cha.* This is not normal. *Bai•sigathangmaran tyi dukungokno. Na•do ramramanchakno.* The friends dammed up the water. There was an unusual quantity of fish. *Nang• garido ramramchagaba kyryngwa.* Your vehicle is making an abnormal sound.

**-ramram**$_2$ *evsp. V* normally, *V* usually, *V* naturally, *V* commonly. *Umigymyn te•ew na•nang ha•gylsakchiba phorenchiba, Igylanchiba, Kalkata, Delhichi, Bambechiba, Badrichiba, Khychuchiba man•ramram khymcha.* Therefore, now, we don't usually get and marry someone from a foreign place, from England, Kolkata, Delhi, Mumbai, Badri or Khychu.

**ran-** *adj1.* dry. *Kha•di nokha•palchi ramaimu ran•ok.* After drying the clothes outside in the sun, they are dry.

**randai** *n.* FOOD/BODY. (1) meat, flesh (2) body

**rang₁** *n.* ART. type of traditional brass drum (see Photo 120) or gong

**rang₂** *n.* GEO. rain. *Rang waaidong.* It's raining. *Rang nemok.* The rain has stopped. "*Rang nemchengama na•nang chyw jamchenga*" *noai rangmu chyw ryngsusaie range san chi byri wawano.* "Will the rain stop first or will we finish our liquor first?" they said and while competing in drinking with the rain, the rain fell for fourteen days.

**rang•set-** *v.* to breathe

**rangbyrym₁** ~ **rangbrym** *n.* GEO. cloud

**rangbyrym-₂** *v.* to be shrouded in clouds, to be blocked by clouds. *Rangsang rangbyrymaidonga, rang wanikhon.* The sun is blocked by the clouds, it might rain.

**rangchinek** *n.* GEO. cloud

**rangdylekpa** *n.* GEO. lightning

**ranggorai** *n.* ANIM. macaque. Monkey with a long tail, brown body and a red face.

**ranggyl** *n.* CORP. fungus infection

**rangkha** *n.* ART. type of traditional metal gong or drum

**rangra** *n.* GEO. sky

**rangrai** *n.* PLANT. species of tree

**rangrengchongcheng rongrengchangcheng** *adv.* swaying from one side to the other

**rangsan** *n.* GEO. sun. *Te•ew una rangsando saniarokno.* Now the sun was setting. *Rangsan dang•angwachian, taw• dang•kyrymanga.* At sunset, all the birds rest. *Rangsan jamangaidok.* The sun is setting. *Rangsan ma•angaidok.* The sun is setting. *Rangsandi•mai phai rewetangwachian, Raka Motbandaaw byletwa.* When the sun was setting, he slew Strong Motbanda.

**rangsyl** *n.* ART. type of traditional metal gong or drum

**rani** *n.* PERS. queen, also used to call one's daughter when she is a little child, like in English 'little princess'

**rap-** *v.* to thatch, to roof

**=rara** *encl.* (1) exclusively, only. *Chanchia ang nang•awrarasa.* I think only of you. (2) among, amongst. *Ge•thengthengdo bobarara myng• ni golpho kha•rukokno.* They, all the fools amongst themselves, started fabricating stories. (3) all. *Uchian Bildo thyi•rara pharirara takyi rai•aidongano.* Then, Bil was coming, all blood and wounds. *Utyme morote gawigababa biphagababa bobirara bobararanowa.* They, these people, the wives and husbands, are all crazy women and crazy men. (4) intensifier enclitic. *Thengthone bylongen chalakno, morot. denggurararano ue.* Thengthon was very cunning, a pure scoundrel of a man.

**rasi** *n.* star sign, good fortune

**rasong** *n.* ACT. (1) praise. *Jisuna rasong* Praise Jesus. (2) blessing. "*De achudyrang, ytykchido re•enganchyibo. Nang•tymba rasong dong•naba ganang. Re•engari*" *nookno.* "Very well then, grandsons, in that case, try to go. You have my blessing." (3) compliment. *Amakan rai•ai takaimungna* "*Mykhangba syma dymbrubru. Di•maiba raw•a dymbrubru*" *noaimuna balai takokno. Rasong man•ai takarokno haiaw daw•khaaw, amakansega.* The monkey came and said: "Your face is sweet and shiny. Your tail is long and shiny", he said. He gave it many compliments, to the crow, the monkey. (4) boasting. *Rasong man•ai takokno usa, chungchunggarangsa, udo phylgymawdo jonggaba kawwano.* The eldest ones boasted a lot (about themselves) although it had been the younger brother who had shot the eagle.

**rasong... gal...** *coll, n.* praise and pride. *Bakbak rasong taknado thapthap galchanado man•chawa.* It's not easy to quickly get praise, to get pride.

**rasun** *n.* PLANT. onion. *rasun pibok* garlic *rasun pisak* red onion *rasun tyisuk* species of onion

**rasuntyisuk** *n.* PLANT. spring onion

**rat-₁** *v.* to throw *Matsa rong ratwa.* A tiger threw a stone.

**-rat₂** *evsp.* (1) V downward. *Mai bytwamungdo pungchina songchina khairata.* Having harvested the rice, it is carried down to the granaries, to the village. (2) V downstream.

Ie gorialdo khengna nanga ido Chaw•ratai-donga ido". This crocodile has to be alive. It is swimming downstream.

**ratat-** *v.* to take out

**ratsok-** *v.* to miss the mark

**raw•-₁** *adj1.* tall, long.

**raw•-₂** *v.* to catch, to grasp. *Changba bydyi myng• sa khen• raw•arong* Somebody, an old man, is catching river crabs. *Uchie songmyng morotmyng jyrym thymaimyng Thengthonaw raw•okno pyi•goropokno.* Then the people of the village, having quietly lain in ambush, caught Thengthon, they grasped him all together.

**raw•reng-₁** *adj1.* slender and long

**raw•reng₂** *n.* PERS someone who is slender and long

**raw•soksok-** *v.* to fail to catch

**-rawraw** *evsp.* continue to V, continuously V, increasingly V. *Kynsangdo rai•wachie napitdo mongma matsana nekarawrawna kyrethyngaimyng phepmyng gal•syrango-kno napitdo.* Later, when the animals were coming, he feared the tigers, the elephants, the ones that were continuously/increasingly getting closer, so much, he fell out of the banyan tree, the barber.

**rawsykot-** *v.* to slip out of the hand

**re•eng-** *v.* (1) to go (away), to leave. *Garu bal-agachi ramai tanaimuna, ha•basang ha•kamna re•engokno.* Having put the mustard leaves outside to dry, she went to the *ha•ba* to weed. *Kynsange nygyltyi ni re•engwachi thik thak jahas kanachina dong•angok.* Later, when she had been going for two weeks, she arrived exactly at the ship and the harbour. *Ytykyimu manapmi jinmamu songsangmyng re•engokno.* So then, the group left the village in the morning. (2) to come from. *Bisang re•engwa na•a?" "Usang nalsasang re•engwa."* "Where do you come from?" "I come from the other side of the river."

**re•eng... jok...** *coll, v.* to go back. *Re•engphi-naribo dada, jokangphinaribo phaw•jong.* Go back, elder brother; run back, elder brother

**re•eng... taw...** *coll, v.* to go away, to leave. *Beanbebe Bandiba re•engbebeaidokno,* *tawangbebeaidokno.* Bandi truly went, he truly left.

**re•koksi** *n.* PLANT. species of plant

**redio** *n.* ART. (< English) radio

**reel** *n.* ART. (< English) train, rail, stud of a fence (see Photo 9)

**reelgari** *n.* ART. (< English and Indic) train

**rek** *n.* PLANT. banana tree

**rekhep** *n.* PLANT. species of huge beans

**rekhep-** *v.* dry (of plants), wrinkled (of person)

**rekkun** *n.* PLANT. banana flower

**rekphang** *n.* PLANT. banana tree

**rekphul** *n.* PLANT. non-edible banana flower

**rekthai** *n.* PLANT. banana

**rens** *n.* ART. (< English) wrench

**repa chepa** *khjyks. adv.* in various places. *Raka Gryngrangba rangsetwa, biba jokgaba dam bangaakno repa chepa matgaba charanga dong•okno.* Strong Gryngrang breathed, [and] had gotten wounded in five places, [and he] had fifteen cuts in various places.

**reprep-** *v.* to rub the clothes while doing the laundry

**ret** *n.* ACT. children's game played in a grid. There are hunters who may only move along the lines of the grid. The other children have to try to cross the grid without being touched by a hunter.

**rewet** *n.* PLACE. riverside, riverbank

**=ri ~ =ryi** *encl.* (privative, indicates a referent, usually kin, that that was lost). without. *Ha•gylsakchi anga jykri mu•waba, uanari-naka, sa•ri parangwaba, uanarinaka.* I lived in the world having lost a wife, and it will just be like that, and I wandered around having lost my children, it will just be like that.

**ri** *n.* BODY. penis. Clf. *goi•. ri• goi• ni.* two penises

 **ri• gang-** to have an erection

 **ri• sa•-** to suck, to perform fellatio

 **ri• selsoksok-** to masturbate

 **ri• sepsep-** to masturbate

**ri•ambanthai** *n.* BODY. glans penis

**ri•baw** *n.* PERS. person with one testicle bigger than the other

**ri•gan•thong** *n.* BODY. erect penis, erection, hard-on. *Nangchi ri•gan•thong ganang.* You have an erection. / You have a hard-on.

**ri•gol** *n.* PERS. penis (used as swearword for men), dick

**ri•karan ~ ri•keren** *n.* BODY. testicle, balls, scrotum. Clf. *rong. ri•karan rong ni.* two balls, testicles

**ri•khu•chul** *n.* BODY. foreskin

**ri•kun** *n.* BODY. glans penis

**ri•myn** *n.* BODY. pubic hair of a male

**ri•pan** *n.* ART. a short dress that women wear around the waist

**ri•ros** *n.* CORP. cum, sperm, semen

**ri•sokop** *n.* BODY. scrotum

**ri•tyi ~ ri•ti** *n.* CORP. cum, sperm, semen

**rijap** *n.* PLACE. (< English) forest reserve

**rimirimi ~ mrimri** *adv.* squinting. *Hyiawe morot re•enggabawdo jan•dugaaimung, rimirimi nuketariokte.* That person going way over there is too far, I can only see him when I squint. *Mykren rimirimi takaidong.* My eyes are almost closed because I'm so tired.

**rimyl-** *adj1.* slippery

**rin-** *v.* to keep as domestic animal

**ring** *n.* PLANT. taro, species of edible tuber with green stems (see Photo 62)

**ringaba** *n.* PLACE. (1) place where domestic animals are kept (2) fishery

**ringgong** *n.* PLANT. species of plant that looks like *ring*, but is not edible.

**ringgythyng** *n.* PLANT. species of plant that looks the same as *ring* but has black stems and is not edible. (see Photo 63)

**riphi- ~ ryphi-** *v.* to plaster (with a mix of clay and cow dung). *Chula riphiaidong.* I'm plastering the cooking place. (see Photo 17)

**riprip-** *v.* to rub

**=ro ~ =aro** *encl.* (declarative modality). I'm telling you! *Ningdo thomungdo jaw•chaaro.* We don't fry with mustard oil. So then, he forgetfully wound his loincloth around his head, I'm telling you! The use of this enclitic in Atong does not always need to be translated into English. (see also =*aro*$_2$)

**roal** *n.* PLACE. (lower) primary school

**robol** *n.* ART/ACT. football

**robolphil ~ robolpil** *n.* PLACE. (< English) football field, playground

**rochok ~ rotchok** *n.* ART. stud, vertical beam that forms part of the side of a house to which the *damdyl* can be attached (see Photo 11)

**rochong** *n.* PLANT. tree stump

**=rogoi** *encl.* (reciprocity). *baju<u>rogoi</u>* friends

**rok-** *v.* to shave. *Ka•myn• rokai matok.* I cut myself while shaving my beard.

**rokhom** *n.* ABSTR. shape, type

**rokset-** *v.* to wipe off

**romrom-** *v.* to roll

**romthom-** *v.* spherical. *Robol romthoma.* A football is round.

**rong**$_1$ *clf.* classifier for small round objects, money, small stones, seeds, stones in a game (when they have a value) and fruits, default classifier for counting. *buchuot rong sa.* one mango. *tangka rong chek.* ten rupees

**-rong**$_2$ *evsp.* usually V. *Thawgaba symgaba phangnan sa•<u>rong</u>chagaba jilami bostudyrangaw raai hyn•aimung kha•sin kha•sin gumukawan palyngchi jalgabadyrangaw jykthangthangaw jumuphynetaaknowa.* Having brought and given tasty and sweet things from the district which are usually never eaten, they slowly recollected all their husbands who had run away into the jungle.

**rong**$_3$ *n.* ABSTR. colour

**rong-**$_4$ *v.* (see *cha•su rongaimu mu•a*)

**rong•** *n.* GEO. stone. *rong• thut sa* one stone *rong• rong sa* one small round stone in a game

**rong•baram** *n.* GEO. type of rock

**rong•cheret ~ rong•chyret** *n.* GEO. pebble-size stone

**rong•chun** *n.* SUBST. lime stone

**rong•chung** *n.* GEO. big rock. *rong•chung thut sa* one rock, one big stone

**rong•chyret ~ rong•cheret** *n.* GEO. pebble-size stone

**rong•dep-** *v.* to crush with a stone

**rong•gyryn ~ rong•rymrym** *adj2.* GEO. being full of big rocks, stony land. *Ie ram ronggyrymrara, angdo rai•chawa.* This road is full of big stones, I will not go.

**rong•han•cheng** *n.* GEO. sedimentary rock

**rong•ka** *n.* PLACE. cliff

**rong•khal** *n.* PLACE. (1) space under a stone, cave. *Ichi rong•khalchi khen• ganangthelnaba ganang.* Here in the spaces under the stones there are river crabs for sure. (2) cave

**rong•khobok ~ rong•bok** *n.* SUBST. chalk

**rong•khol ~ rong•khal** *n.* GEO. cave

**rong•misi** *n.* GEO. very small stone

**rong•patal** *n.* GEO. big rock

**rong•phek** *n.* GEO. a grain of sand or very small stone

**rong•rymrym ~ rong•gyrym** *adj2.* GEO. being full of big rocks, stony land. *Ie ram ronggyrymrara, angdo rai•chawa.* This road is full of big stones, I will not go.

**rong•sa** *n.* ART. whetstone, flat stone for sharpening knives or edged tools (see Photo 26)

**rong•syl** *n.* GEO. flint stone

**rong•syrek** *n.* GEO. small stone

**rong•thai₁** *n.* ART. base stone on which a house is built (see Photo 9 and Photo 10)

**rong•thai₂** *n.* GEO. a rock

**rong•thyk** *n.* GEO. a big rock

**rongkhym** *n.* ANIM. species of yellow beetle

**rongmesak** *n.* SUBST. uranium-containing mineral

**rongmyng-** *v.* to shuffle cards

**-rongreng** *evsp.* V while spinning around. *Ytykyimuan rung chawchiba rung bytrongrengarinowachie, sangkhyning noaimung.* So, when you want to row the boat, and the boat is just driven round and round, it is the water dragon.

**rongrengchangcheng** *adv.* swaying from one side to the other

**rongrong-** *v.* to slide over something

**rongthal-** *v.* to clean, to clarify, to explain

**rongthala-** *adj1.* clean

**rongtyk** *n.* ART. large clay pot to keep rice in, rice pot

**ronok-** *adj1.* smooth

**ronronok** *expr.* eyes almost closed. Used in the expression *Mykren ronronok.* My eyes are almost closed because I am so tired

**rophil- ~ rophyl-** *v.* to joke

  **rophyl... khele...** *coll, v.* to joke around, to play around. *Angai tai•sami ytykni nochido,* *rophylchawachym khelechawachym.* If I would have told you what just happened, you would not have joked around, you would not have played around.

**roprop-** *v.* to crumble

**ros** *n.* PLANT/FOOD/CORP. (1) sap, juice (of meat and fruit) (2) cum, sperm, semen

**rot-** *v.* to boil (something in water). *Sa•na dakang alu rotchengbote!* Boil the potatoes before eating them!

**rotchok ~ rochok** *n.* ART. stud, vertical beam that forms part of the side of a house to which the *damdyl* can be attached (see Photo 9, Photo 11)

**rothop** *n.* PLANT. species of plant, whose seeds can be used to make a type of popcorn (see Photo 55)

**rubibal** *n.* TIME. (< Bengali) Sunday

**ruchut- ~ ruchu-** *v.* to join, to connect. *Sa•gyraiaw ga•dakchichiaimuna singsingkholongsang typsetyi tanangokno, typsetyi tanangokno. Ytykyimung, kynsangdo ruchuphinaimu noksang rai•akno.* Having cut the child up into pieces, they threw him into a deep hole in the ground and left him there. But then it came home after it had joined together again.

**ruda** *n.* PLANT. species of cactus (see Photo 70)

**rugung** *n.* PLACE. edge. *Te•do hawchi chiakol rugungchi jywsawaidongnote, magachakdo.* Now, he was lying at the edge of a deep well, the deer.

**-ruk** *sfx.* (reciprocal). *Na•nange song jan•rukok.* Our villages are very far apart from each other. "*Nang•tyme goi• byisyk man•phawa ie bylsie?*" *noai syng•rukthoka.* "How many did you get this year?" everybody asks each other.

**rukchok ~ lukchokchok** *n.* ANIM. species of frog

**rukpek ~ lukpekpek** *n.* ANIM. species of small frog which says *pekpekpekpek*

**rukwak ~ lukwak** *n.* ANIM. toad (see Photo 104)

**-rum** *evsp.* all, everyone, everybody, everything V. This suffix indicates that the action is done or undergone collectively as a group, at the same place and time. *Myng•phekphekan bunduk ra•angrumokno.*

They all took a gun.(-*rum* cross-references the Agent *Myng•=phekphek=an* (CLF:HUMANS=DISTR=FOC.) *Phylgym chunggaba monokrumokno myng•korokawan.* (-*rum* cross-references the Patient *myng•+korok=aw=an* (CLF:HUMANS+six=ACC=FOC 'the six of them'.) The big eagle had swallowed all six of them together. *Wa• bai•rumok.* All the bamboo is broken. (-*rum* cross-references the Theme *wa•* 'bamboo'.)

**rum•** *num.* twenty. This word is only used in compound numerals *rum• tham* sixty.

**rumal** *n.* ART. head band

**rung** *n.* ART. logboat, dugout boat. river boat made out of a hollowed tree trunk. *rung khan sa* one boat

**rungkhut** *n.* FOOD. broken rice

**runi** *n.* BODY. brains

**-rura** *evsp.* V up and down, V back and forth, V and the opposite motion. *Te•do rai•sotwae tawangaimu ue Grip nok hamgabatykyisa wylangthiriokno. Ytykyimu rai•ruraai rai•ruraaimu noai.* Now, having gone up the shortcut via the Grip House, he went down again. So he went up and down, up and down.

**rurong-** *v.* GEO. to landslide. *ha• runonga* there is a landslide

**-ruru₁** *evsp.* more and more, V around, V all over the place, V through. *Tam•ai chaichie te•do byirakhem hongkhotruruaimu kaksyrangokno pheruawdo.* When he tried to beat it, the bees swarmed out and stung the fox all over.

**ruru-₂** *v.* to make liquid come out

**rychup-** *v.* to fall on one's face, to fall head first.

**rydym-** *v.* to sprout leaves. *Ie nang•myng gore angmyng aluchak rydymgabaaw kobichak rydymgabaaw sa•jyrynga.* This horse of yours eats my sprouting potato leaves and sprouting cabbage leaves every day.

**rygyn** *n.* PLACE. side. *rygynchi* near, next to *Ang choki rygynchi chapaidonga.* I'm standing next to the chair.

**=ryi ~ =ri** *encl.* (privative, indicates a referent, usually kin, that that was lost). without. (see =*ri*)

**ryk₁** *n.* ART. necklace (see Photo 114)

**ryk-₂** *v.* (1) to chase. *"Tai•ni kakai sa•chong•motnaka" noaimyng rykaidokno magachakaw banggale.* "Today I will really devour it", the Bengal said and chased the deer. (2) to herd. *Sa•gyrai ma•su rykarok.* The children are herding the cows. (3) to run to meet someone. *"O ie ang sa•banthai chong•motan bebe" nookno. Ytykyimu de•thenge rykangaimu khabakokno, khu•tymokno.* "Oh! This is really and truly my son", he said. So then, he ran to meet him, and embraced him, and kissed him.

**rym-** *vtr.* to cook, to prepare food. *Mai ja•bek rymna sapama?* Do you know how to cook food?

**rymkhap-** *v.* to cook without *mai•tyi ~ mai•ti. Angdo ja•bek rumkhapni.* I will cook curry without *mai•tyi ~ mai•ti.*

**rymreng rymreng** *adv.* dazed

**rymrym-** *v.* to roll. *Jemi sanchi rong• rymrym dapetaimung Warma sep nogaba jagysimi chak bai•thongokno.* One day, a rock rolled down and broke so-called Warma sahib's hand/arm.

**rymyl** *n.* KIN. (1) marriageable female cousin (2) the relation of female cousins from intermarriageable families (3) the relation of the parents of a married couple (4) girlfriend, lover, sweetheart

**rymyt** *adj2.* yellow, orange

**ryng-** *v.* (1) to drink *Tyi ryngbo.* Drink water. (2) to smoke *Sigyret ryngbo.* Smoke a cigarette. (3) to celebrate (by drinking). *Dakangdo, mamung khem ni•wachido, dymchyrangsangsa chywgyn ryngwano, achu ambi niamdo.* Long ago, where there were no drums, our ancestors used to celebrate *chywgyn* only with string instruments.

**ryng•-** *v.* to sing. *Dilsengdo git ga•sugaba ryng•wa.* Dilseng sings awesome songs.

**ryng•chyw** *n.* FOOD. flat-rice. **ryng•chyw sa•-** to celebrate the flat-rice festival, or, to eat flat rice.

**ryng•chyw syw•ai sa•-** to celebrate the flat-rice festival

**ryngchyw ~ ryngchu** *n.* FOOD. flattened rice
**ryngkhaw-** *v.* to drink sneakily
**ryngkhele-** *v.* to drink for fun
**ryngreng-** *v.* to shake one's head
**ryngring-** *v.* to wiggle, to move back and forth, up and down
**ryp-** *v.* to dive, to be/stay under water, to be submerged. *Thorokangaimyng hawtyi rypokno magachake. Bewal rypaimyng phetaakno.* Having jumped in, he stayed under water for some time, the deer. Having stayed under water for some time, he emerged.
**ryphi- ~ riphi-** *v.* to plaster (with a mix of clay and cow dung). *Chula riphiaidong.* I'm plastering the cooking place. (see Photo 17)
**ryt-** *v.* to pick up, to collect, take back, pick back up. *Ge•theng garu ramgabaaw rytokno.* She had taken the sun-dried mustard leaves back in.

# S

**sa**$_1$ *num.* bound. (1) one. *Ue gawichi sa• myng• korok ganangnoro aro de•theng pipukchi ganangkhua myng• sa.* That woman had six children, and in her belly she had one more. (2) a/an. *Uchie ramchi pheru mang sa gorongwano.* Then they met a fox on the road. **sagaba** (1) first. *Unasa boba myng• sagaba te•ew abun boba nukaisigaakno.* Then the first fool saw another fool. (2) one... another/the other... *Song sagabaaw Songmong myngwanowa, song sagabaaw Songgadal myngwanowa.* One village was called Songmong, the other village was called Songgadal.
**sa-**$_2$ *v.* (1) to be ill, to be sick. (2) to hurt, to be in pain. *Nawengawmu Kumiriawba khamoknowa, tyi. Ytykyimu gumukan thaphuoknowa sawa man•oknowa.* Naweng and Kumiri, got scalded, by the water. They had blisters everywhere, and they (the blisters) hurt. / They had blisters everywhere, and they (Naweng and Kumiri) were in pain.
**dykym sa-** to have malaria. *De•theng dykym saaidong.* She has malaria.
**han•dykmai sa-** to be ill with jungle fever
**sa•ba•na sa-** to go into labour. *Jyw•gado noksangdo oganangarok nookona, sa•ba•na saarokno.* The mother was at home, pregnant, and she was going into labour.
**sa-**$_2$ *vtr.* to put in place, to set as a trap, to do. *Ja•ga saakno uchie, taw• pang•ai banokno.* They set traps and then caught many birds.
**gool sa-** to score a goal
**ra•ai sa-** *expr.* to marry off, to marry someone to someone else. *Mamathanggaba sa•mynchykgana khyrethangaw ra•ai sa•naka.* Mother's brother will marry his daughter to her cousin
**=sa**$_3$ *encl.* (delimitative). *Nang•mi jorado nang•mi madamsate!* Your lover is no one other than your teacher! *Morot myng• sasa bytangwano.* Only one man led him away, they said. *Dakanggabado jineralmitingchengni. umungsa songgumuk thom•aimung ha•ba ha•ryn ha•rynaw sowalni.* In the beginning they begin with a general meeting. Only then, after the whole village has gathered together, they will divide the rice field plot by plot. *Songgumukan ue mongmawana waikhurutaisa boli hyn•aisa man•ai sa•thokwano.* The whole village, precisely because they prayed to the elephant tusk and precisely because they gave offerings, they all became rich.
**sa**$_4$ *interj.* interjection to chase away a chicken
**sa•**$_1$ *n.* KIN. *ref.* child, offspring. Clf. *myng•*. *Ue gawichie sa• myng• korok ganangno aro de•theng pipukchi sa• myng• sa ganangkhuano.* That woman had six children, and in her belly, there was another child.
**sa•-**$_2$ *v.* (1) to eat. *Maijyreng sa•cha, wakna.* You don't eat dried rice, it's for the pigs. (2) to celebrate (by eating). *Nang•e bichi krismas sa•nima?* Where will you celebrate Christmas?
**man•ai sa•-** *expr.*(1) to eat in great amounts. *Uchian magachakdo biskutaw man•ai man•ai man•ai man•ai sa•aidokno.* Then the deer ate the biscuits in great

amounts. (2) rich, wealthy. *Song dam sachi morot man•ai sa•gaba ganangnochym.* In a village supposedly lived a rich man.

**haw•ai kamai sa•-** *expr.* to work hard to survive. *Ue songmi morot haw•ai kamai sa•gaba gumukan.* The people of that village are all people who work hard and struggle to survive.

**wa• sa•gaba** *expr.* though (of persons). *Ge•thengdo morot wa• sa•agaba.* He is a tough person. (Literally 'He is a person who eats bamboo').

**warem sa•-** *expr.* to rust. *Darai warem sa•ak.* The sword has rusted.

**sa•a siwa** *khjyks, v.* famine, starvation. *Ue songchi mu•aiphachi jalangpha•chiba songmi nokmi morotdarange kanggal dongwana sa•a siwana morot myngpha•chano, chonnykariano, che•ephaariano.* While living in that village, having run away, the villagers did not respect them because they were poor and starving, and they just looked down on them.

**sa•banthai** *n.* KIN. *drel, c, ref.* (1) son. (2) nephew: male's brother's son or female's sister's son

**sa•burung jyw•burung ~ sa•byrung jyw•byrung** *khjys, n.* KIN. *set.* a mother and her children

**sa•daiburung** *n.* PERS. child born out of an incestuous relationship

**sa•dap-** *v.* to spill, to take more and more

**sa•gyrai** *n.* PERS. child. *Sa•gyraiwana khymchawa.* Because she's a child I will not marry her.

**sa•gyrai odek** *n.* PERS. baby

**sa•khaw-** *v.* to steal. *"Hai bai•siga biskut sa•khawna" noaidongano.* "Come on, my friend, let's steal the biscuits", he said.

**sa•khele-** *v.* to eat for fun

**sa•lak-** *v.* to lick

**sa•ma** *n.* PLANT. species of tree

**sa•mynchyk** *n.* KIN. *drel, c, ref.* (1) daughter (2) niece: male's brother's daughter or female's sister's daughter

**sa•nal- ~ sa•nyl-** *vgoal.* to be jealous of. *Ge•theng angna sa•nala.* He's jealous of me. *Ge•theng angna jama sa•nyla.* He's jealous of my shirt.

**sa•rong** *adj2.* of the same age

**sa•thup** *n.* BODY. uterus, womb

**sa•thyra** *n.* KIN. stepchild

**sa•wynja** *n.* TIME. (archaic) July

**sabisi** *n.* CORP. disease

**sabun** *n.* ART. (< Hindi) soap. Clf. *thut. sabun thut ni.* two bars of soap

**sadai** *n.* PLANT. species of tree

**sadu** *n.* KIN. *d, ref, a.* brother-in-law: the relation of men whose wives are sisters.
**sadu chunggaba** the elder brother of a *sadu*.
**sadu mylgaba** the younger brother of a *sadu*

**saduthangmaran** *n.* KIN. *set.* two or more men whose wives are sisters

**sagal** *n.* GEO. (< Indic) sea

**sagaltyisam** *n.* PLACE. beach, seaside

**sai**$_1$ *n.* PERS. husband

**sai-**$_2$ *v.* to choose, to select, to elect. *Morot sengbatgabaaw saiok.* The most intelligent person got chosen.

**sai-**$_3$ *v.* to write. *Ang nang•na pang•gaba khathadrangaw saiai baletna.* I want to write a few words to you.

**saido** *n.* ART. fishing line

**saigon** *n.* PLANT. teak tree

**saigyn** *n.* ACT. the third weeding of the *ha•ba*. *Mai kai•manwamungsa ha•jagara kama. Ha•jagara kamaisa kamaimung kynsange jakun kama. Jakun kamaimungsa nobembyl, oktobylsomaichi saigyn khan•a. Umungdo mai mynokodo maidan syla thoka.* Having planted the rice, we weed the land for the first time. Having cleared the weeds for the first time, we will clear them for a second time. Having weeded the land for a second time, in October or November we do a third weeding. Then, when the rice is ripe, we celebrate the new rice festival.

**saikhiribudu** *n.* PLANT. species of creeper

**sainokgaba** *n.* PERS. author

**saip ~ saep** *n.* PERS. (< Hindi) European, white person, British military commander

**sajin** *n.* CORP. illness that makes everything taste bitter

**sak-₁** *adj.* red

**-sak₂** *evsp.* V appropriately, V adequately, V well. *Takrukangwa man•sakchaaimyng, Relaragondi balaidongano. Warasakangaribo" noaimu, baletaidongano.* Not being able to fight appropriately, Relaragondi spoke. "Defend yourself well", he said, and talked. *Unasa Ketketa Burae pheruna jai•sakna chol man•cha cho•motaimyng [...].* Because Ketketa Bura could really not come up with an idea to adequately oppose the fox [...]

**sak-₃** *v.*(1) to bear, to persevere, to endure, to hold out. *Rangsan sakna man•chaaimyng nokchi dang•ok.* Not being able to bear the sun anymore, he went into the house. *Na•a ie sastiaw sakna man•chido jokangni.* If you can endure this punishment, you will go free. (2) to suffer. *Aiaw! Biskynba bylsi nidyrang dong•phinai duk sakwachido de•thengna mamyng tangka poisa, de•thengmi duk sakwana, wak rakhiganaba, tangka poisa hyn•chano.* Oh! Approximately two years have passed in which he suffered from sorrow, he got no money for his suffering and for the pig keeping they gave him no money either. (3) to be patient. *Choi•sa sakkhubo. Ang rai•aphinnaka.* Be patient a little longer. I will come back. **balwa sak-** *expr.* to enjoy the wind. *"Gongwanasa balwa sakai mu•arong" noatakokno amakba.* "I'm just sitting here enjoying the wind because I want to", said the monkey.

**sak-₄** *v.* to fit. *Angdo maja 13 tarikchi chaphang phang 99 ang nok rygynchi kai•ok. Te•ew phang 150-darang sakkhunichym.* A few days ago, on the 13ᵗʰ, I planted 99 tea plants near my house. Now I might fit another 150 or so more.

**sak-₅** *v.* to make a rope by rubbing thread between one's hands

**sak-₆** *vgoal.* to depend on. *Ang maharina sakaidong.* I'm depending on my family.

**sakchyk-** *v.* to endure, to hold out, to be patient, to have patience, to behave well. *Ie sa•gyrai sakchykna man•cha.* These children cannot behave well.

**saket-** *v.* to insert, to plug in. *Waiyraw saketbo.* Insert the wire. *Pluk saketbo.* Plug in the plug.

**sakhap** *n.* PLANT. species of tree

**sakhapnathyng** *n.* PLANT. species of tree

**sakhi** *n.* PERS. witness

**sakhyna-** *v.* wounded

**saknaram ~ salnyram** *n.* PLACE. east

**sakrem-** *v.* to twist

**sal•tareng** *n.* broom (made from the veins of coconut leaves) (see Photo 33)

**sal•wareng** *n.* PANT. species of plant to make brooms to sweep the compound outside the house

**sal•wek** *n.* ART. broom

**salam-₁ ~ selem- ~ serem- ~ saram-** *v.* to break/tear easily, to be easily damaged. *Ie mudupan serema.* This papaya tree breaks easily.

**salam₂** *interj.* (< Arabic) hello

**salam₃** *n.* (< Arabic) greeting. *Barbara, nang•na Miksrang salam baletwa.* Barbara, Miksrang greets you.

**salgypeng** *n.* PLACE. south

**salgyro** *n.* PLACE. north

**Saljong** *n.* SUPER. sun god

**salniram** *n.* PLACE. west

**salnyram ~ saknaram** *n.* PLACE. east

**sam₁** *clf.* classifier for hands, arms, legs, feet, ears and tires. *nakhal sam sa.* one ear. *cha•sam sa.* one leg/foot. *taiyr sam ni.* two tires

**sam₂** *n.* PLANT/FOOD. weed, medicine

**sam-₃** *v.* to wait. *Mosa na•a sambota. Hyn•niba nang•na te•en.* Hey friend, wait! I will give it to you, ok, later.

**samalmaisirong** *n.* ANIM. very small species of ant

**samanggri ~ samanggyri** *n.* PLANT. species of plant

**sambanggyri ~ sambanggri** *n.* PLANT. medicinal plant that stops bleeding. *U•ching cha•aw kakaimu, sambanggyri tokdepdepaimu pha•wa.* Because a leech bit him on the foot, he crushed *sambanggyri* and put it on the wound.

**sambarat** *n.* PLANT. touch-me-not, *mimosa pudica.* Species of plant that closes its leaves when you touch it (see Photo 67)

**samchak** *n.* FOOD/PLANT. vegetable
**samkong** *n.* PLANT. high grass
**sampattar** *n.* PLANT. species of plant
**samphat** *n.* ART. fee paid to a medicine man (*oja*) for his services
**samsai** *n.* PLANT. low grass
**samsi** *n.* PLANT. grass
**samsin** *n.* CORP. big boil, abscess
**samsin maiphara** *n.* CORP. small boil
**samthai** *n.* PLANT. species of plant
**samtokjang** *n.* PLANT. species of plant
**samycheng ~ sasyri** *n.* CORP. bladder infection
**san**$_1$ *n. autoclf.* TIME. day. *Range san chi bri wawano.* The rain fell for fourteen days. **sansan** every day, daily. *Kynokholthanggabado sansanan Dabatwarisang dinggarai sana re•engronganoro.* The son-in-law went to Dabatwari every day to set fish traps.
**san-**$_2$ *v.* to heal
**san-**$_3$ *v.* to put in a bag
**sanarai ~ sanyrai** *n.* ANIM. centipede
**sandi-** *v.* (1) to search (for). *"Abu, angdo dadaparaaw sandiedongachym" nookno. "Nang• dadaparado usang phylgym chunggaaw kawna re•engwanote."* "Grandma, I am searching in vain for my elder brothers" he said. "Your elder brothers went that way to shoot the big eagle!" (2) to inquire (about). *Phulisophischi nang•dadaw sandiaidokno.* People are saying that they are inquiring about your brother at the police station.
**sanglas** *n.* ART. sunglasses
**sang**$_1$ *bound.* side, place (see also *sangphak*). *cha•masang* downstream, bottom of a hill *khambaisang ~ khambaisang* upstream, top of a hill
**sang-**$_2$ *v.* to burn. *Ie pan nemai sangni.* This wood will burn well.
**=sang**$_3$ *encl.* (mobilitative/instrumental/locative ). (1) mobilitative interpretation. *Songsang re•engnima?* Will you go to the village? (2) instrumental interpretation. *Ang rong•sang depywaw ratwa.* I hit the snake with a stone. *Ang ie biskutaw tangkasang ra•wa.* I bought the biscuits with money. (3) locative interpretation. *Uchi rupeke hyiawe rong•ka otyknyng•sang "pekpek pekpek" noai parawaidoknowa.* Then, the frog, way over there at the bottom of the cliff, is calling "pekpek, pekpek". *Dakang ha•haw•a Baljongsangaw.* In the past, they were tilling the soil in Baljong.
**sangkhyning ~ sangkhyni** *n.* SUPER. mythical water dragon that lives in the Symsang river.
**sangori** *n.* GEO. fog
**sangphak ~ samphak** *bound.* side. *isangphak* this side. *usangphak* that side. *ha•byrisangphak* the side of the mountain. *gasamsangphak ~ sagasamsamphak* evening, afternoon, evening, later part of the day.
**sangwal-** *v.* to forget. *Ang nang•aw sangwalchawa.* I won't forget you.
**sanmaji** *tw.* noon, midday
**sanyrai ~ sanarai** *n.* ANIM. centipede (see Photo 96)
**sap-**$_1$ *vsec.* to know a skill. *Bildo te•ewba gore dungna sapchanotyi.* Bil does not know how to ride a horse to my surprise. *Ie ja•bek nemen rymna sapa.* She knows how to cook a good curry.
**sap-**$_2$ *v.* to swoop down (of birds of prey)
**saphairam** *n.* PLANT. species of medicinal plant
**saphang** *n.* ACT. first rice harvest (in August)
**saphaw** *n.* ANIM. rabbit
**sapset-** *v.* to drain. *piseri sapsetbo* drain the fish-tank
**sarai** *n.* CORP. a cut, a wound
  **sarai mat-** to wound someone. *Thalasang satetok. Chybym thyi• hongkhotokno. Sarai matok.* She hit him with a plate. Blood came out of his forehead. She had wounded him.
**saram**$_1$ *n.* ACT. new rice offering festival in which the first rice is offered to the gods or spirits
**saram**$_2$ *n.* FOOD. dry rice grains. *Saram syw•ai sa•a.* Dry rice grains are flattened (by pounding them with an *asam* in an *aman*, and eaten.

**sarat** *n.* PLANT. species of plant

**saraw-** *v.* to borrow. *Bengmi tangka sarawni angdo.* I will borrow money from the bank. *Ang nang•na tangka sarawai hyn•ni.* I will lend you the money.

**sare** *bound.* (< Indic) half past (only used in telling the time). *Sare das baji dong•ok.* It's half past ten.

**sari-**$_1$ *vgoal.* to shun, to ignore someone (out of shame or hatred). *Ang ge•thengna saria.* I ignore him.

**sari-**$_2$ *vtr.* to hide something, to keep something secret

**sasep** *n.* CORP. chicken pox, smallpox

**sason** *n.* ACT. reign, rule

**sasti** *n.* (< Indic) ACT. punishment. *Sa•khawchido na•a sasti man•ni.* If you steal, you will be punished.

**sastro** *n.* PERS. (< Indic) student

**sasyk sasyk tak-** *expr.* feeling unwell, feeling a small pain, feeling an urge. *De•thengdo disunaba sasyk sasyk takarong.* He needs to piss.

**sasyri ~ samycheng** *n.* CORP. bladder infection

**sat**$_1$ *clf.* classifier for bundles. *garu sat tham.* three bundles of mustard leaves

**sat-**$_2$ *v.* to hit with something and wound, to cut with a sword. *Thalasang satetok. Chybym thyi• hongkhotokno. Sarai matok.* She hit him with a plate. Blood came out of his forehead. She had wounded him.

**sat-**$_3$ *v.* to spill, to flush out

**satha ~ sytha** *n.* ART. umbrella. Clf. *khung. satha khung byryi* four umbrellas

**sathup** *n.* PERS. sick person

**satkhap-** *v.* to box, to slap

**satpyret-** *v.* to hit with the open hand

**-saw**$_1$ *evsp.* V expectantly, V and wait, keep V-ing, V and stay, V patiently, V certainly, V persistently, be busy V-ing, fully occupied/preoccupied, engrossed. *Abu bydyi parang khan•sawarongno.* An old woman was busy cutting trees. *Beanbebe usangdo So•redo Relwakmadareaw khymsawaimyng nokthaichi chairatai balwa ryngai mu•aimyng, khyryk chaisawaidongano.* Truly, So•re had married Relwakmadare, and in their small house, they were looking down, enjoying the breeze, and were fully occupied spotting lice.

**saw-**$_2$ *v.* rotten

**saw-**$_3$ *v.* to curse at (use bad words)

**saw•-**$_1$ *v.* to dig. *Nokdanggumuk gopram saw•wa habyri nalsasang.* The whole family dug graves at the other side of the hill. *Thengthon khudalsang ha•aw saw•aidongano. Saw•aidongano, thyw•angaidokno, cha•kyw chyigykdarangdo.* Thengthon is digging in the ground with a chopper. He is digging and he is getting deep, about ten knees deep.

**saw•-**$_2$ *vtr.* (1) to burn, to rage (of fire). *Nokphandaidyrangaw saw•aimung nok pha ndai do•khakhuchi khachapai tangaba mongma wa dora byryi dong•gabaaw ra•ai jalangokno.* Having burnt the bachelors' houses, they took the 20 kg weighing elephant tusks which were tied to the *do•khakhu* and ran away. *Ramchi agal saw•gaba ganang.* On the road is a raging forest fire. *Ja•ryt saw•ai, mantaw saw•ai, mai•chengmung na•lammung thiksa berengai sa•a.* (2) to roast on the hot ashes of the fire. We roast the chilli pepper, we roast the brinjal and cook it until it is well done with *mai•cheng* (a species of leafy green) and *na•lam* (a species of fish) in a bamboo tube and eat it. *ang nok saw•ok* I have burnt the house down.

**saw•khyn ~ sawkun** *n.* ANIM. vulture

**saw•myk-** *v.* to smell rotten, to smell foul

**saw•saw-** *v.* be able to cause a burning sensation

**sawel ~ sawyl ~ jingka ~ jingkha** *n.* PLANT. species of luffa or loofah vegetable (see Photo 51)

**sawkun ~ saw•khyn** *n.* ANIM. vulture

**sawn** *n.* GEO. sound

**sawthal** *n.* PERS. dirty person, person who never washes

**=sega ~ =siga**$_1$ *encl.* (alternative modality). (1) the other, each other, one another. *Morotmi morotsigaaw jongmi jongsigaaw bai•sakrara kakrukok.* Fellow men and brothers fought with each other among friends. (2) next. *Sunibal sanchi Jadi re•engwano biphagabae. Uchie sansegachi phetano.* On Sunday, the lad went to Jadi market. Then, he arrived the next day. *Song sami songsigachina nawrukok tan•rukok.* From one village to the next, people scolded each other and slew each other. (3) in turn. *Matsa kherengwachido, wa•chung byryidarangdo thangaaidongano Bandiba. Bandi kherengwachido, wa•chung byryi wawa wawa thangasigaaidongnote.* When the tiger made an effort, it threw Bandi four bamboo lengths away. When Bandi made an effort, [he] threw [it] four bamboo lengths in turn! (4) instead of. *Nepaldo chepgabaaw dengaimyng, Thengthon hongkhotokno. Ytykyimyng Thengthondo Nep alaw khaaimyng koksepchi chepetsigaaknoro.* The Nepali untied the prisoner, and Thengthon came out. So then, Thengthon tied up the Nepali, and locked him up in the cage instead of himself.

**-sek** *evsp.* to *V* and steal

**sekari** *n.* ART. pin lock

**sekyn** *n. autoclf.* TIME. second

**sel-** *v.* to leak. *Tenkimi mobil selarong.* Oil is leaking from the tank.

**sel•-** *v.* to pour into

**selsoksok-** *v.* to masturbate, to wank. *Ri• selsoksokni angdo.* I'll wank.

**selu** *n.* ANIM. cockroach

**-sem** *evsp.* certainly V. *Gorialdo iskyn san somai jarawachina duk man•aidokno. Ni•wa ido pherudo. "Ang sa•awdo poresemetchagaba kakai sa•na nangok ido" noaimung chanchichypai gorialba.* The crocodile was sad for a long time. The fox was gone. "He must have eaten my children, who are certainly not studying", the crocodile thought.

**-seme** *evsp.* V reluctantly

**sendel ~ sendyl** *n.* ART. (< English) sandal. *sendel jora sa* a pair of sandals

**sene** *num.* seven

**seneng** *n.* ANIM. species of red beetle with black dots, black wings with white tips, black legs with red joints and black antennae (see Photo 90)

**seng-**$_1$ *adj1.* clever, intelligent

**seng-**$_2$ *v.* to stay awake **walseng-** *v.* to stay awake all night

**seng•-**$_1$ *v.* to bother by misbehaving

**seng•-**$_2$ *v.* to shine, to dawn, to become light. *Wal seng•wachi Sijunygylsang re•engni.* At dawn we will go to Siju market.

**seng•khi** *n.* ART. traditional belt made of ivory beads

**seng•sot-** *v.* to abbreviate. *Ue ha•byriawe seng•sotai Matsa Chang•kui myngsigaariok.* That mountain is just called Matsa Chang•kui for short.

**sengki** *n.* PLANT. type of fruit

**sengsyp** *n.* ANIM. species of small fish

**sentimityr** *clf.* (< English) centimetre

**sep-**$_1$ *v.* to be stuck. *Wang•ai sa•chido abongrandai wa•chi sepni.* If I eat the cob by turning it, the corn will get stuck between my teeth.

**sep-**$_2$ *v.* to wring, to squeeze out

**sepjyrot-** *v.* to wring

**sepsep-** *v.* to masturbate, to wank. *Ri• sepsepni angdo.* I will wank.

**septembyl** *n.* TIME. (< English) September

**serabera** *n.* SUBST. dirt

**serabera tak-** dirty. *Ang longpen rai•tyngmi gal•aimu serabera takthiriok.* Because my trousers have fallen off the clothes line they have become dirty again.

**serek** *n.* PLACE. surface, balcony of a rice field house

**serekmyk** *n.* MSRE. the length of one forearm

**serem- ~ salam- ~ selem- ~ saram** *v.* to break/tear easily, to be easily damaged. *Ie mudupan serema.* This papaya tree breaks easily.

**serembut** *n.* ANIM. species of fish

**-set** *evsp.* V and do away with, V so as to dispose of, V away. *Palyngchi songsetokno, ha•thapyra hanggaldarangawdo.* In the jungle, he stored the ashes and cinders away. *Tokkyreng tan•thongaimungna kynsangdo dykymawdo jytsetetokno.* After cutting its throat, he moved the head out of the way. *Sa•gyraiaw ga•dakchichiaimuna* *singsingkholongsang thyps̲etai tanangokno.* They cut the child in pieces, and threw them away in a deep hole in the ground, where they left them.

**si-₁** *v.* to starve. *Ang pi•sachi amapara babapara kanggal dong•wana sa•a siwa.* When I was a child, because my mother and father and their families were poor, we starved for food.
 **kha•thong si-** to feel pity. *Angdo ue sa•gyraina kha•thong sia.* I feel pity for that child.

**si-₂** *v.* to peel

**-si₃** *evsp. V* uncomfortably. *nasi-* to be irritating to listen to, *mu•s̲i-* to stay somewhere uncomfortably

**=si₄ ~ =syi ~ =thai ~ =tyi** *encl.* (mirative modality, indicates an emotion of surprise, annoyance or anger on the part of the speaker). (see *=syi₂*)

**si•-** *v.* to sharpen (a pointy object)

**si•wil-** *v.* to carve, to sharpen a pointy object

**sial** *n.* ANIM. (< Indic) jackal

**sidai** *n.* PLANT. species of tree

**sidikeset** *n.* ART. CD, compact disc

**=siga ~ =sega₁** *encl.* (alternative modality). the other, each other, one another, next, in turn, instead (of), see *=sega ~ =siga*.

**sigyret** *n.* FOOD. (< English) cigarette
 **sigyret ryng-** to smoke a cigarette. *Sigyret ryngnaan thawano.* He said that smoking is tasty. *Sigyret rynga bai.* Don't smoke.

**Sijyw** *n.* PLACE. Siju

**sik-** *v.* to scratch, to pinch

**sikol** *n.* ART. a chain

**siksik-** *v.* to scrape, to rub

**silongket** *n.* PLANT. Shillong tree

**sima** *n.* PLACE. boundary, limit

**simen** *n.* SUBST. (< English) cement

**sin- ~ sin• ~ syn•- ~ syn-** *v.* to lay, to lay out, to spread out on something. *Thymanggami manggisiaw synai tanok.* We have lain the dead person's corps on the bed. *Ang nono jywna synai tanok.* I have lain my younger sister on the bed to sleep.

**singho** *n.* ANIM. (< Assamese or Bengali) lion

**singsingkholong** *n.* PLACE. deep hole in the ground

**singsip** *n.* ANIM. species of fish

**sinthong•-** *v.* to cut/break in two pieces, to cut/break in half, to sever

**sintongtong-** *v.* to cut up in many pieces

**sip-** *v.* to smell (to use one's nose to sense smells)

**sipai** *n.* PERS. (< Indic) soldier

**sipyling ~ spyling ~ spling** *n.* ACT. (< English) spelling. *Atong khu•chukmyng sipyling/spyling/spling rakancha.* The spelling of the Atong language is not difficult.

**siri₁ ~ suri** *n.* GEO/SUBST. snow

**siri-₂** *v.* this is the decorative part of the collocation *jyw- siri-*, see *jyw-*

**sirimynmyn** *tw.* at the break of dawn, at the creak of dawn, very early in the morning, at daybreak

**sirong** *n.* BODY. scrotum

**sisawkhyli** *n.* PLANT. species of tree

**sisawmotgram** *n.* PLANT. species of tree

**sit-₁ ~ syt-** *v.* to clean out the shit from an animal's intestines. *Angdo ma•su pipuk sytaidong.* I'm cleaning out the shit from the cow's intestines. *Nang•do na• pipuk sitbo.* Clean the shit out of the fish's intestines. (see Photo 44)

**sit₂** *interj.* interjection to chase a cat away

**sitbyryt-** *vtr.* to scratch someone or something

**sithi** *n.* SUBST. fermented rice from which *chyw* is drawn by adding water (see Photo 2)

**siwi ~ gylarong** *n.* PLANT. species of sea bean or its pod (see Photo 69)

**siwyl-** *v.* to carve

**ski- ~ syki-** *v.* to learn, to teach

**skrin ~ sykrin** *n.* ART. screen

**skul** *n.* PLACE. (< English through Garo) school

**so•ot-** *v.* to kill, to murder. *Chigachakchi Dibangkongdang Umangchalmangsa mongmaaw so•otai matsaaw so•otai mu•tynwano.* At Chigachak Dibangkogdang Umangchalmang, killing the tigers and killing the elephants, lived as the leader.

**so•re** *n.* SUBST. mica, precious stone

**so•sorot-** *v.* to slip. *Ramchi so•sorotok.* I slipped on the road.

**soal- ~ sual-** *v.* (1) to divide. *Songgumuk thom•aimyng ha•ba ha•ryn ha•rynaw sowalni.*

The whole village gathers and will divide the ha•ba parcel by parcel. (2) to share. *Je ha•ryn ni•gababado uan soalrukai haw•a.* As for those who do not have a plot, they share the land and clear it together.

**sojana** *n.* PLANT. species of long thin vegetable, probably a species of ridged luffa or loofah. (see Photo 52)

**sok**$_1$ *n.* PLANT. the new young leaves of a plant (but not a tree) or vegetable, a shoot, sprout

**sok-**$_2$ *v.* (1) to succeed, to win. *Rang sokchawanasa te•ewchinakhyngkhyng rangmu chyw ryngsusawanasa Mykha Badri bimung myngwanowa.* As the rain hadn't won, because they still drink and compete with the rain, they call it Mykha Badri, it is said (2) to hold out. *Phalthang sokwa dabatdo tyinyng•chi rong•chi pyi•aimyng wa khu•chengphinai sakchykaidokno.* Until he could not hold out any longer, he sat under water as long as he could bear it, holding on to a stone and biting his teeth firmly together.

**sokchuman ~ sokchuwan** *n.* PLANT. species of tree with rotten-smelling white flowers

**sokhop** *n.* ART. cover, sheath

**sokjong** *n.* PLANT. species of tree

**sokrop ~ sokyrop** *n.* BODY. lung

**soksek-** *v.* to shake something without picking it up

**soksok** (1) *v.* to masturbate, to wank (2) *n.* wanker, someone who masturbates

**sokwa** *n.* endurance

**soldi** *n.* CORP. a cold, common cold. *Soldi man•ok.* I have caught a cold.

**sombal** *n.* TIME. (< Bengali) Monday

**somphi** *n.* ACT. a joke, a riddle

**sona** *n.* SUBST. (< Indic) gold

**song**$_1$ *n.* PLACE. village, area that can comprise several *gythym ~ gythum ~ guthum*. *Song dam sachi alsia raja myng• sa gananagnochym.* In a village, there supposedly was a lazy king.
  **song... nok...** *coll, n.* village. *Ue songchi mu•aiphachi jalangpha•chiba songmi nokmi morotdarange kanggal dongwana sa•a siwana morot myngpha•chano, chon-* *nykariano, che•ephaariano.* While living in that village, having run away, the villagers did not respect them because they were poor and starving, and they just looked down on them.

**song-**$_2$ *v.* to elect, to appoint. *Songchi nokchi raja songna angawtara nukariokno. Angaw bytangaidonga, song dam sachi angaw raja songnino.* In the village the people wanted to elect/appoint a king and they just saw only me. They are carrying me away; they will elect/appoint me king in a certain village.

**song-**$_3$ *v.* to keep, to store

**song-**$_4$ *v.* to set up post, to dig a hole and stick something in it so that it keeps standing up, to raise. *Wa•sung ha•bykungchi songbo.* Dig a hole in the sand, and put the bamboo stick in it.

**mykren wa•thok song•phin-** *expr.* gazing in amazement. (see *mykren*)

**song•khot- ~ songkhot** *v.* to come out of a small opening or narrow space, to squeeze out of

**Songdu** *n.* PLACE. the Brahmaputra River. *Symsangdo Gohatichigaba Songduna mylkhala.* The Symsang is smaller than the Brahmaputra in Guwahati.

**songga**$_1$ *n.* PLACE. another village. *songgamyng morot* a person from another village

**songga**$_2$ strange

**songkhamphek** *n.* ART. forked branch or post

**songkhel** *v.* to roll head first

**songkhot- ~ song•khot** *v.* to come out of a small opening or narrow space, to squeeze out of

**songmong** *n.* PLACE. main village

**songrai- ~ songre-** *v.* to travel

**songrat-** *adj1.* bent

**songre- ~ songrai-** *v.* to travel

**songsal** *n.* ABSTR. society

**songsykhep** *n.* ART. big pincher

**songsyrek ~ songsarek** *n.* ABSTR/PERS. animism, an animist (someone who practices animism); pagan, heathen. *Ning songsyrekdo, ning Atongdo, dakangdo mamyng thorom ni•wami somaichido waiaw mania.* We heathens, we the Atongs, in the past, in times when there was no religion, we worshipped spirits.

**sorea ~ soraia** *n.* ART. big metal tub used to wash one's hands before and after eating, and to put dirty dishes into (see Photo 43)

**sorkar** *n.* PERS. (< Indic) government

**sorok-**$_1$ *n.* PLACE. road, path, way. Clf. *chol*. *sorok chol ni.* two ways/roads/ paths

**sorok-**$_2$ *v.* to re-pound the rice

**sorong-**$_1$ *adj1.* straight *khaw soronga* straight hair

**sorong**$_2$ *adj2.* straight

**sorot-** *v.* to hold a ceremony or celebrate in commemoration of a dead person one year after they died. *Ning achuaw sorotaidong.* We are celebrating/holding a ceremony in commemoration of our dead grandfather.

**sosila** *n.* PLANT. plant of the Arum family with a pink inflorescence consisting of an elongate or ovate spathe (a sheathing bract) which envelops the pink spadix (a flower spike with a fleshy axis). This plant looks remarkably like the *Amorphophallus bulbifer*.

**-soso** *evsp.* V to/on the ground *Neng•dugaaimyng mu•sosoangokno.* Having gotten tired, he sat down on the ground.

**-sot**$_1$ *evsp.* V directly, V straight. *Cha•aw re•engwae hyiawe maibachi Kol India hapchi rai•sotgabachi ge•thengdo nekkhala noaimyng rai•sotwanochym.* He went on foot, way over there, maybe at Coal India, via a shortcut, he could have gone straight.

**sot**$_2$ *n.* ANIM. species of very small fly that comes out in the evening and at night and cause itchiness

**sot-**$_3$ *v.* to spit. *Ainachi sa•gyrai khu•ti sotjaak.* The child has spat on the mirror again.

**sot**$_4$ *num.* ten. this word is only used in compound numerals. *sot bri* forty, *sot bonga* fifty, *sot dok* sixty, *sot syni* seventy, *sot chet* eighty, *sot sykhu* ninety.

**sothonthara** *n.* CORP. cancerous swellings all over the body

**sotkat** *n.* ABSTR. (< English) (1) shortcut (2) a short version. *Sotkat anga baletariok.* I just told a short version.

**sotmai** *n.* ANIM. housefly

**spiit** *n.* GEO. (< English) speed. *De•theng gari bytbutungchi bylongen spiit dong•ok.* When he was driving, he went very fast.

**sping ~ spyling ~ sipyling** *n.* ACT. (< English) spelling. *Atong khu•chukmung spyling/sipyling/spling rakancha.* The spelling of the Atong language is not difficult.

**spun** *n.* ART. (< English) spoon

**-srang ~ -syrang** *evsp.* intensifier suffix. *V* very much, *V* strongly, *V* completely, wholly *V*, *V* till the end. This suffix is also used to make things sound more emphatic or slangy. (see -*syrang*)

**ss** *interj.* interjection to chase away a chicken

**stel** *n.* EMO. (< English) haughtiness, arrogance. *Ge•thengchi stel pang•a.* He/she is very haughty. This word probably comes from English 'style'.

**stem** *n.* ART (< English) stamp. Clf. *rong*. *Stem rong chi dok tanaimu chiti wateta.* You put sixteen rupees worth of stamps and post the letter.

**stulkhabar** *n.* ART. (< English) tablecloth

**su-** *v.* to scold

**su•**$_1$ *n.* BODY. vagina

**su•-**$_2$ ~ **syw•-** *v.* to pound, to punch, to prod, to inject, to crush. *Ang khawchi mu•gaba khyrykaw su•bone.* Crush the lice in my hair, will you? *Hospytylsang biji su•na re•engnima?* Are you going to the hospital to get an injection?

**-su**$_3$ *evsp.* intensifier suffix. very. *"Awamyng nakhaldo thawsu thawsu nukwa" noaidonganowa amak sa•ai.* "Father's ears looks very tasty indeed", the monkey's children were saying.

**su•bylok-** *v.* to mash, to beat to a pulp. *Beringwa mynwachido su•byloka.* When the food cooked in the *wa•sung* is ready, we mash it.

**su•gol** *n.* PERS. vagina (used as a swearword for women), bitch

**su•kherek-** *v.* to crash down. *Jariaimu su•kherekaimu wa khaw• sa bai•okno, mongmaba.* Because he was startled, the elephant crashed down and broke one tusk.

**su•myn** *n.* BODY. pubic hair of a female

**su•nadylep** *n.* BODY. clitoris
**su•pyrong-** *v.* to punch a hole through something
**su•that-** *v.* to prod, to poke
**su•ut-** *adj1.* damp
**sua** *n.* ACT. profanation
**sual- ~ soal-** *v.* to divide, to share. *Songgumuk thom•aimyng ha•ba ha•ryn ha•rynaw sowalni.* The whole village gathers and will divide the *ha•ba* parcel by parcel. *Je ha•ryn ni•gababado uan soalrukai haw•a.* As for those who do not have a plot, those mutually share and clear the land.
**suis** *n.* ART. (< English) switch
**suit** *n.* PLANT. species of tree
**suitbipha** *n.* PLANT. species of tree
**suk**$_1$ *n.* (< Indic) happiness, comfort. *Baba, ha•gylsakchi angmi sukdo ni•ok.* Father, I have no happiness in this life. *Ichi mu•ai suk dong•ama?* Is your stay here comfortable? *Juwna suk dong•ancha.* We did not sleep enough.
**suk-**$_2$ *v.* (< Indic) well, comfortable, to enjoy. *Phalthangthangrara golpho kha•rukai takarokno, jywai sukai takaimu.* They were talking amongst themselves, and comfortably asleep. *Na•a gawi khu•thymchido sukama?* Do you enjoy it when you kiss a girl?
**suk-**$_3$ *v.* to insert, to stitch
**sukrung ~ sykrung ~ sukyrung ~ sykurung** *n.* ANIM. river snail
**suksai** *n.* PLANT. species of plant of which traditional umbrellas are made (see Photo 61)
**suksak** *adv.* uncomfortably, tossing and turning
**suksuk** *adv.* comfortably. *Myng• korokawan monokaimu kynsangdo utymdo suksuk jywarokno pipuknyng•chi, phylgym pipuknyng•chi.* After it had swallowed all six of them, they had fallen asleep comfortably inside its belly, inside the belly of the giant eagle.
**suksyrui** *n.* PLANT. species of tree of which the fruits can be eaten
**sukulbal ~ sykulbal ~ sykubal** *n.* TIME. (< Bengali) Friday
**sukyrung ~ sykyrung ~ sukrung ~ sykrung** *n.* ANIM. river snail

**sul**$_1$ *adj2.* next, neighbouring *song sul* the next/neighbouring village
**sul-**$_2$ *v.* to stretch out, to stick out, to extend. *Chak raw•khalai suletkhubo.* Stretch your arm out further.
**thylamphak sul-** to stick out one's tongue
**sun**$_1$ **~ sundul** *n.* PLANT. tree trunk
**sun-**$_2$ *v.* to move, to shift. When many people are sitting on a bench, and someone wants to join them, they can say: *Choi•sa sunbo.* Move a little.
**sung**$_1$ *clf.* classifier for hollow cylindrical objects or tubes. *wa•sung sung tham.* three bamboo tubes
**sung**$_2$ *n.* ACT/ABSTR. remembrance, thought, mind, brain, intelligence, spirit, life. *Ge• thengchi sung ganang.* He is intelligent.
**sung ra•-** *v.* (1) to remember, to think (of, about), to keep in mind. *Dykhimi balgabaaw sung rai•aisa jalphetphetangaidongano.* Remembering what Dykhi had said, he ran quickly. (2) to praise (when used with God). *Isolaw sung ra•ai je Kristen donggabado Isol phi•aia sa•chengna.* Thinking of God, anyone who is a Christian will pray to God and start eating.
**sung•-** *adj1.* short (of time, person, thing)
**sungchal-** *v.* to support (a structure)
**sungman-** (1) *vgoal.* to remember. *Nokchi ang dyngdang mu•chiba, sungmaneta anga nang•na.* (Aristo J Momin) Whenever I'm sitting at home alone, I think of you. (2) *v.* to fantasise (about). *Nokchi ang dyngdang mu•chiba, sungmaneta anga nang•aw.* When I'm sitting at home alone, I fantasise about you.
**sunibal** *n.* TIME. (< Bengali) Saturday
**suri ~ siri** *n.* SUBST. snow
**-susa** *evsp.* V competitively, compete in V-ing. *Ue hapchi mu•wachi rangmu chyw ryngsusawanasa Ha•chykkhu•chuksang mykha badri noyi myngwano.* When they stayed in that place, because they held a drinking competition with the rain, they called it *mykha badri* in Garo.
**suset- ~ susut- ~ susyt-** *v.* to wash
**susu** *n.* BODY. penis

**suthul** *n.* BODY. comb of a rooster

**suting** *n.* ACT. taking pictures, photo shooting. *Balphakramchi suting kha•akma?* Did you take pictures in Balphakram?

**sutuk-** *v.* to put over, to cover, to hide. *Mykhang baketchi sutuka.* She's hiding her face in the bucket.

**swis** *n.* ART. a switch

**syi**$_1$ *n.* KIN. *drel, ref.* aunt: mother's younger sister (addressed as *asyi ~ asi* or *ama*)

=**syi**$_2$ ~ =**si** ~ =**thai** ~ =**tyi** *encl.* (mirative modality, indicates an emotion of surprise, annoyance or anger on the part of the speaker). *Atongtykyi tai•ni ja•bek thawoksyi?* Why is the curry so tasty today? I'm surprised. *Nukoknotyi phylgym chungga•aw.* So then, he suddenly saw the giant eagle! *Rang nemchie ataknakasyi?* Now that the rain has stopped, what the heck shall we do? *De•thengba re•engokthai!* To my surprise, she also went!

**syimaran** *n.* KIN. *set.* my *syi* (mother's younger sister) and her elder sister's child

**syithai-** ~ **syithyi-** ~ **syithi-** *v.* to hang. *Syithai tanarong raityngchi.* It's hanging on the washing line.

**syk-**$_1$ *v.* to insert, to be inserted, to press, to push

**syk-**$_2$ *vsec.* to want. *Jywna sykaidongkhua.* I still want to sleep. *Nang• syka* It's up to you.

**sykdep-** *v.* to press (with one's finger)

**syket-** *v.* to insert

**sykhathang** *adv.* disorderly, carelessly, simply, for nothing, for free, wherever.

**sykhu** *num.* nine. this number is only used in the compound numerals *chi sykhu* nineteen and *sot sykhu* ninety.

**sykhym-** *v.* to moan, to complain, to feel sorrow, to mourn

**syki-** ~ **ski-** *v.* (1) to learn. *Atakna Atongkhuchukaw sykiaidonga na•a?* Why are you learning Atong? (2) to teach. *Angmyng sastrona Atongkhu•chukaw sykia.* I teach my students Atong.

**sykjyret-** *v.* to crush with one's hand

**sykrin** ~ **skrin** *n.* ART. screen

**sykrom-** *v.* to grasp someone. *Thot thyng•thot takwachina dabat sykromaimyng khanetsigaaidongno.* He (Bandi) grasped her (So•re) and poured the liquor into her mouth to the last drop.

**sykrung** ~ **sukrung** ~ **sykurung** ~ **sukyrung** *n.* ANIM. river snail

**sykulbal** ~ **sykubal** ~ **sukulbal** *n.* TIME. (< Bengali) Friday

**sykup-** *v.* to fold

**syl-**$_1$ *adj1.* beautiful, pretty, handsome. *Atongnawmyl sylate.* Atong girls are pretty! *Ang nang• dadaaw sylai nukwa.* I think that your elder brother is handsome.

**syl**$_2$ *n.* SUBST. iron

**syl•et-** *v.* to pour into

**syladyn** *n.* ART. traditional necklace

**sylasyng** *n.* ART. necklace

**syldangkhep** *n.* ART. big pliers to take pans off the fire

**sylet-** *v.* to make beautiful

**sylgythym** *n.* SUBST. iron

**sylkeng** *n.* ART. hoop, ring

**sylkengkun** *n.* ART. stick to drive a hoop

**syltyi** *n.* GEO/SUBST. hail, ice

**sym**$_1$ *adj1.* sweet

-**sym**$_2$ *evsp.* V and follow, consequently, imitate in *V*-ing, *V* attentively. *Amakba rukpekmyng bai•sigathanggaba budiaw tyngsymai takwa.* The monkey was following/imitating his friend the frog's idea. *Bandi paianggabaaw mykren wa•thok song•phinai Gyrynggyrang chaisymaidongano.* Gyrynggyrang is attentively watching the carrying Bandi, gazing in amazement.

**sym-**$_3$ *v.* to follow. *Ang nang•aw kyn kyn symni.* I will follow you closely.

**sym•**$_1$ *n.* SUBST. salt

**sym•-**$_2$ *v.* to soak, to make wet, to make *chyw* by pouring water on the *sythi ~ sithi*. *Chyw sym•ai ryngwa.* We sat around and drunk *chyw*.

**symgong** *n.* PLANT. species of plant of which the red flowers are edible and produce a lot of nectar which you can shake out (see Photo 76)

**symjin-** *v.* to have the sour-sweet taste of a half-ripe fruit

**sympak** *n.* PLANT. species of tree

**symphak** *n.* ART. type of blanket

**symsak-** *vgoal.* to care for/about, to be careful about. *Phalthangthangna symsakaribo, jalthikaribo!* Just care for yourselves, just run away! *Nang•mi amana kha•dangai symsakbo.* Take care of your mother.

**Symsang** *n.* PLACE. the Symsang River, also called Someswari, the Standard-Garo spelling is Simsang.

**-symsym** *evsp.* *V* continuously

**syn-**$_1$ ~ **syn•-** ~ **sin-** ~ **sin•-** *v.* to lay, to lay out, to spread out on something. *Thymanggami manggisiaw synai tanok.* We have lain the dead person's corps on the bed. *Ang nono jywna synai tanok.* I have lain my younger sister on the bed to sleep.

**syn**$_2$ *n.* GEO. a smell. *Syn man•aidong.* He smells something.

**syng•-** *v.* to ask. *Edinchi syng•bo.* Ask Edin.

**syng•gaba** *n.* ACT. question

**synggera** *n.* BODY. handle moustache (a moustache that sticks out from the face to the sides)

**synggi** *n.* ANIM. species of fish

**syngsyngkhol** *n.* PLACE. deep hole in the ground

**syngsyngkholong** *n.* PLACE. deep hole in the ground

**syni** *num.* seven. this word is only used in the compound numerals *chi syni* seventeen and *sot syni* seventy.

**synthi-** *v.* to suffer, to regret, to repent, to lament, to moan, to whine. *Phepchi synthibutungchi te•ewe napit myng• sa rai•phaknoro.* While he was suffering in the banyan tree, a barber came by.

**sypsak-** *v.* to be scratched. *Ha• khamaimu chak sypsakarok.* After working in the field my arm is scratched.

**syraksyrak** *adv.* exactly, just. *"Nang• syraksyrak takgaba, cha•ba nang• takwa, bimangba nang• takwa, chakba nang• takwa, tyi khaw•etwa" nowano.* "Someone who looks exactly like you, with legs like you, with a body like you, with arms like you is getting water", he said.

**-syrang** ~ **srang** *evsp.* intensifier suffix. *V* very much, *V* strongly, *V* completely, wholly *V*, *V* till the end. *V* very much, *V* strongly, *V* completely, wholly *V*, *V* till the end. *Jebadong anga takruksyrangarinaka.* Somehow I will just fight to the end. This suffix is also used to make things sound more emphatic or slangy. *Hai! Sigyret hyn•etsyrang na•a, uaw!* Come on! Give the cigarettes, those ones!

**-syret** *evsp.* to *V* wrongly, mistakenly

**syrong-** ~ **syryng-** *v.* to stretch (out) (rope etc.), to reach out, to build a bamboo bridge

**-syruk** *evsp.* to *V* secretively

**syrup-** *v.* to suck

**syryng-**$_1$ ~ **syrong-** *v.* to stretch (out) (rope etc.), to reach out, to build a bamboo bridge

**syryng**$_2$ *adv.* not clearly. *Angdo ge•thengmi balgaba syryng syryng nawa.* I did not hear clearly what he said.

**syryng**$_3$ *n.* BODY web (of spider)

**syt-**$_1$ ~ **sit-** *v.* to clean out the shit from an animal's intestines. *Angdo ma•su pipuk sytaidong.* I'm cleaning out the shit from the cow's intestines. *Nang•do na• pipuk sitbo.* Clean the shit out of the fish's intestines. (see Photo 44)

**syt**$_2$ *interj.* interjection to chase away a cat

**sytha** ~ **satha** *n.* ART. umbrella

**syw•-**$_1$ ~ **su•-** *v.* to pound, to crush, to punch, to prod, to inject. *Ang khawchi mu•gaba khyrykaw syw•bone.* Crush the lice in my hair, will you? *Hospytylsang biji syw•na re•engnima?* Are you going to the hospital to get an injection?

**syw•**$_2$ *n.* KIN. *drel, d, ref.* grandchild

**syw•muri** *n.* KIN. *c, ref.* great-grandchild

# T

**ta₁** (prohibitive word) don't. *Na•a ta dykyryngto!* Don't make noise! *Ta ie nok dyngdang ham.* Don't build this house alone. *Ta bong!* Don't lie!

**=ta₂ ~ =to** *encl.* (emphatic imperative modality, used when the speaker is impatient, or wants to beseech the addressee for something). *Na•a abundyrangtykyi kam kha•ai chaiphabota!* Try to do some more work like other people! *Rai•aboto!* Come here already! *Haito mosa! Na•a angna hynchakama?* Come on, buddy! Aren't you going to give me some?

**tai•nep** *dtw.* this morning

**tai•ni** *dtw.* today *Ge•theng tai•nidarang rai•anikhon.* He might come sometime today.

**tai•sa** *dtw.* (1) a little while ago (today), just now, just. *Uan tai•saba anga uaw golphochiba baletok.* That's what I also just told in the story. *Awangdo tai•samian ichi mu•aidong.* Uncle was just sitting here. (2) (a little while) before, earlier. *Tai•sa phaw•jonggadarang goronggano, ue achu bydyiaw gorongokno.* He met the old grandpa whom his elder brothers had met before. **tai•satykyi** just like before. *Uan tai•satykyi kantaraaw kyryk kyryk re•engaimyng syng•etthiriokno.* Just like before, he hurried away to an empty place and asks again.

**tai•symphak** *n.* PLANT. species of tree

**taia-** *v.* to pull

**taija** *dtw.* last night. *Taija walchi jywwachi jywmangsang banggirigaba nukwa.* Last night at night when I was sleeping, I saw an earthquake in my dream.

**tainalap** *n.* PLANT. algae

**tairakrak** *adv.* not too big and not too small. *Amakdo ge•theng bai•sigathanggabamyng kynaw rongpatal syltengbigabachi kepreprep bamai hyn•butungchi pantong myk sa donggabasang tairakrak takgabasang tep tep tep tep tokaidoknoa.* The monkey hit his friend who was lying flat on his belly on the very beautiful flat stone with a stick of one *myk*, which was made not too long not too short, tap, tap, tap, tap on the back.

**taisympak** *n.* PLANT. species of tree

**taiyr** *n.* ART. (< English) tire. Clf. *sam/rong/goi•*. *taiyr sam/rong/goi• tham.* three tires

**tak-₁** *vB.* (1) to do. *Ie khamaw krymkraw takna nangni.* We will have to do this work together. (2) to make, to build. *Uchi song dam ni takwano.* There they built two villages. (3) to pretend, to act like. *Te•do magachakan khora takaidongano.* Now the deer is pretending to be lame. (4) to be like, to resemble. *Pipuke moinachongchang takariokno.* His belly was like a bird cage. (5) support verb (see van Breugel 2014: 362–363) "*Ang denggu takni na•a. Na•a paiai jalbone*" *noaidongano magachakan.* "I will do some extortion. You carry the biscuits and run away, ok?" said the deer. *Ha! wen•ni rypwachian miniksuru takokno sylokno magachakmi nyn•do.* So then, ha! when he had bathed twice, his fur was flat, it was beautiful, the deer's fur. (3) to fight

**takwa rukwa** *khjyks, n.* activities, customs, traditions

**=takai ~ =tykyi** *encl.* (perlative/similative). (see *=tykyi*)

**takal.~ dakal** *n.* SUPER. witch, demon

**takap-** *v.* to stick

**takbewal** *n.* ABSTR. tradition

**Takgaba Rywgaba Phathigaba Ra•runggaba** *n.* SUPER. supreme god

**takruk-₁** *v.* (1) to fight. *Rongdyngmi olthoe, dakang somai Jaksongram matsa nok phandaimi matsamu Rongdyng maharimu takrukwanowa.* As for the meaning of Rongdyng, in times long ago, the tigers of Jaksongram's tiger's bachelors' house fought with (alongside?) the Rongdyng clan. (2) to have an orgasm (for a woman)

**taksak-** *v.* to help. *Angaw taksakboto!* Help me! *Ama, baba, angdo byl ni•pha•chiba taksakai hyn•aini*" *noaimung, re•engphaaidongano.* "Mother, father, although I have no strength, [I] will offer to help", [he] said and went.

**takwa rukwa** *khjyks, n.* activities, customs, traditions

**tala** *n.* ART. a lock (< Hindi) *Tala thekbo.* Lock the lock.

**tam-**₁ *v.* to trim, to prune

**tam-**₂ *v.* to wait, to stop

**tam•-** *v.* to beat a drum, to play an instrument

**tam•a toka** *khjyks, v. t*o play an instrument

**tam•o ~ tam•aw** *procl.* Wait!

**tama** *n.* SUBST. (< Bengali) copper

**tan-** *v.* (1) to put. *Ie garuaw ramai tanbo.* Put these mustard leaves outside to dry in the sun. (2) to leave. *Thengthon tangkaaw ra•aimyng, uaw kerengaw palyngchi gopai tansigaakno.* Thengthon took the money and left those bones buried in the jungle. (3) to keep. *So•reba morot man ra•ai takmykai, bylsi sene abek akankhambaichi tangaba...* So•re, is pretending to respect the man i.e. Bandi, and with a seven-years-old *abek*, which is kept on top of a rack above the cooking fire... (4) to spare. *Dugaphinok bai•sigaba angaw thogiwaba tam•ai ge•thengawba tanchawa.* This is too much, friend; you have betrayed me, when the opportunity presents itself, I will not spare him. (5) to leave V. *Kelkhi dawai tanaimu, daw•kha nokmi ruti sa•khawokno.* Because somebody had left the window open, a crow had stolen bread from the house. (6) to put off. (7) to pass (a law). *Ain niamawtakai tanangaimung Mongri chara Thokhang Sangmaaw loskor songai tanangokno.* A law was made and passed that the elder brother of Mongri, Thokhang Sangma, would be appointed *loskor* it is said. (8) completive-verb. *Balai tanangok.* I have already said it. *Angdo ytykyi balaimyng tanarinaka.* As for me, having spoken like this, I will just stop now.

**tanang-** *v.* to leave behind, to leave alone

**tan•-**₂ *v.* to slaughter, to slay, to cut, to cut up, to chop, to chop up. *Una phalthangmyng ma•suthangthangaw tan•aimyng nygylsang khaiangokno.* Then, having slaughtered their own cows, they carried them to the market.

**tan•-**₁ *v.* (see *mykren tan•-*)

**tan•chekchek-** *v.* to cut into small pieces

**tan•choleng-** *v.* to cut a piece out of something

**tan•pat-** *v.* oblique

**tan•pyrak-** *v.* to cut open. *Tokkyreng tan•thongaimungna kynsangdo dykymawdo jytsetetokno, ytykyimungna pipukaw tan•pyrakokno.* After having decapitated (the eagle), he pushed the head out of the way. Then, he cut its belly open.

**tan•set-** *v.* to cut out, to cross out

**tan•thong-** *v.* to decapitate, cut off the head; to cut off. *Gal•aimuna kynsangdo phylgymaw uan rykjolaimuna kukuri byk hotaimuna tokkyrengaw tan•thongokno.* After the eagle had fallen to the ground, he ran and unsheathed his knife and cut off its head.

**tanang-** *v.* to leave behind, to leave alone

**tandap-** *v.* to be on top, to cover

**tang** *clf.* classifier for *koktang* baskets. *Ytykyimyng te•do ge•thengthengdo na• khynaimyng bai•sigathangmaran rukpekba tang sa, amakba tang sa• khaiaknowa.* So then, now, as for them, having collected the fish, the frog and the monkey carried one basket each.

**tang•dap-** *v.* to splash on. *Tyi angchi tang•dapwa.* The water splashed on me.

**tangka** *n.* ART. (< Assamese or Bengali) money. Clf. *phel/khung/rong. tangka phel sa* one coin *tangka khung sa* one banknote *tangka rong chykhyw* nine rupees

**tangka poisa** *khjyks, n.* ART. (< Assamese or Bengali) money

**tankynyng-** *v.* to cut up in many pieces

**tannet** *n.* ART. measure basket

**tanset-** *v.* to abandon, to leave behind. to throw down. *Jaksongrammi matsadu chaw•kyi wa•khu tansetai jalangwanasa, ue ha•byriawe te•ewchinakhyngkhyng Atongkhu•chuksang Matsachaw•kyiasetram myngwanowa.* Because the *matsadu* of Jaksongram threw down their knives and choppers and ran away, that hill is still now still called Matsachaw•kyiasetram in Atong.

**tap** *n. autoclf.* time, turn. *tap sa.* once. *tap ni.* twice. *tap tham.* three times

=**tara** *encl.* (1) only, exclusively, alone. *Anga mamungawan dywanchate. Uan phywramu garutara dywariwate*. I did not add anything, I'm telling you! I put in only rice powder and muster leaves. *Morot myng•sanoromo. jyw•taraanokmo, wa• ni•ok*. There once was a person. She had become a single mother, there was no more father. (2) myself, yourself, himself, herself, itself, ourselves, yourselves, themselves. *Bil phalthangdo goreawtara watetaimyng phalthangdo thyigabamyng degaldyrangaw ra•akno*. Bill sent the horse away by itself and collected the weapons of those who had died.

**tarai** *dtw.* this year. *Angdo tarai Shillongsang re•engchawa*. I won't go to Shillong this year.

**tarak-** *adj1.* quick, fast. swift. *Anga noksang tarakai rai•na nangaidong*. I need to go home quickly. *Tarakbo na•a!* Hurry up!

**tarang** *n.* PLACE. layer. *Angdo ie rajami khemna jywthumaidonga, damana. Ramramchagaba kyryngwa ido. Ha•nyng•tarang chinina imyng kyryngwado rajami dama"*. I am lying here guarding the royal drum. It has an unusual sound, this thing. The sound of it reaches twelve layers inside the earth.

**tarik** *n.* TIME. (< Indic) date. *Angdo maja 13 tarikchi chaphang phang 99 ang nok rygynchi kai•ok*. A few days ago, on the 13$^{th}$, I planted 99 tea plants near my house.

**tas** *n.* ACT. (< Assamese or Bengali) cards (the game). *Tas keleni ningdo*. We are going to play cards.

-**tat**$_1$ *evsp.* compulsorily, inevitably, bound to V. *Uchie jyw•changna nangwachie, matsa gorongtatoknotyi, maikapchi jywwachi*. Then, when he suddenly had to sleep, he was bound to meet that tiger again, which was sleeping on the hay.

**tat-**$_2$ *v.* to drive in (as with a nail in wood)

**tatkhapgaba** *n.* stud (used in the construction of a house, see Photo 9)

**taw-**$_1$ *v.* to go up, to ascend **re•eng... taw... coll**, *v.* to go away, to leave (see *re•eng-*)

-**taw**$_2$ *evsp.* V upward, V upstream. *Sijyw song mu•aidonggaba hapchi bondykaw paiai panaw jap kha•gabaaw kawtawna thymokno*. Those who lived in the village of Siju got their guns, made a defence wall out of wood and lay in ambush to shoot upward. *Hare! Phangnan dakang ang nukgaba gorialdo chaw•sa chaw•tawwachym*. Huh?! In the past, the crocodiles I saw always swam upstream.

=**taw**$_3$ ~ =**aw** *encl.* (accusative). (see =*aw*)

**taw•** *n.* ANIM. chicken, bird

**taw•cha•si** *n.* PLANT. species of liana

**taw•di•mai** *n.* BODY. tail feathers

**taw•gurung** *n.* ART nest

**taw•gylyk** *n.* ANIM. species of jungle bird

**taw•karang** *n.* BODY. bird's wing

**taw•khasi** *n.* ANIM. capon, castrated rooster

**taw•kurung** *n.* PLACE. the nest of a chicken, a chicken that is about to lay an egg

**taw•myn•** *n.* BODY. fluff, body feathers

**taw•nok** *n.* ART. chicken cove, coop

**taw•pachi ~ to•pachi** *n.* ART. the triangular bamboo mat under the roof of a house between the *khyntyri*, *byrym* and *bylbang* (see Photo 10)

**taw•pachi** *n.* ANIM. swallow (Family of *Hirundinidae*)

**taw•pak** *n.* ANIM. bat, butterfly, moth

**taw•pakcha** *n.* PLANT. species of tree

**Taw•pakkhal** *n.* GEO. Bat Cave (in Siju)

**taw•paktyi** *n.* ANIM. caterpillar

**taw•pal** *n.* CORP. freckles

**taw•palyng** *n.* ANIM. jungle fowl

**taw•puk** *n.* ANIM. the innards of a chicken

**taw•pynchyrep** *n.* ANIM. common tailorbird *Orthotomus sutorius* (see Photo 112)

**taw•reksyrup** *n.* ANIM. banana bird, if translated literally its name is 'banana tree sucking bird'

**taw•sa•gyrai** *n.* ANIM. chick

**taw•thup** *n.* ART. nest, bird's nest

**taw•ti ~ taw•tyi** *n.* BODY. egg

**tawa** *n.* ART. frying pan

**tawel** *n.* ART. towel

=**te** *encl.* (declarative modality). I'm telling you! really! (The declarative enclitic is ususally not translated with a specific word or phrase in English, but just with a declarative sentence, or statement sentence.) *Tai•nido ja•bek thawokte. Na•a phangnado thawai rymcha. Atongtykyi tai•ni ja•bek thawoksyi?" nookno. "Atong dywwa ama?" nookno. "Hy? Ni•wate baba. Anga mamungawan dywanchate.* "The curry is so tasty today. You don't usually cook tasty. Why is the curry so surprisingly tasty today?" he said. "What did you put in it, mother?" he said. "Huh? Don't mention it, son. I did not add anything." *"Thol•aidongkhonne babado. Awangdo tai•samian ichi mu•aidong. Re•engkhucha te•ewba." "Nang• awang myng• sagami nokchisate!"* "Dad might be lying. Uncle was just here. In fact, he hasn't left yet." "I was at you other uncle's house, I'm telling you!"

**te•aw ~ te•ew** *dtw.* now. **te•awmangmang ~ te•ewmangmang** just now, only now **te•awrawraw ~ te•ewrawraw** (1) nowadays *Ytykaria, te•ewrawrawmi gawido.* They do just like that, the girls of nowadays. (2) up till now, still *Te•ewrawraw morot nemchabatsyranggaba.* People are still extremely bad. *Angdo te•ewrawrawdo isykynmi dukna khyngkhyngdo nemai mu•phaaidonga.* I am still well, despite my continuing suffering. (3) not yet. *Angdo sawamigymyn te•ewrawrawdo re•engna man•khuchaaidonga.* As for me, I cannot yet go to school because of my illness.

**te•en** *dtw.* later today

**te•ew ~ te•aw** *dtw.* now. (see *te•aw*)

**tebyl** *n.* ART. (< English) table

**tek-** *v.* to tie

**telephon** *n.* ART. (< English) telephone

**-teng₁** *evsp.* still too *V "Balbo, atakna re•engwa. Balchachido tokni", madame. "Jora chaiwa." "Sala! mylteng te•euan jora chaina, roalan jokkhucha!"* "Tell me, why did you go away? If you don't tell me, I'll beat you", said the teacher. "I saw my lover." "Damn! You are still too small now to see your lover, you have not even finished primary school yet!"

**teng₂** *ideo.* the sound of falling coins. plink! clang! plunk! ching ching! *Thengthondo nokchina dongwachie rongtykchi "rong sa rong ni" noai tangkaaw khithyriokno, "teng, teng" noaimyng.* When Thengthon reached his house, he counted the money again in a rice barrel, saying "one rupee, two rupees"; clang! clang! it went.

**tengchypchyp-** *v.* to shine, to glitter. *Ue sona bi•chamchymaw nok ryphiokno. Nok ryphiwamyng kynsangdo te•ew ge•theng nokawan alaga morotdyrangdo tengchypchypai nukariokno.* He plastered his house with the golden flakes. After plastering his house now, other people saw how his house was shiny.

**-tengteng** *evsp.* still much too. *V Sala! Na•a myltengteng. Ang isykyn madam kam kha•phinok, nawmyl dong•phinok, nang•do mylaidongkua.* Damn! You are still much too small. I work as teacher, mind you, I am a marriageable girl, you are still small.

**tenki** *n.* ART. (< English) tank

**teraka** *dtw.* last year. *Na•a teraka ie songchi mu•wama?* Did you stay in this village last year?

**tha•gat-** *v.* to carry a child on one's back. *Sa• angchi tha•gatetbo.* Give me the baby to carry on my back.

**tha•gythyng** *n.* PLANT. species of vegetable

**tha•let-** *v.* to explain. *Ang nang•na Atong khu•chuk saina tha•letni.* I will explain to you how to write the Atong language.

**tha•makhu ~ tha•mykhu** *n.* PLANT. tobacco

**tha•malang ~ tha•mylang** *n.* PLANT. sweet potato

**tha•malangchak ~ tha•mylangchak** *n.* PLANT. leaf of the sweet potato plant

**thaba** *n.* ANIM. bedbug

**thaba•-** *vtr.* to make someone carry a child (in a cloth on the body). *Odek nang•chi thabani* I give you this baby to carry.

**thabai•-** *vtr.* to break

**thabarat-** *vtr.* to make someone feel ashamed

**thabisi** *n.* ART. amulet, antidote

**thagal•-** *v.* to drop. *Ang chabiaw bichiba thagal•ok.* I have dropped my keys somewhere.

=**thai**₁ ~ =**tyi** ~ =**syi** ~ =**si** *encl.* (mirative modality, indicates an emotion of surprise, annoyance or anger on the part of the speaker). (see =*syi*₂)
**thai•**₁ ~ **thai**₂ *n.* PLANT. fruit
**thai•**₂ *clf.* classifier for receptacles. *boiom thai• sa.* one jug. *khap thai• sa.* one cup.
**thai•gundai** ~ **thai•ma•thaigundai** *n.* PLANT. species of bright orange fruit that grows in creepers high in the jungle trees in the rainy season. The round fruits are about seven centimetres in diameter. The outside consists of a thick, uneven leathery rind while inside there are about eight sweet orange carpels each containing a smooth stone. (see Photo 82)
**thai•rokron** *n.* CORP. swollen lymph nodes in the arm pits
**thai•symphak** *n.* PLANT. species of plant
**thai•thuka** ~ **thaikhungka** *n.* PLANT. species of tree
**thai•thuka** *n.* PLANT. species of plant
**thaikhungka** ~ **thai•thuka** *n.* PLANT. species of tree
**thaikuka** *n.* PLANT. species of plant
**thajyri-** *v.* to make trouble
**thal-** *adj1.* clear, explicit *thalai hyn•-* to explain
**thali** *n.* ART. (< Indic) (1) plate (for eating)
**thali** *clf.* plateful
**tham**₁ *num. bound.* three
-**tham**₂ *evsp.* barely V
**thama**₁ *n.* ACT. divination
 **thama chai-** to see the future, to practice divination. *Kamalchi thama chaia.* At a priest's house, divination is practiced.
**thama-**₂ *vtr.* to make lost
-**thamak** *evsp.* V barely, V excessively
**thamat** ~ **thamot** *n.* PLANT. green plant that grows in the jungle and of which the side of the leaves, the young leaves, and the fruits cause irritation when touched
**than•khoana-** *v.* to gut lengthwise, to cut longitudinally
**thang-**₁ *v.* to fall down on
**thang-**₂ *v.* to throw away with great force, to come out with great force. *Matsa kherengwachido wa•chung byryida-rangdo thang•aaidonga Bandiba. Bandi kherengwachido wa•chu byryi wawa wawa thangasigaaidoknote.* When the tiger makes a great effort, he throws Bandi four bamboo lengths away. When Bandi makes a great effort, he throws the tiger whoooosh! four bamboo lengths away, I'm telling you!
=**thang**₃ *encl.* (possessive). (1) marks noun phrases to indicate that the referents are someone's own. *Amakdo nokthangchina dongangwachido na• ni•oknowa.* When the monkey reached his own house, he had no more fish. *Magachakmyng pherumynge bai•sigathangmaran nowa.* The fox and the deer are friends. (2) When used on kinship terms, this enclitic presumably indicates that the kin is biological or consanguineal rather than classificatory, although more fieldwork research is needed to confirm this hypothesis. *Ang abithang uchi mu•a.* My elder sister lives over there. (3) This enclitic occurs on verbs nominalised with the factitive enclitic =*wa*. *Anga nawathangtykyisa balaiwa.* I told it like I myself heard it. (4) Reduplication of this enclitic indicates that the referent is plural. *Gumukawan palyngchi jalgabadyrangaw jykthangthangaw jumuphynaakno.* They recollected all their husbands who had run away into the jungle.
**thang•chichat-** *v.* to drain
**thangguduk** *adv.* suddenly
**thanglang** *n.* PLANT. species of tree
**thangphytphyt-** *v.* to spatter, to splash. *Na•lam gudukwachie te•ewdo tyi thangpytpytaimyng jyksaiaiawan Nawengawmu Kumiribaawma• khamoknowa.* When the *na•lam* (species of fish) wiggled, water spattered on the married couple Naweng and Kumiri and burned them.
**thangtaw-** *v.* to squirt out
**thanthong-** *adj1.* blunt (of pointed things)
**thanyng** *n.* BODY. brain
**thap-**₁ *adj1.* savoury, umami
**thap**₂ *ideo.* hit! slap! a hitting sound

**thap-**₃ *v.* to beat, to beat up, to destroy. *"Phangnan ning nokaw thaparonga" noai balokno. Ytykyimyng myng• tham re•engokno mongma mathaiaw thapna.* "He always destroys out houses" he said. So then the three of them went on their way, to beat up the elephant.

**thaphu**₁ *n.* CORP. blister, sore

**thaphu-**₂ *v.* to blister, to be blistered or to have a sore. *Khu•chul thaphua.* My tongue has a sore.

**thapthap** *adv.* quickly

**thapyra** *n.* SUBST. ashes

**tharai-** *v.* to change, to exchange, to swap. *Ge•theng chola ga•chawana nygylmyngaw tharaiok.* He changed the bad shirt for a new one from the market.

**tharam** *n.* PLANT. species of tree

**tharamdaw•phit** *n.* PLANT. species of tree

**-tharap**₁ *evsp.* as soon as

**tharap-**₂ *v.* to catch up with, to be on time. *"Aia! Udo magachakdo khorate" noaimyng rykoknowa. Tharapna guduk takwachiba tarakai jalariano magachake.* "Hey, this deer is lame!" he said and chased after it. When he almost caught up with the deer, it ran away fast, the deer.

**thari-** *v.* (1) to prepare. (2) to arrange. *Uaw wa•phekgumuk, wa• pangumuk tharithylongaimusa, san sa dyngthangmancha thariaisa, kamalna rykaisa, wai khuruta.* Only after having prepared all that small bamboo, only after having nicely prepared the bamboo and the fire wood, only on an especially arranged day, you call for a priest, and perform the spirit incantation. (3) to repair. *Ang baik nosto dong•aimu, Turachi tharietwa.* Because my bike was damaged, I got it repaired in Tura.

**thasa-** *vtr.* to wake somebody up. *Thasabo uaw.* Wake him up!

**-that** *evsp.* V excessively

**thatthongthong-** *v.* to tear to pieces

**thaw-**₁ *adj1.* tasty

**thaw**₂ *ideo.* bang! pow! the sound of a gun firing

**thaw•jyw** *n.* PLANT. type of fruit

**thawal** *n.* CORP. scab

**the•met-** *v.* to fold

**the•myt** *n.* PLANT. cucumber (see Photo 59)

**thebajaw-** ~ **thebejaw-** *v.* to tickle. *Nang• angaw thebajawwa, ang bejawok.* You tickled me and I feel tickled.

**thek-**₁ *v.* to block off, to lock *Tala thekbo.* Lock the lock.

**thek-**₂ *v.* to insert. *Waiyr karenchi thekbo.* Put the wire into the electric socket.

**-thel** *evsp.* surely V. *Ichi rong•khalchi khen•ganangthelnaba ganang.* Here in the spaces under the stones there are river crabs for sure.

**thel•-** *v.* to tie

**them** *ideo.* pow! bang! the sound of a gunshot. *Myng• sagado them! kawokno.* The first one shot, pow! it is said.

**them•-** *v.* to fold up (clothes, blankets etc.)

**them•taw-** *v.* to roll up. *Longpen them•tawbo.* Roll up your pants.

**theng** *clf.* classifier for pieces food (especially meat)

**theng•** *n. autoclf.* knot. *Sa•mung sa•gyraichie kanwani chunwani kalai kharutchugabaan theng• chidokno.* In his childhood, he walked around half naked in a loincloth tied together by sixteen nots.

**-thengtheng** *evsp.* still too V *mylthengtheng* still too small

**thep** *clf.* classifier for heaps and small packets

**thep gaw** *ideo.* stamp, stamp! the sound of many animals stampeding through the forest

**thep thup** *ideo.* clog clog. tap tap tap.

**-theri** ~ **-thiri** *evsp.* V again. *Myng• sa them! kawoknotyi. khi•anchano. Sagaba: "Angdo nemkhalancha" nochido, aro kamalsang thama chaithiria.* If the sick person says: "I am not better", they will practice divination again at the place of another priest.

**thet-** *v.* to pull, to pull out. *Ang phakwalmyn• theta.* I pull out the hair in my armpit.

**thetchot-** *v.* to tear or break by pulling. *Ning kara thetchotok.* We broke our rope.

**thik**₁ ~ **kuythik** *n.* ANIM. lice (on dogs)

**thik**₂ *adv.* (< Indic) (1) exactly, precisely. *Kawbutungchi thik thokyrengaw man•okno.*

When he shot [the giant eagle] he got it exactly in the neck. (2) well cooked, well done. *Ja•ryt saw•ai, mantaw saw•ai, mai•chengmung na•lammung thiksa berengai sa•a.* We roast the chilli pepper, we roast the brinjal and cook it with *mai•cheng* and *na•lam* in a bamboo tube until it is well done, and then eat it.

**thik dong•- ~ dong-** *v.* (< Indic) correct. *Morot chanchichypai thik dongokodo, uchian rajaan uaw ajot nosawnaka.* Suppose someone gets it right, then the king will tell him *ajot*.

**thik kha•-** *v.* to fix a date and time. *Takrukna san somai thik kha•wachym.* They supposedly fixed a time and a day to fight.

**thik thak** *khjyks, adv.* (< Indic) exactly, precisely. *Thik thak phangnado chykhyw bajichi sa•aidonga.* He always eats at exactly nine o'clock. *Thik thak kawoknotyi ue sa•gyraie. Kawbutungchi thik thokyrengaw man•okno.* That child shot very precisely. When he shot [the giant eagle] he got it exactly in the neck.

**thikthik-** *v.* (< Indic) to instruct. *Uchie kyi•kutukaw pheruaw kaket sa•khalna Ketketa Bura thikthikokno.* Then, Ketketa Bura instructed his dog to devour the fox.

**thimimi-** *vtr.* to make someone smile

**thin-** *v.* to climb a rope that is either vertically hung or horizontally strung.

**kara thin-** to climb a rope

**thing thing thing** *ideo.* pling, pling, pling! clang, clang, clang! the sound of something falling made of metal

**thintaw-** *v.* to climb up

**-thiri₁ ~ -theri** *evsp.* V again. *Myng• sa them! kawoknotyi. khi•anchano. Aro myng• sa kawtherioknom.* Them! One of them, pow! took a shot. And another one shot again. Pow! (see also =*theri*)

**thiri₂** *n.* ART. bow (of a bow and arrow)

**thirikun•** *n.* ART. arrow (of bow and arrow)

**thiriphong** *n.* ART. part of an elephant trap

**-thirithiri** *evsp.* V again and again, V once again. *"Sala! ue sa•gyrai te•ewba rai•athiriok" noaimuna, gyniga phaw•-* *jongga. Uawbathiri ga•dakchichiaimuna thypsetthiriokno. Ytykma•chiba uba sa•gyraiba jumu kha•thirithirioknotyi. Ruchuthirithiriaimu rai•arongno.* "Damn! That child is coming back again", the second elder brother said. He again cut it into pieces and disposed of it. But the child reassembled once again!

**tho₁** *n.* FOOD. mustard oil

**tho-₂** *v.* to compare

**tho•ma** *n.* PERS. group

**tho•theng** *n.* SUPER. forest creature that looks like a person with his feet pointing backwards, so that it looks like you are following his footsteps in the direction he is going, while in fact he was walking the other way.

**thogi-** *v.* to betray, to cheat (on), to deceive. *Gawi angaw thogiok.* The girl has betrayed me. / The girl has cheated on me.

**thojekjek-** *v.* to shake a fixed object. *Ie panaw thojekjekchido thai• gal•khalni.* If you shake this tree, fruit will fall down.

**-thok** *evsp.* everybody V, all of them V. This suffix indicates that a group of individuals all do or undergo the same thing, but not necessarily together at the same place or at the same time. *Ytykyisa dyngthang-dyngthang songchina hapchina jalthokna ga•akok.* So they were all forced to run away to different villages and places. *Nang•tyme goi• byisyk man•phawa ie bylsie? noai syng•rukthoka.* "How many [baskets full of rice] did you get altogether this year?" everybody asks each other.

**thokbyrang** *adj2.* multi-coloured, many coloured

**thokbyrym** *adj2.* multi-coloured, many coloured

**thokthok** *adv.* precisely

**thol** *n.* COPR wrinkles

**thol•-** *v.* to lie, to tell lies. *"Anga Ketketa Bura nogaawan tyngkhucha" noaimyng Ketketa Burae phalthangawan pheruna thol•okno.* "I don't know this so called Ketketa Bura yet", lied Ketketa Bura about himself to the fox.

**thol•wami thol•-** *vpan.* to lie, to tell lies. *Umigymyn te•ew na•nang ha•gylsakchiba phorenchiba, Inglanchiba, Kalkata, Delhichi, Bambechiba, Badrichiba, Khychuchiba man•ramram khymcha, thol•wami thol•ancha.* That's why, now, we, in the whole world, in England and in Kolkata and in Delhi, and in Mumbai and in Badri and in Khychu, do not marry just anyone, I'm not lying.

**thol•am** *n.* PERS. liar

**thom** *clf.* classifier for things in heaps or piles. *jyw• thomsa.* a pile of flattened bamboo used to make mats

**thom•-** *v.* (1) to gather, to come together *Songgumuk thom•aimung ha•ba ha•ryn ha•rynaw sowalni.* The whole village comes together and they will divide the *ha•ba* plot by plot. (2) to make a heap. *Rong• thom•aidonga.* He's making a heap of stones.

**-thong•₁ ~ -thong** *evsp. V* in half (crosswise), *V* off. *Wa• bai•thongok.* The bamboo is broken in half (crosswise).

**thong•₂** *clf.* classifier for cylindrical objects. *betyri thong• byryi.* four batteries

**thong•₃** *n.* SHAPE. half which is the result of a cross section or a cut across the width or a crosscut

**thongmatchang** *n.* ANIM. snake with many colours like the rainbow, when someone sees this snake, they know that someone in their family will die

**thongthong** *adv.* straight

**thop₁** *ideo.* thok! thunk! a hitting sound:

**thop-₂** *v.* to gang up on. *"Watna bai iaw alagaaw!" noaimyng rykathokaidongano, Bandiaw thopna.* "Don't let this stranger go!" they said and they were chasing him, they wanted to gang up on Bandi. **thop... tung...** *coll, v.* to mob, to gather around. *Bawen sene bawenaimung thopangaidongano, tungangaidongano Ban diaw.* They gathered around Bandi in seven circles.

**thore-** *v.* to cry out the name of the *mahari* of one's enemy, for example *Kha Marak!*

**thorok-** *v.* to jump (down from/out of/into). *Alsia raja phe•pmyng thorokokno.* The lazy king jumped out of the banyan tree. *Magachakdo biskutaw tyisamchi tanaimyng chaw! thorokangokno.* H having put the biscuits by the side of the water, the deer splash! jumped into the water.

**thorom** *n.* ABSTR. (1) religion. *Ning songsyrekdo ning atongdo dakangdo mamyng thoromaw ni•wami somaichido waiaw mania.* We pagans, we the Atong, in the past, in times when there was no religion, we worshipped spirits. (2) denomination.

**thot thyng•thot** *adv.* to the last drop. *Thot thyng•thot takwachina dabat sykromaimyng khanetsigaaidongno.* He (Bandi) grasped her (So•re) and poured the liquor into her mouth to the last drop.

**thot-** *v.* to hit, to bump into something or against something. *Cha• rong•chi thotwa.* I hit my foot on a stone.

**thothak** *n.* QUANT/MSRE. a drop, classifier for drops *myktyi thothak ni* two tears/two teardrops *mykrensam thothak ni* two drops of eye medicine.

**thotphyret-** *v.* to smash by hitting against or on something

**thu•-** *v.* to put in one's mouth

**thuk-₁** *v.* to enclose with a fence.
**nol thuk-** to fence (off)

**thuk-₂** *v.* to weave a bamboo mat. *Damdyl thukaimu nok hama.* Having made the bamboo mats, they build the house.

**-thum** *evsp. V* on behalf of someone else, instead of someone else, instead. *"Atakna kyrewa, morotma•dyrangna? Hai ang ganang. Angan raja. Angan balthumni" nooknoro pherue.* Why are you afraid for these people? Come on! I am here. I am the king. I will speak on your behalf", said the fox. *Ue alsia rajae jykba myng• ni khymanoro. Ytykyimyng sa•naba jyk paithumna nangano, jywnaba jyk paina nangano.* That lazy king had married two wives. So, to eat, his wives had to carry him, and to sleep, his wives had to carry him too.

**thumu•-** *vtr.* to sit someone down (used for children)

**thunuk-** *v.* to show

**thup₁** *ideo.* thunk! slap! a beating sound

**thup₂** *n.* PLACE. nest.

**thup-₃** *v.* to nest. *Te•edo ue mongmaai rekchi thupai thupai mu•gabaaw phangnan phai•ai pha•ai sa•ronga.* Now this elephant always breaks and eats the place in which the bird nests in the banana tree. *Ne•kat pankhambaichi thupaidonga.* Bees are nesting in the top of a tree. (see Photo 108)

**thup-₄** *v.* thick (of fog or mist). *Guri thupa.* The fog is thick.

**thup thup** *ideo.* pitapat! the sound of footsteps

**thurung** *n.* ANIM. species of flying insect that comes out after the first rain and fills the air in big swarms like a mist. They come out at the same time as the species of ant called *hang•kyn*

**thut ~ thun** *clf.* classifier for big, lumpy things like stones, bricks, rocks, heads, hills, mountains and bars of soap. *rong thut tham.* three rocks. *ha•byri thut sene.* seven hills, mountains. *sabun thut sa.* one bar of soap. *dykym thut sa.* one head

**thy•yk-** *v.* to have the hiccups

**thyi-** *v.* to die

**thyi•** *n.* CORP. blood

**thyikhop** *n.* ART. dried fruit in which water is stored for consumption

**thyiwami** *n.* ACT. death

**thyk-** *v.* to be fixed sideways

**-thyl** *evsp. V* and avoid, *V* ahead

**thyl•₁** *postp.* (takes genitive-marked complement) up to, until (spatial). *Ie cha·masangmi wai khurutchido ue hyisangmiaw Banggladesmi thyl• Kongosmi jaria ha•gyrsakgumukawan myngani.* When he summons the downstream spirit, the [priest] will call upon the influence of all those faraway places up till Bangladesh and the area of Kongos, all of them.

**thyl•-₂** *v.* to go very far *Ga•thyngaimuna thyl•angok.* Because I kicked it, it went very far.

**thylampak ~ thylapak** *n.* BODY. tongue

**thylamphak sul-** to stick out one's tongue

**-thylong** *evsp. V* nicely. *Uaw wa•phekgumuk, wa• pangumuk tharithylongaimusa, san sa dyngthangmancha thariaisa, kamalna rykaisa, wai khuruta.* Having prepared those small bamboo stalks, having nicely prepared all the bamboo and wood, one special day you need to prepare that, search for a priest and perform the incantation.

**thym-₁** *v.* to lie in ambush, to lie hidden, to hide so that you can still watch what is happening.

**thym-₂** *v.* to take revenge

**thymbylong** *adj2.* to have a hole in it, damaged (of roads, bridges and wooden planks)

**thymyn** *v.* to ripen. *Panchung thymytetbo.* Keep the jackfruit so that it can ripen.

**thymyt-** *v.* to put out (fire), to switch off, to extinguish

**-thyn** *evsp.* intensifier suffix. *Kyn jungthynai mu•arokno.* She was sitting with her back turned away from him

**-thyng₁** *evsp.* intensifier suffix. *V* so much. *Khu•thipna harataimyng ka•wakai mu•gabamyng manamthynggabana sotmai dumaai pang•phinokno.* Because he was so lazy that he did not want to close his mouth, and because he sat there with his mouth open, which stank so much, flies swarmed into it.

**thyng-₂** *v.* to kick

**thyngel-** *v.* to tilt

**thyngpyret-** *v.* to kick

**-thyngthyng** *evsp.* intensifier suffix. *V* so much, *V* continuously, *V* all the time. *Aiaw! Thajyrithyngthyng!.* Oww! You make trouble all the time!

**thyp-** *v.* to throw (sidearm), to throw into. *"Ha• kerengaw sa•bo!"* noai thyprateto-kno. "Take this, eat the bones!" he said and threw them down. *Ge•theng nokhapchina tangkaaw thypai thypai khiaidongano, "Rong sa, rong ni, rong tham" noaimyng.* He is counting the money, throwing it on the ground saying: "One rupee, two rupees, three rupees". *Ga•dakchichiaimuna sings-*

*ingkholongsang thypsetyi tanangokno.* Having cut him into pieces, they threw them in a deep hole in the ground, and left them there.

**thyrgyryw** *v.* to shake something large and unmovable

**thyw•-** *adj1.* deep. *Dabatwari thyw•ama?* Is Dabatwari deep?

**thywkhong** *adj2.* globular, protruding, bulging

**Tibet** *n.* PLACE. Tibet

**tibi** *n.* ART. television

**ticher ~ tichyr** *n.* PERS. teacher

**tiiititi** *interj.* interjection to call a chicken

**tika** *n.* ART. payment

**tiktik-** *v.* to make last. *Tangkaawba tiktika. Na•nangachido tiktikcha. Sa•bongbong ryngbongbong.* They make their money last. As for us, we don't make it last. We are gluttons and drunkards.

**tin** *n.* ART. (< English) corrugated iron sheet used to make roofs. *tin kap sa* one sheet of corrugated iron (see Photo 9)

**tintyrin** *n.* PLANT. tamarind

**tiup** *n.* ART. (< English) tube

**=to ~ =ta** *encl.* (emphatic imperative modality, used when the speaker is impatient, or wants to beseech the addressee for something). (see *=ta*)

**toilet ~ toilyt** *n.* PLACE. (< English) toilet

**tok-** *v.* (1) to beat, to beat up. *Kyi•aw tokbo!* Beat the dog! *Taw•reksyrup mang sa ge•thengmyng thup phangnan mongma phai•ai sa•rongwana, mongma mathaiaw tokna re•engaidongano.* Because the nest of a banana bird always gets broken and eaten by an elephant, it is on its way to beat up the bachelor elephant. (2) to crush, to grind. *Amakmyng rukpekmyng: "Hai bai•siga, na•ru tokna" noaidonganowa, myng• ni.* The monkey and the frog, both of them, said: "Come on, friend, let's grind some fish poison". (3) to play an instrument.

**tokbaw** *n.* CORP. goitre ~ goiter

**tokdepdep-** *v.* to crush, to grind. *Sambanggyri akaiokno, tokdepdepaimu pha•atokno.* He plucked *sambanggyri,* crushed it and put it on the wound.

**tokdyl** *n.* BODY. vocal cords

**tokgepgep-** *v.* to beat to a pulp

**tokkhynyng-** *v.* to smash into pieces

**tokkhyphu ~ tokyphu ~ tokybu** *n.* BODY. gullet, oesophagus, throat

**tokkyreng** *n.* BODY. neck

**tokphyrong-** *v.* to take a powdered substance in the palm of one hand and softly tap on it with the other hand

**tokpyret-** *v.* to crush by hitting

**tokset-**$_1$ *v.* to cough, to have a cold

**tokset-**$_2$ *v.* to pull loose

**tokta** *n.* PLANT. type of wood

**toktai-** *v.* to hang oneself

**tokthining ~ tokthynyng** *n.* BODY. neck

**tokthong•-** *v.* to smash in half

**tokthynyng ~ tokthining** *n.* BODY. neck

**toktokkylek ~ kokkylek** *n.* PLANT. species of plant (see Photo 65)

**tong-** *v.* to copulate, to fuck

**tota** *n.* ART. (Assamese or Bengali) plank. *tota kap sa* one plank *tota khaw• sa* one plank

**totyp** *adj2.* bent. *Bandi mu•etwachian dakhamba ha•china chaksi ni dong•na guduk totyp totyp takaidonganote.* When Bandi sits on the *dakham* it bent almost completely to but two fingers from the ground.

**tu•- ~ ty•-** *v.* to feed (by putting food or drink into the mouth). *Sa•gyraina mai tu•wa.* I fed the child rice.

**tuk-** *v.* overgrown, dense (of vegetation). *Ram tuka.* The road is overgrown. *Palyng tuka.* The jungle is dense.

**tum** *clf.* classifier for places and packets. *Hap tumbyisyk?* How many places?

**tun- ~ tyn-** *v.* to lead, to guide to lead, to guide. *Na•a ang ma•su mang rajasaaw tynangsegabone.* You lead my hundred cows away, OK?

**tung-**$_1$ *adj1.* hot, warm

**tung**$_2$ *clf.* classifier for objects like bridges. *dolong tung ni.* two bridges

**tung-**$_3$ *v.* to gather

**thop... tung...** *coll, v.* to mob, to gather around. *Bawen sene bawenaimung thopangaidongano, tungangaidongano Bandiaw.* They gathered around Bandi in seven circles.

**tungbul-** *v.* to have a warm body (not of fever)

**tungkyryi ~ tyngkyryi** *n.* TIME. hot season

**tupi** *n.* ART. (< Indic) cap, hat

**ty•- ~ tu•-** *v.* (see *tu•-*)

**tyi** *n.* SUBST/CORP. water, (fruit) juice, sweat
  **tyi hung-** *vpan.* to swim. *Na•a tyi hungna sapama?* Do you know how to swim?
  **tyi karan-** *vpan.* thirsty. *Aia! Tyiba karanok bai•siga angdo, ang tyi cho•sa ryngna.* Jeez! I am certainly thirsty, friend. I want to drink some water.

**=tyi ~ =thai ~ =syi ~ =si** *encl.* (mirative modality, indicates an emotion of surprise, annoyance or anger on the part of the speaker). (see *=syi₂*)

**tyi•₁** *n.* BODY egg

**tyi•-₂** *v.* to lay an egg

**tyibal** *n.* GEO. wave

**tyibasal** *n.* GEO. whirlpool

**tyibek** *n.* ART. traditional bottle used to drink water out of, and made of a dried vegetable also called *tyibek*

**tyichabakram** *n.* GEO. waterfall, cascade

**tyichang** *n.* PLACE. island

**tyichaw- ~ tyichaw•** *v.* to drown. *Ie morot tyi hungna sapchaaimu tyichaw•wa.* Because this person did not know how to swim, he drowned.

**tyigat** *n.* PLACE. place in a river or at the end of a water pipe where the people get drinking water, take a bath and wash their clothes and dishes. (see Photo 43)

**tyigum** *n.* ART. waterpot, water container usually made of metal and shaped like a big vase used to store water in the kitchen. Its place in the house is in the *tyinok*. (see Photo 43)

**Tyihanggal** *n.* PLACE. Stream on the way to Balphakram. When the spirit of a dead person takes his bath in that river, he forgets everything about his life. This stream is also called *Chidymak* or *Tyitykmak*.

**tyikhal** *n.* GEO. river. Clf. *chol. tyikhal chol ni.* two rivers

**tyikhop** *n.* ART. dried fruit used for the storage of water

**tyikhyrep-** *v.* to be wrinkled because of being in the water for a long time. *ang chaksi tyikhyrepok* my fingers are wrinkled because they were in the water for a long time *chak tyikhyrepok* dry leaves

**tyimong** *n.* GEO. main river

**tyimuk** *n.* GEO. source, spring (of a stream)

**tyinala** *n.* PLANT. algae

**tyinok** *n.* ART. place in the kitchen where the water pots (*tyigum*) and other utensils like plates, cups and glasses are stored. (see Photo 16)

**tyiphek** *n.* GEO. tributary river, the smaller one of two rivers that flow together

**tyiribok** *n.* PLANT. species of creeper poisonous to cows

**tyis ~ tys ~ yis ~ hyits ~ hys ~ hyis** *interj.* Ugh! Yikes! What the...?! Expresses disapproval or indignation.

**tyisam** *n.* PLACE. river bank, water's edge.

**tyisang** *n.* CORP. piss, urine. *tyisang rai•khuna* I need to piss

**tyisi-** *adj1.* wet. *Magachakmi myn•do tyisiwachian miniksuru takjolarianoro.* When the deer's fur is wet, it just quickly gets flat-haired.

**tyisuk** *n.* CORP. pneumonia.
  **tyisuk ra•-** to have a pneumonia

**tyisurung** *n.* GEO. rainwater that streams over the ground

**tyithai** *n.* ART. water scoop made of a hollow, dried gourd.

**Tyitykmak** *n.* PLACE. Stream on the way to Balphakram. When the spirit of a dead person takes his bath in that river, he forgets everything about his life. This stream is also called *Tyihanggal* or *Chidymak*.

**tyk** *n.* ART. pot, barrel

**tykha** *n.* SUBST. white clay not useful to make pots

**tykhal** *n.* PERS. person who goes around eating in lots of other people houses

**tyksyl ~ dyksyl** *n.* ART. metal pot for cooking rice (see Photo 43)

**=tykyi ~ =takai** *encl.* (perlative/similative). (1) perlative interpretation. *Na•nang Bagmaratykyi re•engni.* We will go via/through Baghmara. *Na•nang itykyi ma utykyi re•engni?* Shall we go this way or that? (2) similative interpretation. *Phulistykyi nukramphinokno bunduk paigana.* They inadvertently looked like the police because they were carrying guns.

**tykyw** *n.* ART. water pot

**tym** *clf.* classifier for fields. *ha•ba tym ni.* two dry rice and vegetable fields on the slope of a mountain

**-tym** *sfx.* (personal pronoun plural). used to form the second person plural exclusive personal pronoun from the second person singular: *nang•tym* you (plural exclusive). Also used to form the third person plural from the proximal and distal demonstrative: *itym, utym* 'they'.

**tyn₁- ~ tun-** *v.* to lead, to guide. *Na•a ang ma•su mang rajasaaw tynangsegabone.* You lead my hundred cows away, OK?

**-tyn₂** *evsp.* lead/bring to *V*, be the leader of the action. *Sipaidyrangaw jaltynokno.* He led his soldiers in running away. *Chigachakchi Dibangkongdang Umangchalmangsa, mongmaaw so•otai matsaaw so•otai mu•tynwano.* At Chigachak, Dibangkongdang Umangchalmang, having killed the elephants and tigers, stayed as the leader.

**tyng₁** *clf.* classifier for long thin objects like ropes, chains, hairs etc. *kara tyng sa.* one rope

**tyng-₂** *v.* (1) to know (a fact or person). *Ketketa Bura? Anga Ketketa Bura nogaawan tyngkhucha.* Ketketa Bura? I don't know this so called Ketketa Bura yet. *Ang ie khata dakangdo tyngchachym, te•ewdo nemen tyngok.* I did not know this word before but now I know it well. (2) to understand. *Angmi balgaaw tyngama?* Do you understand what I say? (3) to recognise. *Nang•baletgaba morote atongtykyi angawe tyng-sawnaka?* How will the person you talk about certainly recognise me?

**-tyng₃** *evsp.* intensifier suffix. *Ge•theng pyi•tynggaba pandala wa•cheksi pang•aimyng morotaw nukphinanchak, pan wa•balwana balphakgabatykyi nuksawphinokno.* Because of the many things he had held onto, twigs and bamboo, you could not see the person at all anymore, because of the blown-away plants, [he] totally looked like something that had been blown away [by the wind].

**tyngcheng-** *v.* to know first, to discover

**tyngen** *adv.* very. *Ue raja kam kha•naba tyngen haratachym.* That king was supposedly very reluctant to do work.

**tyngetwami ~ tyngetwamyng** *n.* ART. announcement, notice

**tyngkarang** *adv.* in one go

**tyngkhalang** *adv.* with one blow

**tyngkyryi ~ tungkyryi** *n.* TIME. hot season

**-tyngtang** *evsp.* *V* all over the place

**tyngtet-** *v.* to hang someone. *Tyng•tet kha•ai tan•!* Hang him up! / Kill him by hanging him!

**tyngwami** *n.* ABSTR. knowledge, understanding

**tyret-** *v.* to bathe someone else

**tyru- ~ tyiru- ~ tyiryw-** *v.* to bathe, to take a bath, to wash oneself

**tys ~ tyis ~ yis ~ hyits ~ hys ~ hyis** *interj.* Ugh! What the...?! Expresses disapproval or indignation. *Tys sala!* Damn!

**tyt-** *v.* to pour. *Ue tyigummi tyi tytbo dipotchi.* Pour water from that *tyigum* into the teapot.

# U

**u•ching ~ ukching** *n.* ANIM. leech

**u•chingrawri ~ ukchingrawri ~ batro** *n.* ANIM. species of brown leech that lives in the soil and mud

**uchi₁** *dem.* there

**uchi₂** *disccon.* then

**uchiba** *disccon.* but then. *Una myng• sagaba sa•banthai sa•banthai myng• sagaba bychymokno, uchiba patangphaariok, dang•angphaariokno.* Then one son pulled the other out from the water, but then they just crossed and they all just drowned.

**ue ~ u-** *dem.* distal demonstrative. (1) modifier function. that, those. *Ue rong•khalaw Durakhal myngwa.* That cave is called *Durakhal*. *Ue song dam niaw Songma Songgyni Khychu Badri myngwa.* Those two villages were called Songma Songgyni Khychu Badri. (2) pronoun function. that (one), those (ones). *Ue Symsangtyikhal.* That is the Symsang River. *Ue Rongsumyng komyla.* Those are oranges from Rongsu. *Bie baik ra•na nemkhalni? Ie ma ue?* Which bike is better to buy? This one, or that one? *Bie biskut ra•ni? Ie ma ue?* Which biscuits shall I buy? These ones, or those ones? (3) he, she, it Only with this meaning is there a plural form *utym* 'they'.

**umi ~ umido ~ umisa ~ umyng ~ umung ~ umyngdo ~ umyngsa** *disccon.* then. *Phasgaba ha•haw•chenga. Umungsa ha•haw•aimungsa wa•cham tan•a.* First we clear the jungle. Then, having cleared the jungle, we cut the old rice stalks.

**umigymynchi ~ umynggymynchi** *disccon* for that reason, therefore, because of that. *Umigymynchi iawdo Dabatwari myngwano.* That's why it's called Dabatwari.

**umyng ~ umung~ umyngdo ~ umyngsa ~ umi ~ umido ~ umisa** *disccon.* then. (see *umi*)

**una** *disccon.* therefore, then. *Jetakai patangchiba rung bytrongrengangariano, sangkhynian. Unasa rung chawna dakangan ytykyi rung dykymaw ga•tyngaimuna "kha Dawa!, kha Dawa!" noaimusa rung chawaimu patronganoro.* Whatever you do whenever you cross, the boat will spin. That is the water dragon. Therefore, before you cross by boat, because you stamp on the head of the boat saying "Kha Dawa! Kha Dawa!" and then having gone by boat, you usually cross.

**uph** *ideo.* oof!

**utyk udong** *expr.* let it be

**utykwachido** *expr.* when it was like that. *Dakangmi pichammi kamdyrangdo gum ukan songsyrek dong•butungchido bylongen han•senga. Dukba ni•wa. Chywgyn rynga, wai khuruta, kana teka. Utykwachido gumukan, bai•damdo, man•ai sa•a kamai sa•a, pungchi phaka, rang ra•a, ryk ra•a.* As for how things were in the past, when everybody practiced animism, we were very happy. There was no sorrow. We celebrated *chywgyn*, performed spirit incantations, there was food and drink. When it was like that, everyone, some people, where rich, had full granaries, got brass drums and necklaces.

**utykyi** *dem.* like that. *Una sa•banthaigababa wa•gaba utykyi noetwana, noksangsamsang khudal paiaimyng, bulai hyn•etokno.* Thereupon, because the father had spoken like that to his son, he went into the house carrying a spade, dug it up, and gave it. *Ha•aw pang•ai sa•na man•chano ha•nyng• khan•syruiba. Phalthangchi wa•churek sa•na man•arinoa. Na•nangba utykyi.* Earthworms cannot eat a lot of soil. They can only eat as much lies within their capacity. We are also like that.

**utym ~ ytym** *ppron.* third person plural personal pronoun, they, them, their. *Utymdo sima ganang, waiba, nang•mi jol thokthoktykyi.* As for them, spirits also have boundaries, just like our places.

# W

**wa**$_1$ *n.* BODY. tooth, tusk (of elephant). Clf. *khaw•*. *wa khaw ni.* two teeth, two tusks

**wa-**$_2$ *vs1.* to rain. *Rang waaidok.* It's raining. This verb can only take the noun *rang* as its argument.

**=wa**$_3$ *encl.* (factitive). (1) on independent-clause predicates. *Angmyng bimung Nykseng myngwa.* My name is Nykseng. *Bisang re•engwa?* Where have you been? *Uchian rajado khusi dong•aimyng: "Na•a* *atongaw nangni?" nookno. Uchie Bildo: "Angna mamyngawan nangcha<u>wa</u>."* Then the king was happy. "Do you need anything?" he asked? Then Bil said: "I don't need anything." *"Nang•mi tangka angna hyn•bo!" "yhy• hyn•cha<u>wa</u>!""* Give me your money!" "No, I won't give it!" (2) on the predicates of Complement Clauses. *Chalman<u>wa</u> machotwamungsa mai kai•chenga.* Only after the sowing is finished, rice is planted. *Ningba ytykyi tak<u>wa</u> ga•nima?* Will it be good if we do it like this? *Mai sa•<u>wa</u> jamkhucha.* I have not finished eating rice yet. *Morot dykymchi tok<u>wa</u> ga•cha.* Hitting a person on the head is not good. *Mungma angawa ga•phynekni<u>wa</u>na kyryia.* I'm afraid that an elephant will stamp me to death. (3) on Type 1 adjectives. *Ooo! Mykgythaldo dong•cha jywmangsama, ma nang• bimang syl<u>wa</u>ai.* (Wilseng S. Marak) Ooo! Is it real or is it a dream, but your body is really beautiful! (4) on adjunct-clause predicates before semantic-role enclitic. (a) Reason Clause. *Sa•gyrai<u>wa</u>na kymchawa.* Because she is a child, I will not marry her. (b) Temporal Location Clause. *Turasang re•eng<u>wa</u>chi angna topi ra•bone.* When you go to Tura, buy me a hat, okay? (c) indicating simultaneous events. *Uchian anga nang•aw nukjyryng<u>wa</u>chian nang•na kha•galwa dang•ok.* Then, when I saw you every day, I started loving you. (d) Similative Clause. *Ian maja nang•bal<u>wa</u>tykyi de•theng gam jamok.* This is like you said in the past, his wealth is finished. (5) indicating an action nominalisation. *Thogidugagabaaw ra•chie ra•chido thyi<u>wa</u>myng ga•akanowa.* If you believe people who lie to much, you will die. *Ian ha•ba haw•<u>wa</u>mung itihas machotaido bytwagumuk khairataisa machota.* Finishing this story about the cultivation of the rice field, they finish by carrying the whole harvest down. *Umigymynchi anga nang•aw khymana daiaido aset<u>wa</u>an nemkhalnaka.* Therefore it will certainly be much better to divorce you than to be married to you.

(6) indicating object nominalisation. *Ian ha•ba haw•<u>wa</u>mung itihas machotaido byt<u>wa</u>gumuk khairataisa machota.* Finishing this story about the cultivation of the rice field, they finish by carrying the whole harvest down. P*ankhambaichi pywaimuna ytykyi sa•khuchano sa•<u>wa</u>awdo.* Having flown up in the treetop like this, he had not yet eaten any food. (7) marking complement clauses of postpositions. *Jaraw jaraw ge•theng sok<u>wa</u> dabatdo sakchykaidongano pheruba.* For a long time, until he did not hold out any longer, he was holding out as long as he could, the fox.

**wa•**₁ *n.* KIN. *drel, d, ref.* biological father (addressed as *awa* or *baba*)

**wa•**₂ *n.* PLANT. bamboo. *wa• dot sa* one culm of bamboo *wa•khaw• sa* one long half of a bamboo *morot wa• sa•gaba* a strong and tough person

**pan... wa•...** *coll, n.* plants, vegetation, plants and trees (see *pan*)

**wa• sa•gaba** *expr.* though (of persons). *Ge•thengdo morot wa• sa•agaba.* He is a tough person. (Literally 'He is a person who eats bamboo').

**wa•byrek** *n.* ART. horizontal beam under the length of the roof underneath the *wa•khaw*. Together, the *wa•byrek* and *wa•khaw* form part of the structure of the roof of a house that keeps whatever covers the roof in its place. (see Photo 11)

**wa•cham** *n.* PLANT. stubble, old rice stalk which is left over after harvesting the rice

**wa•chan** *n.* PLANT. species of fungus that glows in the dark

**wa•cheksi** *n.* PLANT. bamboo stalk or twig

**wa•chu ~ wa•chun** *n.* MSRE. the length of a bamboo pole

**wa•churek** *n.* ABSTR. capacity, capability

**wa•chyrik-** *v.* to be startled. *Gari horn kha•wanasa wa•chyrikok, ge•thenge.* Because the car blew its horn he was startled.

**wa•da** *n.* PLANT. species of bamboo of which each culm comes out of the ground individually instead of in a bush. It is used in the construction of houses. (see Photo 68)

**wa•daweng** *n.* GEO. star sign of three stars in a straight line

**wa•dokolong** *n.* ART. water pipe made of bamboo

**wa•gaba** *n.* PERS. someone else's biological father

**wa•gat** *n.* ART. bamboo shoulder yoke

**wa•gatram** *n.* BODY. shoulder, literally: 'the place where you put the bamboo shoulder yoke'

**wa•gydok** *n.* ART. water pipe made of bamboo

**wa•gylok** *n.* ART. a cut-off piece of bamboo

**wa•jong** *n.* PLANT. species of bamboo

**wa•jongmagal ~ wa•jongmagar** *n.* PLANT. species of plant (see Photo 79)

**wa•kai** *n.* PLANT. species of big bamboo

**wa•khal** *n.* ANIM. grasshopper-like insect

**wa•khaw₁ ~ wa•khu** *n.* ART. purlin, horizontal beam on the outside over the length of the roof, that has the *wa•byrek* as its inside counterpart. Together, the *wa•khaw* and *wa•byrek* form part off the structure of the roof that keeps whatever covers the roof in its place Other speakers call a purlin a *khyntyri*. (see Photo 11)

**wa•khaw₂** *n.* ART. one long half of a bamboo split lengthwise. *wa•khaw sa* one long half of a bamboo

**wa•khel-** *v.* to be stuck in one's teeth

**wa•khelsep-** *v.* to be stuck in one's teeth

**wa•kholchik-** *v.* to show one's teeth

**wa•khu ~ kudal ~ khudal** *n.* ART. chopper (tool used in agriculture) (see Photo 26)

**wa•khu** *n.* ART see *wa•khaw₁*

**wa•khyntha** *n.* PLANT. species of bamboo

**wa•lai** *n.* PLANT. species of bamboo that grows in the jungle, is very long and thin and can be more than one hundred joints long

**wa•lung** *n.* PLACE. place where stuff is burnt

**wa•maran** *n.* KIN. set. a father and his child (son or daughter)

**wa•mychym** *n.* ANIM. fire fly

**wa•phek** *n.* PLANT. bamboo, branch, small bit of bamboo

**wa•phuk** *n.* ART. white half of a strip of bamboo used to make rope

**wa•puk** *n.* PLANT. the inside of a bamboo tube

**wa•rap** *n.* ART. bamboo strip that runs underneath the bamboo floor of a house, and has *engsyri* as its counterpart on top of the floor to keep the bamboo strips that make up the floor in place

**wa•ri ~ wa•ryi** *n.* PERS. child who lost its father

**wa•rok** *n.* ART. bamboo raft. Different verbs used to say that you are going by *wa•rok* are: *phakweng-, gebeng-, rongret-, geching-* (see Photo 124)

**wa•rung** *n.* PLANT. young bamboo

**wa•sung** *n.* ART. bamboo tube used as container, and used to cook *bering* in. Clf. *sung. wa•sung sung ni.* two bamboo tubes

**wa•syl** *n.* ART (1) green half of a strip of bamboo used to make rope (2) the outside of a bamboo tube

**wa•tana** *n.* ART. part of an elephant trap

**wa•thai** *n.* PLANT. species of bamboo smaller than *wa•thyrai*

**wa•thaibok** *n.* PLANT. species of bamboo that is white from the ground a little up

**wa•thok** *n.* PLANT. hollow bamboo stick.

**wa•thyrai** *n.* PLANT. species of bamboo that grows one by one, not in a bush

**wa•tyng** *n.* ART. bamboo strip used to make baskets, and other woven utensils as well as rope. Clf. *tyng/jyw•. wa•tyng tyng tham / wa•tyng jyw• tham.* three bamboo strips. (see Photo 46)

**wach** *n.* ART. watch

**wachyw** *n.* BODY. incisors (the four front teeth used for biting)

**wadi•** *n.* CORP. plaque

**wagyleng ~ wagylok** *n.* PERS. person who is missing one or more teeth

**wai₁** *n.* SUPER. spirit

**wai khurut-** *vpan.* to perform an incantation. *Ie cha•masangmi wai khurutchido ue hyisangmiaw Banggladesmi thyl•Kongosmi jaria ha•gyrsakgumukawan myngani.* When you perform an incantation for the Downstream Spirit, you call upon the whole area up to Bangladesh way over there, upon the influence of Kongos, you call upon the whole area, everything.

**wai-**₂ *v.* to plough. *Bydyi myng• sa ma•su-sang ha•pal waiaidongano.* An old man is ploughing the field with a cow.

**wai-**₃ *v.* to return, to go/come back. *Sikharba kha•chypanchakno, ytykthyngai somai jamchypaimuna jyksang sa•sang waiangokno.* The hunting had failed and having done all this, having wasted time, he went back to his wife and children.

**wai•-** *v.* to scoop (of liquid). *Angna chyw gylaschi wai•bo.* Scoop some liquor into the glass for me. *Tyikhalmi tyi wai•aimu ge•theng ryngok.* He scooped out some water from the river and drank it. (see Photo 2)

**wai•cheng** *n.* ART. longest type of knife

**wai•seng** *n.* ART. very big knife traditionally used to kill tigers and men. Clf. *mang. wai•seng mang sa* one *wai•seng*

**Waimong ~ Waimongha•byri ~ Waimong Ha•bri** *n.* PLACE. most prominent and highest hill in the Atong-speaking area. The Standard-Garo name for this hill is Chutmang A•bri. (see Photo 24 and Photo 25)

**waiphin**₁ *n.* ACT. return. *Waisa waiphin lak sa nanga.* To go and come back you need one hundred thousand rupees.

**waiphin-**₂ *v.* to go back, to return. *Nang•mi sa•banthai waiphinaakte.* Your son has returned, I'm telling you.

**waisa** *n.* ACT. the going (to somewhere). *Waisa waiphin lak sa nanga.* To go and come back you need one hundred thousand rupees.

**waiset-** *v.* to drain a little bit of water, to scoop out water. *Chamussang waisetbo.* Scoop it out with a spoon.

**waiyr** *n.* ART. (< English) wire

**wak** *n.* ANIM. pig, pork

**wakam** *n.* BODY. molar (tooth)

**wakeng** *n.* ART. axe

**wakhol ~ wakholong** *n.* PERS. person who is missing one or more teeth

**waknok** *n.* ANIM. domestic pig

**waknol** *n.* ART. pigsty

**wakpalyng** *n.* ANIM. wild pig

**wakpuk** *n.* BODY. the innards of a pig

**wal**₁ *tw.* night.

**wal-**₂ *vs1.* to be night *Walangaidok.* It's getting night. *San walok.* It has become night. (Literally: The day has become night.)

**wal•** *n.* GEO. fire, torch. *Nokha•palchi wal• chakbo.* Light a fire outside the house. *Na•nang walchi wal• netaimu khen• rawna re•engni.* We will light a torch and go catch river crabs at night.

**wal• chak-** to make fire. *Kumirian wal• chakthiriaimyng te•do na•lam garanawan rymthiriaidongano.* Kumiri rekindled the fire, and cooked the dried fish again.

**wal• kham-** *v.* to burn down. *nok wal• khamgabamyng hanggal ha•thapyra* ashes and charcoal from a burnt-down house

**wal•bek** *n.* FOOD. burnt curry. *Mai sa•naan ja•bek wal•bek thawchawanaan.* As far as eating rice is concerned, the curry was burnt, it was not tasty.

**wal•byt** *n.* ART. match (to make fire)

**wal•cham** *n.* ART. bamboo torch

**wal•di•** *n.* SUBST. ambers, glowing pieces of burnt wood

**wal•khu**₁ *n.* SUBST. smoke

**wal•khu-**₂ *v.* to produce smoke

**wal•kungki** *n.* SUBST. black ashes

**wal•mak** *n.* PLACE. future rice field where the jungle has just been burnt

**wal•sam** *n.* PLACE. fireside

**wal•tum•** *n.* fire that is burned outside the house during winter to sit around and keep warm.

**wala-** *v.* to arrive at night, to be late so that it is already night. *"Ma•, baba, atykyimu walawa?" nookno amakaw, amakmi sa•dyrange.* "But daddy, why are you so late? It is already night", the monkey's children said.

**walchak-** *v.* to kindle the fire with one's breath by blowing

**walmykrak-** *v.* to hold a vigil or watch over the body of a dead person.

**walseng-** *v.* to stay awake all night

**walsymsym** *n.* GEO. twilight, dusk. *Walsymsym takok.* It's dusk.

**wang** *n.* KIN. *drel, d, ref.*(addressed as *awang*) (1) uncle: father's younger brother (2) stepfather (3) the inverse relation of *biawthang*: wife's mother's brother

**wang•-₁** *v.* to bite a bit out of something, to take a mouthful. *"Jebadong anga takruk-syrangarinaka" noaimyng, matsami cha•phungaw wang•joloknoaro, khabakaimyng.* "Somehow I will just fight to the end", he said, and he quickly bites the tiger on the thigh, having grasped him tightly.

**wang•-₂** *v.* to turn, to wind

**wang•kok-** *v.* to eat without using one's hands, with one's mouth. *Ha•aw pang•ai sa•na man•chano ha•nyng• khan•syruiba. Na•nangba utykyi. Ha•awe wang•kokai sa•ancha.* Earthworms cannot eat a lot of soil. We are also like that. We don't eat mouthfuls of soil.

**wanggaba** *n.* KIN. *ref.* (1) the inverse relation of *biawthang*: wife's mother's brother (2) the derelationalised form of *wang* (someone else's *wang*)

**wanggala** *n.* ACT. (< Garo) biggest Garo festival

**wangmaran** *n.* KIN. *set.* my *wang* (father's younger brother) and his elder brother's child

**wara-** *v.* to defend (oneself), to shield (oner-self), to protect (oneself). *"Bawbyl chambyl nochiba, kha•sinai re•engna bai. Waras-akangaribo" noaimu, baletaidongano.* If there are any enemies, don't go slow. Defend yourself well", he said, and talked.

**warem** *n.* SUBST. rust. *Darai warem sa•ak.* The sword has rusted.

**wari** *n.* PLACE. deep place in the river where you can swim or take a bath. *Tyi ga•gaba wari thyw•gaba tyisamchi hap sylgabachi myng• ni bai•sigathangmaran "chang tyry-wchengnaka" noaidongano.* At the waterside of a place in the river where there was nice and deep water, in a beautiful place, the two friends are arguing about who will take a bath first.

**waribul-** *v.* to fish at the festival of *waribula*

**waribula** *n.* ACT. the Siju fishing festival in the Symsang river at Dabatwari

**warung** *n.* PLANT. young or immature bamboo

**wasam** *n.* PLANT. species of plant

**wat-₁** *v.* (1) to send away, to banish, to let get away, to get rid of. *Ge•thengmi nokaw ge•theng watok, ytykyimu dynthang nokchi mu•arok.* His family sent him away, so now he is staying in another house. *Ramchi hampyi na•nangdo watchaka ge•thengawdo, sala! Ge•thengaw wat-khuna so•otthelarinaka.* This evening, on the way, we will certainly not let him get away, the bastard! We will kill him. (2) to let go. *Byiraaw watbo!* Let go of the cat! (3) to avoid. *Joraawdo watna man•cha.* You cannot avoid your match in love. (4) to switch on an electrical appliance like a radio, TV, computer etc. *Tibi watbo.* Switch on the TV. (5) to play music on an electrical device (on the radio / a tape / a CD etc.). *Git watbo.* Play some music. (6) to cum. *Su•nyng•chi ri•tyi watchawa.* I will not cum inside her vagina.

**wat-₂** *v.* to weave things from reed or bamboo, to make a mat or basket from bamboo or reed. *Ge•theng koksep watna sapa.* He knows how to weave a bamboo cage.

**watbyrak-** *v.* shameless. *Ge•theng tawel wat-byrakai kana* He was wearing his towel shamelessly (i.e. not caring to cover up his private parts). *Ge•theng watbyrakai mimirongwa.* He was laughing shamelessly (i.e. without covering his mouth with his hand).

**watet-** *v.* to send (away), to post, to mail. *Ang songthangchina dong•angwachi nang•tymna chiti watetni.* When I have arrived in my own country I will send you letters.

**watwa watwa** *adv.* scattered all over the place

**watyi** *n.* TIME. rainy season

**wawa** *ideo.* the sound of someone throwing something. *Bandi kherengwachido wa•chu byryi wawa wawa thangasigaaidoknote.* When Bandi resisted, he was thrown four bamboo lengths, I'm telling you!

**wek₁** *ideo.* squeal! oink! the sound a pig makes

**wek-₂** *v.* to sweep

**wekwak-** *adj1.* very soft (like mud), sloppy. *Ram wekwakok.* The road is very soft like mud.

**wel-** *v.* to burn (as a sensation). *Ja•ryt wela* chillies burn. *Mykren wela wal•khumigymyn.* My eyes burn because of the smoke.

**wel•-** *v.* to turn left and right, to zigzag

**wel•ang ~ wel•ang wel•ang** *adv.* quickly, fast

**welet-** *v.* to flash

**wen•-**₁ **~ wen-** *v.* to wind around, to wrap around, make as a coil

**wen•**₂ **~ wet** *n. autoclf.* MSRE. time, turn. *wet sa ~ wen• sa.* once. *wen• ni* twice. *wen• tham.* three times

**wen•**₃ **~ wyn• ~ wyt- ~ wot** *v.* to sharpen, to whet

**weng•** *n.* PLANT. node (of bamboo), joint

**wengwang ~ gorweng** *n.* ANIM. species of cicada that makes the noise of a screaming baby or a woman being murdered

**wenphak-** *v.* to wind around something. *Dypyw ang chakaw wenphakwa.* The snake wound itself around my hand.

**wenwen- ~ winwin-** *v.* to wind around. *Panchi kara wenwena.* to wind a rope around the tree.

**wet ~ wen•** *n. autoclf.* MSRE. time, turn. The allomorph *wet* only occurs before the numeral *sa* 'one', and according to many speakers, it is the only appropriate morph in this environment. According to other speakers, the allomorph *wen•* can also appear before the numeral *sa* 'one' The allomorph *wen•* also occurs in all other environments. *wet sa ~ wen• sa* once. *wen• ni.* twice. *wen• tham.* three times

**wetanchian ~ wetantian** *adv.* every time

**wil-**₁ **~ wyl-** *v.* to go down, to descend, to get off

**-wil**₂ *evsp. V* around. *Kynsangdo matsado morotsyn man•aimyng rai•wilokno alsiado. Rai•wilwilokno.* Later, the tiger got the scent of the human and walked around the king. He walked round and round.

**-wilwil** *evsp. V* round and round, *V* around. *Kynsangdo matsado morotsyn man•aimyng rai•wilokno alsiado. Rai•wilwilokno.* Later, the tiger got the scent of the human and walked around the king. He walked round and round.

**winwin- ~ wenwen-** *v.* to wind something around something

**wongong-** *v.* to stir

**wongwet-** *v.* to dangle

**wot- ~ wyt- ~ wen•- ~ wyn•-** *v.* to sharpen, to whet

**wungwung-** *v.* to stir

**wuuuuk** *ideo.* wooosh! the sound of something big falling down

**wyi•** *n.* KIN. *drel, c, ref.* grandmother (archaic, addressed as *abu*)

**wyiset-** *v.* to wipe off

**wyl- ~ wil-** *v.* to go down, to descend, to get off

**wylang-** *v.* to go down, descend

**wyn• ~ wen• ~ wyt- ~ wot-** *v.* to sharpen, to whet

**-wyng** *evsp.* with a swinging motion

**wynget-** *v.* to dangle

**-wyngwang**₁ *evsp. V* in a confused way

**wyngwang-**₂ *v.* to wag. *Kyi• di•mai wyngwangaidong.* The dog is wagging its tail.

**wyngwet-** *v.* to swing, to move back and forth

**wyt- ~ wot- ~ wen•- ~ wyn•-** *v.* to sharpen, to whet

# Y

**yh** *interj.* interjection of hesitation, er…, uhm…, uhhh… . *Isangdo morotdo napit – yh – maibado alsia rajado morottaraanno.* Here however, as far as the species of humans is concerned, the barber – uh, what's it? – the Lazy King is the only human.

**yhy• ~ hy• ~ hy•y** *procl.* I don't agree, no, yes. "*Una tanka hyn•na bai!*" "*Yhy•, hyn•ni.*" "Don't give her the money" "Yes, I'll give her." "*Nang•do re•engchawa?*" "*Yhy•, re•engniba angba.*" "You're not going?" "Yes, I am going." "*Nang•ba re•engni?*" "*Yhy•, re•engchawa angdo.*" "You're going?" "No, I'm not."

**yis ~ hyits ~ hys ~ hyis ~ tyis ~ tys** *interj.* (interjection that expresses disapproval or indignation) Hey! Ugh! What?! Tsk-tsk!

*"Hai bai•siga, biskut sa•khawna." "Hyt man•cha nang•ba atong budi".* "Come on, friend, I want to steal those biscuits." "What?! No! What kind of idea is that?!"

**ym ~ am** *procl.* affirmative, okay, sure, yes, no. This proclause is used to acknowledge another person's statement. *"Ichi taw• banok" nookno. "Ym. Raw•bo" nookno.* "There's a bird trapped here" he said. "Yes, catch it", he said it is said. *"Ang jywcheng-nine" "Ym."* I will go to bed now." "Okay." *"Na•nang myng•ni re•engni, mo?"* "Ym." "The two of us will go together, right?" "Sure." *Nang•do sigyret ryngchamo?* "You don't smoke, do you?" "No, I don't."

**ymbuuu** *ideo.* moo! the sound a cow makes

**ymbyng** *n.* ART. bamboo flute

**ympong** *adj2.* lopsided, convex, having a surface or boundary that curves or bulges outward, as the exterior of a sphere. *Tyibekan ympong.* A traditional water bottle is lopsided.

**ymyi** *interj.* huh?! what?! interjection of surprise

**yndyn** *adv.* in vain, for nothing, for nought, for free, simply.

**ytyk-** *v.* to do like this/that. *Ytykaria, te•ewrawrawmi gawido.* They do just like that, the girls of nowadays.

**ytykchiba** *disccon.* but, however, in that case. *"Acha babaji, angmi joraaw chaina man•nima?" "Man•niba. Ytykchiba raja sa nangnine.* "Ok, fortune teller, can you see my match in love?" "I can, but I will need one hundred rupees." *Angna mamyngawan nangchaw. Ytykchiba na•a angna aro angmyng jykna nang• khengwa dabat ang thyicha dabat angaw mu•ai sa•na hyn•bo"* nookno. "I don't need anything. However, you keep giving me and my wife something to eat as long as you live until I die", he said.

**ytykchido** *disccon.* in that case, this/that being the case, so, but. *"Nang• dada re•engok" "Ama, angba re•engni ytykchido."* "Your elder brother has left". "So, I'll also go, mom."

**ytyken** *adv.* like this/that. *"Atakna rai•awa?" "O, gylgylarong ytyken, haratwanasa."* "Why have you come?" "Oh, I am just roaming like this, just because I'm lazy."

**ytykgaba** *n.* QUAL this kind of, like this, such. *Angba ytykgaba kha•di ra•nichymte.* I would also buy clothes like these.

**ytykkhal** *expr.* it doesn't matter

**ytykma•chiba** *disccon.* but, however. *Ga•dakchichiaimuna thypsetthiriokno. Ytykma•chiba uba sa•gyraiba jumu kha•thirithirioknotyi.* He cut it into pieces again and disposed of it again. But that child reassembled once again!

**ytykram•phinai** *expr., intens.* so (much). *Uchi Ketketa Burae pheruna balsakwano: "Aia anga kan•jotate. Ang ytykram•phinai kan•jota na•a atykyi kakai sa•na?"* Then Ketketa Bura answered the fox: "Jeez, I'm thin. I am so thin, why would you eat me?"

**ytykyi** *adv.* like this/that. *"Ma•, man•ni dongchido ie parang kun• sa•aw kawanchyi" noai hyn•okno. Ytykyi songtawai hyn•okno.* "Well, you could do it, try to shoot this culm of reed", she said and gave one. She chose one and put it upright like this.

**ytykyimyng ~ ytykyimu ~ ytykyimuna ~ ytykyimung ~ ytykyimungna** *disccon.* so then. *Te•ewe alsia rajano song dam sachi. Ytykyimyng jykba myng• ni khymanoro.* Now, there is a lazy king, in a certain village. So then, he is married to two wives.

**ytykyisa** (1) *disccon.* therefore, then, that's why, so. *Arong nokma chaikhawwachi Arong nokmami mukhangaw khiemu thyiokno. Ytykyisa ue Arong nokma thyiwamisa saepe bondyk paiaimu sipaidyrang dang•na man•okno.* When headman Arong took a peek, he was hit in the face, and died. That's why, because of headman Arong's death, the gun-carrying white soldiers were able to enter the village. (2) *adv.* like this/that. *Ytykyisa Bandie balaidongano* [...] Bandi spoke like this [...]

**ytym ~ uytm** *ppron.* third person plural personal pronoun, they, them, their. (see *utym*)

**yyy** *interj.* uhm, eh. Expresses hesitation.

# PART 2: ENGLISH – ATONG DICTIONARY

## What do we see in the English-Atong dictionary?

When we read the English-Atong dictionary we see the headwords, followed by an abbreviation in *italics,* and an Atong translation in **bold face**. The abbreviation in italics indicates the word class of the English headword (see below). These abbreviations are listed in the Prologue. For example:

| yesterday, | *adv.* | **myia** |
|---|---|---|
| headword | word class | translation |

Indication of the word class of English entry words permits the reader to choose the most appropriate translation in cases where two words have the same spelling, but belong to a different word class. For example, the English word *close* has a different translation as a verb, than as an adjective, as we can see in Example (1).

(1) Different word classes may prompt different translations:
close, *adj.* **nek-**
close, *v.* (1) **chep- ~ chyp-** (2) **kap-** (3) **kak-** 'close with a lid' (4) **chugup-** 'close with a lid' (5) **khup- ~ khep- ~ khyp-** 'close, cover, put on clothes, spread out' (6) **khu•tip-** 'close one's mouth' (7) **buthu-**'seal, close a receptacle by putting something in the opening'

The reader should be aware that an English word and its Atong translation do not necessarily belong to the same word class. For example, the English word *accidentally* is an adverb, whereas its Atong translation *-gak* is an event specifier (see van Breugel 2014: 376–385); the English word *strong* is an adjective, whereas its Atong translation *rak-* belongs to the sub-class of verbs within the predicative word classes of Atong (see van Breugel 2014: 65–69). The word classes of Atong translations are not indicated in the English-Atong dictionary, but can be found in the Atong-English dictionary.

When an English headword can be interpreted as one word class or another without consequences for its translation into Atong, the word classes are separated with a slash (/), for example:

| complete, | *v/adj.* | **jam-** |
|---|---|---|
| headword | word classes | translation |

We can read this example as follows: the headword *complete*, in its meaning as verb or adjective is translated as *jam-*.

When there is more than one Atong translation of an English headword, the different translations are numbered, and, if possible, accompanied by an English gloss between single inverted commas, specifying the meaning of the different translations if this meaning deviates from the meaning of the headword, for example:

| bamboo shoot, | n. | (1) **mai•wa** (generic) (2) **mai•wakhyi** 'fermented bamboo shoots' |
|---|---|---|
| headword | word class | (1) **first translation** (2) **second translation** 'gloss' |

Atong words separated by the tilde ( ~ ) represent variations in pronunciation and hence in the spelling of the same lexical item, for example:

| banyan tree, | n. | **phe•ep ~ phep** |
|---|---|---|
| headword | word class | **translation with variation in spelling** |

This dictionary only contains Atong words that the author has recorded. If a particular word of variation of a word does not occur in this dictionary, it does not necessarily mean that this word or variation does not exist, but only that it is not recorded. For example, in this dictionary, the translation of the expression *scoop out* is recorded as *waiset-*. It is possible that the pronunciation *wyiset-* also exists. This pronunciation is not given in this dictionary, because it has not been recorded, and therefore, the author cannot be certain of its existence. Continuing research will hopefully add more words to the dictionary, as well as remedy the many shortcomings in the precision with which variation is presented.

Example sentences are not provided in the English-Atong dictionary, because the use of most Atong words is exemplified in the Atong-English dictionary. Only when an English word needs to be translated with a grammatical construction in Atong, instead of with a lexical expression which can be found in the Atong-English dictionary are examples provided to exemplify the grammatical construction. All examples are followed by a morphemic analysis and glosses. For example, because Atong does not have articles, the English indefinite article *a ~ an* cannot be translated lexically into Atong. However, Atong does have a grammatical or syntagmatic construction to mark a noun phrase as indefinite. This construction, marked in **bold face**, in addition to some useful pragmatic information about the interpretation of noun phrases as indefinite, is proved in the article of the headword *a ~ an*, as can be gauged from the copy of the entry here below.

a ~ an *art*. To mark the indefiniteness of a noun phrase in Atong, the construction **CLASSIFIER+sa** can be used. This construction is usually used when the referent of the noun phrase is introduced for the first time in the discourse, e.g. *Song dam sachi alsia raja myng• sa ganangnochym. Song dam+sa=hi alsia raja myng•+sa ganang=no=chym.* (village CLF:VILLAGES+one lazy.person king CLF:HUMANS exist=QUOT=IRR) 'In a village supposedly lived a lazy king.' Otherwise, an unmarked noun phrase can also be interpreted as being indefinite. *Sympak chunggaba nukoknotyi. Sympak chung=gaba nuk=ok=no=tyi.* (SPECIES. OF.TREE big=ATTR see=COS=QUOT=MIR) 'They saw a big *sympak* tree, to our/their surprise.' (See van Breugel 2019: 239, sentence 119)

The English-Atong dictionary does not translate any English bound morphemes, such as the possessive apostrophe s ('s), the comparative suffix *-er*, or labels of grammatical notions like *plural*, or *instrumental*. Grammatical categories found in Atong are listed in PART 4, §9 of the Prologue. In order to know how Atong functions grammatically, the reader is referred to *A grammar of Atong* (van Breugel 2014).

## How to use the English-Atong dictionary

In order to translate an English word into Atong, first find the English word you wish to translate, and see what translations are available. When in doubt which translations to use, when there is more than one, look in the Atong-English dictionary for more information about the Atong words. This will not always solve the problem, because the author was not always able to ascertain the differences in meaning or usage of different Atong words with the same English translation. An example of a situation where consulting the Atong-English dictionary does help to provide more information about different translations of an English headword is the following.

take, *v.* (1) **ra•-** 'take from' (2) **ra-** 'take to'

The English headword is the verb *take*, and there are two translations in Atong, viz. *ra* and *ra•*. Some extra information is given to each of the translations, in the form of English glosses between single inverted commas. This extra information already provides some minimal differentiation between the two translations. When looking up each Atong word in the Atong-English dictionary, the reader will see the following.

**ra-** *v.* (1) to bring. *Nukwachie phalthangmi gawigaba cha raaknoro.* When he looked, his wife had brought tea. (2) to take to. *Ang ie bostuaw te•en nygylsang raangaimyng phalni.* I will take these things to the market later, and sell them.

**ra•-** *v.* (1) to get, to buy. *"Nang• ie tupi bimi ra•ak?" "Turami ra•ak."* "Where did you buy that cap?" "I bought it in Tura.". *Kha Dawa nochachido phu•chul ra•ariano.* When you don't say "Kha Dawa!", the monitor lizard will just get you. (2) to take (from). *Ge•thengdo uaw thymai chaiaimyng ue morot re•eng-man•wachi, uaw tangkaaw ra•akno.* Having hidden and watched him, when that person left, [he] took that money.

Examples sentences illustrating the use of each word with its relevant meanings can be found. In addition, it can be seen that the words *ra* and *ra•* are polysemous. All the information the articles of the Atong headwords provide, can help the reader choose which Atong word to use to translate the English word *take* in the context in which it occurs.

An example where the Atong-English dictionary does not offer any help in deciding which Atong word to use in a certain translation, is *grind*. The English entry is as follows, with two Atong translations.

grind, *v.* (1) **tok-** (2) **tokdepdep-**

No extra information to each of the two Atong words is given here. When consulting the entries of each of the Atong words in the Atong-English dictionary, the reader will see the following.

**tok-** *v.* (1) to beat, to beat up. *Kyi•aw tokbo!* Beat the dog! *Taw•reksyrup mang sa ge•thengmyng thup phangnan mongma phai•ai sa•rongwana, mongma mathaiaw tokna re•engaidongano.* Because the nest of a banana bird always gets broken and eaten by an elephant, it is on its way to beat up the bachelor elephant. (2) to crush, to grind. *Amakmyng rukpekmyng: "Hai bai•siga, na•ru tokna" noaidonganowa, myng• ni.* So then, now, the monkey and the frog, both of them, said: "Come on, friend, let's grind some fish poison". (3) to play an instrument.

**tokdepdep-** *v.* to crush, to grind. *Sambanggyri akaiokno, tokdepdepaimu pha•a-tokno.* He plucked *sambanggyri*, crushed it and put it on the wound.

The word *tok* is polysemous, and its second meaning is the same as the meaning given for the word *tokdepdep*. However, the examples sentences for the meaning 'to grind, to crush' under *tok-* (2) and the example sentence given for *tokdepdep* do not provide enough information to know if there is a difference in meaning or usage between the two verbs. In other words, the two words could be exact synonyms, or not, but the reader cannot know. Moreover, if one assumes that *tokdepdep* consists of the root *tok* and a suffix *-depdep*, and one searches for this suffix in order to see if its meaning can differentiate the verbs *tok* and *tokdepdep*, one sees that no such suffix is recorded, and again, not enough information exists in the dictionary to know whether or not the two verbs are exact synonyms.

Some English expressions consisting of more than one word are often given under the most important headword in that expression. For example, the Atong translation of the English expression *cut in half* can be found under the English headword word *cut*.

When an English word is not in this dictionary, the reader is advised to consult the appendices. If that fails, the desired word does probably not occur in the English-Atong dictionary. Continuing research in the field of Atong lexicography will hopefully add many more word to future editions of this dictionary.

# English-Atong Dictionary

## A

a ~ an *art.* To mark the indefiniteness of a noun phrase in Atong, the construction **CLASSIFIER+sa** can be used. This construction is usually used when the referent of the noun phrase is introduced for the first time in the discourse, e.g. *Song <u>dam</u> <u>sa</u>chi alsia raja <u>myng• sa</u> ganangnochym. Song dam+sa=hi alsia raja myng•+sa ganang=no=chym.* (village CLF:VILLAGES+ one lazy.person king CLF:HUMANS exist= QUOT=IRR) 'In <u>a</u> village supposedly lived <u>a</u> lazy king.' Otherwise, an unmarked noun phrase can also be interpreted as being indefinite. <u>Sympak chunggaba nukokno-tyi.</u> *Sympak chung=gaba nuk=ok=no=-tyi.* (SPECIES.OF.TREE big=ATTR see=-COS=QUOT=MIR) 'They saw <u>a</u> big *sympak* tree, to our/their surprise.' (See van Breugel 2019: 239, sentence 119)

a few, *det.* **pang•cha**
a few days ago, *adv.* **maja**
a little, *det.* **pang•cha**
a little bit, *adv.* (1) **choi•sa** (2) **alamyla**
a little while ago today, *adv.* **tai•sa**
a long time ago, *adv.* **dakang**
a lot of, *det.* **pang•-**
abandon, *v.* **tanset-**
abbreviate, *v.* **seng•sot-**
abdominal membrane, *n.* **bichylap**
abduct, *v.* **bytsek-**
about, *prep.* **gymyn**
about, *adv.* **=darang ~ =dyrang**
about to, *adv.* **-dykdyk**
about to lay an egg, *adj.* **khurung**
abrasion, *n.* (1) **gusylak** (2) **khosylak**
abscess, *n.* (1) **samsin** (2) **nangthaigaba**
absentmindedly, *adv.* (1) **awan awan** (2) **-parang**
abundance, *n.* **pang•wami**
abundantly, *adv.* **-bongbong**
accept, *v.* **ra•sak-**
accidentally, *adv.* (1) **-gak** (2) **-jol**
according to, *adv.* **kri**

accuse of adultery, *v.* **jykrat-**
across, *adv.* **-pat**
act like, *v.* **tak-**
activity, *n.* (1) **kam ~ gam** (2) **takwa rukwa**
add, *v.* **dyw-**
address, *n.* **edres**
advantage, *n.* **nemgyni**
aeroplane, *n.* **alupren**
afraid, *adj.* **kyryi-**
affirmative, *n.* (1) **ym ~ am**
Africa, *n.* **Efrika**
after, *prep.* **kynsang**
afternoon, *n.* (1) **gasam** (2) **gasamphang ~ gasamphak**
afterwards, *adv.* **kynsang**
again, *adv.* **-theri ~ -thiri**
again and again, *adv.* **– thirithiri**
ago, *adv.* **dakang**
agree, *v.* **bam-**
ahead, *adv.* **-thyl**
aim, *v.* **nisan-**
aimlessly, *adv.* **-parang**
air, *n.* **balwa**
algae, *n.* (1) **tainalap** (2) **tyinala**
alive, *adj.* (1) **khengaidong** (2) **kheng-** (3) **pikheng**
all, *det.* (1) **=gumuk** (3) **gumukan** (3) **janggal** (4) **janggalan** (5) **-thok** (6) **ha•gyrsak ~ ha•gylsak** (7) **=khakhet**
all over the place, *adv.* (1) **dymbyra dymbyra** (2) **watwa watwa** (3) **byldyng byldang** (4) **-tyngtang** (5) **-ruru**
all the time, *adv.* (1) **-jyryng** (2) **-thyngthyng**
alligator, *n.* **mejakbal**
allowed, *adj.* **man•-**
almost, *adv.* **guduk tak-**
alone, *adv.* (1) **dyngdang** (2) **-chep** (3) **=tara**
along with, *adv.* (1) **-chap** (2) **-pha**
alphabet, *n.* **oikor**
already, *adv.* **-man ~ -man•**
also, *adv.* (1) **aro** (2) **-dap** (3) **-khan** (4) **-pha** (5) **=ba**
alternatively, *adv.* (1) **=sega ~ =siga**
aluminium, *n.* **elmoni**

always, *adv.* (1) **phangnan** (2) **jyryngnam** (3) **jyryng jyryng** (4) **-barai**
ambers, *n.* (1) **wal•di•**
America, *n.* **Amerika**
amulet, *n.* **thabisi**
anaconda, *n.* **dypywnokma**
ancestors, *n.* (1) **achuambi**
and, *conj.* (1) **aro** (2) =**myng ~ =mung ~ =mu ~ =mungna ~ =muna**
angle, *n.* **gyching ~ giching**
angry, *adj.* **kha•pet-**
animal, *n.* (1) **mat** (2) **matburung ~ matpalyng**
anime, *n.* **jokal**
announcement, *n.* **tyngetwami ~ tyngetwamyng**
annoyed, *adv.* **chaisi-**
another, *det.* (1) **alaga** (2) **abun**
another village, *n.* **songga**
answer, *v.* **khu•sak-**
ant, *n.* **kabin, butsa, kalthek, gompara ~ gompyra, samalmaisirong, melanggaw**
antenna, *n.* **kakmyn•** (of an insect)
antidote, *n.* **thabisi**
anus, *n.* **di•khal**
any, *det.* **je**
anybody *pron.* **darangba**
anyone, *n.* (1) **darang ~ dyrang** (2) **changba** (3) **darangba**
anyway, *adv.* (2) =**ari** (2) **dymdym damdam**
anywhere, *adv.* **jechiba**
appearance, *n.* **bimang**
apple, *n.* **epyl**
apply, *v.* (1) **nong-** (2) **pha•at- ~ pha•et-**
appoint, *v.* **song-**
appreciate, *v.* **mythel-**
appropriate, *adj.* (1) **ke-** (2) **mili-**
appropriately, *adv.* (1) =**darang ~ =dyrang** (2) **-sak** (3) **chacha**
April, *n.* (1) **epril** (2) **boisaja**
area, *n.* **jol**
arm, *n.* (1) **chak** (2) **chakphong ~ chakphung** 'arm, upper arm'
armful, *n.* **khabak**
armpit, *n.* **phakwal**
around, *adv.* (1) **-phak** (2) **-ruru** (3) **-wil** (4) **-wilwil** 'around and around'

arrange, *v.* **thari-**
arrive, *v.* (1) **phet-** (2) **phetang-** (3) **pheta-** (4) **dong•-** (5) **dongang-** (6) **wala-** 'arrive at night'
arrogance, *n.* (1) **gal** (2) **bawra** (3) **stel**
arrogant, *adj.* **bawra tak-**
arrow (of bow and arrow), *n.* **thirikun•**
arse, *n.* **di•khal**
arse crack, *n.* (1) **di•sep** (2) **di•sepra**
as best you can, *adv.* **-mangmang**
as long as you can, *adv.* **-chik ~ -chyk**
as soon as, *adv.* **-tharap**
as well, *adv.* (1) **aro** (2) **-dap** (3) **-khan** (4) **-pha**
as well as, *adv.* **aro**
ascend, *v.* **taw-**
ashamed, *adj.* **barat-**
ashes, *n.* (1) **thapyra** (2) **ha•thapyra** (3) **wal•kungki** 'black ashes'
Asia, *n.* **Esia**
ask, *v.* (1) **syng•-** 'ask something' (2) **pi•-** 'request, beg'
aslant, *adj.* (1) **ching•pheng** (2) **gyching ~ giching** (3) **khingcheng**
asperity, *n.* **mukthai**
assemble, *v.* **mili-**
astral projection, *n.* **jasyri** 'the experience that one sees oneself in a different place while one is asleep as if one's soul leaves one's body'
Atong *n./adj.* **Atong**
attempt, *n.* **joton**
attempt, *v.* **joton kha•-**
attend, *v.* (1) **ganang** (2) **chairok-**
attentively, *adv.* **-sym**
August, *n.* (1) **agos** (2) **badolja** (archaic)
aunt, *n.* (1) **akai** (2) **ama** (3) **syi** (4) **asyi ~ asi** (5) **nyng** (6) **anyng** (7) **mani** (8) **nai•** (9) **anai**
Australia, *n.* **Ostrelia**
author, *n.* **sainokgaba**
auto rewind, *n.* **otorewain**
auto rickshaw, *n.* **oto**
averrhoa carambola, *n.* **galdai**
avoid, *v.* (1) **jok-** (2) **kyl-** (3) **-thyl**
away, *adv.* **-ang**
axe, *n.* (1) **kulal ~ kular ~ kural ~ kurar** (2) **wakeng**

## B

baby, *n.* (1) **odek** (2) **sa•gyrai odek** (3) **babu** (term of address) (4) **chame** 'lover, sweetheart' (5) **gogyrek** 'baby with its neck bent sideways while it is being carried on the back'
bachelor, *n.* **banthai**
bachelor elephant, *n.* (1) **mathai** (2) **mongmamathai ~ mungmamathai**
Bachelor's degree, *n.* **Bechylyrdygri**
bachelors' house, *n.* **nokbanthai**
back, *n.* **kyn**
back and forth, *adv.* **-rura**
back of the hand, *n.* (1) **chakphakhung** (2) **chaksikhum**
backside of the ear, *n.* **nakhong**
backwards, *adv.* (1) **kynbyret** (2) **-phin ~ -phyn**
bad smell, *n.* **manam**
bag, *n.* (1) **dokra** (2) **bosta** (3) **bek**
bagful, *n.* **chatom**
balcony of a rice field house, *n.* **serek**
ball, *n.* (1) **robol** 'football' (2) **ri•karan ~ ri•keren** 'testicle'
bam, *ideo.* (1) **dam ~ dym** (2) **gyp**
bamboo, *n.* (1) **wa•** 'generic' (2) **wa•phek** 'small bit of bamboo'
bamboo flute, *n.* **ymbyng**
bamboo shoot, *n.* (1) **mai•wa** (generic term) (2) **mai•wakhyi** 'fermented bamboo shoots'
bamboo stalk or twig, *n.* **wa•cheksi**
bamboo strip, *n.* (1) **wa•tyng** (2) **engsyri** (3) **jyw•** (4) **khiil** (5) **kyryw** (6) **wa•syl** (7) **wa•phuk**
bamboo torch, *n.* **wal•cham**
bamboo tube, *n.* **wa•sung**
banana, *n.* **rekthai**
banana bird, *n.* **taw•reksyrup**
banana flower, *n.* (1) **rekkun** (edible) (2) **rekphul** (not edible)
banana tree, *n.* (1) **rek** (2) **rekphang**
bang, *ideo.* (1) **thaw** (2) **them**
Bangladesh, *n.* **Banglades**
Bangladeshi, *n.* **Banggal**
banknote, *n.* **tangka khung sa**
banyan tree, *n.* **phe•ep ~ phep**

barber, *n.* **napit**
barebina-xariegata, *n.* **palengma**
barefoot, *adv.* **cha• kantara**
barely, *adv.* (1) **-tham** (2) **myryng myryng**
barely, *adv.* (1) **-mangmang** (2) **-thamak**
barf, *v.* (1) **chisat-** (2) **kha•rekrek-** (3) **khawakwak-**
bark, *v.* **khong•-**
bark (of a tree), *n.* (1) **mawkhol** (2) **pankhol**
barking deer, *n.* **magachak**
barrel, *n.* (1) **tyk** (2) **dram**
base stone on which a house is built, *n.* **rong•thai**
basket, *n.* (1) **kok** (2) **asok** (3) **chokhoi** (4) **katha** (5) **khugyri ~ koksi** (6) **kokbal** (7) **kokcheng** (8) **kokdam ~ koktang** (9) **koksep** (10) **nakamai ~ namakai** (11) **net** (12) **pai•ra ~ phai•ra** (13) **tannet**
bat, *n.* **taw•pak**
bathe, *v.* (1) **tyru- ~ tyiru- ~ tyiryw-** 'take a bath, wash oneself'' (2) **tyret- ~ tyiret-** 'bathe/wash someone else'
battery, *n.* **betyri**
be, *v.* (1) **dong•- ~ dong-** (2) **ganang** 'be' (3) **ni•-** 'not to be, there isn't/aren't'
be able, *v.* **man•-**
be an orphan, *v.* **ma•chot-**
be at, *v.* **mu•-**
be in pain, *v.* **sa-**
be like a fist, *v.* **mym•-**
be on top, *adj.* **tandap-**
beam, *n.* (1) **gandai** (2) **han•dyng** (3) **jagybeng ~ jagebeng** (4) **khyntyri** (5) **pakrai ~ pakri**
bean, *n.* (1) **biins** (2) **kha•rek** (3) **bai•khop** (4) **rekhep** (5) **siwi ~ gylarong**
bear, *n.* **makbul**
bear, *v.* (1) **sak-** (2) **nang-** 'bear fruit'
beard, *n.* (1) **ka•myn** (2) **khu•symang** 'beard, moustache'
beat, *v.* (1) **thap-** 'beat, beat up' (2) **tok-** 'beat, beat up' (3) **tam•-** 'beat a drum' (4) **tokgepgep-** 'beat to a pulp' (5) **su•bylok-** 'beat to a pulp'
beautiful, *adj.* (1) **syl-** (2) **chaithawa-**
beautifully, *adv.* (1) **sylai** (2) **-nang**

because (of), *conj.* (1) **genitive-marked noun phrase + gymyn** (2) **=ok=ona** (3) **=ai=mu ~ =ai=mung ~ =ai=mungna ~ =ai=myng** (4) **=tykyi=myng**
become light, *v.* **seng•-**
become more and more, *v.* **dairukruk-**
become uprooted, *v.* **phuruk-**
bed, *n.* **palong**
bedbug, *n.* **thaba**
bedroom, *n.* **din**
bee, *n.* **ne•** (generic), **ne•kat, byirakhem, ne•wal**
bee's nest, *n.* **ne•katthup**
beer, *n.* (1) **beer** (2) **chyw** 'rice beer, liquor'
beetle, *n.* (1) **gogak** (2) **ambisuthyk** (3) **rongkhym** (4) **gongchit**
before, *adv.* **dakang**
beg, *v.* **pi•-**
begin, *v.* **ha•bacheng-**
beginning, *n.* **ha•bachenggaba**
behave well, *v.* **sakchyk-**
behind, *prep.* (1) **kynchi**
believe, *v.* **bebe ra•a**
belly, *n.* **pipuk**
bellybutton, *n.* **gandyrui**
belong to, *v.* **=mi ~ =myng**
belongings, *n.* **chakra**
beloved person, *n.* **daldi**
below, *prep.* (1) **cha•machi** (2) **hama ~ nokhama**
belt, *n.* (1) **seng•khi** (2) **khalpak** 'belt of a basket'
bend, *v.* (1) **gom-** (2) **bam-** 'bend one's head' (3) **bamkhup ~ bangkhylok-** 'bend one's head' (4) **gonggong-** 'bend over' (5) **khomchuk-** 'bend over'
Bengali, *n.* **Banggal**
bent, *adj.* (1) **kompyl ~ kongpyl** (2) **gongdang** (3) **totyp** (4) **songrat-** (5) **chogop-** 'fully bent but not touching the ground (used only with plants)'
beside, *adv.* **-phak**
bestow upon, *v.* **phathi-**
betel nut, *n.* (1) **goi** (2) **goichara** 'betel nut sapling'
betel nut and paan/pan, *n.* **goilapan**
betray, *v.* **thogi-**

between, *n.* **matji ~ maji**
Bible, *n.* **Bailyl**
biceps, *n.* (1) **biambong** (2) **jagydok**
biconcave, *adj.* **jang•jot**
bicycle, *n.* **baisykyl**
big, *adj.* **chung-**
Big Dipper, *n.* (1) **ma•suchawkhol** (2) **do•jenjok**
bigger, *adj.* (1) **chungkhal-** (2) **dai-**
bike, *n.* **baik**
bird, *n.* **taw•** (generic), **taw•gylyk, taw•pynchyrep, ma•rek,**
bird cage, *n.* **chongchang**
bird's nest, *n.* (1) **thup** (2) **taw•thup**
biscuit, *n.* **biskut**
bit, *n.* (1) **choi•sa** 'a little bit' (2) **bi•chamchym** 'fragment' (3) **alamyla** 'a little bit, somewhat, ordinary, normal'
bitch, *n.* (1) **kyi• gawi** (2) **su•gol** (swearword)
bite, *v.* (1) **kak-** (2) **kakdep-** 'bite on something' (3) **kakkhap-** 'bite down on, firmly hold between the teeth or in the beak (4) **wang-** 'bite a bit out of something'(5) **khu•cheng-** 'bite one's teeth firmly together'
bitter (taste), *adj.* **ka•- ~ kha•-**
bitter gourd, *n.* **kolachita**
bitter-sweet, *adj.* **kha•sym-**
blabbermouth, *n.* **khuchylep**
black, *adj.* (1) **pinak** (2) **nak-**
bladder infection, *n.* (1) **samycheng** (2) **sasyri**
blame, *n.* **dosi**
blanket, *n.* (1) **kombol** (2) **symphak**
blarp, *ideo.* **phong**
bless, *v.* **phathi-**
blessed, *adj.* **pathigaba**
blessing, *n.* **rasong**
blind person, *n.* **kana**
blink (with one's eyes), *v.* (1) **phylyp-** (2) **mykphylyp-**
blister, *n.* (1) **chichugaba** (2) **thaphu** 'blister, sore'
blister, *v.* (1) **chichu-** (2) **thaphu-**
blistered, *adj.* **thaphu-**
block, *v.* (1) **ping-** (2) **thek-** (3) **bythyw-**
blond, *adj.* **pisak**
blood, *n.* **thyi•**

blood relative, *n.* (1) **bai•** (2) **bai• tyng** (3) **bai•siga ~ bai•sega**
bloom, *v.* **pal-**
blow, *v.* (1) **haphu-** (of a person) (2) **balwa-** '(of the wind) (3) **balphak-** 'blow away'
blow one's nose, *v.* **het-**
blue *n./adj.* **bylu**
blue, *adj.* (1) **khengchek** (2) **blu**
blunt (of pointed things), *adj.* (1) **thanthong-** (2) **kek-**
boast, *v.* (1) **galcha-** (2) **rasong dong•- ~ rasong dong-**
boasting, *n.* **rasong**
body, *n.* (1) **bimang** (of a human) (2) **kan•** (of a human) (3) **randai** 'meat, flesh, body'
body hair, *n.* **myn•**
body odour, *n.* **dil**
body smell, *n.* **dil**
boil, *v.* (1) **buthu- ~ buthyw- ~ bythyw-** (of water) (2) **rot-** 'boil something in water'
boil, *n.* (1) **samsin** 'large boil' (2) **samsin maiphara** 'small boil'
bold person, *n.* **kokalang ~ khokalang**
bond, *n.* **ha•khym**
bone, *n.* **kereng**
book, *n.* (1) **lekha** (2) **kitap**
border, *n.* **bai**
born, *adv.* (1) **achi-** (2) **ba•-**
borrow, *v.* **saraw-**
bosom, *n.* (1) **mu•thai** (2) **chungthai** 'large breast of a woman' (3) **mylthai** 'small breast of a woman' (4) **chel** 'breast or bosom of a man'
both, *det.* CLASSIFIER+**ni**
bother by misbehaving, *v.* **seng•-**
bottle, *n.* (1) **botol** (2) **tyibek**
bottom, *n.* (1) **di•khal** (2) **oktyk** 'bottom of ravine or cliff'
bougainvillea, *n.* **getphul ~ lekhaphul**
bouncily, *adv.* **-gangang**
boundary, *n.* **sima**
bow, *n.* (1) **thiri** (2) **pawai**
bowels, *n.* **pipuk**
bowl for curry *n.* **pawai**
bowl of rice, *n.* **powa ~ pywa**
box, *v.* **satkhap-**
boy, *n.* **bipha**

boyfriend, *n.* **chame ~ chamai**
bracelet, *n.* (1) **chaksan**
bracket, *n.* **breket ~ brekyt**
Brahmaputra, *n.* **Songdu**
brain, *n.* (1) **runi** (2) **thanyng** (3) **sung**
branch, *n.* (1) **dala** 'not directly attached to the trunk' (2) **phek** (classifier for smaller branches of trees)
brap, *ideo.* **phong**
brawler, *n.* **gunda**
breadth, *n.* **gebeng**
break, *v.* (1) **bai•-** (2) **thabai•-**(3) **phai•-** 'break, translate'(4) **gyrym ~ gyrum-** 'break off and fall down' (5) **phuruk-** (for plants and trees) (6) **khorop-** (only used for bamboo) (7) **okhynyng-** 'break a round hollow object in half'(8) **phai•thong-** 'break a solid object in half'(9) **chogyp** 'break off and fall down (for branches and big leaves)'(10) **salam- ~ selem- ~ serem- ~ saram-** 'break/tear easily, be easily damaged'
break of dawn, *n.* **sirimynmyn**
breakfast, *n.* **maimanap**
breast, *n.* (1) **mu•thai** 'breast or bosom of a woman' (2) **chel** 'breast or bosom of a man'
breastfeed, *v.* **mu•thai hal-**
breath, *n.* **biba**
breathe, *v.* **rang•set-**
brick, *n.* **itha ~ ita**
bridge, *n.* **dolong**
bright, *adj.* **chyng•-**
bring, *v.* **ra-**
British *n./adj.* **Britis**
British military commander, *n.* **saip ~ saep** (from *sahib ~ saheb*)
broken rice, *n.* **rungkhut**
bronchitis, *n.* **leng**
brood, *v.* **bam-**
broom, *n.* (1) **nokwek** (2) **cha•wek** (3) **sal•wek** (3) **sal•tareng** (4) **nogek**
brother, *n.* (1) **dada** (2) **phaw•jong ~ phawjong** (3) **jong** (4) **jojong**
brother-in-law, *n.* (1) **gumi** (2) **jongsyri** (3) **bonyng** (4) **biawthang** (4) **sadu**
brown, *adj.* **ha•mangrong**
brush, *v.* **nat-**
bucket, *n.* (1) **baket** (2) **baltin**

budge, *v.* **guduk-**
bug, *n.* **chong•** (generic)
bubulcus ibis (cattle egret), *n.* **alabok**
build, *v.* (1) **ham-** (2) **chanpat-**'build a bamboo bridge'
building, *n.* (1) **bilding** 'house made with masonry, building' (2) **dolan** 'big house, big building'
building ground, *n.* **nokhap**
bulbul bird, *n.* **daw•blok**
bulging, *adj.* **thywkhong**
bulky, *adj.* **chat-**
bullock cart, *n.* **ma•sugari**
bum, *n.* **di•khal**
bump, *v.* (1) **cha•duk-** (2) **godot-**
bumpily, *adv.* **-gangang**
bunch, *n.* (1) **bada** (2) **chok**
bundle, *n.* (1) **khasot** (2) **chok** (3) **sat**
burn, *v.* (1) **kham-** (2) **saw•-chyng•-** (3) **sang-** (3) **ha• kham-** 'burn the land' (4) **chem•-** 'burn up'
burn (as a sensation), *v.* (1) **wel-** 'like chillies' (2) **bol-** 'like the sensation of a being stung by a stinging nettle' (3) **basak-** 'burn and cause a rash'
burned rice, *n.* **maikhyt**
burnt curry, *n.* **wal•bek**
burst, *v.* **pheret- ~ peret-**
bury, *v.* **gop-**
bus, *n.* **bas**
bus stop, *n.* **basneng•thakgaba**
bush, *n.* (1) **gyrym** (2) **burung**
but, *conj.* (1) **ytykchiba** (2) **utykwachido** (3) **ytykma•chiba**
but then, *conj.* **uchiba**
butterfly, *n.* **taw•pak**
buttock, *n.* **di•phathai**
button, *n.* **baton**
buy, *v.* **ra•-**
by accident, *adv.* **-ram•**
by which way?, *interr.* **bitykyi**

# C

cabbage, *n.* (1) **kobi** (2) **laisak**
cable, *n.* **kebyl**
cake, *n.* **kek**
calf (part of the leg), *n.* **cha•pathai**
call, *n.* (1) **kol** 'telephone call' (2) **phone ~ phoon** 'telephone call'
call, *v.* (1) **hok** (2) **myng-** 'to call someone a name' (3) **kol kha•-** 'to make a phone call' (4) **phone ~ phoon kha•-** (5) **mynga-** 'to call upon' (6) **myngkhylek-** 'to call someone by a nickname'
camera, *n.* **kemyra**
campfire, *n.* **wal•tum**
can, *v.* **man•-**
cane, *n.* **raima**
cancer, *n.* **sothonthara** 'cancerous swellings all over the body'
candle, *n.* **kendyl ~ kendel**
candy, *n.* **choklet**
canine teeth, *n.* **kyi•wa**
cannibal, *n.* **dykyl**
cap, *n.* **tupi**
capability, *n.* **wa•churek**
capacity, *n.* **wa•churek**
capon, *n.* **taw•khasi**
capture, *v.* **watcha-**
car, *n.* **gari**
carambola, *n.* **galdai**
carapace, *n.* **khung**
cards (game), *n.* **tas**
care, *n.* **khel**
care, *v.* (1) **kha•dang-** (2) **symsak-**
careful, *v.* **symsak-**
carelessly, *adv.* (1) **dymdym damdam** (2) **sykhathang**
caretaker, *n.* **rakhigaba**
carrot, *n.* **gajol**
carry, *v.* (1) **khai-** 'carry on one's back with a strap tied around the head '(2) **pai-** 'carry by hand'(3) **ba•-** 'carry a child in a cloth on the body'(4) **gogat-** 'carry on one's shoulders'(5) **tha•gat-** 'carry on one's back'(6) **rai•byt-** 'carry around'
cartoon, *n.* **jokal**
cartoon figure, *n.* **jokal**
carve, *v.* (1) **ry-** (2) **si•wil-**
cascade, *n.* **tyichabakram**
cassava, *n.* **khan**
cassette, *n.* **keset ~ kheset**
castrate, *v.* **khasi-**

castrated rooster, *n.* **taw•khasi**
cat, *n.* **byira** (generic), **byira amanthong** (jungle cat)
catch, *v.* (1) **pyi•khap-** 'to catch with one's hands' (2) **pun-** 'to catch with a fishing rod and fishing hook' (3) **na• pun-** 'to fish, to catch fish' (4) **raw•-** 'to catch' (5) **kap-** (6) **joi•- ~ doi•-** 'to catch by dragging a net through the water' (7) **khaw•-** 'to catch water in the palms of one's hands' (8) **man•-** 'to catch a disease' (9) **wal• kham-** 'to catch fire'
catch, *v.* (1) **raw•-** 'catch, grasp' (2) **khap-** 'catch in one's hands' (3) **pyi•khap-** 'catch with one's hands' (4) **dirikhap-** 'catch and clasp with one's hands' (5) **ban-** 'catch in a trap (6) **doi•- ~ joi•-** 'catch in a fishing net' (7) **pun-** 'catch fish' (8) **na• pun-** 'catch fish' (9) **khaw•-** 'catch water in the palms of one's hands' (10) **wal• kham-** 'catch fire' (11) **kap-** 'catch, close' (12) **raw•soksok-** 'fail to catch'
catch up with, *v.* **tharap-**
caterpillar, *n.* **chong•su** (generic), **mongma-chong•, taw•paktyi, asalchong•, pan-chungchong•su**
catfish, *n.* **galjak ~ kaljak**
cattle egret, *n.* **alabok**
caught, *adj.* **chep- ~ chip- ~ chup- ~ chyp-**
cauliflower, *n.* **phulkobi**
cause, *n.* **gymyn**
cause a burning sensation, *v.* **saw•saw-**
cause irritation, *v.* **basak-**
cause itching, *v.* **basak-**
cave, *n.* (1) **rongkhol ~ rong•khal** (1) **hang•khal** (3) **kep**
caw, *ideo.* **aak**
CD, *n.* **sidikeset**
celebrate, *v.* (1) **ryng-** (2) **sa•-** (3) **chywgyn-** 'celebrate the festival of the dead'
cement, *n.* **simen**
centimetre, *n.* **sentimityr**
centipede, *n.* **sanyrai ~ sanarai**
central government, *n.* **jyw•mrong solkari ~ jyw•myrong sorkar**
centre strap of a sandal, *n.* **pakyl**
certainly, *adv.* (1) **-chong•mot ~ cho•mot** (2) **-saw** (3) **-sem**
chaff, *n.* (1) **cha•wek** (2) **kokpylak**

chain, *n.* (1) **chen** (2) **sikol**
chair, *n.* **choki ~ chuki**
chalk, *n.* **rong•khobok ~ rong•bok**
change, *v.* (1) **bodol-** (2) **tharai-** 'change, exchange, swap' (3) **phyl•-** 'transform, change into'
change into, *v.* **phyl•-**
chaos, *n.* **golmal ~ gormal**
chase, *v.* (1) **bak-** (2) **ryk-**
cheat (on), *v.* **thogi-**
cheek, *n.* (1) **phaithawa ~ phaithopa** (2) **natheng** 'cheek, cheekbone'
cheekbone, *n.* **natheng**
chest, *n.* (1) **chelbak** (2) **kha•phak**
chew, *v.* **chym•-**
chick, *n.* **taw•sa•gyrai**
chicken, *n.* (1) **taw•** (2) **taw•kurung** 'chicken about to lay an egg'
chicken coop, *n.* **taw•nok**
chicken cove, *n.* **taw•nok**
chicken heart, *n.* **kha•rongthai**
chicken pox, *n.* **sasep**
chicken's nest, *n.* **khurung**
child, *n.* (1) **sa•** 'offspring' (2) **sa•gyrai** (3) **babu** (term of address) (4) **sa•daiburung** 'child born out of an incestuous relationship' (5) **bakdongmi sa• ~ bakdongmyng sa•** 'child from a forbidden marriage' (6) **wa•ri ~ wa•ryi** 'child who lost its father' (7) **jyw•ri ~ jyw•ryi** 'child who lost its mother'
childhood, *n.* **pi•sa**
chilli pepper, *n.* **ja•ryt**
chin, *n.* **ka•dymbai**
Chinese rose, *n.* **joba**
ching ching, *ideo.* **teng**
chip (of wood), *n.* **pantiki**
chirp, *ideo.* **chipchip**
chocolate, *n.* **choklet**
choose, *v.* **sai-**
chop, *v.* (1) **khan•-** (2) **kek-** 'chop wood'
chopper, *n.* (1) **khudal ~ kudal** (2) **wa•khu**
Christian, *n.* **kristan ~ kristen**
Christmas, *n.* **krismas**
church, *n.* (1) **gyljanok** (2) **gylja** (3) **mondoli** 'the Christian Church, congregation, church community'

church community, *n.* **mondoli**
cicada, *n.* (1) **ganthi ~ ganthai** (2) **gorweng ~ wengwang**
cigarette, *n.* (1) **sigyret** (2) **biri**
circle, *n.* **bawen**
circular, *adj.* **dawel-**
clam, *n.* **kapangsi**
clan, *n.* **mahari**
clang, *ideo.* **teng**
clang, clang, clang, *ideo.* **thing thing thing**
clarify, *v.* **rongthal-**
class, *n.* **klas**
claw, *n.* **khen•khorong** 'crab's claw, crab's pincher'
clay, *n.* (1) **ha•mang** (2) **tykha** 'white clay'
clean, *adj.* **rongthala-**
clean, *v.* (1) **rongthal** (2) **het-** 'clean an orifice or hole' (2) **sit- ~ syt-**‘ clean out the faces from an animal's intestines'
cleaning cloth, *n.* **pha•lak**
clear (be free from doubt or confusion), *v.* **rongthal-**
clear (free from doubt or confusion), *adj.* (1) **kyryk-** (2) **thal-**(3) **rongthal-**
clear (remove vegetation), *v.* (1) **kam-** (2) **ha•kam-** *both (1) and (2) mean* 'clear the field, cut the jungle make a field, tear out or cut weeds' (3) **haw•-** 'clear or cut the jungle to make a rice field'
clever, *adj.* (1) **seng-** (2) **chalak**
cliff, *n.* **rong•ka**
climb, *v.* (1) **ga•kat-** (2) **man-** (3) **dung•-** (4) **thin-** 'climb a rope that is either vertically hung or horizontally strung' (5) **thintaw-**
clip, *n.* **kilip ~ kylip**
clitoris, *n.* **su•nadylep**
cloaca, *n.* **di•chongkhamai**
clog, clog, *ideo.* **thep thup**
close, *adj.* **nek-**
close, *v.* (1) **chep- ~ chyp-** (2) **kap-** (3) **kak-** 'close with a lid' (4) **chugup-** 'close with a lid' (5) **khup- ~ khep- ~ khyp-** 'close, cover, put on clothes, spread out' (6) **khu•tip-** 'close one's mouth' (7) **buthu-** 'seal, close a receptacle by putting something in the opening'
close to, *prep.* (1) **gychingchi** (2) **nek-**

close together, *adv.* **chapchap**
close one's mouth, *v.* **khu•tip-**
closely, *n.* **kyn kyn**
cloth, *n.* (1) **gamsa** (2) **ba•sek** 'cloth to carry a baby in'
clothes, *n.* **kha•di**
clothes line, *n.* **raityng**
cloud, *n.* (1) **rangbrym ~ rangbyrym** (2) **rangchinek**
coal, *n.* **koila**
coal truck, *n.* **koilagari**
coalmine, *n.* **jul**
cob, *n.* **borong**
cobra, *n.* **dypywpoda**
cock, *n.* (1) **gogylek** 'rooster' (2) **taw•khasi** 'castrated cock/rooster' (3) **ri•** 'penis'
cockroach, *n.* **selu**
cockscomb, *n.* (1) **chyngmat** (2) **suthul**
coconut, *n.* (1) **daba** (2) **narykel ~ narykhel**
coconut water, *n.* **narykeltyi**
coffee, *n.* **kopi**
coil up, *v.* **wen•- ~ wen-**
coin, *n.* (1) **bisyl** (2) **tangka phel sa**
cold, *n.* **soldi** (disease: common cold)
cold, *adj.* **chek- ~ chyk-**
cold rice, *n.* **maichek ~ maichyk**
cold season, *n.* **chekkyryi**
collapse, *v.* (1) **pyi•ru-** (2) **gurum- ~ gyrum-** (3) **dangkhym- ~ dangthym-** (for roads and bridges)
collect, *v.* (1) **jumu-** (2) **ra•gat-** (3) **gyl-** (4) **ryt-** (5) **khyn-** 'collect, gather, pick up' (6) **ha•khyn-** 'collect the remaining cinders after burning the field make it ready for agriculture'
colony, *n.* **kolani**
colour, *n.* **rong**
comb, *v.* (1) **chok-** (2) **chyngmat** 'comb of a rooster' (3) **suthul** 'comb of a rooster'
come, *v.* (1) **rai•-** (2) **rai•a-** (3) **thom-** 'come together' (4) **mili-** 'come together' (5) **wai-**'come back' (6) **dang•-**'come in' (7) **kholthyrai-**'come off (of skin) ' (8) **hongkhot-** 'come out' (9) **jok-**'come out' (10) **song•khot- ~ songkhot-**'come out of a small opening or narrow space, squeeze out of'

come on **hai**
comfort, *n.* **suk**
comfortable, *adj.* **suk-**
comfortably, *adv.* **suksuk**
comic strip, *n.* **jokal**
command, *v.* **hit-**
commit adultery, *v.* **dari-**
commonly, *adv.* (1) **-ramram** (2) **-rong**
compact disc, *n.* **sidikeset**
compare, *v.* **tho-**
compelled, *adj.* **ga•ak-**
compete, *v.* **-susa**
competitively, *adv.* **-susa**
complain, *v.* **sykhym-**
complete *v./adj.* **jam-**
completely, *adv.* (1) **-chyp** (2) **-phin ~ -phyn** (3) **-syrang ~ -srang**
complicated, *adj.* **man•dyk-**
compliment, *n.* **rasong**
compulsorily, *adv.* **-tat**
computer, *n.* **kompiutyr**
confusedly, *adv.* **-wyngwang**
confusion, *n.* **jajyreng**
congregation, *n.* **mondoli**
connect, *v.* **ruchut- ~ ruchu-**
consequently, *adv.* **-sym**
conspire, *v.* **khu•mong-**
construct, *v.* (1) **ham-** (2) **chanpat-**'construct a bamboo bridge'
continent, *n.* **kontinen**
continue, *v.* **-rawraw**
continuing, *adj.* (1) **=khu** (2) **-khyngkhyng**
continuously, *adv.* (1) **-thyngthyng** (2) **-rawraw** (3) **-symsym**
contractor, *n.* **kontrektyr**
convenient, *adj.* **dong•- ~ dong-**
convex, *adj.* **ympong**
cook, *v.* (1) **rym-** (2) **bering- ~ bereng-** 'cook in a bamboo tube' (3) **rymkhap-** 'cook without *mai•tyi ~ mai•ti*' (4) **pyn•-** 'cook in a banana leaf'
cooked, *adj.* **myn-**
cooked without *mai•tyi ~ maiti*, *adj.* **khap-**
cooking place, *n.* **chula**
cooking pot, *n.* **dyksyl ~ tyksyl**
cooking stone, *n.* **phong•khal**
coolish, *adj.* **chekjyrym-**

coop, *n.* **taw•nok**
co-operatively, *adv.* **krymkraw**
copper, *n.* **tama**
copulate, *v.* (1) **pel-** (2) **tong-** (3) **hat-** (4) **chom-**
cord, *n.* **pakara ~ pakyra**
corn, *n.* **abong**
corner of the eye, *n.* **myksep**
corner of the mouth, *n.* **khu•sep**
corps, *n.* **manggisi**
correct, *adj.* **thik dong•- ~ thik dong-**
corrugated iron, *n.* **tin**
cotton, *n.* **kylchap**
cough, *v.* (1) **gusu-** (2) **tokset-**
could, *v.* **=chym**
count, *v.* **khi-**
courageous, *adj.* **kha•dong-**
court, *v.* **bot-**
courtyard, *n.* **nukhu**
cousin, *n.* (1) **chamai ~ chame** (2) **mawsa ~ mosa** (3) **dada** (4) **ja•naw** (5) **abi** (6) **naw** (7) **nono** (8) **jong** (9) **jojong** (10) **phaw•-jong ~ phawjong**
cover, *n.* (1) **kabal ~ kabar** (2) **sokhop**
cover, *v.* (1) **pyn- ~ phyn-** (2) **dap- ~ dep-** (3) **pyndap ~ phyndap** (4) **tandap-** (5) **gyryp-** (6) **chugup-** 'cover with a lid' (7) **gop-** 'cover up, bury'
cow, *n.* **ma•su**
cow fly, *n.* **begyri**
cow shed, *n.* **kuti**
crab, *n.* (1) **hen•** (river crab) (2) **khen•jasyri** 'river crab that is walking on the road'
crabs, *n.* **kakhirok** (head and pubic lice)
crack, *v.* **pheret- ~ peret-**
crack, *n.* (1) **pheret** 'crack in the skin' (2) **cha•pheret** 'crack in the callous skin of the heel' (3) **mai•byram** 'cracks in the skin of the cheek' (4) **sirimynmyn** 'at the crack of dawn'
crash, *v.* (1) **su•kherek-** (2) **pywgak-** (in flight)
crawl, *v.* (1) **man-** (2) **manram-**
crazily, *adv.* **kingreng kingcheng**
crazy person, *n.* (1) **boba** (male) (2) **bobi** (female) (3) **bobylawthok** (male or female) (3) **jabyra**
credit, *n.* **baki**
creep, *v.* **man-**

cricket, *n.* **mangkung, achepchep, mangkhung, che•et**
crime, *n.* **denggu**
criminal, *n.* **gunda**
crippled, *adv.* **lengla**
croak, *ideo.* **pekpek**
crocodile, *n.* **gorial**
crore (2) **kror** (2) **krorsa**
cross, *n.* **chisol**
cross, *v.* (1) **pat-** 'go to the other side' (2) **badai-** 'cross a boundary'
crosscut, *n.* **thong•**
cross-eyed, *adj.* **mykpeng-**
cross-eyed person, *n.* **mykpeng mykpeng**
crosswise, *adv.* **-thong• ~ -thong**
crow, *n.* **daw•kha**
crow bar, *n.* **jomphol**
crowd, *n.* **duma**
crown feathers, *n.* **basu**
crumble, *v.* **roprop-**
crumpled, *adj.* **cho•chep-**
crush, *v.* (1) **kakpyret-** 'crush by biting' (2) **tok-** (3) **tokpyret-** 'crush by hitting' (4) **mu•pyret-** 'crush by sitting on something' (5) **dep- ~ -dap-** (6) **rong•dep-** 'crush with a stone' (7) **ga•jyret-** 'crush with one's foot' (8) **sykjyret-** 'crush with one's hand' (9) **pyi•khyrep-** 'crush with one's hand' (10) **tokdepdep-** 'crush by grinding' (11) **syw•- ~ su•-** 'pound, crush' (12) **ga•kynyng-** 'to trample on, to crush, to destroy' (13) **ga•jonong-** 'to trample on, to crush, destroy' (14) **ga•pyret-** 'crush with one's foot'
cry, *v.* (1) **khep-** 'shed tears when sad' **thore-** 'cry out the name of the lineage of one's enemy'
cubit, *n.* **myk**
cucumber, *n.* **the•myt**
cunning, *adj.* **chalak**
culture, *n.* (1) **ain niam** (2) **takwa rukwa**
cup, *n.* **khap**
cupboard, *n.* **jaljeng**
curl, *v.* **kunremrem-**
curly, *adj.* (1) **kun-** (2) **kunremrem-**
curry, *n.* (1) **ja•bek** (2) **wal•bek** 'bunt curry'
curry juice, *n.* **mai•tyi ~ mai•ti**
curse, *v.* (1) **saw-** 'use bad words' (2) **peng•-** put a curse on someone'

custom, *n.* (1) **ain** (2) **niam** (3) **ain niam** (4) **takwa rukwa**
cut, *n.* (1) **sarai** 'flesh wound' (2) **mat** 'a wound a cut' (3) **matgaba** 'a wound, a cut'
cut, *v.* (1) **tan•-** 'cut, cut up, chop, chop up, slay, slaughter' (2) **ga•dak-** 'cut up' (3) **a•haw•-** 'cut or clear the land make a rice field' (4) **khan•chot-** 'cut hair' (5) **khan•phyt-** 'cut a solid object in half lengthwise' (6) **khan•peret- ~ khan•pyret** 'split, cut open' (7) **khanpyrak-** 'cut a hollow object in half lengthwise' (8) **khan•thong•-** 'cut in half' (9) **khan•tongthong•-** " (10) **pan•pyrak-** 'cut breadthwise' (11) **sinthong•-** 'cut/break in two pieces, cut/break in half, sever' (12) **sintongtong-** 'cut up in many pieces' (13) **byl-** 'cut and kill, slay' (14) **tan•chekchek-** 'cut into small pieces' (15) **tan•choleng-** 'cut a piece out of something' (16) **tan•pyrak-** 'cut open' (17) **tan•set-** 'cut out, cross out' (18) **tankynyng-** 'cut up in many pieces' (19) **mat-** 'sharp, cut' (20) **khan•-** 'to slaughter, to chop, to mince, to cut' (21) **chot-** 'tear, cut' (22) **choka-** 'cut off' (23) **khep-** 'cut with scissors' (24) **sarai mat-** 'to cut someone'
cyclone, *n.* **balmundyri**

# D

dagger, *n.* **kukuri**
daily, *adv.* **-jyryng**
dam, *v.* **dukung-**
damage, *v.* **benek-**
damaged, *adj.* (1) **thymbylong** (2) **nosto dong• ~ nosto dong-**
dammed up, *adj.* **kung-**
damp, *adj.* **su•ut-**
dance, *v.* **byisa- ~ bysa- ~ bywsa-**
dandruff, *n.* **khawkhirok**
danger, *n.* (1) **kyrewami ~ kyryiwami** (2) **jajyreng**
dangle, *v.* (1) **dyngdai-** (2) **wongwet-** (3) **wynget-**
dare, *v.* **pha•-**
dark, *adj.* **manak-**
dark green, *adj.* **khengsyryk**

darling, *n.* (1) **chame ~ chamai** (2) **daldi**
date, *n.* **tarik**
daughter, *n.* (1) **sa•mynchyk** (2) **ama** (3) **rani**
'daughter-in-law, *n.* **namnokhol**
dawn, *v.* (1) **seng•-** (2) **sirimynmyn** 'at the crack of dawn'
day, *n.* (1) **san** (2) **rangsan**
day after tomorrow, *adv.* **cheknai**
daily, *adv.* **sansan**
dazed, *adv.* **rymreng rymreng**
dead body, *n.* **manggisi**
deaf, *adj.* **nagok**
deaf person, *n.* **kala**
deal, *v.* (1) **badyng-** (2) **chiwal-**
death, *n.* **thyiwami**
debt, *n.* **karaw**
decapitate, *v.* **tan•thong-**
deceive, *v.* **thogi-**
December, *n.* (1) **disembyl** (2) **mai•raj** (archaic)
decide, *v.* **myksong-**
decoration, *n.* **dekoresyn**
deep, *adj.* (1) **thyw•-** (2) **nyng•thyw-**
deep place in the river where you can swim or take a bath, *n.* **wari**
deer, *n.* **machok, magachak**
defecate, *v.* **di•it- ~ de•et- ~ di•et**
defective, *adj.* **nosto dong• ~ nosto dong-**
defend, *v.* (1) **chanpheng-** (2) **wara-**
definitively, *adv.* **-chong•mot ~ cho•mot**
deflate, *v.* **peleng- ~ pe•leng-**
degree, *n.* **dygri**
deity, *n.* **myte**
delay, *n.* **dile**
demand, *n.* **dabia**
denomination, *n.* **thorom**
dense, *adj.* **pyn•-**
dense, *adj.* **jyngjang**
dense (of vegetation), *adj.* **tuk-**
depend on, *v.* **sak-**
deplete, *v.* **jam-**
descend, *v.* (1) **wil ~ wyl-** (2) **wylang**
despise, *v.* **mykchep-**
destroy, *v.* (1) **ga•kynyng-** (2) **ga•jonong-** (3) **thap-**
detached, *adj.* **doksylok-**
determinedly, *adv.* **-chong•mot ~ cho•mot**

detrimentally, *adv.* **-phet**
devour, *v.* (1) **kakai sa•-** (2) **monok-**
diagonal, *adj.* **gyching ~ giching**
diamond, *n.* **hira**
diaper, *n.* (1) **di•thap** (2) **dempharai**
dick, *n.* (1) **ri•** (2) **ri•gol** (used as swearword for men)
dictionary, *n.* **diksyneri**
die, *v.* (3) **thyi-** (2) **janggi thyi-**
different, *adj.* **dyngthang**
difficult, *adj.* (1) **rak-** 'mentally difficult' (2) **man•dyk-** 'physically difficult'
dig, *v.* (1) **gat-** (2) **saw•-** (3) **bul-** 'dig up' **jul-** 'dig up'
dining room, *n.* **dainingrum**
dinner, *n.* **maigasam**
directly, *adv.* **-sot**
dirt, *n.* (1) **moila** (2) **serabera** (3) **chinik** 'dirt on the body'
dirty, *adj.* **serabera tak-**
dirty person, *n.* (1) **khalput** (2) **sawthal** 'person who never washes'
disassemble, *v.* (3) **choka-** (2) **dok-**
disco, *n.* **disko**
discover, *v.* **tyngcheng-**
disease, *n.* (1) **sabisi** (2) **gawak**
disembowel, *v.* **phe-**
dish out/up, *v.* **chok-**
dislike, *v.* (1) **khu•bisi-** (2) **chaisi-**
disorderly, *adv.* (1) **koksi kongdang** (2) **sykhathang**
dispose of, *v.* (1) **aset- ~ asyt-** (2) **-set**
dissolve, *v.* **jonong- ~ jorong-**
distant relatives, *n.* (1) **bai•maran** (2) **bai•maran chingmaran**
district, *n.* **distrik**
disturb, *v.* **hala kha•-**
ditch, *n.* **digi**
dive, *v.* **ryp-**
divide, *v.* (1) **soal- ~ sual-** (2) **hanthi-**
divination, *n.* **thama**
divorce, *v.* **jyk aset-**
dizzily, *adv.* **chuwyng chuwang**
do, *v.* (1) **tak-** (2) **kha•-** (4) **sa-** (5) **ytyk-** 'do like this/that' (6) **atak** 'do what?'
do business, *v.* (1) **badyng-** (2) **chiwal-**
dog, *n.* **kyi•**

doll, *n.* **dalibibi**
don't, *v.* **ta•**
donkey, *n.* **gada**
door, *n.* (1) **nokchol** (2) **nokkhap**
down, *adv.* **-rat**
downstream, *n.* **cha•ma**
downstream, *adv.* **-rat**
downward, *adv.* **-rat**
doze off, *v.* **mykjyw-**
drag, *v.* (1) **bytruru-** (2) **doi•-** ~ **joi•-** 'drag through the water like a fishing net'
dragonfly, *n.* (1) **maijyk** (2) **maimijyk**
drain, *v.* (1) **jokset-** (2) **thang•chichat-** (3) **sapset-** (4) **waiset-** 'drain a little water, scoop water out' (5) **khyw-** 'drain, shake fluid out'
draw a knife, *v.* **hot-**
draw a line, *v.* **bytchirit-**
draw or write scratchily, *v.* **bytjekjek-**
dream, *n.* **jumang** ~ **jywmang**
dress, *n.* (1) **dakmanda** (2) **ri•pan** (3) **phatsai**
dress someone else, *v.* (1) **dakan-** (2) **dukhup-** ~ **dykhyp-**
dressed, *adv.* **chup** ~ **chyp**
dried rice grains, *n.* **saram**
drink, *v.* (1) **ryng-** (2) **ryngkhaw-** 'drink sneakily' (3) **ryngkhele-** 'drink for fun' (4) **ka•syrak-** 'drink from the bottle' (4) **kana theka** 'drink liquor'
drip out, *v.* **chokchok-**
drive, *v.* (1) **byt-** (2) **bytganggang-** 'drive a vehicle over a bumpy road' (3) **tat-** 'drive in (as with a nail in wood)'
drop, *n.* **thothak**
drop, *v.* (1) **baphai-** (2) **thagal•-** 'lose'
drown, *v.* (1) **chaw•** (2) **-tyichaw-** ~ **tyichaw•**
drum (barrel), *n.* **dram**
drum (instrument), *n.* (1) **dama** (2) **khem** (3) **rangkha** (4) **rangsyl**
drunk, *adj.* (1) **phek-** (2) **rekhep-** (of plants)
dry, *adj.* **ran•-**
dry, *v.* (1) **ram-** 'dry in the sun' (2) **baw•-** 'make jerky, dry vegetables' (3) **hang-** 'dry meat or other things near the fire'
duck, *n.* **daw•gep**
dud jackfruit, *n.* **chichot**
dugout boat, *n.* **rung**

dumb person, *n.* **khu•ma**
dumbass, *n.* (1) **boba** (male) (2) **bobi** (female) (3) **bobylawthok**
dung beetle, *n.* **di•but**
duplicate, *n.* **dupliket**
dusk, *n.* **walsymsym**
dust, *n.* **busi**
dustpan, *n.* **huksetgaba** ~ **huksetga**
Dutch, *n.* (1) **Nedyran** (2) **Holen**
Dutchman, *n.* (1) **Nedyranmorot** (2) **Holenmorot**

# E

each, *det.* **-phek**
each other, *adv.* **=ruk**
eagle, *n.* (1) **daw•gamdot** (2) **phylgym** (3) **daw•phylgym**
ear, *n.* **nakhal**
ear infection, *n.* **nakhal ruru-** 'to have an ear infection'
earlier, *adv.* **dakang**
earlobe, *n.* **nadekaram**
earring, *n.* **na•thek**
earth, *n.* (1) **ha•** (2) **ha•mang** (3) **ha•gylsak** ~ **ha•gyrsak** 'the Earth'
earthquake, *n.* (1) **banggyri-** ~ **banggiri-** (2) **banggyrigaba** ~ **banggirigaba**
earthworm, *n.* **khansyrui, ha•mai**
easily, *adv.* **kawrawraw**
east, *n.* **saknaram** ~ **salnyram**
easy, *adj.* (1) **nom•-** (2) **althu•-**
easy to deal with, *adj.* **mal-**
eat, *v.* (1) **sa•-** (2) **mai sa•-** (3) **wang•kok-** 'eat without using one's hands, eat with one's mouth' (4) **sa•khele-** 'eat for fun'
eczema between the toes, *n.* **cha•sitokkyreng**
edge, *n.* (1) **rugung** (2) **dareng** (3) **tyisam** 'edge of the water'
edition, *n.* **edisyn**
eel, *n.* **khuchina, na•nyl** 'electric eel'
effortlessly, *adv.* **kawrawraw**
egg, *n.* (1) **tyi•** (2) **taw•ti** ~ **taw•tyi**
egg shell, *n.* **koplak**
egret, *n.* **alabok**
eight, *num.* **chatgyk**
eighteen, *num.* **chi chat**

eighty, *num.* (1) **sot chet** (2) **rum• byryi** (3) **khol chang byryi**
ejaculate, *v.* (1) **hongkhot-** (2) **wat-**
elbow, *n.* **chakchuk**
elbow pit, *n.* **chakkhawak**
elect, *v.* (1) **sai-** (2) **song-**
election, *n.* **ileksyn**
electric eel, *n.* **na•nyl**
electricity, *n.* (1) **karen** (2) **ilektrisiti**
elephant, *n.* **mongma ~ mungma**
elephant tusk, *n.* **mongmawa• ~ mungmawa•**
eleven, *num.* **chitsa**
ell, *n.* **myk**
embrace, *v.* **khabak-**
emerge, *v.* **phet-**
emptiness, *n.* **kantara**
empty, *adj.* **bangbang**
empty, *v.* **batpyret- ~ papret**
encircle, *v.* **bawen-**
enclose (with a fence), *v.* **nol kha•-**
enclose with a fence, *v.* **thuk-**
enclosure, *n.* **nol**
encourage, *v.* **jut-**
end, *v.* (1) **machot-** (2) **jam-**
end of a pointy object, *n.* **chokdeng**
endure, *v.* (1) **sak-** (2) **sakchyk-**
endurance *n.* **sokwa**
enemy, *n.* (1) **bawbyl** (2) **bawbyl chambyl**
engagement, *n.* **inggeech**
engine oil, *n.* **mobil**
England, *n.* **Inggylan**
English, *adj.* **Ingglis ~ Inglis**
engrossed, *adj.* **-saw**
enjoy, *v.* **sak-**
enjoy, *v.* (1) **suk-** (2) **han•seng-** (3) **han•saw-** (4) **suk dong•- ~ suk dong-**
enough, *det.* **dong•- ~ dong-**
enter, *v.* (1) **dang•-** (2) **dangkhym- ~ dangthym-** 'enter a hole'
entrance, *n.* **nokchol**
envious, *v.* (1) **mykbyryw- ~ mykbryw- ~ mykbyru-**
er..., *interj.* (1) **yyy** (2) **yh** (3) **ah** (4) **ma•** (5) **ba• ~ ba** (6) **hai•e**
era, *n.* **chasong**
erect, *adj.* (1) **gang-** (2) **chenggang**

erect something, *v.* **song-**
erection, *n.* **ri•gan•thong**
escape, *v.* **jok-**
especially, *adv.* **dyngthangmancha**
eternal, *adj.* **khengkhang**
Europe, *n.* **Iyrop**
European, *n.* (1) **Iyropmorot** (2) **saip ~ saep**
even, *adv.* **=mangmang**
evening, *n.* (1) **gasam** (2) **gasamphang ~ gasamphak**
event, *n.* (1) **dong•wa** (2) **obosta**
ever, *adv.* **pang•aiba**
every, *det.* **=gumuk**
every time, *adv.* **wetanchian ~ wetantian**
everybody, *n.* (1) **gumukan** (2) **=gumuk** (3) **janggal** (4) **janggalan**
everybody *pron.* (1) **gumukan** (2) **=gumuk** (3) **=khakhet** (2) **-thok**
everyone, *n.* **darang ~ dyrang**
everyone *pron.* (1) **gumukan** (2) **=gumuk** (3) **=khakhet** (2) **-thok**
everything, *n.* (1) **jaggal** (2) **ha•gyrsak ~ ha•gylsak**
everything *pron.* (1) **gumukan** (2) **=gumuk** (3) **=khakhet** (2) **-thok**
everywhere, *adv.* **gumuksangan ~ gumuksang**
exactly, *adv.* (1) **thik thak** (2) **thik** (3) **syrak syrak** (4) **chacha**
exam, *n.* **porika**
excessively, *adv.* (1) **-duga** (2) **-thamak** (2) **-that**
exchange, *v.* **tharai-**
exclusively, *adv.* (1) **=rara** (2) **-mangmang** (3) **=mangmang** (4) **=tara** (5) **khali**
exclusively, *adv.* (1) **=rara** (2) **=tara**
exist, *v.* (1) **ganang** 'exist' (2) **ni•-** 'not exist'
exit, *v.* **hongkhot-**
expectantly, *adv.* **-saw**
expenses, *n.* **koros**
expensive, *adj.* **damrak-**
experience the sensation of being tickled, *v.* **bejaw-**
experience the sensation of irritation or itching, *v.* **bosok-**
explain, *v.* (1) **rongthal-** (2) **thalai hyn•-**
explain, *v.* **tha•let-**

explicit, *adj.* **thal-**
explode, *v.* **pheret-** ~ **peret-**
extinguish, *v.* **myt-**
extortion, *n.* **denggu**
extract, *v.* **hot-**
eye, *n.* **mykren** ~ **mykyren**
eyebrow, *n.* **myksymyl** ~ **myksmyl**
eyelash, *n.* **myksyram** ~ **myksram**
eyelid, *n.* **kykgul**

# F

face, *n.* **mykhang**
face, *v.* **mykhang-**
fail, *v.* (1) **-soksok** (2) **peel dong•** ~ **pheel dong•-** ~ **peel dong-** ~ **pheel dong-** (3) **neng•-**
fake, *n.* **dupliket**
fall, *v.* (1) **galat-**(2) **gal•-** (3) **thang-** 'fall down on something' (4) **chygyp-** 'fall face down on the ground'(5) **rychup-** 'fall on one's face, fall head first' (6) **chabak** 'fall (of water in a waterfall) ' (7) **da•rat-**'fall down (for persons) ' (8) **gal•ruru-** 'fall through something'
famine, *n.* (1) **sa•a siwa** (2) **akal**
familiar, *adj.* **mal-**
family, *n.* **mahari**
fan, *n.* (1) **bangka** ~ **bangkha** (2) **ajip** (3) **motorajip**
far, *adj.* **jan•-**
fart, *n.* **diphu**
fart, *v.* **diphu-**
fast, *adv.* (1) **joljol** (2) **-jol** (3) **bykbyk** ~ **bakbak** (4) **dykdyk** (5) **thapthap** (6) **bakrukrak** ~ **bakrukylak** (7) **rabak** ~ **rabak rabak** (8) **wel•ang** ~ **wel•ang wel•ang**
fast, *adj.* **tarak-**
fat, *n.* **bytym**
fat, *adj.* **mel•-**
fat, *adj.* (1) **mel•-** (person) (2) **chat-** (thing)
father, *n.*
father, *n.* (1) **awa** 'biological father' (2) **baba** (3) **wa•** 'biological father' (3) **wa•gaba** 'someone else's biological father'
father-in-law, *n.* (1) **haw•nokhol** (2) **mama**
fear, *n.* (1) **kyryi** (2) **kyryiwa**

February, *n.* (1) **phebuari** (2) **pargunj** (archaic)
fee paid to a medicine man, *n.* **samphat**
feed, *v.* (1) **hal-** (2) **tu•-** ~ **ty•-**'feed by putting food or drink into someone's mouth' (3) **haldun-** 'feed, provide for'
feel, *v.* (1) **tyng-** (2) **pyi•ram-** 'feel for, search by feeling' (3) **khonok-** 'search by feeling' (4) **kom-** 'feel like a loser' (5) **komok-** 'feel insulted' (6) **kha•thong si-** 'feel pity' (7) **han•tung-** 'feel secure' (7) **sykhym-**'feel sorrow'
feeler, *n.* **kakmyn•**
female, *n.* **gawi**
fence, *n.* (1) **bera** (2) **nol**
fence, *v.* **nol kha•-**
fermented bamboo shoot, *n.* **mai•wakhyi**
fermented fish, *n.* **na•saw**
fermented rice, *n.* **sithi**
few, *det.* **pang•cha**
fidget, *v.* **jot-**
field, *n.* **ha•pal**
fifteen, *num.* **charanga** ~ **chi banga**
fifty, *num.* (1) **sotbonga** (2) **rum•ni chyigyk**
fight, *n.* **golmal** ~ **gormal**
fight, *v.* (1) **takruk-** (2) **dyng-** (3) **tak-**
fighter, *n.* **gunda**
fighting spirit, *n.* **kha•**
fill, *v.* **diphing-**
film, *n.* **philm** ~ **philym** ~ **philim**
filter, *n.* **janti**
filth, *n.* (1) **moila** (2) **serabera**
fin, *n.* **gangthai**
find, *v.* **nuk-**
finger, *n.* **chaksi**
fingernail, *n.* **chaksikhol**
finish, *v.* (1) **jam-** (2) **machot-**
fire, *n.* (1) **wal•** (2) **wal•tum**
fire place (for cooking), *n.* **phong•**
fire place for cooking, *n.* **phong•thu**
firefly, *n.* (1) **wa•mychym** (2) **na•chan**
fireside, *n.* **wal•sam**
firewood, *n.* (1) **pan** (2) **panbai** (3) **panju**
firmly, *adv.* **-khep**
first, *num.* (1) **phas** (2) **phasgaba** (3) **dakang-gaba**
first *adv.* **-cheng**
firstly, *adv.* (1) **dakanggaba** (2) **phasgaba**

fish, *n.* **na•**
fish, *v.* (1) **pun-** (2) **na• pun-** (3) **waribul-** 'fish at the festival of *waribula*'
fish trap, *n.* (1) **dinggarai** (2) **asok** (3) **jolpi** (4) **jonggi**
fishbone, *n.* **asu**
fishery, *n.* **ringaba**
fishery, *n.* **piseri**
fishing basket, *n.* **chokhoi**
fishing hook, *n.* **apun**
fishing line, *n.* (1) **apunkara** (2) **saido**
fishing net, *n.* **chek**
fishing rod, *n.* **apunphong**
fist, *n.* **mym•**
fit, *v.* (1) **sak-** (2) **kyp-** 'fit tightly, fit and close off'
five, *num.* **banga**
fix, *v.* (1) **thik kha•-** 'fix a date and time' (2) **thyk-**' fixed sideways' (3) **chat-** 'fixed together (like a stapled pile of paper or a pile of wood etc.)'
flash, *v.* **welet-**
flat, *adj.* (1) **pa•-** (2) **pylang ~ pelang ~ peleng ~ pyl•eng**
flat piece of hard material, *n.* **khap**
flat rice, *n.* **ryngchyw ~ ryngchu**
flatbread, *n.* **barata**
flat-haired (of animals), *adj.* **miniksuru-**
flatten, *v.* **dypyleng-**
flatter, *v.* **bot-**
flea, *n.* **kaithuk**
flesh, *n.* **randai**
flint stone, *n.* **rong•syl**
flip, *v.* **phekphek-** 'flipping and turning (like fish do on dry land or in a dammed-up fishing place)'
float, *v.* **chaw•-**
floater organs of a fish, *n.* **phe•phong**
floor, *n.* **nokweng**
flour, *n.* **ata**
flow, *v.* (1) **ban-** (2) **chaw•-** (3) **jokruru-** 'flow into'
flower, *n.* **pal**
fluff, *n.* **taw•myn•**
flush out, *v.* **sat-**
flute, *n.* (1) **bangsi** (2) **ymbyng**
fly, *n.* **sot**
fly, *v.* (1) **pyw-** (2) **pywram- ~ poram-** 'fly over'

fog, *n.* (1) **sangori** (2) **goira guri**
fold, *v.* (1) **dem.•-** (2) **sykup-** (3) **the•met-** (4) **khatdep-** (5) **them•-** 'fold up clothes, blankets etc.'
follow, *v.* **sym-**
following, *adj.* **abun**
food, *n.* (1) **sa•wa** (2) **sa•gaba ~ sa•ga** (3) **bering** (4) **beringwa ~ berengwa** 'food cooked in a *wa•sung*'
food pipe, *n.* (1) **tokkhyphu ~ tokyphu ~ tokybu** (2) **nadanggorot**
fool, *n.* (1) **boba** (male) (2) **bobi** (female) (3) **bobylawthok** (male or female) (3) **jabyra**
foot, *n.* (1) **cha•** (body part) 'foot/leg' (2) **cha•pha** (body part and measurement) 'foot' *cha•pha tham* three feet long
foot of a tree, *n.* **ja•phang**
football, *n.* **robol**
football field, *n.* **robolphil ~ robolpil**
footprint, *n.* **cha•man**
footstep, *n.* **cha•gyl**
for a short while, *adv.* **dykdyk**
for a while, *adv.* (1) **bewal** (2) **-phak**
for example, *adv.* **jekhai**
for free, *adv.* (1) **sykhathang** (2) **yndyn**
for fun, *adv.* **-khelek**
for instance, *adv.* **jekhai**
for no reason, *adv.* **ha•sel**
for nothing, *adv.* (1) **sykhathang** (2) **yndyn**
for some time, *adv.* **hawtyi**
forbidden marriage, *n.* **bakdong**
forced, *adj.* **ga•ak-**
forehead, *n.* **chybym**
foreign country, *n.* (1) **phoren** (2) **phoren ha•song**
foreigner, *n.* **phoren**
foreskin, *n.* **ri•khu•chul**
forest reserve, *n.* **rijap**
forget, *v.* (1) **awan-** (used in Badri area) (2) **sangwal-** (used in Siju area)
forgetfully, *adv.* **awan awan**
forgive, *v.* **khema kha•-**
forgiveness, *n.* **khema**
forked branch or post, *n.* **songkhamphek**
fortune teller, *n.* **babaji**
forty, *num.* (1) **sotbyri** (2) **rum•ni**
four, *num.* **byryi**

fourteen, *num.* **chi bri**
fox, *n.* **pheru**
fragment, *n.* **bi•chamchym**
freckle, *n.* (1) **dak** (2) **taw•pal**
freed, *v.* **jok-**
freeloader, *n.* **pokotia**
freely, *adv.* (1) **sykhathang** (2) **yndyn**
fresh, *adj.* **pikheng**
Friday, *n.* **sukulbal ~ sykulbal ~ sykubal**
friend, *n.* **baju**
frog, *n.* (1) **ong ang, kangkang** (2) **rukchok ~ lukchok, rukpekpek ~ lukpekpek ~ rukpek ~ lukpek** (3) **gandalak** (4) **me•mangkyi ~ mi•mangkyi,**
from, *prep.* (1) **=mi ~ =myng** (2) **dabat**
from time to time, *adv.* **gisep gisep ~ gysep gysep**
front, *n.* (1) **mykhang** (2) **okma** 'front of the body'
fruit, *n.* **thai• ~ thai**
fruit fly, *n.* **kyimang**
fry, *v.* **jaw•-**
frying pan, *n.* **tawa**
fucker, *n.* **butang**
full, *adj.* (1) **phing-** (2) **okha-** (after eating)
full moon, *n.* **jarambong**
fully, *adv.* **-phin ~ -phyn**
fungus, *n.* **wa•chan** 'species of fungus that glows in the dark'
fungus infection, *n.* **ranggyl**
fur, *n.* **myn•**
furthermore, *adv.* **aro**
future, *n.* **mykhang**

# G

gain, *n.* (1) **lap** (2) **man•dapami ~ mandapwami**
gain, *v.* (1) **lap-** (2) **man•dap-**
gall, *n.* **ke•ret**
gall bladder, *n.* **piryt**
game, *n.* **khelegaba**
gang up on, *v.* **thop-**
garbage, *n.* **jabol**
garbage heap, *n.* **jaboldam**
garden, *n.* (1) **bagan** (2) **bari**

garlic, *n.* **rasun pibok**
Garo *n./adj.* **Ha•chyk**
gather (around), *v.* (1) **duma-**(of people) (2) **dum-** 'gather, swarm'
gather (something), *v.* (1) **jumu-** (2) **gyl-** (3) **khyn-**(4) **ryt-**
gay person, *n.* **hijra**
gear, *n.* **geer**
gecko, *n.* (1) **pape ~ baphe** (2) **keko ~ koke** (3) **lukchokchok** (4) **noktapa**
generation, *n.* **chasong**
get, *v.* (1) **man•-** 'obtain, get' (2) **ra•-** 'get, take, buy'
get better, *v.* **nem-**
get up, *v.* **jasa-**
gibbon, *n.* **hu•raw**
gift, *n.* **bot**
gill, *n.* (1) **kaw•warai** (2) **chengkhana**
ginger, *n.* **cheng•khu ~ cheng•khyw**
girl, *n.* (1) **gawi** (2) **nawmyl**
girlfriend, *n.* **chame ~ chamai**
give, *v.* **hyn•-**
give an injection, *v.* **biji su•-**
give birth, *v.* **ba•-**
gizzard, *n.* **di•thom**
glans penis, *n.* (1) **ri•ambanthai** (2) **ri•kun**
glass, *n.* **gylas ~ gilas**
glitter, *v.* (1) **chyng•chet-** (2) **tengchypchyp-**
globular, *adj.* **thywkhong**
glottal stop, *n.* **raka**
glove, *n.* **chakkhop**
glutton, *n.* **khalbong**
go, *v.* (1) **rai•-** 'go/come' (2) **ram rai•-** (3) **re•eng-** 'go away, leave' (4) **dangkhym- ~ dangthym-** 'go into a hole' (5) **waiphin-** 'to back, return' (6) **chaw-** 'go by boat' (7) **wil- ~ wyl-** 'go down, descend, get off' (8) **wylang-** 'go down, descend' (9) **rai•phak-** 'go through' (10) **taw-** 'go up, ascend' (11) **thyl•-** 'go very far' (12) **re•eng-** 'go, go away, leave' (13) **rai•ganggang-** 'go/drive/ride over on a bumpy road' (14) **hongkhot-** 'go/come out' (15) **sa• ba•na sa-** 'go into labour'(16) **parang** 'wanter, go astray' (17) **wai-** 'go back' (18) **dang•-** 'go in'
goal, *n.* **gol ~ gool** (in football)
goat, *n.* **purun**

God, *n.* (1) **Isol** (2) **Nokgaba**
god (pagan), *n.* (1) **myte** (2) **Rabuga** 'god who created the world (according to ancient religion)'
going (to somewhere), *n.* **waisa**
goitre ~ goiter, *n.* **tokbaw**
gold, *n.* **sona**
gollop, *adv.* **lyp tak-**
gong, *n.* (1) **rang** (2) **rangkha** (3) **rangsyl**
good, *adj.* (1) **ga•-** (2) **nem-**
good fortune, *n.* (1) **nemgyni** (2) **rasi**
good-for-nothing, *n.* **lapchagaba ~ lapchaga**
gorge, *v.* **nal-**
gossip, *v.* **golpho kha•-**
gourd, *n.* **mantaw**
government, *n.* (1) **gobormen ~ golmen** (2) **sorkar**
grab, *v.* **pyi•-**
grain of sand, *n.* **rong•phek**
gram, *n.* **grem**
grammar, *n.* (1) **gremyr** (2) **khu•hamgaba**
granary, *n.* **pung**
grandchild, *n.* (1) **syw•** (2) **achu** (3) **chuchu** (4) **abu**
granddaughter, *n.* (1) **syw•** (2) **abu**
grandfather, *n.* **achu**
grandmother, *n.* (1) **abu** (2) **awyi** (3) **wyi•**
grand-nephew / great-nephew, *n.* (1) **khyryithangsyw•** (2) **syw•**
grand-niece / great-niece, *n.* (1) **namchyksyw•** (2) **syw•**
grandson, *n.* (1) **syw•** (2) **chuchu**
grape, *n.* **drakha**
grasp, *v.* (1) **sykrom-** (2) **pyi•-**
grass, *n.* **samsi**
grasshopper, *n.* (1) **aguk** (generic) (2) **gukchepchep** (3) **gukmadym** (4) **ha•chepchep** (5) **aguk** (6) **bukylek** (7) **gugyrengsa•** (8) **aganggi** (9) **gugyreng**
grate, *v.* **morot-**
gratuitously, *adv.* **dymdam**
grave, *n.* **gopram**
grease, *n.* **bytym**
greater, *v.* **dai-**
great-granddaughter, *n.* **syw•muri**
great-grandfather, *n.* **achumuri**
great-grandmother, *n.* **abumuri**
great-grandson, *n.* **syw•muri**
green, *adj.* **khengchek**
grind, *v.* (1) **tok-** (2) **tokdepdep-**
groin, *n.* **cha•phak**
group, *n.* (1) **jinma** (2) **dol** (3) **tho•ma** (4) **-burung**
grow, *v.* (1) **chungtaw-** (2) **kek-** (3) **dym-** (of plants) (4) **buk-** 'grow like a creeper or liana' (5) **chunggalgal-** 'grow up, become an adult'
grunt, *ideo.* **•hm**
growl, *ideo.* **gulgul galgal**
guard, *v.* (1) **chaithum-** (2) **rakhi-**
guava, *n.* (1) **pangkol** (2) **pangkywal**
guide **tun- ~ tyn-**
gullet, *n.* (1) **tokkhyphu ~ tokyphu ~ tokybu** (2) **nadanggorot**
gun, *n.* **bonduk ~ bondyk ~ byndyk**
gush out, *v.* **batpyret- ~ papret**
gut, *v.* (1) **phe-** (2) **than•khoana-** 'gut lengthwise/longitudinally'

# H

habit (1) **bewal** (2) **takbewal**
hail, *n.* **syltyi**
hair, *n.* (1) **khaw** 'hair of the head' (2) **myn•** body hair (of humans), fur (of animals) (3) **ri•myn** 'pubic hair of a male' **su•myn** 'pubic hair of a female' (4) **chachura** 'hair on top of the head' (5) **khawra** 'hair that has fallen out of the head' (6) **myn•symok** 'small body hair' (7) **khawchi ~ khawkhai ~ khawkhi** 'grey hair' (8) **khawchyryng** 'scalpel hair'
hairdresser, *n.* **napit**
hairy with small hairs, *adj.* **myn•sym-**
half, *n./det./adj.* (1) **atha** (2) **di•thap** 'half of a volume' (3) **thong•** 'half which is the result of a cross section or a cut across the width' (4) **phak** 'half which is the result of a longitudinal section or a cut along the length' (5) **hap** (6) **sare** 'half past' (only used to tell the time)
half-brain, *n.* **nawang**
halo of the moon, *n.* **baguriwa•**
hand, *n.* (1) **chak** 'hand/arm' (2) **chak chok khjyks**, *n.* hand/arm

hand over, *v.* **ra-**
handful, *n.* **chakwak**
handkerchief, *n.* **engkal ~ ingkal**
handle, *n.* (1) **gan•thong** (2) **phong**
handle, *v.* **kha•at-**
handwriting, *n.* **henraiting**
hang, *v.* (1) **syithai- ~ syithyi- ~ syithi-** (2) **nang-** (3) **bijyrang-** 'hang when dry' (4) **ram-** 'hang in the sun to dry' (5) **nang-** 'hang down from' (6) **toktai-** 'hang oneself' (7) **khaithyi-** 'hang oneself' (8) **tyngtet-** 'hang someone'
hanging root, *n.* **kyrydyl**
happiness, *n.* **suk**
happy, *v.* (1) **khusi dong•- ~ khusi dong-** (2) **han•seng-**
harbour, *n.* **khana**
hard-on, *n.* **ri•gan•thong**
harelip, *n.* **khu•chit**
harvest, *n.* (1) **bytwa ~ bytwami ~ bytwamyng** (2) **saphang** 'first rice harvest' (3) **maikung** 'second rice harvest'
hat, *n.* **tupi**
hate, *v.* (1) **khu•bisi-** (2) **chaisi-**
haughtiness, *n.* **stel**
have, *v.* (1) **ganang** 'have. there is/are' (2) **ni•-** 'not have, there isn't/aren't'
have a cold, *v.* **tokset-**
have a hole in a cloth or paper as the result of burning, *v.* (1) **khampyryw-** 'have a hole in a cloth or paper as the result of burning' (2) **khamthymbylong-** 'have a hole in a road or bridge as the result of burning'
have a hole in it, *adj.* (1) **phyryw** (2) **thymbylong** (of roads, bridges and wooden planks)
have a nutty taste, *v.* **jyryk-**
have a sore, *v.* **thaphu-**
have a warm body (not of fever), *v.* **tungbul-**
have an orgasm, *v.* **takruk-**(of a woman)
have an ulcer on one's skin, *v.* **matchirit-**
have bronchitis **leng man•-**
have diarrhoea, *v.* (1) **di•chyrak-** (2) **di•pyru-**
have holes in it (of clothes), *adj.* **bukalang**
have itchy eyes, *v.* **mykbyryw- ~ mykbryw-**
have malaria, *v.* **dykym sa-**

have rabies, *v.* **kong•-**
have the hiccups, *v.* **thy•yk-**
have the sour-sweet taste of a half-ripe fruit, *v.* **symjin-**
have to, *v.* **nang-**
hawk, *n.* **daw•reng**
hay, *n.* **maikap**
he *pron.* (1) **ge•theng ~ de•theng** (2) **ue** (3) **ie**
head, *n.* **dykym**
headmaster, *n.* **hetmastel**
headmistress, *n.* **hetmadam**
head band, *n.* **rumal**
head cloth, *n.* (1) **khawkhuthuk** (for men) (2) **khophynga** (for women)
headman, *n.* **nokma**
heal, *v.* (1) **nem-** (2) **san-**
heap, *n.* (1) **ali** (2) **chok** (3) **thep** (4) **thom**
heap up, *v.* **thom•-**
hear, *v.* (1) **na-** (2) **nakhal na-**
heart, *n.* **kha•thong**
heat, *v.* (1) **tunget-** (2) **chawarai-** 'heat meat on a frying pan without salt or water in order preserve it'
heat up, *v.* **tunget-**
heathen, *n.* **songsyrek ~ songsarek**
heavy, *adj.* **chyrym**
heel, *n.* (1) **cha•dok ~ cha•tok** (2) **cha•pakithyk** (3) **hil** (of a shoe)
heiress, *n.* (1) **nokchina** (1) **nokna** (2) **nokrom**
helicopter, *n.* **alukotar**
hell, *n.* **norok**
Hello? Is someone there?, *interj.* **huhu**
help, *v.* **taksak-**
her, *det.* (1) **get•theng ~ de•theng** (2) **get•thengmi ~ ge•thengmyng ~ de•thengmi ~ de•thengmyng**
herd, *n.* **jinma**
herd, *v.* **ryk-**
here, *procl.* **ha•**
here, *adv.* **ichi**
heron, *n.* **alabok**
hers, *pron.* (1) **get•theng ~ de•theng** (2) **get•thengmi ~ ge•thengmyng ~ de•thengmi ~ de•thengmyng**
hey, *interj.* (1) **ha** (2) **chys** (3) **hyt** (4) **oi** (5) **o**
hibiscus sabdariffa, *n.* **dachang**

hide, *n.* **khol**
hide, *v.* (1) **kel-** 'hide behind or in something' (2) **kyl-** 'hide, avoid' (2) **sari-**'hide something, keep something secret'(3) **gop-** (4) **thym-**'lie in ambush, lie hidden'
high, *adj.* **chyw•-**
high ground, *n.* **ha•kha**
hill, *n.* **ha•byri**
hillslope, *n.* **ha•kha**
hinder, *v.* (1) **peng ~ peng•-** (2) **khapeng-**
hinge, *n.* (1) **eskrup** (2) **khopja**
hip, *n.* **cha•pungdym**
his, *det./pron.* (1) **get•theng ~ de•theng** (2) **get•thengmi ~ ge•thengmyng ~ de•thengmi ~ de•thengmyng**
history, *n.* **itihas**
hit, *n.* **byl•**
hit, *v.* (1) **khi•-** 'hit a target, hit the mark'(2) **sat-•-**'hit with a stick or bat, cut with a sword' (3) **satpyret-**'hit with the open hand' (4) **chympyret-**'hit with one's fist, crash head-on'(5) **ga•phak-**'hit with one's foot while walking' (6) **tap-**'hit, beat-up' (7) **thot-**'hit, bump into something or against something' (8) **chagak-**'hit, crash into' (9) **rai•phak-**'elbow, hit with one's elbow while walking'
hit, *ideo.* **thap**
hive, *n.* **ne•katthup**
hoe, *n.* (1) **khudal ~ kudal** (2) **wa•khu**
hoist, *n.* **jomphol**
hold, *v.* (1) **diri- ~ dyri- diritat-**'hold firmly' (2) **pyi•khep-**'hold firmly' (3) **bala-**'hold in the beak' (4) **pyi•thyng-**'hold on, grasp' (5) **pheng•chang-**'hold something in front of something else'
hold a ceremony or celebrate in commemoration of a dead person one year after they died, *v.* **sorot-**
hold a meeting, *v.* **miting**
hold a vigil or watch over the body of a dead person, n., *v.* **walmykrak-**
hold a wake, *v.* (1) **mykrak-** (2) **wal mykrak-**
hold out, *v.* (1) **sak-** (2) **sakchyk-** (3) **sok-**
hole, *n.* (1) **khal** (2) **singsingkholong ~ syngsyngkhol ~ syngsyngkholong** 'hole in the ground'

Holland, *n.* **Holen**
hollow, *adj.* **gopgylang**
hollow, *v.* **phyryw-**
hollow between the roots of a tree, *n.* **cha•kok**
hollow part of the elbow, *n.* **chakkhawak**
hollow side of the knee, *n.* **cha•khawak ~ cha•khok**
holy, *adj.* **rongthalgaba**
home, *n.* **nok**
homosexual, *n.* **hijra**
hoop, *n.* **sylkeng**
hope, *v.* **kha•dong-**
hopeful, *adj.* **kha•dong-**
hopscotch, *n.* **aiding**
horn, *n.* (1) **korong ~ kyron** (of an animal) (2) **ka** (traditional instrument)
hornbill, *n.* **hynggek**
horse, *n.* **gorai ~ gore**
horseback riding, *n.* **gore dung•-**
hot, *adj.* **tung-**
hot season, *n.* **tungkyryi ~ tyngkyryi**
hour, *n.* **khanta ~ khantha**
house, *n.* (1) **nok** (generic) (2) **bilding** 'house built with masonry' (3) **dolan** 'very big house' (4) **nokthai** 'separate, small house on a premises'
house builder, *n.* **mistyri**
house owner, *n.* **nokgaba**
housefly, *n.* **sotmai**
how?, *adv.* (1) **atakai ~ atykyi ~ atakai** (2) **atongtykyi**
how come?, *interr.* **atongtykyi**
how many?, *interr.* (2) **biskyn** (2) **=byisyk**
how much?, *interr.* (1) **biskyn** (2) **=byisyk**
however, *adv.* (1) **ytykchiba** (2) **ytykchido**
however many/much, *adj.* **jesykyn**
huh?, *interj.* **ymyi**
human, *n.* **morot**
hundred, *num.* (1) **rajasa** (2) **rum•banga**
hunger, *n.* **ok**
hungry, *adj.* **okhi-**
hurt, *v.* **sa-**
husband, *n.* (1) **jyk** (2) **biphagaba** (3) **sai**
husband and wife, *n.* **jyksai**
husk, *n.* (1) **khoppalak ~ khoppylak** (2) **maikhol** (of rice)

# I

I *pron.* (1) **ang** (2) **anga**
I agree. **ho•ong**
I don't agree. (1) **hy•** ~ **hy•y** ~ **yhy•** (2) **hm•m** ~ **m•m** ~ **mm**
I don't care. **chungai rai•cha**
I don't know. **haida**
ice, *n.* **syltyi**
idea, *n.* (1) **chol** (2) **aidia**
idiot, *n.* (1) **burbok** ~ **bulbok** (2) **ga•tha** ~ **gatha** (male) (3) **gathi** (female) (4) **boba** (male) (5) **bobi** (female) (6) **bobylawthok**
if, *conj.* **=chido**
ignite, *v.* **chak-**
ignoramus, *n.* (1) **boba** (male) (2) **bobi** (female) (3) **bobylawthok**
ignore, *v.* **sari-** 'ignore someone out of shame or hatred'
ill, *adj.* **sa-**
illness that makes everything taste bitter, *n.* **sajin**
imaginary, *adv.* **-chyp**
imagine, *v.* **chanchichyp-**
imitate, *v.* **-sym**
important, *adj.* (1) **gamchat** (2) **nangchomot-**
imprisoned, *adj.* **chep-** ~ **chip-** ~ **chup-** ~ **chyp-**
in a group, *adv.* (1) **-kyrym** (2) **-thok**
in a pile, *adv.* **japrukruk**
in accordance with, *adv.* **kri**
in addition, *adv.* (1) **aro** (2) **-dap** (3) **-khan** (4) **-pha**
in co-operation, *adv.* **krymkraw**
in front of, *prep.* **mykhangchi**
in half, *adv.* **-thong•** ~ **-thong** (crosswise)
in one blow, *adv.* **-phak**
in one go, *adv.* **tyngkarang**
in reverse motion, *adv.* (1) **kynbyret** (2) **-phin** ~ **-phyn**
in succession, *adv.* **-damdam**
in that case, *adv.* (1) **ytykchiba** (2) **ytykchido**
in the evening, *adv.* **hampyi**
in the far future, *adv.* **naija**
in the future, *adv.* **hambun**
in the late afternoon, *adv.* **hampyi**
in the past, *adv.* **dakang**
in the recent past, *adv.* **maja**
in turn, *adv.* **=sega** ~ **=siga**
in unison, *adv.* **krymkraw**
in vain, *adv.* (1) **yndyn** (2) **magyna** (3) **=chym**
in various places, *adv.* **repa chepa**
inadvertently, *adv.* **-ram•**
inattentively, *adv.* **-parang**
incisors, *n.* **wachyw**
inclination, *n.* **gyching** ~ **giching**
inclined, *adj.* **gyching** ~ **giching**
increase, *v.* **jel-**
increasingly, *adv.* (1) **-ruru** (2) **-rawraw**
indeed, *adv.* **=ba**
index finger, *n.* **chaksijotram**
India, *n.* **India**
infection of the inner ear, *n.* **kholjisop**
influence, *n.* **jaria**
inherit, *v.* **man•symrukruk-**
inject, *v.* **syw•-** ~ **su•-**
injection, *n.* **biji**
injection needle, *n.* **biji**
innards, *n.* (1) **pipuk** (2) **taw•puk** (of a chicken) (3) **wakpuk** (of a pig)
inquire, *v.* **sandi-**
insect, *n.* **chong•** (generic), **buna, me•mesi, wa•khal, ganthai, thurung**
insert, *v.* (1) **syket-** ~ **saket-** (2) **thek-** (3) **suk-** 'insert, stitch'
inside, *n.* **nyng•**
inside out, *adj.* **bykphyl**
insipid, *adv.* **dapet**
inspect, *v.* **chai-**
instead, *adv.* **=sega** ~ **=siga**
instead of, *adv.* (1) **-thum** (2) **phal**
instep, *n.* **cha•bykung**
instruct, *v.* **thikthik-**
insult, *v.* **chonnyk-**
insurance, *n.* **insuren**
intelligence, *n.* **sung**
intelligent, *adj.* **seng-**
intend, *v.* **myksong-**
intensely, *adv.* **-phetphet**
interest, *n.* (1) **lap** (2) **man•dapami** ~ **man-dapwami**
interval, *n.* **gesep** ~ **gysep** ~ **gisep**
intestines, *n.* (1) **pipuk** (2) **taw•puk** (of a chicken) (3) **wakpuk** (of a pig)

into pieces, *adv.* (1) -**chichi** (2) -**khynyng** (3) -**pyrak**
invite, *v.* **phi-**
iron, *n.* (1) **syl** (2) **sylgythym**
irritated, *adj.* **bosok-**
island, *n.* **tyichang**
it doesn't matter **ytykkhal**
itch, *v.* **bosok-**

## J

jack up, *v.* **jul-**
jackal, *n.* **sial**
jackfruit, *n.* (1) **panchung** (2) **chichot** 'dud jackfruit'
January, *n.* **januari**
jaundice, *n.* **holdiasop**
jaw, *n.* **chengkhyna ~ chengkana**
jealous, *adj.* (1) **mykbyryw- ~ mykbryw- ~ mykbyru-** (2) **sa•nal- ~ sa•nyl-**
jerk, *v.* (1) **bytjekjek-** 'give short jerks' (2) **soksok-** 'jerk off, masturbate)
jerkily (over a rough road), *adv.* **phyltawtaw**
jerky, *n.* **garan**
jerry can, *n.* **galon**
join, *v.* **ruchut- ~ ruchu-**
joint, *n.* **weng•**
joke, *n.* (1) **somphi** (2) **mimiwami ~ mimiwamyng**
joke, *v.* (1) **rophil- ~ rophyl-** (2) **chikyrak- ~ chikarak-**
joyful, *adj.* (2) **han•seng-**
judge, *n.* **loskor**
jug, *n.* **boiom**
juice, *n.* (1) **tyi** (2) **ros**
July, *n.* (1) **julai** (2) **sa•wynj** (archaic)
jump, *v.* (1) **ho-** (2) **thorok-** 'jump down from/out of/into' (3) **hochorokchorok-** 'jump like a deer' (4) **hopat-** 'jump like taking a step, i.e. with one's legs apart, not with both legs together' (5) **pywtaw-** 'jump over' (6) **hojokjok-** 'jump up and down' **jok-** 'jump because something startled you'
June, *n.* (1) **jun** (2) **asalj** (archaic)
jungle, *n.* **palyng**
jungle fever, *n.* **han•dykmai**
jungle fowl, *n.* **taw•palyng**
jungle goat, *n.* **matrong**
jungle thicket, *n.* **pangyrym**
just, *adv.* (1) =**ari** (2) **dymdym damdam** (3) **syrak syrak** (4) -**mangmang**

## K

kangaroo, *n.* **kangguru**
keep, *v.* (1) **song-** 'keep, store' (2) **rin-** 'keep as domestic animal' (3) **mu•ten-** 'keep company, look after, watch' (4) **rakhi-** 'keep, guard, look after' (5) **sung ra•-** 'keep in mind' (6) **rakhi-** 'keep, guard' (6) **mu•-** 'keep doing something (durative)' (7) **dep- ~ -dap-** 'keep together by force' (8) -**saw**
kernel, *n.* **karan**
key, *n.* **chabi**
Khasi *n./adj.* (1) **Khasi** (2) **Dykyl** (pejorative)
kick, *v.* (1) **ga•thyng-** (2) **ga•ryngreng-** (3) **thyng-** (4) **thyngpyret-**
kidney, *n.* **kha•rongthai**
kill, *v.* (1) **so•ot-** (2) **dythyi-**
kilogram, *n.* **keji**
kilometre, *n.* **kilomytyr**
kind, *n.* **rokhom**
kindle, *v.* **walchak-** 'kindle the fire with one's breath by blowing'
king, *n.* **raja**
king stud, *n.* **manjuri ~ manjyri**
kiss, *v.* **khu•thym-**
kitchen, *n.* **babelsi ~ babylsi**
kite, *n.* (1) **lekadaw•reng** (2) **daw•reng**
knee, *n.* **cha•kyw ~ cha•ku** (both the body part and the measurement: the length from the knee to the foot)
knee pit, *n.* **cha•khawak ~ cha•khok**
kneel, *v.* **cha•gywgyw-**
knife, *n.* (1) **chang•kui ~ cheng•kui ~ chaw•kyi ~ chaw•ki** (2) **chang•khui nagap ~ chang•khui kaldap** (3) **chang•kuikatri ~ cheng•kuikatri ~ chaw•kyikatri ~ chawkikatri** (4) **churi** (5) **mongreng ~ mongyreng** (6) **wai•cheng** (7) **wai•seng**
knot, *n.* (1) **hen•** (2) **nabak** (3) **theng•**

know, *v.* (1) **tyng-**'know a fact or person' (2) **sap-**'know a skill'
knowledge, *n.* **tyngwami**
knuckle, *n.* **chaksiweng**

## L

labour, *v.* **sa• ba•na sa-** 'go into labour'
labyrinthitis, *n.* **kholjisop**
lace, *n.* **barat**
lack, *v.* **neng•-**
lad, *n.* **bipha**
ladder, *n.* (1) **guchung** (2) **dojanggre**
lady's finger, *n.* **dorai**
lakh (1) **lak** (2) **laksa**
lament, *v.* **synthi-**
land, *n.* (1) **ha•** (2) **kyndam** 'land behind a village' (3) **ha•khyng** 'land that belongs to a *nokma*'
landlord, *n.* **nokgaba**
landslide, *n.* **rurong-**
language, *n.* **khu•chuk**
last, *adj.* (1) **las** (2) **lasgaba** (3) **kynpha-** (4) **jamkhamwa**
last, *v.* **kan•-**
last night, *adv.* **taija**
last year, *adv.* **teraka**
late, *adj.* **kynpha-**
later, *adv.* (1) **kynsang** (2) **te•en** 'later today' (3) **hambun** 'later, but not today'
latex, *n.* **myn•tyi**
laugh, *v.* **mimi-**
law, *n.* (1) **ain** (2) **niam**
lay (1) **syn- ~ syn•- ~ sin- ~ sin•-** 'lay, lay out, spread out on something' **dan-** 'spread out, lay out (mats etc.) ' (6) **tyi•-**'lay an egg'
layer, *n.* **tarang**
lazy person, *n.* (1) **alsia ~ halsia** (2) **halsia kongtoksi**
lead, *v.* (1) **dyl-** (2) **-tyn** (3) **tun- ~ tyn-**
leader, *n.* **dylgaba**
leaf, *n.* (1) **panchak** (2) **chak** (3) **chyw•** 'young leaf'
leak, *v.* (1) **sel-** (2) **chep-** (3) **jok-** 'leak out'
lean, *v.* (1) **dandan-** (2) **chanchok-**
learn, *v.* **ski- ~ syki-**

leave, *v.* (1) **re•eng-** (2) **tanang-**'leave alone' **tanset-** 'leave behind' (3) **jagat-** 'experience that one's soul leaves one's body and temporarily enters an animal'
leech, *n.* (1) **uching ~ u•ching ~ ukching** (2) **gomga ~ gomgaba** (3) **nadanggap** (4) **batro** (5) **u•chingrawi**
left, *n.* **jagysi**
leftovers of cooked rice dried in the sun used to feed the pigs, *n.* **maijyreng**
leg, *n.* **cha•** 'leg/foot'
lemon, *n.* (1) **chinara** (2) **gakji**
lengthwise, *adv.* **-phak**
leprosy, *n.* **khonchi**
let it be **utyk udong**
Let's go. **hai**
letter, *n.* **chiti**
liana, *n.* (1) **durymytdyl** (2) **karydyl** (3) **taw•cha•si** (4) **me•mangguchung** (5) **na•lamsusyrakdyl**
liar, *n.* (1) **bokbok ~ bongbong ~ bong** (2) **thol•am**
licence, *n.* **laisen**
lick, *v.* **sa•lak-**
lid, *n.* **kak**
lie, *v.* (1) **jyw-** 'lie down (both the movement and the position) ' (2) **kapkap-** 'lie flat on one's belly' (3) **thym-**'lie in ambush, lie hidden' (4) **jywdap-**'lie on' (5) **jywkarang-** 'lie on one's back' (6) **jywbythyn-** 'lie on one's belly' (7) **jywkapkap-** 'lie on one's belly' (8) **jywgebeng-** 'lie on one's side'
lie (tell lies), *v.* **thol•-**
life, *n.* (1) **janggi** (2) **janggi khengwa** (3) **sung**
lift up, *v.* (1) **paitaw-** (2) **jul-** 'lift up, uproot, swell' (3) **phok-** (4) **gadaw-**'lift one's chin up'
light, *n.* **lait**
light, *v.* **chak-**
light (not heavy), *adj.* **cheng•-**
light green, *adj.* **pibok**
lightning, *n.* **rangdylekpa**
like, *adv.* **=tykyi ~ =takai**
like, *v.* (1) **nemnuk-** (2) **mykcha-** 'fancy'
like that, *adv.* (1) **ytykyi** (2) **ytyken**
like this, *adv.* (1) **itykyi**
limb, *n.* **sam**

lime stone, *n.* **rong•chun**
limestone, *n.* **chun**
limit, *n.* **sima**
line, *n.* **raityng**
lion, *n.* **singho**
lip, *n.* **khu•chul**
lipstick, *n.* **lepstik**
liquor, *n.* **chyw**
listen, *v.* **natym-**
litre, *n.* **lityr**
little bit, *adv.* **choi•sa ~ cho•sa**
little finger, *n.* **chaksirengma**
live somewhere, *v.* **mu•-**
livelihood, *n.* (1) **chol** (2) **chol chal**
liver, *n.* **bi•thyn ~ pi•thyn**
lizard, *n.* (1) **kangkylek** (2) **phu•chul** 'monitor lizard'
lock, *n.* **tala**
lock, *v.* **thek-**
locked up, *adj.* **chep- ~ chip- ~ chup- ~ chyp-**
log, *n.* (1) **gandi** (2) **panchoka** 'small log'
logboat, *n.* **rung**
loin cloth, *n.* **kalai**
long, *adj.* (1) **raw•-** 'long, tall' (persons, things) (2) **jaraw-** (long time) (3) **-kham** (long time)
long pants, *n.* **longpen**
look, *v.* (1) **chai-** 'look at' (2) **mu•ten-** 'look after, watch, keep company' (3) **chairuru- ~ chairura-** 'look around' (4) **chonnyk-** 'look down on, mock, scorn, insult' (5) **mykchep-** 'look down on, despise, scorn' (6) **mysepai** (7) **chai-**'look/watch with one eye' (8) **gopjyrujyru-** 'look down with one's head bent down' (9) **rakhi-** 'guard, look after, keep'
look like, *v.* **nuk-**
lopsided, *adj.* **ympong**
lose (fail to win), *v.* **magana-**
lose (unable to find), *v.* (1) **ma-** (2) **thama-** 'make unable to find'
loosen *v.* **nom•-**
lotus, *n.* **mongnal**
loud, *adj.* **rak-**
louse, *n.* **kakhirok, thik ~ kuythi** (on dogs), **khyryk**
love, *n.* (1) **chame ~ chamai** (2) **daldi**

love, *v.* **kha•gal-**
lover, *n.* (1) **chame ~ chamai** (2) **kha•wa**
low, *adj.* **pa•-**
low ground, *n.* (1) **ha•dawak** (2) **ha•rongrong**
lower primary school, *n.* **roal**
lower side of a hill, *n.* (1) **ha•dawak** (2) **ha•rongrong**
luck, *n.* **nemgyni**
luggage, *n.* **peking ~ pheking**
lunch, *n.* **maisan**
lung, *n.* **kha•sop ~ sokrop ~ sokyrop**
lychee, *n.* **lechu**
lying on one's back, *v.* **bathan**
lying on one's belly, *v.* **byryp**
Lyngam *n./adj.* **Megam**

# M

macaque, *n.* (1) **amak** (2) **ranggorai**
magic, *n.* **jadu**
magic spell, *n.* **muni**
magistrate, *n.* **mejistret**
magnet, *n.* **chymbuk**
main river, *n.* **tyimong**
main village, *n.* **songmong**
maize, *n.* **abong**
make, *v.* (1) **tak•-** (2) **-et**
make a great effort, *v.* **khereng-**
make a hole in a road or bridge by stamping, *v.* **ga•thymbylong**
make a rope by rubbing thread between one's hands, *v.* **sak-**
make a sound, *v.* **kyryng-**
make an about turn, *v.* **kynjung-**
make an angle, *v.* **gychingching mu•-**
make barren, *v.* **bak-**
make beautiful, *v.* **sylet-**
make fire, *v.* **wal• chak-**
make jerky, *v.* **baw•-**
make last, *v.* **tiktik-**
make liquid come out, *v.* **ruru-**
make noise, *v.* (1) **kyryng-** (2) **jykjak-** (3) **dykyryng-** 'make noise on purpose'
make profit, *v.* **lap-**
make someone be quiet, *v.* **jyrypet-**
make someone carry a child, *v.* **thaba•-**
make someone cry, *v.* **dykhep-**

make someone feel ashamed, *v.* **thabarat-**
make someone smile, *v.* **thimimi-**
make wet, *v.* **sym•-**
male, *n.* **bipha**
man, *n.* (1) **bipha** 'man, male' (2) **morot** 'person, man, woman' (3) **me•apha** 'married man'
mango, *n.* **bu•chot**
mango tree, *n.* **buchotpan**
many, *det.* **pang•-**
many coloured, *adj.* **thokbyrang ~ thokbyrym**
map, *n.* **maip ~ mep**
March, *n.* (1) **mars** (2) **choi•etj** (archaic)
market, *n.* **nygyl**
marriage, *n.* (1) **bia** (2) **bakdong** 'forbidden marriage, marriage between two people from the same *mahari*'
marriageable boy, *n.* **banthai**
marriageable girl, *n.* (1) **gawi** (2) **nawmyl**
married couple, *n.* **jyksai**
married man, *n.* **me•apha**
married woman, *n.* **me•ama**
marry, *v.* (1) **khym-** (2) **bia kha•-** 'have a marriage ceremony'
mash, *v.* (1) **jotkhyngkhyng-** (2) **su•bylok-**
mason, *n.* **mistyri**
Master's degree, *n.* **Masteldygri**
masturbate, *v.* (1) **soksok-** (2) **selsoksok-** (3) **sepsep-** (4) **ri• selsoksok-** (5) **ri• sepsep-**
mat, *n.* (1) **dam** (2) **damplak** (3) **damthol** (4) **dengdyl** (5) **damdyl** (6) **dala** (7) **beraberi** (8) **taw•pachi ~ to•pachi**
match (to make fire), *n.* **wal•byt**
match in love, *n.* **jora**
matter, *n.* **kam ~ gam**
mature, *adj.* **pyryi-**
May, *n.* (1) **mai ~ mei** (2) **jetj** (archaic)
maybe, *v.* (1) **=khon** (2) **khon** (3) **maiba**
me *pron.* **ang**
mean, *v.* **myksong-**
meaning, *n.* **oltho ~ ortho**
meat, *n.* **randai**
medial malleolus, *n.* **cha•muk**
medicine, *n.* **sam**
medicine man, *n.* **oja**
meet, *v.* (1) **gorong-** (2) **mili-** (3) **ryk-**

meeting, *n.* (1) **miting** (2) **jineral miting** 'general meeting'
Megam *n./adj.* **Megam**
Meghalaya, *n.* **Mekalaia**
melon, *n.* **chin•thai**
melt away, *v.* **chem•-**
meow, *ideo.* **mew**
metre, *n.* **mityr**
mezenga, *n.* **mai•cheng**
mica, *n.* **so•re**
midday, *n.* **sanmaji**
middle, *n.* **matji ~ maji**
middle finger, *n.* **mykthoram**
might, *v.* **=khon**
milk, *v.* **chep-**
millet, *n.* **maisi**
millimetre, *n.* **milimityr**
million, *num.* **wan milion**
millipede, *n.* **manggywak**
mince, *v.* **khan•-**
mind, *n.* **sung**
mine, *pron.* (1) **ang** (2) **angmi ~ angmyng**
mine, *n.* **jul**
minister (in the government), *n.* **montyri**
minute, *n.* **minit ~ minyt**
mirror, *n.* (1) **aina**
miss (feel the absence of someone or something), *v.* **kha•pak-**
miss the mark, *v.* **ratsok-**
mist, *n.* **goira guri**
mistakenly, *adv.* **-syret**
mix, *v.* **pirin-**
moan, *v.* (1) **ma•am-** 'make a moaning noise' (2) **sykhym-** 'moan, complain, feel sorrow, mourn' (3) **synthi-** 'lament, complain'
mobile phone, *n.* **mobail**
mock, *v.* (1) **chonnyk-** (2) **che•e-**
molar, *n.* **wakam**
Monday, *n.* **sombal**
money, *n.* (1) **tangka** (2) **tangka poisa**
monitor lizard, *n.* **phu•chul**
month, *n.* **ja**
moo, *ideo.* (1) **ymbuuu** (2) **baaa**
moon, *n.* (1) **ja** (2) **jajong** (3) **chang•ai** (3) **jarambong** 'full moon'
more, *adv.* (1) **=khu** (2) **-khal** (3) **-dap**
more, *adj.* (1) **-khal** (2) **dai-**

more and more, *adv.* **-ruru**
moreover, *adv.* (1) **aro** (2) **daikhalaisa**
morning, *n.* **manap**
morning star, *n.* **phryngphrang askui**
mortar, *n.* **asam**
mosquito, *n.* **ganggawa**
mosquito net, *n.* **musuri**
moth, *n.* **taw•pak**
mother, *n.* (1) **jyw•** 'biological mother' (2) **ama** (3) **jyw•gaba** 'someone else's biological mother'
mother, *n.* (1) **ama** (2) **jyw•**
mother's house, *n.* **jyw•para**
mother's household, *n.* **jyw•para**
mother-in-law, *n.* (1) **nai•nokhol** (2) **mani**
motor oil, *n.* **mobil**
moulded, *adj.* **dumut-**
mountain, *n.* **ha•byri**
mourn, *v.* **sykhym-**
mouse, *n.* **muchot**
moustache (1) **khu•symang** 'moustache, beard' (2) **synggera** 'handle moustache'
mouth, *n.* **khu•chuk**
mouthful, *adv.* **wang•-** 'to take a mouthful, to bite a bit out of something'
move (transitive/intransitive), *v.* (1) **jit-** ~ **jyt-** 'move' (2) **sun-**'move, shift' (3) **bawen-** 'move in a circle, make a circle around something, encircle, be rolled up' (3) **ryngring-** 'move back and forth, up and down' (4) **hapjyt-** 'move house' (5) **guduk-**'move unstably'
movie, *n.* **philm** ~ **philym** ~ **philim**
much, *det.* **pang•-**
mucus *n.* **nakhung** (from the nose)
mud, *n.* **doba**
multi-coloured, *adj.* (1) **thokbyrang** (2) **byrymbyrym**
multiplied by, *adv.* **-chang**
multiply, *v.* **jel-**
mumbling, *n.* **khu•thikhu•thyraiga(ba)**
murder, *v.* **so•ot-**
murky, *adj.* **bui-**
muscle, *n.* **byl**
mushroom, *n.* (1) **panachol** (not edible) (2) **mairugu** ~ **meringgu** ~ **meringgaw** ~ **merenggaw** (edible) (3) **dumuta** (edible)
(4) **chingchongphyrot** ~ **chomchomphyrot** (edible)
must, *v.* **nang-**
mustard oil, *n.* **tho**
mustard plant, *n.* **garu**
my, *det.* (1) **ang** (2) **angmi** ~ **angmyng**
mythical black amphibian like a salamander which eats people., *n.* **phot** ~ **pho•ot**

# N

nag, *v.* **laklak-**
nah, *interj.* **na**
nail (finger), *n.* **chaksikhol**
nail (iron), *n.* **khiil**
naked, *adj.* **nanggandai**
naked, *adj.* **dymdam**
naked person, *n.* (1) **nanggandai** (2) **nanggodolong**
name, *n.* **bimung** ~ **bimyng**
national highway, *n.* **nesynyl haiwe**
naturally, *adv.* **-ramram**
naughtiness, *n.* **denggu**
nauseating, *adj.* **khalthyng-**
navel, *n.* **gandyrui**
near, *adj.* (1) **nek-** (2) **gycheng**
neck, *n.* (1) **tokkyreng** (2) **tokthining** ~ **tokthynyng** (3) **gyngjangjang**
neck feathers of a chicken, *n.* **mirang**
necklace, *n.* (1) **pi•tyng** (2) **syladyn** (3) **sylasyng** (4) **ryk**
need, *v.* **nang-**
neighbouring, *adj.* (1) **sul** (2) **abun**
Nepal, *n.* **Nepal**
Nepali *n./adj.* **Nepal**
nephew, *n.* (1) **khyryithang** ~ **khyrythang** (2) **sa•banthai**
nervous, *adv.* **nalbas**
nest, *n.* (1) **thup** (2) **taw•thup** (2) **khurung** (3) **taw•gurung**
nest, *v.* **thup-**
net, *n.* **chek**
Netherlands, *n.* **Nedyran**
network, *n.* **netwak**
never, *adv.* (1) **phangnan** (2) **bichiba**
new, *adj.* **pidan**
next, *adj.* **abun**

next, *adj.* (1) **sul** (2) **myksul**
next to, *prep.* (1) **rygynchi** (2) **gychingchi** 'next to, close to, near'
next year, *adv.* **naija**
nice, *adj.* (1) **nem-** (2) **ga•-**
nicely, *adv.* (1) **-nang** (2) **-thylong**
nickname, *n.* **myngkheleka**
niece, *n.* (1) **namchyk** (2) **namgaba** (3) **sa•mynchyk**
night, *n.* **wal**
night blindness, *n.* **mykdaw ~ mykdo**
nine, *num.* **chykhyw**
nineteen, *num.* **chi sykhu**
ninety, *num.* (1) **sot sykhu** (2) **rum• byryi chyigyk** (3) **khol chang byryi chyigyk**
nipple, *n.* **mu•khuchok**
no one *pron.* **darangba**
nobody *pron.* **darangba**
nod, *v.* **gakgu-**
node (of bamboo), *n.* **weng•**
noise, *n.* (1) **kyryngwa** (2) **kyryngwami** (3) **chengchang bengchang** 'loud noise' (4) **sawn**
noisy, *adj.* **jykjak-**
noon, *n.* **sanmaji**
normal, *adj.* (1) **ramram** (2) **alamyla**
normally, *adv.* (1) **-ramram** (2) **-rong**
north, *n.* **salgyro**
nose, *n.* **nakhung**
nose hair, *n.* **nakhungmyn•**
nose piercing, *n.* **nakhungthek**
nostril, *n.* **nakhungkhal**
not be, *v.* (1) **ni•-** (2) **dong•cha ~ dongcha**
not clearly, *adv.* (1) **myrumyru** (2) **syryng**
not exist, *v.* **ni•-**
not have, *v.* **ni•-**
not like and ignore, *v.* **kha•si-**
not too big and not too small, *adv.* **tairakrak**
not yet, *adv.* **te•ewrawraw ~ te•awrawraw**
nothing, *v.* (1) **mamung ~ mamyng** (2) **ni•wa**
notice, *n.* **tyngetwami ~ tyngetwamyng**
November, *n.* (1) **nobembyl ~ nobembol** (2) **agynj** (archaic)
now, *adv.* **te•aw ~ te•ew**
nowadays, *adv.* **te•ewrawraw ~ te•awrawraw**
nowhere *adv.* **bichiba**

number, *n.* **nambal ~ nombol**
numerous, *v.* **jel-**

## O

o'clock, *adv.* **baji**
oar, *n.* **kewal ~ khewal**
obey, *v.* (1) **kata ~ khatha ~ khata ~ katha ra•-** (2) **bam-**
obligation, *n.* **karaw**
oblique, *adj.* **tan•pat-**
obstruct, *v.* **peng ~ peng•-**
obtain, *v.* **man•-**
obviously, *adv.* **-phin ~ -phyn**
occupy, *v.* **khang-**
October, *n.* (1) **oktobyl** (2) **katij** (archaic)
oesophagus, *n.* (1) **tokkhyphu ~ tokyphu ~ tokybu** (2) **nadanggorot**
of, *prep.* **=mi ~ =myng**
offer to the dead, *v.* **chyn-**
offering, *n.* (1) **boli** (2) **chyngaba** 'offering to a dead person'
office, *n.* **opis ~ ophis**
officer, *n.* **opiser ~ ophiser**
offspring, *n.* **sa•**
often *adv.* **pang•ai**
oh, *interj.* (1) **ah** (2) **aia** (3) **aia** (4) **aiu** (5) **atyw ~ atyyyw** (6) **baaa**
oi, *interj.* **oi**
oink, *ideo.* **wek**
OK/okay, *adv.* (1) **ym ~ am** (2) **=ne** (3) **=de** (4) **dong•- ~ dong-**
old, *adj.* (1) **bydy** (for persons) (2) **picham** (for things)
old and crooked, *adj.* **kynkom-**
old couple, *n.* **bydyi badai**
old man, *n.* **bydyi**
older, *adj.* (1) **bydyikhal-** (2) **mykhal-**
omelette, *n.* **mamylet**
on behalf of, *adv.* **-thum**
on hands and knees, *adv.* **kepleplep ~ kepreprep**
on its side, *adj.* **chugup-**
on the ground **-soso**
on top, *adv.* **-dap**
on top of, *prep.* **khambaichi**

one, *num.* **sa**
one after the other, *adv.* **ha•chang ha•chang**
one after the other, *adv.* **-damdam**
one on top of the other, *adv.* **japrukruk**
onion, *n.* **rasun**
only, *adv.* (1) **=sa** (2) **=rara** (3) **-mangmang** (4) **=mangmang** (5) **=tara** (6) **khali**
oof, *interj.* **uph**
open, *v.* (1) **khuli- ~ kuli-** (2) **daw-** 'open, peel' (3) **kha•wak- ~ ka•wak-**'open one's mouth, have one's mouth open'
opium, *n.* **khuli**
oppose, *v.* **jai•-**
or, *adv.* **ma**
orange, *adj.* (1) **rymyt** (2) **konglarong**
orange, *n.* (1) **komyla** (2) **narang**
orchid, *n.* **balgyto•**
ordinary, *adj.* (1) **ramram** (2) **alamyla**
ornament, *n.* **baru**
other, *det.* (1) **alaga** (2) **abun**
other side, *n.* **nalsasang**
otter, *n.* **matdam**
our, *det.* (1) **ning** (inclusive), **na•nang** (exclusive) (2) **ningmi ~ ningmyng** (inclusive), **na•nangmi ~ na•nangmyng** (exclusive)
ours, *pron.* (1) **ning** (inclusive), **na•nang** (exclusive) (2) **ningmi ~ ningmyng** (inclusive), **na•nangmi ~ na•nangmyng** (exclusive)
out-of-body experience, *n.* **jasyri** 'the experience that one sees oneself in a different place while one is asleep as if one's soul leaves one's body'
outside, *n.* (1) **balaga** (2) **ha•pal**
over-, *adv.* **-pyryt**
overflow, *v.* **phingpyryt- ~ phingpurut-**
overgrown, *v.* **tuk-**
overstay, *v.* **daijol-**
overtime, *n.* **-phin ~ -phyn**
owl, *n.* **daw•phaw ~ do•pho**
own, *adv.* **=thang**
ox, *n.* **ma•subolot**

# P

paan leaf, *n.* **lapan**
pack, *v.* **pyn•-**
packed, *adv.* **chapchap**
packet, *n.* (1) **peket** (2) **thep**
packing, *n.* **peking ~ pheking**
packing leaf, *n.* **rai•chak**
paddy, *n.* **maiphang**
paddy field, *n.* **badym**
paddy sprouts (used for planting), *n.* **cha•ri**
pagan, *n.* **songsyrek ~ songsarek**
pain in the lower abdomen, *n.* **oksephang**
painter, *n.* **mistyri**
pair, *n.* **jora**
palate, *n.* **chagak**
palm of the hand, *n.* **chakpha**
pan, *n.* (1) **dyksyl ~ tyksyl** (2) **thyk** (3) **korea** (4) **maityk**
pan leaf, *n.* **lapan**
pants, *n.* **longpen**
papaya, *n.* **mudu**
paper, *n.* **lekha**
paratha, *n.* **barata**
parcel, *n.* (1) **ha•ryn** (2) **ha•gun**
pardon, *n.* **khema**
pardon, *v.* **khema kha•-**
parents, *n.* **jyw• wa**
parrot, *n.* **daw•sik**
part of the head behind the ear, *n.* **nakhalcha•dan**
partner, *n.* **jora**
pass, *v.* **badai**
passed, *adj.* **dong•- ~ dong-**
past, *adj.* **dakangmi ~ dakangmyng**
past, *adv.* **dong•- ~ dong-**
pasta, *n.* **papol**
patch (of vegetation), *n.* **gyrym**
path, *n.* (1) **ram** (2) **sorok**
patient, *adj.* (1) **sak-** (2) **sakchyk-**
patiently, *adv.* **-saw**
pauper, *n.* **kanggal**
pay respect, *v.* **mani-**
payment, *n.* **tika**
peacock, *n.* **do•de**
peddle, *n.* **khewal ~ kewal**
peddle, *v.* **badyng-**
peel, *n.* (1) **khoppalak ~ khoppylak** (2) **khol**
peel, *v.* (1) **si-** (2) **daw-** (3) **kar-** 'peel off of something like corn from a cob'
peep, *v.* **chaikhaw-**

pen, *n.* **pen**
penetrate, *v.* **but-**
penis, *n.* (1) **ri•** (2) **susu** (3) **ri•gol** (used as swearword for men)
people, *n.* (1) **darang ~ dyrang** (2) **morotdarang** (3) **nokdang** 'people who live together in one house'
perch, *v.* **pa•-**
perform an incantation, *v.* (1) **khurut** (2) **wai khurut-**
perl, *n.* **mukta**
persevere, *v.* **sak-**
persistently, *adv.* (1) **-saw** (2) **-dyngdyng**
person, *n.* (1) **morot** (2) **menpart** 'most important or most salient person' (3) **nanggandai** 'naked person' (4) **nanggodolong** 'naked person' (5) **gun montyro man•gaba** 'person who can control the spirits' (6) **khuchylep** 'person who cannot keep secrets and talks a lot' (7) **khu•ma** 'person who cannot speak' (8) **khu•gri** 'person who does not talk much' (9) **khalbong** 'person who eats scandalously much' (10) **tykhal** 'person who goes around eating in lots of other people houses' (11) **kykulwil** 'person who has an ear infection and has lost his balance' (12) **madong** 'person who has married a person from the same mahari as themselves' (13) **wagyleng ~ wagylok** 'person who is missing one or more teeth' (14) **wakhol ~ wakholong** 'person who is missing one or more teeth' (15) **aragong** 'person who is too big for his age' (16) **kaltyk**' person who never washes' (17) **khurutgaba** 'person who performs an incantation to determine which spirit makes someone ill' (18) **pokotia** 'person who takes advantage of the kindness of others' (19) **khu•eng** 'person with a crooked or slant mouth' (20) **po•tolong** 'person with a naked chest (21) **phebaw** 'person with a swollen cheek or swollen tonsils' (22) **chokrek** 'person with a touting mouth' (24) **khu•jylok** 'person with an open mouth' (25) **ja•jol ~ ja•gol** 'person with long legs' (26) **ri•baw** 'person with one testicle bigger than the other' (27) **myn•dyluk** 'person without body hair' (28) **khu•gri** 'quiet person'
pestle, *n.* **aman**
phlegm from the lungs, *n.* **dak**
photo, *n.* **piktiyr**
photo shoot, *n.* **suting**
pick, *v.* (1) **ak** (flowers, plants) (2) **di•itset- ~ di•etset ~ de•etset-** 'pick one's nose'
pick up, *v.* (1) **ryt-** (2) **khyn-**
pick one's nose, *v.* **di•itset- ~ di•etset ~ de•etset-**
picture, *n.* **piktiyr**
piece of meat, *n.* **theng**
pierce, *v.* (1) **pyryw- ~ pyru-** (2) **jotpyryw-** (3) **ga•pyryw-** 'pierce by stamping'
pig, *n.* (1) **wak** (generic) (2) **waknok** 'domestic pig'
pigeon, *n.* **daw•kyru, daw•kruha•sym** 'green pigeon'
pigsty, *n.* **waknol**
pile, *n.* (1) **ali** (2) **chom•** (3) **thom**
pile up, *v.* (1) **jap-** (2) **chom-**
pillow, *n.* **khawkham**
pillow stuffing, *n.* **khul**
pimple, *n.* **gangma**
pin lock, *n.* **sekari**
pincer, *n.* **khen•khorong** 'crab's claw, crab's pincher'
pinch, *v.* (1) **dep- ~ -dap-** (2) **khep-** (3) **sik-**
pincher, *n.* **songsykhep**
pineapple, *n.* (1) **anaros** (2) **kewa**
pipe, *n.* **paip**
piss, *n.* (1) **disutyi** (2) **tyisang**
piss, *v.* (1) **disu-** (2) **disudap-** 'piss on top of'
pitapat, *ideo.* **thup thup**
pitcher plant, *n.* **me•mangkoksi ~ mi•mangkoksi**
place, *n.* (1) **dam** (2) **hap** (3) **ram** (4) **-sang** (5) **nokhap** 'place to build a house' (6) **tyigat** 'place in a river or at the end of a water pipe where the people get drinking water, take a bath and wash their clothes and dishes' (7) **tyinok** 'place in the kitchen where the water pots and other utensils like plates, cups and glasses are stored' (8) **ma•sutan•dam** 'place where cows are slaughtered' (9) **wa•lung** 'place where

stuff gets burnt' (10) **cha•wekdam** 'place where the chaff is thrown after winnowing the rice' (11) **ringaba** 'place where domestic animals are kept'
plain, *n.* **ha•wai**
plain, *adj.* **pa•-**
plan, *v.* (1) **mangsong-** (2) **myksong-** (3) **aidia**
plank, *n.* **tota**
plant, *n.* **sam**
plant, *v.* (1) **kai••-**' plant by putting it into the earth with one's hands' (2) **pot-**'plant by sticking a sprout in the mud' (3) **chal-**'plant or sow by making a hole in the ground with a stick and putting the seed into the hole'
plaque, *n.* **wadi• ~ wakhi**
plaster, *v.* **ryphi- ~ riphi-**
plate, *n.* **thali**
play, *v.* (1) **khele-** (2) **tam•** 'play an instrument' (3) – **tok-** 'play an instrument'
playground, *n.* **robolphil ~ robolpil**
please, *adv.* **=khu**
pliers, *n.* (1) **syldangkhep** (2) **bawili ~ bawyli**
pling, pling, pling, *ideo.* **thing thing thing**
plink, *ideo.* **teng**
plot, *n.* (1) **ha•ryn** (2) **ha•gun**
plot (old), *n.* **ha•gun**
plough, *v.* **wai-**
pluck, *v.* **ak-**
plug in, *v.* **syket- ~ saket**
plunk, *ideo.* **teng**
pneumonia, *n.* **tyisuk**
point, *v.* **jot-**
poison, *n.* **bisi**
poison, *v.* **nisi-**
poke, *v.* **su•that-**
police, *n.* **phulis ~ pulis**
pomegranate, *n.* **dalim**
pomelo, *n.* **jamura**
poor person, *n.* **kanggal**
poorly, *adv.* **-mangmang**
popped rice, *n.* **muri**
porcupine, *n.* **bythyi**
pork, *n.* **wak**
port, *n.* **khana**
possible, *adj.* **man•-**
pot, *n.* (1) **tyk** (2) **gora**
potato, *n.* **alu**

pound, *v.* **syw•- ~ su•-**
pour, *v.* (1) **tyt-** 'pour liquid out of a container' (2) **kha•-** 'pour liquid into a jug'(3) **sel•-** 'pour into' (4) **syl•et-** 'pour into'
pow, *ideo.* (1) **thaw** (2) **them**
powder, *n.* **busi** 'dust', **pawdyr** 'talcum powder'
practice divination, *v.* **thama chai-**
praise, *n.* **rasong**
pray, *v.* **pi•-**
precious stone, *n.* **so•re**
precisely, *adv.* (1) **thik thak** (2) **thokthok**
predict, *v.* **nukcham-**
pregnancy, *n.* **ogynanggaba ~ okgynanggaba**
pregnant, *adj.* **ogynang- ~ oknak- ~ oknang- ~ okgynang**
premises, *n.* **noksam**
preoccupied, *adj.* **-saw**
prepare, *v.* **thari-**
press, *v.* (1) **dep- ~ -dap-** (2) **syk-** (3) **sykdep-** 'press with one's finger' (4) **dandan-** 'press with one's back against something'
pretend, *v.* **tak-**
pretty, *adj.* **syl-**
prevent, *v.* **peng ~ peng•-**
previous, *adj.* **dakanggaba**
previously, *adv.* **dakang**
price, *n.* **dam**
pride, *n.* **gal**
priest, *n.* **kamal**
prisoner, *n.* **chepgaba**
prod, *v.* (1) **jot-**(2) **jothat-** (3) **leklek-** 'prod in an orifice or hole' (4) **laklak-** 'prod in an orifice or hole for pleasure' (5) **ga•duk-duk-** 'prod with one's legs or feet' (6) **su•that-** (7) **syw•- ~ su•-**
profanation, *n.* **sua**
profit, *n.* (1) **lap** (2) **man•dapami ~ mandapwami**
profitable, *adj.* **lap-**
promise, *v.* (1) **khu•rasak-** (2) **chat-**
proper, *adj.* **khe-**
prostitute, *n.* **pha•lap**
protect, *v.* (1) **warasak-** (2) **rakhi-**
protruding, *adj.* **thywkhong**
protrusion, *n.* **mukthai**
prune, *v.* **tam-**

pry bar, *n.* **jomphol**
pry open, *v.* **chel•-**
pubic hair, *n.* (1) **ri•myn** (of a male) (2) **su•myn** (of a female)
puddle, *n.* (1) **ha•thywkong** (2) **ha•tykylok**
pull, *v.* (1) **taia-** (2) **bytjengjeng-**'pull jerkily' (3) **tokset-**'pull loose' (4) **thet-**'pull, pull out' (5) **byt-** 'pull, drag, drive, ride, transport, lead, haul, draw, shock (electricity)' (6) **bytsorok-**'pull out' (7) **bychym** 'pull up/out'
pumpkin, *n.* (1) **gomynda** (2) **akyrudygyl ~ akrudygyl**
punch, *v.* (1) **syw•- ~ su•-** (2) **su•pyrong-**'punch a hole through something'
punishment, *n.* **sasti**
purlin, *n.* (1) **wa•khaw** (2) **khyntyri**
push, *v.* **syk-**
puss, *n.* **myn•tyi**
put, *v.* (1) **tan-** 'put, put down, stop, keep, put off' (2) **bara-** 'put in, load'(3) **dung-** 'put into'(4) **san-** 'put in a bag'(5) **chanet-** 'put on the fire'(4) **dap- ~ dep-** 'put on top, stack'(6) **thu-** 'put in one's mouth' (7) **dukhup- ~ dykhyp-** 'put on clothes, dress'(8) **khawdam-** 'put down'(9) **gatdap-** 'put on top, stack'(10) **khep ~ khup ~ khyp-** 'cover, put on'(11) **nong-** 'put on, apply'(12) **pha•at- ~ pha•et-** 'put on, apply'(13) **thymyt-** 'put out, extinguish'(14) **sutuk-** 'put over, cover, hide'(15) **ram-** 'put in the sun dry'(16) **sa-** 'put in place' (17) **syket-** 'put in, insert'(18) **thek-** 'put in, insert' (19) **sa-** 'put in place'
pygostyle, *n.* **di•chongkhanthyi**

# Q

quack, *ideo.* **gepgep**
quantity, *n.* **pang•wami**
quarrel, *n.* **golmal ~ gormal**
quarrel, *v.* (1) **mangneng-** (2) **mangnengruk-**
queen, *n.* **rani**
question, *n.* **syng•gaba**
quick, *adj.* **jang-**
quickly, *adv.* (1) **joljol** (2) **-jol** (3) **bykbyk ~ bakbak** (4) **dykdyk** (5) **thapthap** (6) **bakrukrak ~ bakrukylak** (7) **rabak ~ rabak rabak** (8) **wel•ang ~ wel•ang wel•ang**
quietly, *adv.* **jyrym ~ jyryp**

# R

rabbit, *n.* (1) **husyring** (2) **saphaw**
race, *n.* **jat**
rack, *n.* (1) **akan** (2) **gan•chang**
racket **chengchang bengchang** 'loud noise'
radio, *n.* **redio**
rafter, *n.* (1) **byrym** (2) **kenchi**
rail, *n.* **reel**
rain, *n.* (1) **rang** (2) **mykha badri** 'long period of incessant, heavy rainfall'
rain, *v.* **wa-**
rainwater that streams over the ground, *n.* **tyisurung**
rainy season, *n.* **watyi**
raise, *v.* **song-**
ramp, *n.* **dalni tatdepgaba**
rape, *v.* **dyra-**
rapidly, *adv.* **-pyl**
rat, *n.* **muchot**
rattan, *n.* **raityng**
raw, *adj.* **githyng ~ gythyng ~ githing ~ gi•thyng**
razor blade, *n.* **bylet**
reach, *v.* (1) **phet-** (2) **phetang-** (3) **pheta-** (4) **dong•-** (5) **dongang-** (6) **wala-** 'reach at night'
read, *v.* **porai- ~ pore-**
ready to eat, *adj.* **myn-**
realise, *v.* **jasa-**
reality, *n.* **mykgythal**
really, *adv.* (1) **cho•mot ~ chong•mot** (2) **asol** (3) **-asol** (4) **-bebe** (5) **-dam**
reason, *n.* **gymyn**
receive, *v.* **ra•sak-**
recognise, *v.* **tyng-**
red, *adj.* (1) **sak-** (2) **pisak**
red, *n.* **pisak**
red, *adj.* **pisak**
redden, *v.* (1) **sak-** (2) **saket-**
reed, *n.* (1) **rai** (2) **parang**

refuse, *v.* **jai•-**
region, *n.* **jila**
regret, *v.* **synthi-**
reign, *n.* **sason**
relative, *n.* (1) **mahari** 'someone of the same clan' (2) **bai•** 'blood relative' (3) **bai•maran** 'two distant relatives' (4) **bai•maran chingmaran** 'two distant relatives' (5) **bai•sakthangmaran** two people who belong to the same lineage (6) **bai• tyng** 'blood relative' (7) **bai•siga ~ bai•sega** 'blood relative'
religion, *n.* **thorom**
reluctant, *adj.* **harat-**
reluctantly, *adv.* **-seme**
remember, *v.* (1) **sung ra•-** (2) **sungman- ~ suman -**
remembrance, *n.* **sung**
remove (skin, bark, peel, dress etc.), *v.* **khok-**
repair, *v.* **thari-**
repeatedly, *adv.* (1) **-chekchek** (2) **-jyryng** (3) **-phetphet**
repent, *v.* **synthi-**
replica, *n.* **dupliket**
reply, *v.* **khu•sak-**
reportedly, *adv.* **=no**
re-pound rice, *v.* **sorok-**
request, *v.* **pi•-**
resemble, *v.* **nuk-**
resin, *n.* (1) **laha** (2) **myn•tyi**
resist, *v.* **khereng-**
respect, *n.* **man**
respond, *v.* **khu•sak-**
rest, *v.* **neng•thak- ~ ning•thak-**
retard, *n.* **nawang**
retract, *v.* **bil•- ~ bel•-**
return, *n.* **waiphin**
return (go back), *v.* (1) **wai-** (2) **waiphin-** (3) **nokphin-** 'return home'
revoke, *v.* **ra•rung-**
rewind, *v.* **bytphin-**
rhinoceros, *n.* **gondu**
rib, *n.* **gawasu ~ gawsu**
rib cage, *n.* **chelku**
ribbit, *ideo.* **pekpek**
rice, *n.* (1) **mai** 'generic, plant and grain, cooked or uncooked) (2) **mairong** 'husked, uncooked rice' (4) **mairongkholnang ~ maikholnang** 'unhusked rice'(3) **sithi** 'fermented rice' (5) **mainyl** 'sticky rice' (6) **maisen** 'sticky rice in a banana leaf' (7) **maichek ~ maichyk** 'cold rice' (8) **maikhyt** 'burnt rice' (9) **maijyreng** 'leftovers of cooked rice dried in the sun used to feed the pigs' (10) **muri** 'popped rice' (11) **phywra** 'rice powder' (12) **rungkhut** 'broken rice' (13) **saram** 'dried rice grains' (14) **ryngchyw ~ ryngchu** 'flat rice' (15) **cha•gang** 'bad rice that is thrown away in the husking process' (16) **maidan** 'new rice (just harvested)' (17) **pijyw** 'unhusked rice that is thrown away when cleaning a portion of rice before cooking it, newly harvested rice' (18) **memaboro** 'species of nice smelling rice' (19) **maiguru** 'species of rice from which beer is made for the *chywgyn* festival' (20) **rungkhut** 'broken rice'
rice beer, *n.* (see *wine (from rice)*)
rice field, *n.* (1) **ha•ba** (2) **ha•byreng** 'old rice field' (3) **wal•mak** (future rice field)
rice pot, *n.* **rongtyk**
rice powder, *n.* **phywra**
rice seeds, *n.* **pijyw**
rice stock house, *n.* **pung**
rice wine, *n.* (see *wine (from rice)*)
rice-field house, *n.* **ha•banok**
rich, *adj.* **man•ai sa•-**
rich man, *n.* **nokma**
riches, *n.* (1) **gam ~ kam** (2) **gam jym**
riddle, *n.* **somphi**
ride a horse, *v.* **gore dung•-**
right, *n.* **jagyra**
right?, *adv.* **=mo**
ring, *n.* **jaksithem**
ring finger, *n.* **myksolkhare**
ringworm, *n.* **khat**
rip, *v.* **chit-**
ripe, *adj.* **myn-**
ripen, *v.* **thymyn-**
rise (of the sun or moon), *v.* **phet-**
river, *n.* (1) **tyikhal** (2) **tyimong** 'main river' (3) **tyiphek** 'tributary'
river bank, *n.* (1) **tyisam** (2) **ha•khung**

river crab, *n.* (1) **hen•** (2) **khen•jasyri** 'river crab that is walking on the road'
river junction, *n.* **para**
river shrimp, *n.* **na•cheng**
river snail **sukrung ~ sykrung ~ sukyrung ~ sykurung**
riverside, *n.* (1) **tyisam** (2) **ha•khung**
road, *n.* (1) **ram** (2) **sorok**
roam, *v.* **gylgyl-**
roast, *v.* **saw•-**
rob, *v.* **daket tak-**
rock, *n.* (1) **rong•** (2) **patal ~ phatal ~ phathal ~ pathal** (3) **rong•baram** (4) **rong•thai** (5) **rong•cheret** 'pebble-sized stone' (6) **rong•chung** 'big rock' (7) **rong•thyk** 'big rock' (8) **rong•patal** 'big rock' (9) **rong•chyret** 'very small rock' (10) **rong•misi** 'very small rock' (11) **rong•phek** 'rock almost the size of grain of sand' (12) **rong•han•cheng** 'sedimentary rock' (13) **rongsyrek** 'very small stone'
rocky, *adj.* **rong•gyrym ~ rong•rymrym**
roll, *v.* (1) **romrom- ~ rymrym** (2) **kyrurua** 'roll by itself' (3) **songkhel** 'roll head first' (4) **jitymryma-** 'roll something for the purpose of transporting it' (5) **byl•-** 'roll something into something' (6) **dol•romrom- ~ dolromrom** 'roll up' (7) **jol-** 'roll up' (8) **themtaw-** 'roll up'
rolled up, *v.* **bawen-**
roller, *n.* **lolal**
roof, *n.* **nokkhung ~ nukkhung**
roof, *v.* **rap-**
roo-koo, *ideo.* **khrukhru**
rooster, *n.* (1) **gogylek** (2) **taw•khasi** 'castrated rooster'
root, *n.* (1) **dyl** (2) **cha•dyl** (3) **cha•dylmorong** 'main root of a tree' (4) **cha•dylsaphek** 'small root'
rope, *n.* **kara**
rose, *n.* **golap**
roselle, *n.* **dachang**
rotten, *adj.* **saw-**
rough, *adj.* **baram-**
roughly, *adv.* **=darang ~ =dyrang**
round, *adj.* **romthom-v**

rub, *v.* (1) **siksik** (2) **riprip-** (3) **reprep-** 'rub the clothes while doing the laundry'
rubber tree, *n.* **rabal**
rule, *n.* **sason**
run, *v.* (1) **jal-**'run, run away' (2) **jalphakang-**'run from one side the other' (3) **bak-**'run after'
run, *v.* **ryk-**
rust, *n.* **warem**

## S

sad, *adj.* (1) **duk ganang** (2) **duk ni•-** 'not sad' (3) **duk man•-** (4) **duk sak-** (5) **duk dong•-**
sadness, *n.* **duk**
salary, *n.* **dorma ~ dolma**
saliva, *n.* **khu•tyi ~ khu•ti**
salt, *n.* **sym•**
salty, *adj.* (1) **kha-** (2) **kyisym-** (3) **parap-**
same, *adj.* (1) **baibai** (2) **gapsan ~ hapsan**
same age, *adj.* **sa•rong**
sand, *n.* (1) **ha•bykung** (2) **han•cheng**
sandal, *n.* **sendel ~ sendyl**
sap, *n.* (1) **tyi** (2) **ros**
sapling, *n.* (1) **chara** (2) **dala** (3) **mochok** (4) **panphek**
Saturday, *n.* **sunibal**
savoury, *adj.* **thap-**
saw, *n.* **khorot ~ khorat**
say, *v.* (1) **bal-** (2) **no-**
scab, *n.* **thawal**
scald, *v.* **kham-**
scale (for weighing), *n.* **paila**
scale (of fish), *n.* **khol**
scandalously much, *adv.* **-phet**
scar, *n.* (1) **dagi** (2) **gal**
scatter, *v.* **gal•ruru-**
scattered about, *adv.* (1) **dymbyra dymbyra** (2) **watwa watwa** (3) **byldyng byldang**
school, *n.* **skul**
scissors, *n.* **kensi ~ kesi**
scold, *v.* (1) **naw-** (2) **su-** (3) **jai-**
scoop, *v.* (1) **choket-**(for solid substances) (2) **wai•-**(for liquid) (3) **chok-** 'scoop, serve up, dish up, dish out, comb' (4) **chokset-**

'scoop away (to dispose of something) ' (5) **waiset-** 'scoop out'
scorn, *v.* (1) **chonnyk-** (2) **mykchep-**
scorpion, *n.* **ha•mangkyrang ~ mangkhrang ~ mangkhram ~ mangkyrang ~ mankyrang**
scour, *v.* **nat-**
scrape, *v.* (1) **siksik-** (2) **bak-** (with a spade or chopper)
scratch, *v.* (1) **khen-** (2) **sik-** (3) **sitbyryt-** (4) **matchirit-**
scratched, *adj.* (1) **sypsak-** (2) **matchirit-**
screen, *n.* **skrin ~ sykrin**
scrotum, *n.* (1) **ri•karan ~ ri•keren** (2) **ri•sokop** (3) **sirong**
scrub, *v.* **nat-**
sea, *n.* **sagal**
sea bean, *n.* (1) **gylarong** (2) **siwi**
seal, *v.* **buthu-**
search, *v.* (1) **ram•-** (2) **sandi-** 'search, inquire'
second, *num.* **nigaba**
second, *n.* **sekyn**
secretively/secretly, *adv.* (1) **-syruk** (2) **-khaw**
sedimentary rock, *n.* **rong•han•cheng**
see, *v.* (1) **nuk-** (2) **mykren nuk-**
seed, *n.* (1) **karan** (of a fruit) (2) **cha•ri** (for planting)
seize, *v.* (1) **pyi• –** (2) **watcha-**
seldom, *adv.* (1) **bichiba** (2) **bichiba bichiba** (3) **gasam gasam**
select, *v.* **sai-**
self *pron.* **phalthang**
sell, *v.* **phal-**
selves, *pron.* **phalthangthang**
semen, *n.* (1) **ri•ros** (2) **ri•tyi ~ ri•ti**
send, *v.* (1) **wat-** 'send away', banish, get rid of, avoid, switch on, let go, squirt' (2) **watet-** 'send (away), post, mail'
separate, *adj.* **ek-**
September, *n.* (1) **septembyl** (2) **asingja ~ asyngj** (archaic)
servant, *n.* **chakol**
serve up, *v.* **chok-**
set (of the sun), *v.* (see *rangsan*) (1) **jamang-** (2) **ma•ang-** (3) **san-** (4) **dang•-**
set as a trap, *v.* **sa-**
set up post, *v.* **song-**
seven, *num.* **sene**

seventeen, *num.* **chi syni ~ chi sene**
seventy, *num.* (1) **sot sene ~ sot syni** (2) **rum•tham chygyk**
sever, *v.* **sinthong•-**
severely, *adv.* (1) **nemen** (2) **bylongen ~ blongen** (3) **tyngen** (4) **-bi**
shade, *n.* **bythyn**
shadow, *n.* **jagyryng** (cast by a person)
shake, *v.* (1) **jingonget-** (2) **mot- -**(fixed objects) (3) **mojekjek-**(fixed objects) (4) **gyryw-** (objects that you can pick up, non-fixed objects) (5) **thojekjek- -**(fixed objects) (6) **jekjek-**'shake from side to side' (7) **chultet-** 'shake off of' (8) **thyrgyryw** (large and unmovable objects) (9) **ryngreng-** 'shake one's head' (10) **soksek-**'shake something without picking it up' (11) **khyw-**'shake out fluid' (12) **mimikakak-** 'shake with laughter'
shame, *n.* **baratwami**
shameless, *adj.* **watbyrak-**
shape, *n.* (1) **rokhom** 'shape, type' (2) **bimang** 'the shape of someone's body'
share, *v.* (1) **soal- ~ sual-** (2) **hanthi-**
share,, *n.* **phal**
sharp, *adj.* (1) **mat-**(of knives) (2) **bu•chok-** (of pointy objects) (3) **bu-** (of pointy objects)
sharpen, *v.* (1) **wyn• ~ wen• ~ wyt- ~ wot-** 'sharpen, whet' (2) **si•-** 'sharpen a pointy object' (3) **si•wil-** 'sharpen a pointy object' (4) **chokchok-** 'sharpen a pointy object'
shave, *v.* **rok-**
she *pron.* (1) **ge•theng ~ de•theng** (2) **ue** (3) **ie**
sheath, *n.* **sokhop**
shed skin, *v.* **kholthyrai-**
sheep, *n.* **mes**
shelf, *n.* **gadang**
shell, *n.* (1) **khung** 'shell of a crab, tortoise etc.' (2) **koplak** 'egg shell'
shield, *n.* **danyl**
shield, *v.* **warasak-**
shift (of work), *n.* **phal**
shift (transitive/intransitive), *v.* **sun-**
shinbone, *n.* **cha•kereng**
shine, *v.* (1) **seng•-** (2) **jarang-** (3) **jaseng•-** (4) **mykchel-** 'shine in the eyes' **tengchyp-chyp-** 'shine, glitter'

shiny, *adj.* **dymbrubru**
ship, *n.* **jahas**
shirt, *n.* (1) **chola** (2) **jama**
shit, *n.* **di•**
shit, *v.* (1) **di•it-** ~ **de•et-** ~ **di•et-** (2) **mu•rong-** 'shit on' **di•pyryw-** 'shit one's pants'
shiver, *v.* **dekdek-**
shoddily, *adv.* **-mangmang**
shoe, *n.* (1) **cha•khop** ~ **ja•khop** (2) **juta**
shoot, *v.* **kaw-**
shooting (of pictures), *n.* **suting**
shop, *n.* **dokhan**
short, *adj.* **sung•-**
shortcut, *n.* **sotkat**
short pants, *n.* **happen**
shortcut, *n.* (1) **rai•sotwa** (2) **rai•sotgaba**
shorts, *n.* **happen**
shotgun, *n.* **bonduk** ~ **bondyk** ~ **byndyk**
should, *v.* **=chym**
shoulder, *n.* (1) **phagongma** ~ **phagungma** (2) **wa•gatram**
shoulder yoke, *n.* **wa•gat**
shout, *v.* (1) **paraw-** (2) **parawchyrik-**'shout loudly'
show, *v.* (1) **thunuk-** (2) **wa•kholchik-**'show one's teeth'
shrimp, *n.* **na•cheng** (river shrimp)
shrouded in clouds, *adj.* (1) **guruchup-** (2) **rangbyrym-** 'shrouded in clouds, blocked by clouds'
shuffle cards, *v.* **rongmyng-**
shut somebody up, *v.* **jyrypet-**
shy *adv./adj.* **barat-**
sick, *adj.* **sa-**
sick person, *n.* **sathup**
sickle, *n.* **khatchi**
side, *n.* (1) **phak** (2) **gycheng** (3) **rygyn** (4) **-sang** (5) **sangphak** ~ **samphak** (6) **ramga** 'side of an object that faces away from the wall' (7) **noksuk** 'side of an object that faces the wall' (8) **kan•peng** 'side of the body' (9) **puksuk** 'side of the body' (10) **myktyiwatram** 'side of the hand under the index finger' (11) **natheng** 'side of the head' (12) **dykymphak** 'side where the head is'
side by side, *adv.* (1) **phakwil phakwal** ~ **phakwyl phakwal** (2) **phakthangthang**

sideburn, *n.* **khawcha•ryng**
sieve, *n.* **cheke**
sign, *n.* **chin**
Siju, *n.* **Sijyw** ~ **Siju**
silently, *adv.* **jyrym** ~ **jyryp**
silkworm, *n.* **khoryndachong**
simply, *adv.* (1) **=ari** (2) **sykhathang** (3) **yndyn** (4) **dymdam**
Simsang, *n.* **Symsang**
since, *prep.* **dabat**
since, *conj.* **gymyn**
since, *adv.* **dabat**
sing, *v.* **ryng•-**
sister, *n.* (1) **abi** (2) **ja•naw** (3) **naw** (4) **nono**
sister-in-law, *n.* (1) **bochi** (2) **anyng** (3) **ja•chung** (4) **nawsyri**
sit (1) **mu•-** 'sit down, be in sitting position' (2) **mu•peng-**'sit and block someone's view' (3) **mu•dap-** 'sit on something' (4) **mu•symbylek-** ~ **mu•symblek-**'sit on the floor (with one's bum touching the floor)' (5) **khom•-** 'sit with one's head in one's lap and one's legs pulled up' (6) **thumu•-** 'sit someone down (used for children)'
six, *num.* **korok**
sixteen, *num.* **chi dok**
sixty, *num.* (1) **sot dok** (2) **rum• tham**
skewer, *n.* **dabogos**
skin, *n.* **khol**
skinny, *adj.* **kan•jot-**
skirt, *n.* **dakmanda**
sky, *n.* **rangra**
slant, *adj.* (1) **ching•pheng** (2) **gyching** ~ **giching** (3) **khingcheng**
slap, *v.* **satkhap-**
slap, *ideo.* (1) **thap** (2) **thup**
slaughter, *v.* (1) **khat-** (2) **khan•-**
slave, *n.* **nokhol** ~ **nokhor**
sleep, *v.* (1) **jyw-** (2) **kynphak-** 'sleep in, sleep late'
slender, *adj.* **raw•reng-**
slice, *v.* **phyt-**
slide over something, *v.* **rongrong-**
slim, *adj.* **kan•jot-**
slime from the eyes, *n.* **mykkhi**
slingshot, *n.* **batdyl** ~ **patyl**

slip, *v.* (1) **so•sorot-** (2) **rawsykot-** 'slip out of one's hand' (3) **ga•syrot-** 'slip and fall'
slippery, *adj.* **rimyl-**
slither, *v.* **rai•ram-**
slope, *n.* **ha•kha**
sloppy, *adj.* **wekwak-** 'soft like mud'
slow, *adj.* **khasin**
slowly, *adv.* (1) **kha•sin** (2) **kha•sin kadym**
small, *adj.* **myl-**
small intestine, *n.* **puktyng**
smallpox, *n.* **sasep**
smash, *v.* (1) **thotphyret-** (2) **batkhynyng-** (3) **tokthong•-**'smash in half' (4) **tokhynyng-** 'smash into pieces'
smell, *n.* (1) **syn** (2) **dil** 'body smell, body odour' (3) **bytym** 'a nice smell'
smell, *v.* (1) **syn man•-** 'to perceive a smell' (1) **sip-** 'to use one's nose to sense smells' (3) **bytym-** 'smell nice' (4) **manam-** 'smell bad' (5) **saw•myk-** 'smell rotten, smell foul'
smile, *v.* **mychym-**
smoke, *n.* **wal•khu**
smoke, *v.* (1) **ryng-** (2) **muk-** (3) **wal•khu-** 'produce smoke'
smooth, *adj.* **ronok-**
snail, *n.* (1) **sukrung ~ sykrung ~ sukyrung ~ sykurung** (river snail) (2) **echaluk, kapkung** (land snail), **chyhyl**
snake, *n.* **dypyw** (generic), **thongmatchang, dypywkaram, dypywha•saw ~ ha•saw, dypywkheng,**
snatch, *v.* **ra•sek-**
sneak, *v.* **jom•-** 'sneak, sneak up on somebody'
sneeze, *v.* **hachi-**
snore, *v.* **hogol ra•-**
snoring, *n.* **hogol**
snort, *v.* **nakhung ra•taw-**
snot, *n.* (1) **nakung** (liquid mucus) (2) **nakhungdi•** 'hard piece of snot'
snow, *n.* **siri ~ suri**
so, *conj.* (1) **ytykyimyng ~ ytykyimu ~ ytykyimuna ~ ytykyimung ~ ytykyimungna** (2) **umi ~ umido ~ umisa ~ umyng ~ umung ~ umyngdo ~ umyngsa** (3) **una** (4) **ytykchido**
so many, *adj.* **isykyn ~ iskyn**

so much, *adv.* (1) **-thyng** (2) **-thyngthyng**
so then, *conj.* (1) **ytykyimyng ~ ytykyimu ~ ytykyimuna ~ ytykyimung ~ ytykyimungna** (2) **umi ~ umido ~ umisa ~ umyng ~ umung ~ umyngdo ~ umyngsa** (3) **una**
soak, *v.* **sym•-**
soap, *n.* **sabun**
soccer, *n.* **robol**
soccer field, *n.* **robolphil ~ robolpil**
society, *n.* **songsal**
sock, *n.* **muja**
soda, *n.* **khaltyi**
soft, *adj.* (1) **nom•-** (2) **demdong-**
soil, *n.* (1) **ha•** (2) **ha•mang**
soldier, *n.* **sipai**
sole of the foot, *n.* (1) **cha•pa** (2) **cha•chok**
solidify, *v.* **khang-**
some, *det.* **pang•cha**
some (people), *pron.* **bai•dam ~ baidam**
someday, *adv.* **bibyrokhon ~ bibakoron**
some time ago, *adv.* **maja**
some time ago today, *adv.* **tai•sa**
somebody *pron.* **changba**
somehow, *adv.* **jenethene**
someone *pron.* **changba**
someone else's, *adj.* **abun**
something *pron.* **atongba**
sometimes, *adv.* (1) **bichiba** (2) **bichiba bichiba** (3) **gasam gasam**
somewhere, *adv.* (1) **bichiba** (2) **bisangba**
son, *n.* (1) **sa•banthai** (2) **baba** (3) **babu**
song, *n.* (1) **git** (2) **chaira** 'type of traditional song'
son-in-law, *n.* **kynokhol**
soot, *n.* **garamak**
sore, *n.* **thaphu**
sorrow, *n.* **duk**
sort, *n.* **rokhom**
sound, *n.* (1) **kyryngwa** (2) **kyryngwami** (3) **chengchang bengchang** 'loud noise' (4) **sawn**
sour, *adj.* **khyi-**
source, *n.* (1) **chiakhol** (2) **khawsuk** (3) **tyimuk**
south, *n.* **salgypeng**
sow, *v.* (1) **pyjyw-** 'sow seeds by scattering them' (2) **khit-** 'sow seeds by sprinkling them'

space, *n.* (1) **gesep ~ gysep ~ gisep** (2) **cholwat** (3) **dykymphak** (4) **chaksigysep** 'space in between the fingers' (5) **rong•khal** 'space under a stone'
spade, *n.* **belcha**
sparrow, *n.* **chanchora ~ chanchura**
sparse, *adj.* **heng•-**
spatter, *v.* **thangphytphyt-**
speak, *v.* (1) **bal-** (2) **ol-** (3) **golpho-** 'tell a story, speak/talk very long' (4) **balsem-** 'talk very long'
spear, *n.* (1) **guthini ~ guthyni** (2) **jatha**
speed, *n.* **spiit**
spell, *n.* **muni**
spelling, *n.* **spyling ~ spling ~ sipyling**
sperm, *n.* (1) **ri•ros** (2) **ri•tyi ~ ri•ti**
spherical, *adj.* **romthom-**
spider, *n.* **gawang ~ guwang, chengchengmachok**
spider web, *n.* **gawangsyryng**
spill, *v.* (1) **sat-** (2) **sa•dap-** 'spill, take more and more' (3) **phakphaklak-**
spine, *n.* **kynkyreng**
spinning, *adv.* (1) **chuwil chuwal** (2) **-rongreng**
spirit, *n.* (1) **wai** (2) **sung**
spirit, *n.* **wai**
spirit house, *n.* **delang ~ dylang**
spit, *n.* **khu•tyi ~ khu•ti**
spit, *v.* (1) **khu•tyisot-** (2) **dak-** 'spit out' **phuset-** 'spit out'
spittle, *n.* **khu•tyi ~ khu•ti**
splash, *v.* (1) **thangphytphyt-** (2) **tang•dap-** 'splash on'
splash, *ideo.* **chaw**
spleen, *n.* **mansylang ~ manthylang**
splendid, *adj.* **ga•su-**
split, *v.* (1) **peret- ~ pheret-** (2) **khan•peret- ~ khan•pyret -**'split, cut open'
spoiled (only used with meals), *v.* **gusum-**
sponger, *n.* **pokotia**
spoon, *n.* (1) **palak** (2) **spun** (3) **abek**
spot, *v.* **chai-**
spouse, *n.* (1) **jyk** (2) **khymgaba**
sprain one's foot, *v.* **ga•sylek-**
spread out, *v.* (1) **khep ~ khup ~ khyp-** (2) **dan-**

spring (of a stream), *n.* (1) **chiakhol** (2) **khawsuk** (3) **tyimuk**
spring onion, *n.* **rasun tyisuk**
sprinkle, *v.* **khit-**
sprout, *v.* (1) **dym-** (2) **rydym-** 'sprout leaves'
spy (on), *v.* **chaikhaw-**
squat, *v.* (1) **chongchyron- ~ choncholon-** (2) **mu•chongchyron- ~ mu•choncholon-** (3) **cha•choron- ~ cho•choron-**
squeak, *ideo.* (1) **chepchap chepchap ~ chepchep chepchap** (2) **chutchut**
squeal, *ideo.* **wek**
squeeze, *v.* (1) **but-** 'squeeze in' (2) **sep-** 'squeeze out'
squinting, *adv.* **mrimri ~ rimirimi**
squirrel, *n.* **karat ~ ka•rat**
squirt, *ideo.* **chyryt chyryt**
squirt out, *v.* **thangtaw-**
stack, *v.* (1) **dep- ~ -dap-** (2) **gatdap-** (3) **chom-**
stalk, *n.* (1) **gantheng** (2) **pakara ~ pakyra** (2) **cheksi** (3) **maikhoppylak ~ maikhoppalak** 'stalk left over after threshing rice' (4) **maipalak ~ maipylak** 'stalk left over after threshing rice'
stamp, *v.* (1) **ga•tyn-** (2) **ga•pyryw-** 'stamp through something'(3) **ga•phynek-** 'stamp to death' (4) **ga•pyret-** 'stamp to death'
stamp, stamp, *ideo.* **thep gaw**
stand (be in standing position), *v.* **chap-**
star, *n.* **aski ~ askhui ~ askui**
star fruit, *n.* **galdai**
star sign, *n.* (1) **rasi** (2) **wa•daweng** 'star sign of three stars in a straight line'
start, *v.* (1) **ha•bacheng-** (2) **dang•-**(3) **-gat**
startled, *adj.* (1) **jari-** (2) **wa•chyrik-**
starve, *v.* (1) **okmyng-** (2) **si-**
starvation, *n.* **sa•a siwa**
station, *n.* **khana**
stay (somewhere), *v.* (1) **mu•-** (2) **mu•si-** 'stay somewhere uncomfortable' (3) **ryp-** ;stay under water'
stay awake, *v.* (1) **seng-** (2) **walseng-** 'stay awake all night'
steal, *v.* (1) **sa•khaw-** (2) **-sek** (3) **bytsek-** 'steal a person, abduct'
steam, *n.* **biba**
steep, *adj.* **chyw•-**

steep slope, *n.* **ha•kha**
stem (of leaf or fruit), *n.* **gan•theng ~ ga•theng**
stench, *n.* **manam**
step by step, *adv.* **gadang gadang**
step on, *v.* (1) **ga•dap-** (2) **ga•reret-**
stepchild, *n.* **sa•thyra**
stepfather, *n.* **wang**
stepmother, *n.* **ade**
stick, *n.* (1) **gan•thong** (2) **kun•** (3) **panthong** (4) **sylkengkun** ;drive a hoop'
stick (adhere to), *v.* **takap-**
stick insect, *n.* **me•mangkereng**
stick into, *v.* **bat-**
stick out one's tongue, *v.* **thylamphak sul-**
sticky rice, *n.* (1) **mainyl** (2) **maisen** 'sticky rice in a banana leaf'
still, *adv.* (1) **=khu** (2) **-khyngkhyng** (3) **te•ewrawraw ~ te•awrawraw**
still too, *adv.* (1) **-teng** (2) **-tengteng ~ -thengtheng**
sting (of a bee etc.), *v.* **dyl- ~ del-**
stink, *v.* **manam-**
stir, *v.*
stir, *v.* (1) **bul-** (2) **wongong- ~ wungwung-**
stirring rod, *n.* **palak**
stitch, *v.* **suk-**
stomach, *n.* (1) **pipuk** (2) **athom**
stomach pain, *n.* **okhuchak**
stone, *n.* (1) **patal ~ phatal ~ phathal ~ pathal** (2) **rong•** See also *rock*
stone (of a fruit), *n.* **karan**
stool (from body), *n.* **di•**
stool (to sit on), *n.* (1) **mura** (2) **dakham**
stop, *v.* **machot-**
store **song-**
story, *n.* **golpho**
straight, *adj.* (1) **pereng-** (2) **sorong-** (3) **sorong**
straight, *adv.* (1) **pering tongtong** (2) **thongthong** (3) **-sot**
straight, *adj.* (1) **pering-** (2) **thongthong**
strainer, *n.* **chek**
strange, *adj.* **songga**
strength, *n.* (1) **byl** (2) **byl chak** (3) **jagydok**
stretch (1) **syryng ~ syrong** 'stretch (out) (used for rope etc.), reach out, build a bamboo bridge' (2) **cha•syrong-** 'to stretch one's leg' (3) **chaksyrong** 'to stretch one's arm' (4) **sul-** 'stretch out, stick out, extend' (5) **kan•ol-** 'stretch one's body'
strike, *n.* **byl•**
strong, *adj.* (1) **rak-** (2) **bylak-** (3) **gyl-**
strongly, *adv.* **-syrang ~ -srang**
struggle, *v.* **khereng-**
strut, *n.* **ges ~ kes**
stubble old rice stalk which is left over after harvesting the rice, *n.* **wa•cham**
stuck, *adj.* (1) **khet-** (2) **sep-** (3) **chang•khet-** (4) **phuk- ~ puk-** (in one's throat) (5) **wa•khel-** (in one's teeth) (6) **wa•khelsep-** (in one's teeth)
stud, *n.* (1) **rochok ~ rotchok** (2) **tatkhap-ga(ba)** (3) **reel**
study, *v.* **porai- ~ pore-**
stuff, *n.* **bostu**
stumble, *v.* (1) **ga•sokhok-** (2) **cha•godot-**
stump (of a tree), *n.* **rochong**
stupid person, *n.* **jada**
submerged, *adj.* **ryp-**
succeed, *v.* (1) **sok-** (2) **chu•sok-** (3) **jam-**
such, *det.* (1) **ytykgaba** (2) **isykyn ~ iskyn**
such as, *adv.* **jekhai**
suck, *v.* (1) **hup-** (2) **mojet- ~ mojot-** (3) **syrup-** (4) **ri• sa•-**
suckle, *v.* **khan-**
suddenly, *adv.* (1) **thangguduk** (2) **-chang**
suffer, *v.* (1) **sak-** (2) **jurimana kam-** 'suffer a penalty' (3) **nasi-** 'suffer from a loud noise or sound' (4) **synthi-** 'suffer, regret, repent, lament, moan, whine'
sufficient, *adj.* **dong•- ~ dong-**
sugar, *n.* **chini**
sugarcane, *n.* **ko•rot**
suitable, *adj.* (1) **khe-** (2) **myngnang-**
summon a spirit, *v.* **khurut-**
sun, *n.* **rangsan**
Sunday, *n.* **rubibal**
support, *n.* **chalgaba**
support, *v.* (1) **chal-** 'support a person' (2) **sungchal-** 'support a structure' (3) **pai-** 'support, tolerate'
supporting post, *n.* **manjuri ~ manjyri**
supporting structure, *n.* **nokhama**

suppose, *v.* (1) **chanchichyp** (2) **chanchichypai**
supposedly, *adv.* (1) **=chym** (2) chym
surely, *adv.* **-thel**
surface, *n.* **serek**
surrender, *v.* **bam-**
surreptitiously, *adv.* **-khaw**
swallow, *n.* **taw•pachi**
swallow,, *v.* **monok-**
swap, *v.* **tharai-**
swarm, *v.* **dum-**
swarming, *adv.* (1) **-chichak chikchak** (2) **wekwak**
swaying, *adv.* (1) **rongrengchangcheng** (2) **rangrengchongcheng rongrengchangcheng**
sweat, *n.* **tyi**
sweat, *v.* **tyi hongkhot-**
sweep, *v.* (1) **wek-** (2) **huk-** 'sweep together'
sweet, *n.* **choklet**
sweet, *adj.* **sym**
sweet potato, *n.* **tha•malang ~ tha•mylang**
sweetheart, *n.* (1) **chame ~ chamai** (2) **mykchagaba** (3) **daldi**
swell, *v.* (1) **pok-** (2) **phet-** (3) **gangphu-** 'blow up (like a chapatti on the fire) '
swelling, *n.* (1) **samsin** (2) **nangthaigaba** (3) **sothonthara** 'cancerous swellings all over the body'
swidden, *n.* **ha•ba**
swift, *adj.* **tarak-**
swim, *v.* (1) **hung-** (2) **tyi hung-**
swing, *v.* (1) **wyngwet-**'move back and forth' (2) **chingchoroi-** 'swing from something'
switch, *n.* **suis**
switch off, *v.* **ni•et-**
swollen lymph nodes in the arm pits, *n.* **thai•rokron**
swoop down (of birds of prey), *v.* **sap-**
sword, *n.* **darai**

# T

table, *n.* **tebyl**
tablecloth, *n.* **stulkhabar**
tadpole, *n.* **na•luk**
tail, *n.* **di•mai**

tail feathers, *n.* **taw•di•mai**
take, *v.* (1) **ra•-** 'take from' (2) **ra-** 'take to'
take a bath, *v.* **tyru- ~ tyiru- ~ tyiryw-**
take a bite, *v.* **khyp-**
take a rest, *v.* **neng•thak- ~ ning•thak-**
take apart, *v.* (1) **dok-** (2) **choka-**
take away, *v.* **ra•ang-**
take back, *v.* **ra•rung-** 'revoke'
take care of, *v.* **chairok-**
take off (clothes), *v.* **dok-**
take out, *v.* (1) **ratat-** (2) **bykot-**
take revenge, *v.* **thym-**
Take this. **ha•**
talk, *v.* (1) **bal-** (2) **ol-** (3) **golpho-** 'tell a story, talk a lot, talk a long time' (4) **balsem-** 'talk a long time' (5) **golpho kha•-** 'talk a lot, gossip'
tall, *adj.* **raw•-**
tamarind, *n.* (1) **tintyrin** (2) **chengcheng**
tank, *n.* **tenki**
tank top, *n.* **genji**
tap, *v.* **tokphyrong-**'take a powdered substance in the palm of one hand and softly tap on it with the other hand'
tap, *ideo.* (1) **gyp** (2) **kak**
tap, tap, tap, *ideo.* **thep thup**
tape, *n.* **keset ~ kheset**
tapioca, *n.* **khan**
taro, *n.* **ring**
tasty, *adj.* **thaw-**
tea, *n.* **cha**
tea leaf, *n.* **chachak**
tea plant, *n.* **chachakphang ~ chaphang**
tea strainer, *n.* **chachek**
teach, *v.* **syki- ~ ski-**
teacher, *n.* (1) **tichyr** (2) **madam** (female) (3) **mastel** (male)
teacup, *n.* **khap**
teak tree, *n.* **saigon**
teapot, *n.* **dipot**
tear (2) **chit-** (3) **chet-**'tear, tear off (clothes paper etc.) ' (4) **chot-** 'tear (off), cut' (5) **bytphyrak-** 'tear apart' (6) **chetpyrak-** 'tear apart' (7) **kanting-**'tear spontaneously' (8) **thatthongthong-**'tear to pieces' (9) **dikirin-** 'tear clothes, paper etc.' (10) **chithong-** 'tear cloths to shreds' (11) **thetchot-**

'tear or break by pulling' (12) **bytphuruk-** 'tear out with the roots'
tear (from one's eye), *n.* **myktyi**
telephone, *n.* **telephon**
telephone call, *n.* **kol**
television, *n.* **tibi**
tell, *v.* (1) **bal-** (3) **no-** (4) **thil•-** 'tell lies' (5) **myk-** 'tell lies' (6) **golpho-** 'tell stories, speak at length'
temple (body part), *n.* **mykkep**
ten, *num.* (1) **chyigyk** (2) **-chek** (3) **chi**
terminal bud, *n.* **pansok**
termite, *n.* **hangkyn, hangkyn raja**
terrific, *adj.* **ga•su-**
testicle, *n.* **ri•karan ~ ri•keren**
thank, *v.* **mythel-**
that *pron./det.* (1) **ue ~ u-** (2) **hawe ~ haw-** (3) **hyiawe ~ hyiaw-**
That's right. **ho•ong**
that's why, *conj.* (1) **ytykyis** (2) **umi ~ umido ~ umisa ~ umyng ~ umung ~ umyngdo ~ umyngsa** (3) **una**
thatch, *n.* **parang**
thatch, *v.* **rap-**
the, *art.* In Atong, noun phrases or nominal clumps (see van Breugel 2014:98–103) preceded by a demonstrative are always definite, e.g. *Morot myng• sa ganangno. Uba jyw•taraanokno, wa• ni•okno. Ue gawichie sa• myng• korok ganangno. Morot myng•+sa ganang=no. U=ba jyw•=tara=an=ok=no, wa• ni•=ok=no. Ue gawi=chi=e sa• myng•+korok ganang=no.* (person CLF:HUMANS+one exist=QUOT DST=EMPH mother=EXCLUSIVELY FOC=COS=QUOT father not. exist=COS=QUOT DST woman=LOC=CT child CLF:HUMANS+one exist=QUOT) 'There was a person. She was a single mother. The woman had six children.' A noun phrase or nominal clump with a topic enclitic =*do* can also be interpreted as being definite, e.g. *Maido mynokno. Mai=do myn=ok=no.* (rice=TOP ready=COS=QUOT) 'The rice was ready.' Noun phrases or nominal clumps marked with the contrastive/new topic enclitic =*e* (CT) can also be interpreted as being definite, e.g. *Sympak chunggaba nukoknotyi. Sympak chunggabachie phylgym pa•ai mu•sawarongno. Sympak chung=gaba nuk=ok=no=tyi. Sympak chung=gaba=chi=e phylgym pa•=ai mu•-saw=arong=no.* (SPECIES.OF.TREE big=ATTR see=COS=QUOT=MIR SPECIES.OF.TREE big= ATTR=LOC=CT eagle perch=ADV=SEQ sit- FULLY.PREOCCUPIED=DUR=QUOT) 'They saw a big *symphak* tree, to our/their surprise. 'In the big sympak tree, the eagle was sitting, fully preoccupied.' Unmarked noun phrases or nominal clumps can also be interpreted as being definite, e.g. the word *phylgym* in the previous example. the fact that the animal has been mentioned before in the story, prompts the definite interpretation of its occurrence here (see van Breugel 2019: 239, sentences 119 and 120). Another example of an unmarked noun phrase is *Dinggarai goi•sagaba chaithiriokno. Dinggarai goi•+sa=gaba chai-thiri=ok=no* (fish.trap CFL:RESIDU+one=ATTR inspect=AGAIN=COS=QUOT) 'He inspected the first fish trap again.' Here, the fact that the referent was mentioned before in the text, as well as the occurrence of the event specifier -*thiri* 'again' on the predicate, both prompt the definite interpretation of *Dinggarai goi•sagaba* 'first fish trap' (see van Breugel 2019: 269, sentence 17).
the day before yesterday, *adv.* **maja**
their, *det.* (1) **ge•thengtheng ~ de•thengtheng** (2) **ge•thengthengmi ~ de•thengthengmyng**
theirs, *pron.* (1) **ge•thengtheng ~ de•thengtheng** (2) **ge•thengthengmi ~ de•thengthengmyng**
then, *adv.* (1) **ytykyimyng ~ ytykyimu ~ ytykyimuna ~ ytykyimung ~ ytykyimungna** (2) **umi ~ umido ~ umisa ~ umyng ~ umung ~ umyngdo ~ umyngsa** (3) **una** (4) **uchie** (5) **ytykyisa**
there is/are, *adv.* **ganang**
there isn't/aren't, *adv.* **ni•-**
thereafter, *conj.* **kynsang**
therefore, *adv.* (1) **ytykyisa** (2) **una** (3) **umigymynchi ~ umynggymynchi**

these *pron./det.* **ie ~ i-**.
they *pron.* (1) **ge•thengtheng ~ de•thengtheng** (2) **utym**
thick, *adj.* (1) **chat-** (2) **pyn•-**
thick (of fog or mist), *v.* **thup-**
thigh, *n.* **cha•phong ~ cha•phung**
thin, *adj.* (1) **kan•jot-** (of person) (2) **pa•-** (things)
thing, *n.* **bostu**
think, *v.* (1) **chanchi-** (2) **sung ra•-** 'think of, remember'
thirsty, *adj.* (1) **karan ~ ka•ran- ~ kha•ran-** (2) **tyikaran**
thirteen, *num.* **chi tham**
thirty, *num.* **kholachi ~ kholechyi**
this *pron./det.* **ie ~ i-**
this many, *adj.* **isykyn ~ iskyn**
this morning, *adv.* **tai•nep**
this much, *adj.* **isykyn ~ iskyn**
this year, *adv.* **tarai**
thorn, *n.* **asu**
those *pron./det.* (1) **ue ~ u-** (2) **hawe ~ haw-** (3) **hyiawe ~ hyiaw-**
thought, *n.* **sung**
thousand *num.* **hajalsa**
thread, *n.* **pi•tyng**
threaten, *v.* **dykyret- ~ dykyryi-**
three, *num.* **tham**
thresh, *v.* (1) **ga•-** (2) **mai ga•-**
throat, *n.* (1) **tokkhyphu ~ tokyphu ~ tokybu** (2) **chokdeng**
through, *prep.* (1) **=tykyi ~ =takai** (2) **-ruru**
throw, *v.* (1) **rat-** (2) **thyp-** 'throw sidearm, throw into'(3) **aset- ~ asyt-** 'throw away,, dispose of'(4) **thang-** 'throw away with great force'(5) **batphai•-** 'throw pieces'(6) **batpyret- ~ papret-**'throw and smash'(7) **phak- ~ phat-** 'throw out'(8) **phakset-** 'throw away, dispose of' (9) **thang-**'throw away with great force, come out with great force'
throw away, *v.* **-set**
throw up, *v.* (1) **chisat-** (2) **kha•rekrek-** (3) **khawakwak-**
thud, *ideo.* **dam ~ dym**
thumb, *n.* **chaksijyw•bydyi**
thunder, *n.* **goira**

thunk, *ideo.* (1) **gyp** (2) **thop** (3) **thup**
Thursday, *n.* **bistibal**
Tibet, *n.* **Tibet**
tick, *n.* **nasengkhet, nathyra**
tickle, *v.* **thebajaw- ~ thebejaw-**
tickly, *adv.* **bejaw-**
tie, *v.* (1) **tek-** (2) **thel•-** (3) **kha-** (4) **khet-** 'tie the cloth in which you carry a baby'
tie beam, *n.* **bylbang**
tiger, *n.* **matsa**
tight, *adj.* (1) **ket-** (2) **kyryng-** (3) **ha•kha-** 'very tight'
till the end, *adv.* **-syrang ~ -srang**
tilling the soil (for a living), *v.* **ha• haw•ai sa•-**
tilt, *v.* **thyngel-**
tilted, *adj.* **gychingching mu•-**
time, *n.* (1) **somai ~ somoi** (2) **wen• ~ wet** (3) **tap**
times, *adv.* **-chang**
tire, *n.* **taiyr**
tired, *adj.* (1) **neng•-** (2) **chyi•-**(3) **nombok** (after eating too much)
to, *prep.* (1) **=sang** (2) **=na**
to be torn (of cloth and paper), *adj.* **kirin**
to the last drop, *adv.* **thot thyng•thot**
toad, *n.* **lukwak ~ rukwak, bengblok**
tobacco, *n.* **tha•makhu ~ tha•mykhu**
today, *adv.* **tai•ni**
toe, *n.* (1) **cha•si** (2) **cha•sijyw•bydyi** 'big toe'
together, *adv.* (1) **gapsan ~ hapsan** (2) **-thok** (3) **-rum** (4) **-gorop** (5) **krymkraw** (6) **=maran**
toilet, *n.* (1) **di•kyntyk** (2) **letrin** (3) **phaikana ~ paikhana ~ phaikhana** (4) **toilet ~ toilyt**
tolerate, *v.* **pai-**
tomato, *n.* **mantawbylati**
tomorrow *n./adv.* **hanep**
tongue, *n.* **thylampak ~ thylapak**
too (excessively), *adv.* **-duga**
too (in addition), *adv.* **=ba**
too much, *adj.* (1) **-duga•** (2) **dugaphinok** (3) **-bongbong** (4) **bylong-** (5) **agre ~ agrai**
too salty, *adj.* **pyrap-**
tooth, *n.* **wa**

top, *n.* (1) **khambai** (2) **dykym** (3) **khutai** 'top of a house'
torch, *n.* **wal•**
torn, *adj.* (1) **choka-** (2) **kirin-**
tortoise, *n.* **khu•sum ~ ku•sum, katua ~ khatua**
tossing and turning, *adv.* **suksak**
total, *n.* **jam-**
totally, *adv.* **-phin ~ -phyn**
touch, *v.* (1) **khi-** (2) **pyi•-**
touch-me-not plant, *n.* **sambarat**
towards, *prep.* **-a ~ -ai**
towel, *n.* **tawel**
trade, *v.* (1) **badyng-** (2) **chiwal-**
tradition, *n.* (1) **ain** (2) **niam** (3) **ain niam** (4) **bewal** (5) **takbewal** (6) **takwa rukwa**
train, *n.* **reelgari**
trample, *v.* (1) **ga•-** (2) **ga•kynyng-** (3) **ga•jonong-**
transform, *v.* **phyl•-**
transparent, *adj.* **kyryk-**
trap, *n.* (1) **ja•ga** (2) **jap**
trap, *v.* **ban-**
travel, *v.* **songrai- ~ songre-**
tread on, *v.* **ga•reret-**
tree, *n.* **pan**
tree house, *n.* (1) **bo•rang** (2) **bandaw ~ bando** (3) **noga**
tree stump, *n.* **gan•thong**
tree stump, *n.* **rochong**
tree trunk, *n.* (1) **japang** (2) **panchong** (3) **sun** (4) **sundul**
tremble, *v.* **dekdek-**
triangular pastry eaten with tea, *n.* **poop**
tribe, *n.* **jat**
tributary river, *n.* **tyiphek**
trim, *v.* **tam-**
trod, *v.* **ga•-**
trouble, *n.* **karaw**
troublesome, *adj.* (1) **ha•sel** (2) **ha•mat** (3) **man•dyk-**
trousers, *n.* **longpen**
truly, *adv.* (1) **asol** (2) **-asol** (3) **-bebe** (4) **bebe** (5) **-dam**
trump, *n.* **chun**
try, *n.* **joton**

try, *v.* (1) **ram•-** (2) **chyi-** (3) **joton kha•-**
tsk, *interj.* **chys**
tub, *n.* **sorea ~ soraia**
tube, *n.* **tiup**
Tuesday, *n.* **monggolbal**
tug-of-war, *n.* **bytwami**
turban, *n.* (1) **khawphyng** (2) **bagukhawa**
turbid, *adj.* **bui-**
turmeric, *n.* **raidi**
turn, *n.* (1) **wen• ~ wet** (2) **tap**
turn, *v.* **wang•-**
turn off, *v.* **ni•et-**
turn over, *v.* **chuduk- ~ chyduk-**
turn upside down,, *v.* **chuduk- ~ chyduk-**
turn one's back to someone, *v.* **kyngjung-**
turtle/tortoise, *n.* **katua ~ khatua**
tusk, *n.* (1) **wa** (2) **mongmawa• ~ mungmawa•**
twelve, *num.* **chi ni**
twenty, *num.* (1) **kholgyk ~ kholgryk** (2) **khol** (3) **rum•**
twig, *n.* (1) **pandala** (2) **panchyksi** (3) **cheksi**
twilight, *n.* **walsymsym**
twin, *n.* **jonja**
twist, *v.* **sakrem-**
two, *num.* **ni**
type, *n.* **rokhom**

## U

ugh, *interj.* (1) **hyits** (2) **hys** (3) **hyis** (4) **tyis** (5) **tys** (6) **yis** (7) **chys**
umami, *adj.* **thap-**
umbilical cord, *n.* **gandurian**
umbrella, *n.* **satha ~ sytha**
unblock, *v.* **dok-**
uncle, *n.* (1) **mama** (2) **haw•** (3) **dytyi** (4) **wang** (5) **awang** (6) **baba**
uncomfortably, *adv.* **suksak**
unconscious, *adj.* (1) **nombok-** (2) **nombok thyibok**
uncooked, *adj.* **githyng ~ gythyng ~ githing ~ gi•thyng**
under, *n.* **hama ~ nokhama**
under water, *adv.* **ryp-**
underarm, *n.* **chakgytok**

underestimate, *v.* **mykchep-**
underneath, *n.* **hama ~ nokhama**
underside, *n.* **okma**
understand, *v.* **tyng-**
understanding, *n.* **tyngwami**
unearth, *v.* **bul-**
unfold, *v.* **badal-**
unintentionally, *adv.* **-ram•**
United States, *n.* **Amerika**
university, *n.* **unibersyti**
unmarried man, *n.* **banthai**
unplug, *v.* **dok-**
unprofitable, *adv.* **lapchaga(ba)**
unripe, *adj.* (1) **githyng ~ gythyng ~ githing ~ gi•thyng** (2) **pibok**
unsheathe, *v.* **bykot-**
unstable, *adj.* (1) **ingjong-** (2) **guduk-**
unsuccessfully, *adv.* **-chyp**
untie, *v.* **deng-**
until, *prep./conj.* (1) **dabat** (temporal) (2) **thyl•** (spatial) (3) **=china**
unwell, *adv.* **sasyk sasyk tak-**
up, *adv.* **-taw**
up and down, *adv.* (1) **-jokjok** (2) **-rura**
up onto, *prep.* **-gat**
up till now, *adv.* (1) **-khyngkhyng** (2) **te•ewrawraw ~ te•awrawraw**
up to, *prep.* (1) **=china** (2) **thyl•**
upper arm, *n.* **chakphong ~ chakphung** 'upper arm, arm'
upper leg, *n.* **cha•phung**
upper side of a hill, *n.* **ha•kha**
upright, *adv.* **-phak**
upright, *adj.* **chenggang-**
uproot, *v.* **phok-**
upside, *n.* **dykym**
upside down, *adj.* **phangphyl**
upstream, *n.* **khambai**
upstream, *adv.* **-taw**
upwards, *adv.* **-taw**
uranium, *n.* **rongmesak**
urge, *n.* **sasyk sasyk tak-**
urinate, *v.* (1) **disu-** (2) **disudap-** 'urinate on top of'
urine, *n.* (1) **disutyi** (2) **tyisang**
urine bladder, *n.* **disutyitup**

Ursa Major (constellation), *n.* **do•jenjok**
use, *v.* **jakhal-**
useless, *adj.* **achok**
use up, *v.* **jam-**
useful, *adj.* (1) **chuli-** (2) **jakhal-**
uselessly, *adv.* **ha•sel**
usually, *adv.* (1) **-ramram** (2) **-rong**
uterus, *n.* **sa•thup**

# V

vagina, *n.* **su•**
valley, *n.* **ha•khong**
valuable, *adj.* **gamchat-**
value, *n.* (1) **lap** (2) **man•dapami ~ mandapwami** (3) **gamchatgaba**
vapour, *n.* **biba**
vegetable, *n.* **samchak**
vegetation, *n.* **pan wa•**
vehicle, *n.* **gari**
vehicle repair man, *n.* **mistyri**
vein, *n.* (1) **cha•dyl** (2) **kara**
vengeful, *adj.* **machak-**
Venus (planet), *n.* **athamphang**
veranda, *n.* **baranda**
vertically, *adv.* **-phak**
very, *adv.* (1) **nemen** (2) **bylongen ~ blongen** (3) **tyngen** (4) **-bi** (5) **=ok** (6) **-thyng** (7) **-thyngthyng**
very early in the morning, *adv.* **manapmi**
very little food, *n.* **churu**
very much, *adv.* **-syrang ~ -srang**
via, *prep.* **=tykyi ~ =takai**
vicinity, *n.* **gyching ~ giching**
video, *n.* **pidio**
village, *n.* (1) **song** (2) **gythym ~ guthum ~ gythum** (3) **songmong** 'main village'
village and surrounding lands, *n.* **ha•song**
village headman, *n.* **nokma**
vine, *n.* **cha•dyl**
vitiligo, *n.* **chong•khobok**
vocal cords, *n.* **tokdyl**
voice, *n.* **khu•rang**
vomit, *v.* (1) **chisat-** (2) **kha•rekrek-** (3) **kha-wakwak-**
vulture, *n.* **saw•khyn ~ sawkun**

# W

wag, *v.* **wyngwang-**
wages, *n.* (1) **dolma** (2) **ha•chak** (3) **hajira**
waist, *n.* (1) **changchon** (2) **puksuk**
waist cloth (for men), *n.* **bagu**
wait, *v.* (1) **sam-** (2) **tam-**
wake someone up, *v.* (1) **hasa-** (2) **hala kha•-**
wake up, *v.* **jasa-**
walk, *v.* (1) **cha•aw rai•-** (2) **rai•-** (3) **re•eng-** 'walk away, leave' (5) **rai•wil-**'walk around something' (6) **gul- ~ jul-** 'walk through the jungle with difficulty'
walking stick, *n.* (1) **me•mangkereng** (animal) (2) **guthini ~ guthyni** (artefact)
wall, *n.* **noksam**
wander, *v.* **parang-**
wank, *v.* (1) **soksok-** (2) **selsoksok-** (3) **sepsep-** (4) **ri• selsoksok-** (5) **ri• sepsep-**
wanker, *n.* **soksok**
want, *v.* (1) **syk-** (2) **ram•-**
warden, *n.* **chokida**
warm, *adj.* **tung-**
warm, *v.* (1) **tunget-** (2) **hang-**'warm one's hands by the fire'
warm up, *v.* **tunget-**
warn, *v.* **mykraket-**
wart, *n.* **me•mangsawdet**
wash, *v.* (1) **suset- ~ susut- ~ susyt-** 'was something' (2) **tyru- ~ tyiru- ~ tyiryw-** 'wash oneself' (3) **myksu-** 'wash one's face' (4) **chaksu-** 'to wash one's hands' (5) **cha•su-** 'to wash one's feet/legs' (6) **dai•-** wash away (as in a landslide)
washing line, *n.* **raityng**
wasp, *n.* **ong**
waste, *v.* (1) **kha•chyp-** (2) **lekat-** 'waste time'
wastefully, *adv.* **-chyp**
watch, *n.* (1) **khori** (2) **wach**
watch, *v.* (1) **chai-** (2) **mu•ten-**'look after, watch, keep company'
watch over, *v.* **chaithum-**
watchman, *n.* **chaitumgaba**
water, *n.* (1) **tyi** (2) **tyisurung** 'rainwater that streams over the ground'
water buffalo, *n.* **chyndyk**
water container, *n.* **tyigum**
water monitor, *n.* **phutsul**
water pipe, *n.* (1) **wa•dokolong** (2) **wa•gydok**
water pot, *n.* **tykyw**
waterfall, *n.* **tyichabakram**
wattle, *n.* **kha•thol**
wave, *n.* **tyibal**
way, *n.* (1) **ram** (2) **sorok**
we *pron.* (1) **ning** (exclusive) (2) **na•nang** (inclusive)
weak, *adj.* (1) **nom•-** (2) **demdong-**
wealth, *n.* (1) **gam ~ kam** (2) **gam jym**
weapon, *n.* **ostro**
wear, *v.* **kan-**
weave, *v.* (1) **dok-** 'weave clothes'(2) **wat-** 'weave things from reed or bamboo, make a mat or basket from bamboo or reed' (3) **khep-** 'weave a bamboo mat (*damdyl*) ' (4) **thuk-** 'weave a bamboo mat (*damdyl*) '
web, *n.* **syryng**
wedding, *n.* **bia**
Wednesday, *n.* **butbal**
weed, *n.* **sam**
weed out, *v.* **bak-**
weeding, *n.* (1) **ha•jagyra** 'the first weeding of the *ha•ba*' (2) **jakun** 'the second weeding of the *ha•ba*' (3) **saigyn** 'the third weeding of the *ha•ba*'
week, *n.* **nygyltyi**
weekly, *adv.* **nygyltyityi**
welcome, *v.* **ra•sak-**
well, *n.* (1) **chiakhol** (2) **digi**
well, *interj.* **=de**
well, *adv.* (1) **nem-** (2) **suk-**
well cooked, *adv.* **thik**
well done, *adv.* **thik**
west, *n.* **salniram**
wet, *adj.* **tyisi-**
what?, *pron./det.* **atong**
whatchamacallit *n./interj.* **hai•e ~ hai•-**
whatever, *pron./adj.* **je**
wheat, *n.* **gom**
wheel, *n.* **chaka**
when, *conj.* **=wachido**
when?, *adv.* **biba**
when, *conj.* **=wachi**
where?, *adv.* (1) **bichi** (2) **bie**
where from?, *interr.* (1) **bisang** (2) **bisangmi ~ bisangmyng**

where to?, *interr.* (1) **bisang** (2) **bisangna**
wherever, *adv.* (1) **bibasa** (2) **jechiba** (3) **jesangba**
whet *v.* **wyt-** ~ **wot-** ~ **wen•-** ~ **wyn•-**
whetstone, *n.* **rong•sa**
which?, *pron./det.* (1) **bie** (2) **bigaba** ~ **biga**
whichever, *pron./adj.* **je**
while, *adv.* **=butung**
whine, *v.* (1) **mangneng-** (2) **synthi-**
whirlpool, *n.* **tyibasal**
whisper, *v.* **balsyruk-**
whistle, *v.* **khu•sylip** ~ **khu•sylyp-**
white, *adj.* **pibok**
white patches, *n.* **chong•khobok**
white person, *n.* (1) **phoren** (2) **saip** ~ **saep**
white radish, *n.* **mula**
who?, *pron.* **chang**
whoever, *n.* (1) **changba** (2) **changgaba**
whole, *adj.* **=gumuk**
wholeheartedly, *adv.* **-nap**
wholly, *adv.* **-syrang** ~ **-srang**
whoosh, *ideo.* **wuuuuk**
whore, *n.* **pha•lap**
why?, *adv.* (1) **atakna** (2) **atongtykyi**
widely spaced,, *adj.* **heng•-**
widow, *n.* (1) **jykri** ~ **jykyryi** (2) **jyknyi**
widower, *n.* (1) **jykri** ~ **jykyryi** (2) **jyknyi**
width, *n.* **gebeng**
wife, *n.* (1) **gawigaba** (2) **jyk** (3) **jyktyi** 'second wife of a man who is already married' (4) **jykmong** ~ **jykmongma** 'first wife of a man who has two wives'
wiggle, *v.* (1) **guduk-** (2) **ryngring-** 'move back and forth, up and down'
wiggly, *adj.* **jingjong-**
wild pig, *n.* **wakpalyng**
wild water buffalo, *n.* **matdi**
will, *v.* (1) **=ni** (2) **=naka**
willing, *adv.* (1) **bam-** (2) **gong-**
willingly, *adv.* **bamai**
wind, *n.* **balwa**
wind, *v.* **wang•-**
wind something around something, *v.* (1) **winwin-** ~ **wenwen-** (2) **wenphak-** (2) **wen•-** ~ **wen-**
winding, *adj.* **kon•**
winding, *adv.* **kongken naken**

window, *n.* **kelki** ~ **khelki**
wine (from rice), *n.* (1) **chyw** (2) **chywbok** (white) (3) **pityi** (4) **mainyl pityi** (5) **dykha** 'wine drunk during the *chywgyn* festival'
wing, *n.* (1) **karang** (2) **taw•karang**
wink, *v.* (1) **uk-** (2) **mykren tan•-**
winnow, *v.* **chaw-**
winnowing basket, *n.* **awan**
wipe off, *v.* (1) **wyiset-** (2) **rokset-**
wire, *n.* **waiyr**
wise person, *n.* **bida**
with, *prep.* (1) **=myng** ~ **=mung** ~ **=mu** (2) **=ba** (3) **=sang**
with a spinning head, *adv.* **chuwyng chuwang**
with open mouth, *adv.* **kha•wak khu•wak**
without, *prep.* (1) **=nyi** ~ **=ni** (2) **=ri** ~ **=ryi**
without hesitation, *adv.* **-parang**
witness, *n.* **sakhi**
wobble, *v.* **guduk-**
woman, *n.* (1) **gawi** (2) **jyw•bydyi** 'woman with children, old woman' (3) **me•ama** 'married woman'
womb, *n.* **sa•thup**
woo, *v.* **bot-**
wood chip, *n.* **pantiki**
wood shed, *n.* **pannok**
woof, *ideo.* **bawbaw**
word, *n.* **kata** ~ **khata** ~ **katha** ~ **khatha**
work, *n.* **kam** ~ **gam**
work, *v.* (1) **kam kha•-** (2) **dairamphin-** 'work overtime' **kha•at-** 'work with ( material) '
world, *n.* **ha•gyrsak** ~ **ha•gylsak**
worm, *n.* **chygyl, krimichong, kyrywkeng**
worry, *v.* **jajyreng-**
worship, *v.* **mani-**
worthless, *adj.* **lapchaga(ba)**
would, *v.* **=chym**
wound, *n.* (1) **mat** (2) **matgaba** (3) **phari**
wounded, *adj.* (1) **sakhyna-** (2) **matok**
wow, *interj.* (1) **aia** (2) **aiu** (3) **atyw** ~ **atyyyw** (4) **baaa** (5) **bapre** (6) **baprebap** (7) **bylongok** (8) **da•nang** (9) **ha•gyrsak** ~ **ha•gylsak**
wrap, *v.* (1) **chu•-** (2) **khatdep-** (3) **pyn•-** (4) **pyt-**'wrap neatly as a present' (5) **wen•-** ~ **wen-** 'wrap around'
wrench, *n.* **rens**

wring, *v.* (1) **sepjyrot-** (2) **sep-** (3) **chaksi phai•-** 'wring one's hands'
wrinkle, *n.* **dol ~ thol**
wrinkled, *adj.* (1) **rekhep-** (of person) (2) **tyikhyrep-** 'wrinkled because of being in the water for a long time'
wrist, *n.* (1) **chakgydok** (2) **jaksan**
write, *v.* **sai-**
wrong, *adv.* **nemcha**
wrongly, *adv.*(1) **-syret** (2) **nemchaai**

# Y

yard long bean, *n.* **kha•rek**
yawn, *v.* **ajam-**
yawn, *v.* **hajam-**
year, *n.* **bylsi**
yeast, *n.* **aphap**
yellow, *adj.* **rymyt**
yellow fever, *n.* **holdiasop**

yesterday, *adv.* **myia**
yonder, *adj./adv.* (1) **hawchi** (2) **hyiawchi**
you *pron.* (1) **nang•** (singular) (2) **nang•tym** (plural)
your, *det.* (1) **nang•** (singular), **nang•tym** (plural) (2) **nang•mi ~ nang•myng** (singular), **nang•tymmi ~ nang•tymmyng** (plural)
yours, *pron.* (1) **nang•** (singular), **nang•tym** (plural) (2) **nang•mi ~ nang•myng** (singular), **nang•tymmi ~ nang•tymmyng** (plural)

# Z

Zanthoxylum oxyphyllum, *n.* **mai•cheng**
zero, *n.* **jero**
zigzag, *adv.* (1) **kongken naken** (2) **wel•-**
zip ~ zipper, *n.* **cheen**
zoo, *n.* **chirokhana**

# PART 3: SEMANTIC LEXICA

This part of the volume contains semantic lexica, i.e. lists of words organised by their meanings. The first four lexica could have been embedded within the semantic lexicon of verbs and nouns; however, given their informational complexity, they require a special layout. Therefore, these four lexica have been given separate primary sections within this part of the book. Each lexicon is preceded by a short introduction of its own.

## 1 Days of the week

The Atong names for the days of the week are nouns of Bengali origin. However, these words are completely integrated into Atong, and adapted to Atong pronunciation, to make them indistinguishable from other Atong words.

| English | Atong | Bengali |
|---|---|---|
| Monday | **sombal** | সোমবার |
| Tuesday | **monggolbal** | মঙ্গলবার |
| Wednesday | **butbal** | বুধবার |
| Thursday | **bistibal** | বৃহস্পতিবার |
| Friday | **sukulbal ~ sykulbal ~ sykubal** | শুক্রবার |
| Saturday | **sunibal** | শনিবার |
| Sunday | **rubibal** | শনিবার |

## 2 Months of the year

Nowadays, Atongs use the names of the months borrowed from English, and adapted to Atong pronunciation. However, some speakers still remembered a distant past, when words of Bengali origin were used. Both the modern and archaic names are presented here, with the names of the months in the languages they are borrowed from. Months of the year are nouns. Those of Bengalis origin are compounds of a Bengali loanword and the Atong noun *ja* 'month'. As far as the author was able to establish, there are no traditional or original Atong words for the months of the year.

| English | Atong Modern Style | Atong Old Style | Bengali |
|---|---|---|---|
| January | jenuari | makja | মাঘ |
| February | phebuari | pargunja | ফাল্গুন |
| March | march | choi•etja | চৈত্র |
| April | epril | boisaja | বৈশাখ |
| May | me | jetja | জ্যৈষ্ঠ |
| June | jun | asalja | আষাঢ় |
| July | julai | sa•wynja | শ্রাবণ |
| August | agos ~ agys | badolja | ভাদ্র |
| September | septembol ~ septembyl | asyngja ~ asingja | আশ্বিনি |
| October | oktobol ~ oktobyl | katija | কার্তকি |
| November | nobembol ~ nobembyl | agynja | অগ্রহায়ণ |
| December | disembol ~ disembyl | mai•raja | পৌষ |

## 3 Lexicon of kinship terms: Atong – English

English has far fewer kinship terms than Atong. It is therefore not possible to translate each Atong word by a different English word. Some Atong kinship terms can be translated into English, but a description needs to be added because the English word is not specific enough. For example, the words *awang*, *dytyi*, *haw•* and *syi* can all be translated by 'uncle' in English, but the Atong words refer to four different relationships. This is why a description is added after a colon, as is done in the following example of the headword *awang*, where the translation is 'uncle' and the explanation is 'father's younger brother'.

**awang** *c, ref, a.* uncle: father's younger brother

For other Atong kinship terms, there is no English word available at all, and only a description can be given in English.

All kinship terms in this lexicon are nouns. For each kinship term, except those denoting pairs or groups family members (labelled *set*), it is indicated whether it can be directly derelationalised (*drel*), whether it is classificatory (*c*) or descriptive (*d*), whether it can be used referentially (*ref*) or as a term of address (*a*), and whether it can be use reciprocally (*rec*). Note that all address terms can also be used referentially, but not vice versa. For an in-depth description of Atong kinship terms, see van Breugel (2020). Words labelled $c_{(1)}$ are classificatory only when used with the meaning of the first provided translation, but descriptive when used with their other meaning or meanings.

**abi** *c, ref, a.* (1) elder sister (2) Mothers-in-law can address each other as *abi* too.

**abu** *c, ref, a, rec\**. (1) grandmother. (2) Also used to address a granddaughter. (3) Also used to address an unrelated elderly woman. *This word is only reciprocal when used as a term of address between grandparent and granddaughter.

**abumuri** *c, ref.* great-grandmother (addressed as *abu*). Can be followed by the terms *jagyra* 'maternal' or *jagysi* 'paternal'.

**achu** *c, ref, a, rec\**. (1) grandfather. (2) Also used to address a grandson. (3) Also used to address an unrelated elderly man. *achu ambi* grandparents, ancestors. (4) Also used to talk about or address an elephant when you are in the jungle. *This word is only reciprocal when used as a term of address between grandparent and grandson.

**achu ambi** *set.* (1) grandparents (2) ancestors

**achumuri** *c, ref.* great-grandfather (addressed as *achu*). Can be followed by the terms *jagyra* 'maternal' or *jagysi* 'paternal'.

**achuthangmaran** *set* a grandfather and his grandchild

**ade** *ref, a.* stepmother

**akai** *c, ref, a.* (1) aunt: mother's elder sister (2) Also used to address an unrelated married woman older than the speaker.

**akaithangmaran** *set* my *akai* (mother's elder sister) and her younger sister's child

**ama** *c, ref, a, rec\** (1) mother (biological or classificatory) (2) Also used to talk about or address a maternal aunt. (3) Also used to address a daughter. *This word is only reciprocal when used as a term of address between parent and daughter.

**ambithangmaran** *set.* a grandmother and her grandchild

**anai** *c, ref, a.* aunt: father's sister

**anyng** *c₍₁₎, ref, A.* (1) aunt: father's sister (2) sister-in-law: husband's elder sister

**asyi ~ asi** *c, ref, a.* aunt: mother's younger sister

**awa** *c, ref, a.* biological father

**awang** *c₍₁₎, ref, a.* (1) uncle: father's younger brother (2) what children call their stepfather

**awyi** *c, ref, a.* grandmother (archaic in Badri and Siju)

**baba** *c, ref, a, rec\*.* (1) father (biological or classificatory) (2) Also used to talk about or address a paternal uncle. (3) Also used to address a son. *This word is only reciprocal when used as a term of address between parent and son.

**bai•maran** *set* two distant relatives.

**bai•maran chingmaran** *set.* two distant relatives

**bai•sakthangmaran** *set.* two people who belong to the same *mahari* 'lineage'

**biawthang** *ref, a.* the relationship between a male and the husband of his sister's daughter (*namgaba*)

**biawthangmaran** *set.* my wife's elder brother and me together

**biphagaba ~ biphaga** *ref.* husband

**bochi** *ref, a.* Only used in the Siju dialect (in the Badri dialect: *ja•chung*) sister-in-law: elder brother's wife

**bochithangmaran ~ buchithangmaran** *set.* my wife and her sister or brother together

**bonyng** *ref, a, rec.* (1) brother-in-law: the reciprocal relation between a man and his younger sister's husband or a man and his wife's elder brother (2) any man of another clan from the same generation as a male speaker

**chamai ~ chame** *ref, a.* (1) female cross-cousin: mother's brother's daughter or father's sister's daughter (2) the relation of female cousins from intermarriageable families (3) the relation of the parents of a married couple

**chamaithangmaran ~ chamethangmaran** *set.* a couple of marriageable cross-cousins, a boy and girl who can marry

**chara** *set.* (1) wife's elder brothers (2) mother's brothers

**charamong** (1) wife's eldest brother (2) mother's eldest brother

**chuchu** *a.* the address term a grandparent uses to their grandson

**dada** *c, ref, a.* (1) elder brother. (2) Also used to speak about or address a related older male relative of one's own generation: cousin. (3) Also used to address an unrelated man older than the speaker.

**dytyi** *c, ref, a.* uncle: father's elder brother

**dytyithangmaran** *set.* my *dytyi* (father's elder brother) and his younger brother's child

**gawigaba ~ gawiga** *ref.* wife

**gumi** *d, ref, a.* brother-in-law: (1) elder sister's husband (2) husband's elder brother

**gumithangmaran** *set.* my husband and my younger sister or younger brother together

**haw•** *drel, c, ref.* uncle: mother's brother (addressed as *mama*)

**haw•maran** *set.* my *haw•* (mother's elder or younger brother) and his (elder or younger) sister's unmarried child

**haw•nokhol** *ref.* father-in-law, deceased father-in-law's heir (addressed as *mama*)

**haw•nokholburung** *set.* a group of fathers-in-law (*haw•nokhol*) and sons-in-law (*kynokhol*)

**ja•chung** *d, ref, a.* Badri dialect: sister-in-law: (1) wife's elder sister (2) elder brother's wife. Siju dialect: wife's elder sister

**ja•chungthanmaran** *set.* my husband and my elder sister together

**ja•naw** *ref, a.* (1) elder sister. (2) Also used to address an older female cousin or a woman older than the speaker (The word *abi* is more respectful as a term of address for both referents.)

**ja•nawburung** *set.* a group of sisters

**ja•nawmaran** *set.* two sisters

**jojong** *c, ref, a.* (1) younger brother. (2) Also used to talk about or address a related younger male of one's own generation: cousin (3) Also used to address a young male unrelated person younger than the speaker.

**jong** *drel, d, ref, a.* (1) younger brother (2) Also used to address a younger male cousin or (3) an unrelated man younger than the speaker.

**jongsyri** *d, ref, a.* (1) brother-in-law: spouse's younger brother (2) female's younger sister's husband (A male's younger sister's husband is *bonyng*.)

**jyk** *ref.* spouse

**jyksai** *set.* a married couple, husband and wife

**jyw•** *c, ref.* biological mother

**jyw•burung** *set.* a group of mothers and daughters

**jyw•mong ~ jyw•morong** *ref.* eldest of a group of sisters

**khyryithang ~ khyrythang** *c, ref, a.* nephew: male's sister's son or female's brother's son

**khyryithangsyw•** *d, ref.* great-nephew: the son of my husband's sister's daughter or the son of my brother's wife's daughter

**kynokhol** *d, ref, a.* son-in-law: deceased testator's son-in-law, or husband of a household's heiress

**machong** woman-founder of a clan

**mama** *c, ref, a.* (1) uncle: mother's brother (2) Also used to address my father-in-law. (3) Also used to address an unrelated man older than the speaker in a respectful way.

**mani** *c, ref, a.* (1) aunt: father's sister. (2) Also used to address one's mother-in-law.

**mawsa ~ mosa** *c, ref, a, rec.* (1) male cross-cousin: father' sister's son or mother's brother's son (2) the relation of male cousins from intermarriageable families (3) a male friend belonging to an intermarriageable family

**mawsathangmaran ~ mosathangmaran** *set.* two boys of different *maharis* (lineages), for example Marak and Sangma, in two possible relationships depending on the gender of the speaker. Male speaker: my elder sister's son and my son. Female speaker: my elder brother's son and my son.

**nai•** *drel, c, ref.* aunt: father's sister. (addressed as *anyng, anai,* or *mani*)

**nai•maran** *set.* my *nai•* (father's elder or younger sister) and her (elder or younger) brother's unmarried child

**nai•nokhol** *d, ref.* mother-in-law (addressed as *mani* or *anai*)

**nai•nokholburung** *set.* a group of mothers-in-law and daughters-in-law

**nai•nokholthangmaran** *set.* my *nai•* (father's elder or younger sister) and her (elder or younger) brother's married child

**namchyk** *d, ref, a.* niece: (1) female's brother's daughter (2) male's sister's daughter (Denotes the same relation as *namgaba*.)

**namchyksyw•** *d, ref.* grand-niece / great-niece: (1) the daughter of my husband's sister's daughter (2) the daughter of my brother's wife's daughter

**namgaba** *d, ref.* niece: (1) male's sister's daughter (2) female's brother's daughter (addressed as *namchyk*)

**namnokhol** *d, ref, a.* daughter-in-law

**namnokholburung** *set.* group of daughters-in-law

**naw** *drel, d, ref, a.* (1) younger sister. (2) Also used to address a younger female cousin or (3) an unrelated woman younger than the speaker.

**nawsyri** *d, ref, a.* sister-in-law: (1) spouse's younger sister (2) younger brother's wife (Elder brother's wife is *ja•chung* or *bochi*.)

**nokchama** *d, ref.* the relationship of the parents of a married couple

**nokchina ~ nokna** *ref.* (Siju dialect) the heiress of a household or her husband.

**nokrom** *ref.* (Badri dialect) the heiress of a household or her husband.

**nono** *c, ref, a.* (1) younger sister. (2) Also used to talk about or address a related younger female of one's generation: cousin or (3) to address a young unrelated female person younger than the speaker.

**nyng** *d, ref.* (1) aunt: father's sister (2) sister-in-law: husband's elder sister (addressed as *anyng*)

**phaw•jong ~ phawjong** *drel, c, ref, a.* (1) elder brother. (2) Also used to address an older male cousin or (3) a man older than the speaker.

**phaw•jongmaran ~ phawjongmaran** *set.* two elder brothers

**sa•** *ref.* child, offspring

**sa•banthai** *drel, c, ref.* (1) son (2) nephew: male's brother's son or female's sister's son

**sa•burung jyw•burung ~ sa•byrung jyw•byrung** *khjys n. ref.* a mother and her children

**sa•mynchyk** *drel, c, ref.* (1) daughter (2) niece: male's brother's daughter or female's sister's daughter

**sadu** *d, ref, a.* brother-in-law: the relation of men whose wives are sisters.

**saduthangmaran** *set.* two or more men whose wives are sisters

**syi** *drel, ref.* aunt: mother's younger sister

**syimaran** *set.* my *syi* (mother's younger sister) and her elder sister's child

**syw•** *drel, d, ref.* grandchild

**syw•muri** *c, ref.* great-grandchild

**wa•** *drel, d, ref.* biological father

**wa•maran** *set.* a father and his child (son or daughter)

**wang** *drel, d, ref.* (1) uncle: father's younger brother (2) stepfather (3) the inverse relation of *biawthang*: wife's mother's brother

**wanggaba** *ref.* (1) the inverse relation of *biawthang*: wife's mother's brother (2) the derelationalised form of *wang*.

**wangmaran** *set.* my *wang* (father's younger brother) and his elder brother's child

**wyi•** *drel, c, ref, a.* grandmother (archaic)

# 4 Lexicon of kinship terms: English – Atong

Since English has much fewer kinship terms than Atong, many English terms can be translated in more than one way in Atong. The appropriate Atong kinship term for use in a particular context can be found by referring to the Atong-English Lexicon of Kinship Terms. For example, the English kinship term *uncle* can be translated into Atong as *mama, haw•, dytyi, wang, baba* or *awang*. Which one can be used to address father's younger brother (a paternal uncle)? To find the answer, look up the precise meaning and use of each of the Atong kinship terms in the Atong-English lexicon, and find that the answer is the words *baba* and *awang*. To facilitate searches for specific kinship relations, this lexicon also contains descriptive entries, e.g. *my elder sister's son and my son (male speaker)*, and *wife's younger sister*.

ancestors **achuambi**
aunt **akai, ama, syi, asyi ~ asi, nyng, anyng, mani, nai•, anai**
blood relatives **bai•, bai• tyng, bai•siga ~ bai•sega**
brother (elder) **dada, phaw•jong ~ phawjong**
brother (younger) **jong, jojong**
brother-in-law **gumi, jongsyri, bonyng, biawthang, sadu**
child **sa•, sa•gyrai**
couple of marriageable cross-cousins, a boy and girl who can marry (pair) **chamethangmaran**
cousin **chamai ~ chame, mawsa ~ mosa, dada, ja•naw, abi, naw, nono, jong, jojong, phaw•jong ~ phawjong**
cross-cousin (female) **chamai ~ chame**
cross-cousin (male) **mawsa ~ mosa**
daughter **sa•mynchyk, ama, rani,**
daughter-in-law **namnokhol**
daughters-in-law (group) **namnokholburung**
distant relatives (pair) **bai•maran, bai•maran chingamaran**
elder brother's wife **bochi** (Siju), **ja•chung** (Badri)
elder sister's husband **gumi**
eldest sister **jyw•mong ~ jyw•morong**
father and his child (son or daughter) (pair) **wa•maran**

father **awa, baba, wa•** (biological)
father-in-law **haw•nokhol, mama**
fathers-in-law (*haw•nokhol*) and sons-in-law (*kynokhol*) (group) **haw•nokholburung**
father's brother (elder) **dytyi**
father's bother (younger) **wang, awang**
female's brother's daughter **namchyk, namgaba**
female's brother's son **khyryithang ~ khyrythang**
father's sister **anyng, nyng, anai, nai•, mani**
female's younger sister's husband **jongsyri**
grandchild **syw•, achu, chuchu, abu**
granddaughter **syw•, abu**
grandfather **achu**
grandfather and grandchild (pair) **achuthangmaran**
grandmother **abu, awyi, wyi•**
grandmother and her grandchild (pair) **ambithangmaran**
grand-nephew / great-nephew **khyryithangsyw•, syw•** (the son of my husband's sister's daughter or the son of my brother's wife's daughter)
grand-niece / great-niece **namchyksyw•, syw•** (the daughter of my husband's sister's daughter, or the daughter of my brother's wife's daughter)
grandson **syw•, chuchu**
great-granddaughter **syw•muri**

great-grandfather **achumuri**
great-grandmother **abumuri**
great-grandson **syw•muri**
husband and wife **jyksai**
husband **biphagaba, jyk**
husband's brother (elder) **gumi**
husband's brother (younger) **jongsyri**
husband's sister (younger) **nawsyri**
male's sister's daughter **namchyk, namgaba**
male's sister's son **khyryithang ~ khyrythang**
married couple **jyksai**
men whose wives are sisters (pair or group) **saduthangmaran**
mother **ama, jyw•** (biological)
mother and her children (group) **sa•burung jyw•burung**
mother's brother **mama, haw•**
mother's brothers (group) **chara**
mother's brother's wife **nai•**
mother's elder sister **akai**
mother's elder sister (*akai*) and her younger sister's child (pair) **akaithangmaran**
mother's eldest brother **charamong**
mother's younger sister **asyi, asi, syi**
mother's younger sister's husband **wang, awang**
mother-in-law **nai•nokhol, mani**
mothers and daughters (group) **jyw•burung**
mothers-in-law and daughters-in-law (group) **nai•nokholburung**
my elder brother's son and my son (female speaker) (pair) **mawsathangmaran ~ mosathangmaran**
my elder sister's son and my son (male speaker) (pair) **mawsathangmaran ~ mosathangmaran**
my father's elder brother (*dytyi*) and his younger brother's child (pair) **dytyithangmaran**
my father's elder or younger sister (*nai•*) and her (elder or younger) brother's unmarried child (pair) **nai•maran**
my father's elder or younger sister (*nai•*) and her (elder or younger) brother's married child (pair) **nai•nokholthangmaran**

my father's younger brother (*wang*) and his elder brother's child (pair) **wangmaran**
my husband and my elder sister together (pair) **ja•chungthanmaran**
my husband and my younger sister or younger brother together (pair) **gumithangmaran**
my mother's elder or younger brother (*haw•*) and his (elder or younger) sister's unmarried child (pair) **haw•maran**
my mother's younger sister (*syi*) and her elder sister's child (pair) **syimaran**
my wife and her sister or brother together (pair) **buchithangmaran ~ bochithangmaran**
my wife's elder brother and me (pair) **biawthangamaran**
nephew **khyryithang ~ khyrythang** (male's brother's son or female's sister's son), **sa•banthai** (male's sister's son or female's brother's son)
niece **namchyk ~ namgaba** (female's brother's daughter or male's sister's daughter), **sa•mynchy** (male's brother's daughter or female's sister's daughter)
offspring **sa•**
pair of sisters **ja•nawmaran**
relatives **bai•maran, bai•maran chingmaran, bai•sakthangmaran, bai•, bai•siga ~ bai•sega, bai• tyng**
sister (elder) **abi, ja•naw**
sister (younger) **naw, nono**
sister-in-law **bochi, anyng, ja•chung, nawsyri**
sisters (group) **ja•nawburung**
son **sa•banthai, baba, babu**
son-in-law **kynokhol**
spouse **jyk**
stepchild **sa•thyra**
stepfather **wang**
stepmother **ade**
two elder brothers **phaw•jongmaran**
two relatives of the same *mahari* (lineage) **bai•sakthangmaran**
two sisters **ja•nawmaran**
uncle **mama, haw•, dytyi, wang, awang, baba**
wife **gawigaba, jyk**

wife's elder brothers **chara**
wife's eldest brother **charamong**
wife's elder sister **ja•chung**
wife's mother's brother **wang, wanggaba**
wife's younger brother **jongsyri**

wife's younger sister **nawsyri**
woman's brother's daughter **namchyk, namgaba**
woman's brother's son **khyryithang ~ khyrythang**
woman's younger sister's husband **jongsyri**

## 5 Semantic lexicon of verbs and nouns

This section presents lists of recorded nouns and verbs, organised by semantic domain. The reader is referred to the Table contents for the listing of all the categories found in this section. Note that nouns and verbs are both members of the predicative word class (see PART 5, Section 15). The lists are incomplete, and thus do not contain all nouns and verbs of the Atong language. There are many words in Atong the author of this dictionary has not yet recorded. The lists will hopefully be expanded through future fieldwork.

This lexicon is inspired by Burling (2004a) and Jose & Kholar et al. (2014).[11] The categories in which the words presented here are divided have been kept rather more broad than in the two volumes just referred to, so as to keep the lexicon easily accessible. The English-Atong dictionary will provide a more refined tool, in many cases, to search for more specific information. For example, if a reader wishes to search all words regarding *cutting* and *piercing*, they can look under the English entries *cut* and *pierce* in the English-Atong dictionary. Likewise, when the reader wants a list of types of necklaces, they can refer to the entry word *necklace* in the English-Atong Dictionary.

### 5.1 Nature and natural phenomena

#### 5.1.1 Heavenly bodies
**aski ~ askhui ~ askui** *n.* star
**athamphang** *n.* the planet Venus
**baguriwa•** *n.* halo of the moon
**chang•ai** *n.* the moon
**do•jenjok** *n.* Big Dipper (star sign), Ursa Major (constellation)
**jajong** *n.* moon
**jarambong** *n.* full moon
**phryngphrang askui** *n.* the morning star
**rangsan** *n.* sun, day
**wa•daweng** *n.* star sign of three stars in a straight line

#### 5.1.2 Parts of the day
**gasam** *tw/vØ.* evening/to be evening
**manap** *tw/vØ.* morning/to be morning
**wal** *tw/vs1.* night/to be night

#### 5.1.3 Seasons
**chekkyryi** *n.* cold season
**tungkyryi ~ tyngkyryi** *n.* hot season
**watyi** *n.* rainy season

---

11 See also van Breugel (2016).

## 5.1.4 Weather

**balmundyri** *n.* cyclone
**balwa** *n.* wind, air
**goira** *n.* thunder
**guri** *n.* mist, fog
**mykha badri** *n.* long period of incessant heavy rainfall
**rang** *n.* rain

**rangbrym ~ rangbyrym** *n.* cloud
**rangchinek** *n.* cloud
**rangdylekpa** *n.* lightning
**rangra** *n.* sky
**sangori** *n.* fog
**balwa-** *vs1.* to blow (of the wind)
**wa-** *vs1.* to rain

## 5.1.5 Fire and related words

**agal** *n.* forest fire
**chak-** *v.* to ignite
**chem•-** *v.* to melt away, to burn up
**chyng•-** *v.* to burn
**garamak** *n.* soot
**ha•thapyra** *n.* ashes
**kham-** *vintr.* to burn, to scorch, to be on fire, to scald
**muk-** *v.* to smoke (to produce smoke, like a fire does)
**sang-** *v.* to burn
**saw•** *v.* to burn, to roast on the hot ashes of the fire

**thapyra** *n.* ashes
**thup-** *v.* to be thick (of fog or mist)
**wal•** *n.* fire, torch
**wal•di•** *n.* ambers, glowing pieces of burnt wood
**wal•khu** *n.* smoke
**wal•khu-** *v.* to produce smoke
**wal•kungki** *n.* black ashes
**wal•tum** *n.* fire that is burned outside the house during winter to sit around and keep warm

## 5.1.6 Other natural phenomena

**balphak-** *tw/vtr.* to blow something away
**ban-** *vintr.* to flow (of rivers)
**banggyri- ~ banggiri-** *v.* earthquake
**bytym-**₁ *adj1.* to smell nice
**bytym**₂ *n.* a good smell
**chabak-** *vintr.* to fall (of water in a waterfall)
**dai•-** *v.* to wash away (as in a landslide)
**gebeng** *n.* width, breadth
**guruchup-** *v.* to be shrouded in clouds
**jamang-** *v.* to set (of the sun)

**ma•ang-** *vintr.* to set (of the sun)
**manam-**₁ *adj1.* to stink, to smell bad
**manam**₂ *n.* bad smell, stench
**rangbyrym-** *v.* to be shrouded in clouds, to be blocked by clouds.
**rurong-** *v.* to landslide
**seng•-** *v.* to shine, to dawn, to become light
**spiit** *n.* speed
**syn** *n.* a smell

## 5.2 The earth, soil, products of the earth

### 5.2.1 Stones and rocks

**bai•wak-** *v.* GEO. to be land of broken-off rocks
**busi** *n.* dust that is stirred up by the wind or by human activity
**chymbuk** *n.* magnet
**doba** *n.* mud
**ha•** *n.* soil, earth, land
**ha•bykung** *n.* sand

**ha•mang** *n.* soil, earth, clay
**han•cheng** *n.* sand
**hanggal** *n.* charcoal, cinder
**patal ~ phatal ~ phathal ~ pathal** *n.* stone
**rong•** *n.* stone
**rong•baram** *n.* type of rock
**rong•cheret** *n.* pebble size stone

**rong•chung** *n.* big rock
**rong•chyret** *n.* very small stone
**rong•gyrym ~ rong•rymrym** *n.* being full of big rocks, stony land
**rong•han•cheng** *n.* sedimentary rock
**rong•khobok ~ rong•bok** *n.* chalk
**rong•khol ~ rong•khal** *n.* cave
**rong•misi** *n.* very small stone
**rong•patal** *n.* big rock

**rong•phek** *n.* a grain of sand or very small stone
**rong•rymrym ~ rong•gyrym** *n.* being full of big rocks, stony land
**rong•syl** *n.* flint stone
**rong•syrek** *n.* small stone
**rong•thai** *n.* a rock
**rong•thyk** *n.* a big rock
**tykha** *n.* white clay not useful to make pots

### 5.2.2 Precious stones
**hira** *n.* diamond
**so•re** *n.* mica, precious stone

### 5.2.3 Metals and minerals
**koila** *n.* coal
**rong•chun** *n.* lime stone
**rong•khobok ~ rong•bok** *n.* chalk
**elmoni** *n.* aluminium
**rongmesak** *n.* uranium
**sona** *n.* gold

**syl** *n.* iron
**sylgythym** *n.* iron
**sym•** *n.* salt
**tama** *n.* copper
**warem** *n.* rust

### 5.2.4 Water bodies and aquatic phenomena
**biba** *n.* breath, vapour, steam
**sagal** *n.* sea
**siri ~ suri** *n.* snow
**syltyi** *n.* hail, ice
**tyi** *n.* water
**tyibal** *n.* wave
**tyibasal** *n.* whirlpool

**tyichabakram** *n.* waterfall, cascade
**tyikhal** *n.* river
**tyimong** *n.* main river
**tyimuk** *n.* source, spring (of a stream)
**tyiphek** *n.* tributary river, the smaller one of two rivers that flow together
**tyisurung** *n.* rainwater that streams over the ground

### 5.2.5 Dirt, filth
**serabera** *n.* dirt
**moila** *n.* dirt, filth
**jabol** *n.* garbage

## 5.3 Physical processes

**chem•-** *v.* to melt away, to burn up
**kheng-** *v.* to be alive
**phyl•-** *vØ.* to transform, to change into
**chyng•-** *v.* to burn

**wal•khu-** *v.* to produce smoke
**chak-** *v.* to ignite
**sang-** *v.* to burn
**jonong- ~ jorong-** *v.* to dissolve

**jokruru-** *v.* to flow into
**jarang-** *vintr.* to shine
**kham-** *vintr.* to burn

**khang-** *v.* to solidify
**muk-** *v.* to smoke (to produce smoke, like a fire does)

## 5.4 Physical development

**dym-** *v.* to grow (of plants), to sprout
**pyryi-** *v.* to be mature
**chungtaw-** *v.* to grow
**chunggalgal-** *v.* to grow up, to become an adult
**badal-** *v.* to unfold (flowers)

**buk-** *vintr.* to grow like a creeper or liana
**dairukruk-** *v.* become more and more, to increase
**jel-** *v.* to increase, to multiply
**kek-** *v.* to grow
**nem-** *v.* to get better, to heal

## 5.5 Humans

### 5.5.1 Human body parts

**athom** *n.* stomach
**bi•thyn ~ pi•thyn** *n.* liver
**biambong** *n.* biceps
**bichylap** *n.* abdominal membrane
**bimang** *n.* body, appearance
**byl** *n.* strength, muscle
**byl... chak...** *n.* arm, hand
**cha•** *n.* leg, foot
**cha•bykung** *n.* instep
**cha•chok** *n.* sole of the foot
**cha•dok ~ cha•tok** *n.* heel
**cha•dyl** *n.* vein
**cha•kereng** *n.* shinbone, shin
**cha•khawak ~ cha•khok** *n.* hollow side of the knee
**cha•kyw ~ cha•ku** *n.* knee, length from the knee to the foot
**cha•muk** *n.* medial malleolus
**cha•myn** *n.* leg hair
**cha•pa** *n.* sole of the foot
**cha•pakithyk** *n.* heel
**cha•pathai** *n.* calf
**cha•phak** *n.* groin
**cha•phong ~ cha•phung** *n.* thigh
**cha•phung** *n.* upper leg
**cha•pungdym** *n.* hip
**cha•si** *n.* toe
**cha•sijyw•bydyi** *n.* big toe

**cha•tok ~ cha•dok** *n.* heel
**chachura** *n.* hair on top of the head
**chagak** *n.* palate
**chak chok** hand
**chak** *n.* arm, hand
**chakchuk** *n.* elbow
**chakgydok** *n.* wrist
**chakgytok** *n.* underarm
**chakkhawak** *n.* hollow part of the elbow
**chakpha** *n.* palm of the hand
**chakphakhung** *n.* back of the hand
**chakphong ~ chakphung** *n.* arm, upper arm
**chaksi** *n.* finger *chaksi goi•banga* five fingers
**chaksigysep** *n.* space in between the fingers
**chaksijotram** *n.* index finger
**chaksijyw•bydyi** *n.* thumb
**chaksikhol** *n.* fingernail
**chaksikhum** *n.* back of the hand
**chaksirengma** *n.* little finger
**chaksiweng** *n.* knuckles
**changchon** *n.* waist
**chel** *n.* bosom of a man
**chelbak** *n.* chest
**chelku** *n.* rib cage
**chengkhyna ~ chengkana** *n.* jaw
**chokdeng** *n.* throat
**chungthai** *n.* big bosom.
**chybym** *n.* forehead

**di•khal** *n.* bottom, arse, anus
**di•phathai** *n.* buttock
**di•sep** *n.* arse crack
**di•sepra** *n.* arse crack
**disutyitup** *n.* urine bladder
**gandurian** *n.* umbilical cord
**gandyrui** *n.* bellybutton, navel
**gawasu ~ gawsu** rib
**jaksan** *n.* wrist
**ka•dymbai** *n.* chin
**ka•myn•** *n.* beard
**kan•** *n.* body (of human)
**kan•peng** *n.* side of the body
**kereng** *n.* bone
**kha•phak** *n.* chest
**kha•rongthai** *n.* kidney
**kha•sop** *n.* lung
**kha•thong** *n.* heart
**khaw** *n.* hair (of the head)
**khawcha•ryng** *n.* sideburn
**khawchi ~ khawkhai ~ khawkhi** *n.* grey hair
**khawchyryng** *n.* scalpel hair
**khol** *n.* skin
**khu•chul** *n.* lip
**khu•sep** *n.* corner of the mouth
**khu•symang** *n.* facial hair, beard, moustache
**kyi•wa** *n.* canine teeth
**kykgul** *n.* eyelid
**kyn** *n.* back
**kynkyreng** *n.* spine
**manggisi** *n.* corps, dead body
**mansylang ~ manthylang** *n.* spleen
**mu•khuchok** *n.* nipple
**mu•thai** *n.* breast (of woman), bosom
**mykkep** *n.* temple
**mykren ~ mykyren** *n.* eye
**myksep** *n.* corner of the eye
**myksolkhare** *n.* ring finger
**myksymyl ~ myksmyl** *n.* eyebrow
**myksyram ~ myksram** *n.* eyelash
**mykthoram** *n.* middle finger
**myktyiwatram** *n.* side of the hand under the index finger
**mylthai** *n.* small bosom
**mym•** *n.* a fist
**myn•** *n.* body hair (of humans), fur (of animals)
**myn•symok** *n.* a small body hair

**nadanggorot** *n.* oesophagus, food pipe, gullet
**nadekaram** *n.* earlobe
**nakhal** *n.* ear
**nakhalcha•dan** *n.* part of the head behind the ear
**nakhong** *n.* backside of the ear
**nakhung** *n.* nose, snot (liquid)
**nakhungkhal** *n.* nostril
**nakhungmyn•** *n.* nose hair
**natheng** *n.* cheek and cheekbone, side of the head
**okma** *n.* the front of the body, belly, underside
**phagongma ~ phagungma** *n.* shoulder
**phaithawa ~ phaithopa** *n.* cheek
**phakwal** *n.* armpit
**pi•thyn ~ bi•thyn** *n.* liver
**pipuk** *n.* belly, intestines, bowels, stomach
**piryt** *n.* gall bladder
**puksuk** *n.* waist, side of the body
**puktyng** *n.* small intestine
**randai** *n.* body, flesh, meat
**ri•** *n.* penis
**ri•ambanthai** *n.* glans penis
**ri•gan•thong** *n.* erect penis
**ri•karan ~ ri•keren** *n.* testicle, balls, scrotum
**ri•khu•chul** *n.* foreskin
**ri•kun** *n.* glans penis
**ri•myn** *n.* pubic hair of a male
**ri•sokop** *n.* scrotum
**runi** *n.* brains
**sa•thup** *n.* uterus, womb
**sirong** *n.* scrotum
**sokrop ~ sokyrop** *n.* lung
**su•** *n.* vagina
**su•myn** *n.* pubic hair of a female
**su•nadylep** *n.* clitoris
**susu** *n.* penis
**synggera** *n.* handle moustache
**thanyng** *n.* brain
**thylampak ~ thylapak** *n.* tongue
**tokdyl** *n.* vocal cords
**tokkhyphu ~ tokyphu ~ tokybu** *n.* gullet, oesophagus, throat
**tokkyreng** *n.* neck
**tokthynyng ~ tokthining** *n.* neck
**wa** *n.* tooth, tusk
**wa•gatram** *n.* shoulder
**wachyw** *n.* incisors
**wakam** *n.* molar

## 5.5.2 Products of the human body

**bytym** *n.* fat
**cha•gyl** *n.* footstep
**cha•man** *n.* footprint
**chinik** *n.* dirt on the body
**dak** *n.* freckle
**dak** *n.* phlegm from the lungs
**di•** *n.* shit
**dil** *n.* body smell, body odour
**diphu** *n.* a fart
**disutyi** *n.* piss, urine
**gangma** *n.* pimple
**jagyryng** *n.* shadow cast by a person
**ke•ret** *n.* gall
**khawchi ~ khawkhai ~ khawkhi** *n.* grey hair
**khawkhirok** *n.* dandruff
**khawra** *n.* a hair that has fallen out of the head
**khu•rang** *n.* voice
**khu•tyi ~ khu•ti** *n.* spit, spittle, saliva. *khu• tyi thandapai bala* to speak with a lot of spittle

**mai•byram** *n.* cracks in the skin of the cheek
**mykkhi** *n.* slime from the eyes
**myktyi** *n.* tear
**myn•tyi** *n.* puss
**myn•tyi** *n.* puss; resin; latex of jackfruit, thick fluid of various fruits
**nakhung** *n.* nose, snot (liquid)
**nakhungdi•** *n.* hard piece of snot
**ri•ros** *n.* sperm, semen
**ri•tyi ~ ri•ti** *n.* sperm, semen
**ros** *n.* sap, juice (of meat and fruit), sperm, semen,
**sarai** *n.* a cut
**taw•pal** *n.* freckles
**thawal** *n.* scab
**thyi•** *n.* blood
**tyi** *n.* sweat
**tyisang** *n.* piss, urine
**wadi• ~ wakhi** *n.* plaque

## 5.5.3 Bodily functions

**chichu-** *v.* to blister
**thy•yk-** *v.* to have the hiccups
**rang•set-** *v.* to breathe
**takruk-** *v.* to have an orgasm (for a woman)
**khawakwak-** *v.* to vomit, to barf, to chunder
**chisat-** *v.* to vomit, to throw up, to barf
**kha•rekrek-** *v.* to vomit, to barf, to chunder
**dekdek-** *v.* to shiver, to tremble
**de•et- ~ di•it- ~ di•et-** *v.* to shit, to do number two, to defecate
**di•chyrak-** *v.* to have diarrhoea

**di•pyru-** *v.* to have diarrhoea
**khep-** *v.* to cry
**disu-** *v.* to piss, to pass urine, to do number one, to urinate
**diphu-** *v.* to fart
**gusu-** *v.* to cough
**hachi-** *v.* to sneeze
**hajam-** *v.* to yawn
**gang-** *v.* to be erect, to have an erection, to have a hard on
**hongkhot-** *v.* to ejaculate, to cum

## 5.5.4 Afflictions of the human body

**cha•pheret** *n.* crack in the callous skin of the heel
**cha•sitokkyreng** *n.* eczema between the toes
**chichugaba** *n.* a blister
**chong•khobok** *n.* white patches, vitiligo
**dagi** *n.* scar
**dol** *n.* wrinkles
**gal** *n.* scar

**gusylak** *n.* abrasion
**khosylak** *n.* abrasion
**khu•chit** *n.* harelip
**mai•byram** *n.* cracks in the skin of the cheek
**mat ~ matgaba** *n.* a wound
**me•mangsawdet** *n.* wart
**mykbyryw- ~ mykbryw- ~ mykbyru-** *v.* to have itchy eyes

**mykdaw ~ mykdo** *n.* night blindness
**mykpeng-** *v.* to be cross-eyed.
**nangthaigaba** *n.* swelling, abscess
**ogynanggaba ~ okgynanggaba** *n.* pregnancy
**okhuchak** *n.* stomach pain
**oksephang** *n.* pain in the lower abdomen
**pheret** *n.* crack in the skin
**sa-** *v.* to be ill, sick, to hurt, to be in pain

**sakhyna-** *v.* to be wounded
**samsin maiphara** *n.* small boil
**samsin** *n.* big boil, abscess
**san-** *v.* to heal
**thaphu** *n.* blister, sore
**thaphu-** *v.* to blister, to be blistered or to have a sore
**tokbaw** *n.* goitre ~ goiter

### 5.5.5 Physical sensations

**basak-** *vintr.* to burn and cause a rash to cause irritation or itching
**bejaw-** *vintr.* to experience the sensation of being tickled
**bol-** *v.* to burn (like the sensation of a being stung by a stinging nettle); to cause burning, irritation or strong itching

**bosok-** *v.* to itch, to be irritated, to experience the sensation of irritation or itching
**byt-** *v.* to shock (electricity)
**jagydok** *n.* strength
**tungbul-** *v.* to have a warm body (not of fever)
**wel-** *v.* to burn (as a sensation)

### 5.5.6 Diseases and infections

**gawak** *n.* disease
**han•dykmai** *n.* jungle fever
**holdiasop** *n.* jaundice, yellow fever
**khat** *n.* ringworm (a fungal infection)
**kholjisop** *n.* infection of the inner ear, labyrinthitis
**khonchi** *n.* leprosy. khonchi man•ok ~ khonchi sa•ak to have leprosy
**kong•-** *v.* to have rabies
**leng** *n.* bronchitis
**matchirit-** *v.* to have an ulcer on one's skin
**ranggyl** *n.* fungus infection
**sabisi** *n.* disease

**sajin** *n.* illness that makes everything taste bitter
**samycheng ~ sasyri** *n.* bladder infection
**sasep** *n.* chicken pox, smallpox
**sasyri ~ samycheng** *n.* bladder infection
**soldi** *n.* a cold, common cold
**sothonthara** *n.* cancerous swellings all over the body
**thai•rokron** *n.* swollen lymph nodes in the arm pits
**tokset-** *v.* to cough, to have a cold
**tyisuk** *n.* pneumonia

### 5.5.7 Emotions and psychological feelings and states

**barat-** *vgoal.* to be ashamed, to be shy
**baratwami** *n.* shame
**bawra** *n.* arrogance
**chaisi-** *vtr.* to hate, to dislike, to be annoyed by something or someone
**duk** *n.* sorrow, sadness
**gal** *n.* pride, arrogance
**gong-** *v.* to agree, to be willing
**han•saw-** *v.* to enjoy
**han•seng-** *v.* be happy, be joyful, to enjoy

**han•tung-** *v.* to feel secure, to feel safe
**harat-** *v.* to be reluctant
**jajyreng** *n.* confusion; danger
**jajyreng-** *vgoal.* to worry
**jari-** *v.* to be startled
**ka•ran- ~ kha•ran-** *v.* to be thirsty
**kha•** *n.* fighting spirit.
**kha•dong-** *v.* to be courageous, to be hopeful
**kha•dong** *v.* to hope
**kha•gal-** *vgoal.* to love

kha•pak- *vgoal.* to miss
kha•pet- *v.* be angry
kha•si- *v.* to shun, to not like and ignore
khu•bisi- *v.* to hate, to dislike
khu•chi- *v.* to dislike
kom- *v.* to feel like a loser
komok- *v.* to feel insulted
kyryi *n.* fear
kyryi- *vgoal.* to be afraid of
machak- *v.* to be vengeful
man *n.* respect
mykbu- *v.* to be jealous, to be envious
mykbyryw- ~ mykbryw- ~ mykbyru- *v.* to be jealous

mykcha- *v.* to like somebody
nemnuk- *v.* to like
nombok- *v.* to be unconscious, to be tired after eating a lot
ok *n.* hunger
okhi- *v.* to be hungry
sa•nal- ~ sa•nyl- *vgoal.* to be jealous of
sak- *v.* to enjoy
stel *n.* haughtiness, arrogance
suk- *v.* to be well, to be comfortable, to enjoy
sykhym- *v.* to feel sorrow, to mourn
synthi- *v.* to suffer, to regret, to repent
wa•chyrik- *vgoal.* to be startled

## 5.6 Human behaviour

### 5.6.1 General behaviour

ain *n.* custom, law, tradition
ain niam *n.* laws, customs, traditions
bam- *vgoal.* to obey, to agree, to surrender, to do willingly
bewal *n.* tradition, habit
bot *n.* a gift
bot- *vtr.* to court, to woo, to flatter, to give a present to a girl after dating her
byisa- ~ bysa- ~ bywsa- *v.* to dance
byl• *n.* a strike, a hit
bytchirit- *vtr.* to draw a line
bytsek- *vtr.* to abduct, to kidnap, to take away a person, to steal a person
chairok- *v.* to attend to, to take care of
chanpheng- *vgoal.* to defend
chip- ~ chep- ~ chup- ~ chyp- *v.* to be imprisoned, to be caught, to be locked up
chol chal *n.* livelyhood, way to make a living
chol *n.* livelyhood, way to make a living
chom- *v.* to copulate, to fuck
chonnyk- *v.* to look down on, to mock, to scorn, to insult
chyi- *v.* to try
daijol- *v.* to overstay
dairamphin- *v.* to work overtime
dakan- *v.* to dress someone else
dari- *v.* to commit adultery
di•it- ~ de•et- ~ di•et- *v.* to shit, to defecate

di•pyryw- *v.* to shit one's pants
de•etset- ~ di•itset- ~ di•etset *v.* to pick one's nose
dekdek- *v.* to shiver
denggu *n.* crime, extortion, naughtiness
di•pyryw- *v.* to shit one's pants
disudap- *v.* to piss on top of
dok- *v.* to take off (clothes), to unplug, to take apart, to disassemble, to unblock
dosi *n.* blame
dukhup- ~ dykhyp- *vtr.* to dress someone, to put clothes on someone else
dykhep- *vtr.* to make someone cry
dykyret- ~ dykyryi- *vtr.* to threaten
dykyryng- *v.* to make noise on purpose
dyng- *v.* to fight
dyra- *v.* to rape
dythyi- *vtr.* to kill (only used for animals)
galcha- *v.* to boast
gat- *v.* to dig
golmal ~ gormal *n.* a fight, a quarrel, chaos
gorong- *v.* to meet
hala kha•- *v.* to wake someone up, to disturb someone
haldun- *v.* to feed, to provide for
hang- *v.* to warm one's hands by the fire, to dry meat or other things near the fire
hanthi- *v.* to divide, to share

**haphu-** *v.* to blow
**hapjyt-** *v.* to move house
**hat-** *v.* to copulate, to fuck
**het-** *v.* to clean an orifice or hole, to blow one's nose
**ja sa-** *v.* to wake up, to get up, to get out of bed
**jai•-** *v.* to oppose, to refuse
**jasa-** *v.* to wake up, to get up, to get out of bed; to realise
**jok-** *v.* to be freed
**jot-** *v.* to prod, to point, to fidget
**joton** *n.* attempt, try
**juk-** *v.* to wink
**jut-** *v.* to encourage
**jykjak-** *v.* to be noisy, to make noise
**jykrat-** *v.* to accuse of adultery
**jyrypet-** *v.* to shut somebody up, to make someone be quiet
**ka•wak- ~ kha•wak-** *v.* to open one's mouth, to open one's mouth widely, to say aaah, to have one's mouth open
**kak-** *v.* to bite
**kan-** *v.* to wear
**kel-** *v.* to hide behind or in something
**kha•-** *v.* to do, to work
**kha•dang-** *vgoal.* to care for with great love
**khaithyi-** *v.* to hang oneself
**khapeng-** *v.* to hinder
**khasi-** *v.* to castrate, to remove the testicles
**khel** *n.* care
**khele-** *v.* to play
**khema** *n.* forgiveness
**khep- ~ khup ~ khyp-** *v.* put on clothes
**khereng-** *v.* to resist, to struggle, to make a great effort
**khu•cheng-** *v.* to bite one's teeth firmly together
**khu•mong-** *v.* to conspire
**khu•thym-** *v.* to kiss
**khu•tip-** *v.* to close one's mouth
**khu•tyisot-** *v.* to spit
**khym-** *v.* to marry
**kyl-** *v.* to hide, to avoid
**kyngjung-** *v.* to turn one's back to someone
**kynphak-** *v.* to sleep in
**laklak-** *v.* to prod in an orifice or hole for pleasure
**lekat-** *v.* to waste time

**leklek-** *v.* to prod in an orifice or hole
**mani-** *v.* to worship, to pay respect to someone
**mili-** *v.* to assemble, to meet, to come together
**mimi-** *v.* to laugh, to laugh at someone
**miting** *v.* to hold a meeting
**mu•rong-** *v.* to shit on
**mu•ten-** *v.* to look after, to watch, to keep company
**mychym-** *v.* to smile at someone
**mykchep-** *v.* to look down upon, to despise, to scorn, to underestimate.
**mykphylyp-** *v.* to blink with one's eyes
**mykrak-** *v.* to hold a wake (often used with the incorporated noun *wal* 'night')
**mykraket-** *v.* to warn
**mykren tan•-** *v.* to wink
**myksu-** *v.* to wash one's face
**mythel-** *v.* to thank, to appreciate
**nakhung ra•taw-** *v.* to snort
**neng•thak- ~ ning•thak-** *v.* to rest, to take a rest, stop for a while
**niam** *n.* custom, law, tradition
**nisan-** *v.* to aim
**pel-** *v.* to copulate, to fuck
**peng ~ peng•-** *v.* to prevent, to hinder, to obstruct.
**pha•-** *v.* to dare
**phal** *n.* share, shift of work, instead of
**phathi-** *v.* to bless, to bestow upon
**phi-** *v.* to invite
**phylyp-** *v.* to blink (with one's eyes)
**ra-** *v.* to bring, to give, to hand over
**ra•-** *v.* to get, to take
**ra•ang-** *v.* to take away
**rakhi-** *v.* to protect, to guard (against), to keep, to look after
**ra•sak-** *v.* to accept, to receive
**ra•sek-** *v.* to snatch
**ram•-** *v.* to search, to want, to try
**rasong** *n.* praise, blessing, compliments, boasting
**rin-** *v.* to keep as domestic animal
**ryng-** *v.* to celebrate (by drinking)
**ryngreng-** *v.* to shake one's head
**sa•-** *v.* to celebrate (by eating)
**sa•khaw-** *v.* to steal
**sakchyk-** *v.* to behave well

**saraw-** *v.* to borrow
**sari-** *vgoal.* to shun, to ignore someone (out of shame or hatred)
**sari-** *vtr.* to hide something, to keep something secret
**sason** *n.* reign, rule
**sasti** *n.* punishment.
**saw•-** *v.* to dig
**selsoksok-** *v.* to masturbate, to wank
**seng-** *v.* to stay awake
**seng•-** *v.* to bother by misbehaving
**sepsep-** *v.* to masturbate, to wank
**sitbyryt-** *vtr.* to scratch someone
**sok-** *v.* to hold out
**soksok-** *v.* to masturbate, to wank
**song-** *v.* to elect, to appoint
**sot-** *v.* to spit
**suset- ~ susut- ~ susyt-** *v.* to wash
**syk-** *vsec.* to want
**symsak-** *vgoal.* to care for/about, to be careful about
**tak-** *v.* to fight
**tak-** *vB.* to do, to make, to pretend, to act like, to be like
**takbewal** *n.* tradition
**takruk-** *v.* to fight
**taksak-** *v.* to help
**taksakgaba** *n.* help
**tam•-** *v.* to beat a drum, to play an instrument

**tam•a toka** *khjyks*, *v.* to play an instrument
**tan-** *v.* to put off
**tanang-** *v.* to leave behind, to leave alone
**tanset-** *v.* to abandon, to leave behind
**thabarat-** *vtr.* to make someone feel ashamed
**thajyri-** *v.* to make trouble
**thangphytphyt-** *v.* to spatter, to splash
**thasa-** *vtr.* to wake somebody up
**thebajaw- ~ thebejaw-** *v.* to tickle
**thik kha•-** *v.* to fix a date and time
**thimimi-** *vtr.* to make someone smile
**thogi-** *v.* to betray, to cheat (on), to deceive
**thom-** *v.* to come together
**thorom** *n.* religion
**thym-** *v.* to lie in ambush, to lie hidden, to hide so that you can still watch what is happening.
**thym-** *v.* to take revenge
**tun- ~ tyn-** *v.* to lead, to guide to lead, to guide
**tyret-** *v.* to bathe someone else
**tyru- ~ tyiru- ~ tyiryw-** *v.* to bathe, to take a bath, to wash oneself
**wa•kholchik-** *v.* to show one's teeth
**wa• nat-** *v.* to brush one's teeth
**walseng-** *v.* to stay awake all night
**wara-** *v.* to defend (oneself), to shield (oneself), to protect (oneself)
**wat-** *v.* to send away, to banish, to get rid of, to avoid, to let go, to squirt

### 5.6.2 Child bearing and raising
See also §5.11.7, Babies and children.
**ba•-** *v.* to be born, to give birth
**ba•-** *v.* to carry a child in a cloth on the body
**hal-** *v.* to feed
**mu•thai hal-** *v.* to breastfeed

**achi-** *vintr.* to be born
**ma•chot-** *v.* to be an orphan
**tha•gat-** *v.* to carry a child on one's back
**thaba•-** *v.* to make someone carry a child (in a cloth on the body)

### 5.6.3 Hunting and fishing
**apun** *n.* fishing hook
**apunphong** *n.* fishing rod
**ban-** *vtr.* to trap, to catch in a trap
**ban-** *vtr.* to trap, to catch in a trap
**batdyl** *n.* slingshot
**bonduk ~ bondyk ~ byndyk** *n.* gun, shotgun
**chek** *n.* net, fishing net

**dinggarai** *n.* fish trap
**dukung-** *vtr.* to dam, to make circular a wall of stones in the water in the river to trap fish
**dukung-** *vtr.* to dam, to make circular a wall of stones in the water in the river to trap fish
**jolpi** *n.* bamboo fish trap

**jonggi** *n.* type of fish trap that has an opening on the op with long spikes pointing inwards
**kaw-** *v.* to shoot
**kaw-** *v.* to shoot
**kukuri** *n.* dagger, type of knife with a blade with an obtuse angle used to survive in the jungle
**pun-** *v.* to catch with a fishing rod and fishing hook
**pun-** *v.* to catch with a fishing rod and fishing hook
**sa-** *vtr.* to set as a trap, to put in place, to do
**sa-** *vtr.* to set as a trap, to put in place, to do
**saido** *n.* fishing line
**thiri** *n.* bow
**thirikun•** *n.* arrow
**wai•seng** *n.* very big knife traditionally used to kill tigers and men

## 5.6.4 Killing
**byl-** *vtr.* to cut and kill a big animal or person, to slay
**khat-** *v.* to slaughter
**khan•-** *v.* to slaughter
**nisi-** *v.* to poison
**tyngtet-** *v.* to hang someone
**so•ot-** *v.* to kill, to murder
**papret-** *v.* to throw to death

## 5.6.5 Death
**chywgyn** *n.* the festival of the dead
**chywgyn-** *v.* to celebrate the festival of the dead
**dylang ~ delang** *n.* small house for the spirit of a deceased person
**gop-** *v.* to burry
**gopram** *n.* grave
**manggisi** *n.* corps, dead body
**okmyng-** *v.* to starve
**si-** *v.* to starve
**sorot-** *v.* to hold a ceremony or celebrate in commemoration of a dead person one year after they died
**thyi-** *v.* to die
**toktai-** *v.* to hang oneself
**tyichaw- ~ tyichaw•** *v.* to drown
**walmykrak-** *v.* to hold a vigil or watch over the body of a dead person.

## 5.6.6 Agriculture
**bak-** *vtr.* to make barren, to weed out all the plants, to scrape with a spade or chopper
**bat-** *vtr.* to stick in, to plant by sticking a seed into a hole in the soil
**bytwa ~ bytwami ~ bytwamyng** *n.* harvest
**chal-** *v.* to plant or sow by making a hole in the ground with a stick and putting the seed into the hole
**chaw-** *v.* to winnow
**chep-** *v.* to milk
**ha• haw•-** *v.* to cut/clear the land to make a *ha•ba*
**ha• kam-** *v.* to clear the field, to cut the jungle to make a field, to tear out or cut weeds
**ha•jagyra** *n.* the first weeding of the *ha•ba*
**haw•-** *v.* to clear/cut the jungle to make a rice field
**jakun** *n.* the second weeding of the *ha•ba*
**kai•-** *v.* to plant
**kam-** *v.* to clear the field, to cut the jungle to make a field, to tear out weeds
**khit-** *v.* to sprinkle, to sow seeds
**mai ga•-** *v.* to thresh the rice with one's feet so as to separate the grains from the ores
**maikung** *n.* second rice harvest
**pot-** *v.* to plant by sticking a sprout in the mud
**pyjyw-** *v.* to sow seeds by scattering them
**saigyn** *n.* the third weeding of the *ha•ba*
**saphang** *n.* first rice harvest in August
**wai-** *v.* to plough

## 5.6.7 Household chores

**bijyrang-** *vtr.* to hang to dry
**huk-** *v.* to sweep together
**kek-** *v.* to chop wood
**nat-** *v.* to scour
**ram-** *v.* to dry in the sun, to put in the sun to dry
**reprep-** *v.* to rub the clothes while doing the laundry
**rongthal-** *v.* to clean
**rym-** *v.* to cook
**ryphi- ~ riphi-** *v.* to plaster (with a mix of clay and cow dung)
**walchak-** *v.* to kindle the fire with one's breath by blowing
**wek-** *v.* to sweep

## 5.6.8 Religion

**boli** *n.* offering to a spirit
**chisol** *n.* a cross
**chyngaba** *n.* offering to a dead person
**chyngaba** *n.* offering to a dead person
**gylja ~ gyljanok** *n.* church
**ha•bachenggaba** *n.* beginning, Genesis
**kethylik** *n.* Catholic
**kristan** *n.* Christian
**mondoli** *n.* The Church, congregation, church community
**phathi-** *v.* to bless
**songsarek ~ songsyrek** *n.* animist, animism, heathen, pagan
**thorom** *n.* religion, denomination

## 5.6.9 Games, toys and jokes

**aiding** *n.* hopscotch (a children's game)
**ajot** *n.* children's game
**ambret bambret** *n.* children's game
**bytwami** *n.* tug-of-war
**chaira** *n.* a traditional song
**dalibibi** *n.* doll
**gol ~ gool** *n.* goal (in football)
**khelegaba** *n.* game
**lekadaw•reng** *n.* ART. kite
**mimiwami ~ mimiwamyng** *n.* a joke
**pido** *n.* game played with small stones
**ret** *n.* children's game
**robol** *n.* football
**somphi** *n.* a joke, a riddle
**sylkeng** *n.* ART. hoop, ring
**sylkengkun** *n.* ART. stick to drive a hoop
**tas** *n.* cards (the game)

## 5.6.10 Festivals, ceremonies and events

**chyn-** *v.* to offer to the dead
**chywgyn** *n.* the festival of the dead
**chywgyn-** *v.* to celebrate the festival of the dead
**dong•wa** *n.* event
**gop-** *v.* to bury
**ileksyn** *n.* election
**istyr** *n.* Easter
**jineral miting** *n.* general meeting
**katha ~ kata ~ khata ~ khatha jyw•khynwa** *expr.* to tell long epic stories during the festival of *chywgyn*
**krismas** *n.* Christmas
**obosta** *n.* event
**porika** *n.* exam, examination
**ryng-** *v.* to celebrate by drinking
**sa•-** *v.* to celebrate by eating
**saram** *n.* new rice offering festival
**sorot-** *v.* to hold a ceremony or celebrate in commemoration of a dead person one year after they died
**walmykrak-** *v.* to hold a vigil or watch over the body of a dead person.
**wanggala** *n.* biggest Garo festival
**waribul-** *v.* to fish at the festival of *waribula*
**waribula** *n.* the Siju fishing festival

## 5.6.11 Things humans and/or animals undergo

**kam-** *v.* to suffer a penalty
**man•symrukruk-** *v.* to inherit
**mykchel-** *v.* to shine in the eyes
**nasi-** *v.* to suffer from a loud noise or sound
**sak-** *v.* to bear, to endure, to enjoy, to hold out, to be patient, to suffer
**neng•-** *vsec.* to lack, to fail to

**pai-** *vgoal.* to support, to tolerate
**peel ~ pheel dong•- ~ dong-** *v.* to fail
**sakchyk-** *v.* to endure, to hold out, to be patient, to have patience
**sam-** *v.* to wait
**sangwal-** *v.* to forget

## 5.7 Language

**bal-** *v.* to speak, to tell, to say
**balsem-** *v.* to talk very long
**balsyruk-** *v.* to whisper
**Bechylyrdygri** Bachelor's degree
**bimung ~ bimyng** *n.* name
**bokbok- ~ bongbong- ~ bong-** *v.* to lie, to tell lies
**breket ~ brekyt** *n.* bracket.
**chat-** *v.* to promise
**che•e-** *v.* to mock
**chikarak-** *v.* to joke
**chiti** *n.* letter
**diksyneri** *n.* dictionary
**dygri** *n.* degree
**edisyn** *n.* edition
**git** *n.* a song
**golpho** *n.* story
**golpho-** *v.* to talk extensively
**gremyr** *n.* grammar
**gremyr** *n.* grammar
**henraiting** *n.* handwriting
**hit-** *v.* to command
**hok-** *v.* to call loudly
**itihas** *n.* history
**jai-** *v.* to scold someone
**jokal** *n.* comic strip, cartoon, anime; a character from one of these categories
**kata ~ khata ~ katha ~ khatha** *n.* (< Assamese or Bengali) word
**khu•chuk** *n.* language
**khu•hamgaba** *n.* grammar
**khu•rasak-** *v.* to promise
**khu•sak-** *v.* to answer, to reply, to respond
**kitap** *n.* book

**laisen** *n.* licence
**laklak-** *v.* to nag
**lekha** *n.* book, paper
**ma•am-** *v.* to moan
**maip ~ mep** *n.* map
**mangneng-** *v.* to whine, to quarrel
**mangnengruk-** *v.* to quarrel
**Masteldygri** *n.* Master's degree
**myk-** *v.* to tell lies
**myng-** *v.* to call someone/somebody a name
**mynga-** *v.* to call upon someone or something
**myngkhelek-** *v.* to call somebody by a nickname
**myngkheleka** *n.* nickname
**naw-** *v.* to scold
**no-** *v.* to say
**oikor** *n.* alphabet, spelling
**ol-** *v.* to speak, talk
**paraw-** *v.* to shout (of animal and human)
**parawchyrik-** *v.* to shout loudly
**peng•-** *v.* to curse
**phai•-** *vtr.* to translate
**pi•-** *v.* to ask, te request, to beg, to pray
**ra•rung-** *v.* to revoke, to take back
**ra•sak-** *v.* to welcome
**raka** *n.* the first letter of the Atong alphabet, glottal stop
**rongthal-** *v.* to clarify, to explain
**rophil- ~ rophyl-** *v.* to joke
**ryng•-** *v.* to sing
**sandi-** *v.* to inquire (about), to search (for)
**saw-** *v.* to curse at (use bad words)
**seng•sot-** *v.* to abbreviate
**sipyling ~ spyling ~ spling** *n.* spelling

**su-** *v.* to scold
**sua** *n.* profanation
**sykhym-** *v.* to moan, to complain
**syng•-** *v.* to ask
**syng•gaba** *n.* question
**synthi-** *v.* to lament, to moan, to whine
**tha•let-** *vgoal.* to explain

**thikthik-** *v.* to instruct
**thol•-** *v.* to lie, to tell lies
**thore-** *v.* to cry out the name of the *mahari* of one's enemy
**tyngetwami ~ tyngetwamyng** *n.* announcement, notice

## 5.8 Food and cooking, eating and drinking

### 5.8.1 Food items and ingredients used for food

**ata** *n.* flour
**barata** *n.* paratha, flatbread *barata phel sa* one paratha
**bering** *n.* food cooked in a *wa•sung*
**beringwa ~ berengwa** *n.* food cooked in a *wa•sung*
**bisi** *n.* poison
**biskut** *n.* biscuit
**bytym** *n.* fat (of human or animal), grease
**cha** *n.* tea
**chini** *n.* sugar
**choklet** *n.* a sweet, chocolate
**chun** *n.* ground limestone
**chyw** *n.* rice beer, rice wine alcohol, wine, liquor
**chywbok** *n.* white alcoholic liquid made from fermented rice, white rice beer, white rice wine
**dykha** *n.* wine drunk during the *chywgyn* festival
**garan** *n.* jerky
**goilapan** *n.* betel nut and paan/pan
**ja•bek** *n.* curry
**kek** *n.* cake
**khuli** *n.* opium
**kopi** *n.* coffee
**mai** *n.* rice
**mai•tyi ~ mai•ti** *n.* juice from *ja•bek*, curry juice, the watery liquid or broth that is the result of cooking *ja•bek*
**mai•wakhyi** *n.* fermented bamboo shoots

**maichek~ maichyk** *n.* cold rice
**maigasam** *n.* meal eaten in the later part of the day or evening, dinner
**maijyreng** *n.* leftovers of cooked rice dried in the sun used to feed the pigs
**maikhyt** *n.* burned rice
**maimanap** *n.* meal eaten in the morning, breakfast
**mainyl** *n.* sticky rice
**maisan** *n.* meal eaten in the middle of the day, lunch
**maisen** *n.* sticky rice in a banana leaf
**mamylet** *n.* omelette
**muri** *n.* popped rice
**na•saw** *n.* fermented fish
**papol** *n.* pasta
**phywra** *n.* rice powder
**pi•ti ~ pi•tyi** *n.* rice beer (gold coloured)
**poop** *n.* triangular pastry eaten with tea
**rungkhut** *n.* broken rice
**ryng•chyw** *n.* flat-rice
**ryngchyw ~ ryngchu** *n.* flattened rice
**sam** *n.* weed, medicine
**samchak** *n.* vegetable
**saram** *n.* dry rice grains
**sigyret** *n.* cigarette
**sithi** *n.* fermented rice from which *chyw* is drawn by adding water
**tho** *n.* mustard oil
**wal•bek** *n.* burnt curry

### 5.8.2 Kitchen furniture

**jaljeng** *n.* cupboard
**tebyl** *n.* table
**tyinok** *n.* utensil storage
**akan** *n.* wooden rack above the cooking fire
**gan•chang** *n.* rack above the *tyinok* for plates and other kitchen utensils; rack under the *akan*, where meat is put to dry above the fire
**phong•** *n.* fire place for cooking
**phong•khal** *n.* stones to put a cooking pot on
**phong•thu** *n.* stones to put a cooking pot on, fire place for cooking

### 5.8.3 Cooking

**baw•-** *vtr.* to dry: to make jerky, to dry vegetables
**bering- ~ bereng-** *vtr.* to cook in a bamboo tube (*wa•sung*) which is sealed with banana leaves and placed in the fire
**buthu- ~ buthyw- ~ bythyw-** *vintr.* to boil (of water)
**chanet-** *v.* to put on the fire
**chawarai-** *v.* to heat meat on a frying pan without salt or water in order to preserve it
**jaw•-** *v.* to fry
**jotkhyngkhyng-** *v.* to mash
**khan•-** *v.* to chop, to mince, to cut
**khap-** *v.* to be cooked without *mai•tyi ~ maiti*
**morot-** *v.* to grate
**phe-** *v.* to disembowel, to gut (fish),
**pyn•-** *v.* to cook in a banana leaf
**rym-** *vtr.* to cook
**rymkhap-** *v.* to cook without *mai•tyi ~ mai•ti*
**saw•-** *vtr.* to burn, to roast on the hot ashes of the fire
**sit- ~ syt-** *v.* to clean out the shit from an animal's intestines
**sorok-** *v.* to re-pound the rice

### 5.8.4 Cooking and eating utensils

**abek** *n.* hollow spoon
**aman** *n.* pestle
**asam** *n.* mortar
**awan** *n.* winnowing basket
**boiom** *n.* jug
**botol** *n.* bottle
**chachek** *n.* tea strainer
**chamus** *n.* spoon
**chang•kui ~ cheng•kui ~ chaw•kyi ~ chaw•ki** type of knife
**chang•kuikatri** *n.* type of knife
**cheke** sieve
**churi** *n.* knife
**dabogos** *n.* skewer
**dipot** *n.* teapot
**dyksyl ~ tyksyl** pan for cooking rice
**gora** *n.* wine pot
**gylas ~ gilas** *n.* glass
**janti** *n.* filter
**kak** *n.* lid
**kaldap** *n.* type of knife
**khap** *n.* cup
**korea** pan
**maityk** *n.* pot to cook rice
**palak** *n.* spoon
**pawai** *n.* curry bowl
**rai•chak** *n.* big leaf used to pack food
**spun** *n.* spoon
**stulkhabar** *n.* tablecloth
**syldangkhep** *n.* plyers
**tawa** *n.* frying pan
**thali** *n.* plate
**thyikhop** *n.* dried fruit used to store water
**thyk** *n.* rice-cooking pan
**tyibek** *n.* bottle made of a dried vegetable
**tyigum** *n.* metal water jug
**tyithai** *n.* spoon made of a dried and hollow gourd
**tyksyl ~ dyksyl** *n.* rice-cooking pan
**tykyw** *n.* water pot
**wa•sung** bamboo tube

## 5.8.5 Eating and drinking

**chym•-** *v.* to chew
**churu** *n.* very little food
**dak-** *v.* to spit out
**hup-** *v.* to suck
**ka•syrak-** *v.* to drink from the bottle
**kak-** *v.* to bite
**kana thetka** *khjyks, v.* to drink (liquor)
**khan-** *v.* to suckle
**khaw•-** *v.* to catch water in the palms of one's hands
**khyp-** *v.* to take a bite
**mojet- ~ mojot-** *v.* to suck
**monok-** *v.* to swallow, to devour
**nal-** *v.* to gorge, to stuff one's face
**okha-** *v.* to be full after eating
**phuset-** *v.* to spit out
**rot-** *v.* to boil (something in water)
**ryng-** *v.* to drink, to smoke, to celebrate (by drinking)
**ryngkhaw-** *v.* to drink sneakily
**ryngkhele-** *v.* to drink for fun
**sa•-** *v.* to eat
**sa•dap-** *v.* to take more and more
**sa•khele-** *v.* to eat for fun
**sa•lak-** *v.* to lick
**sym•-** *v.* to soak, to make wet, to make *chyw* by pouring water on the *sythi ~ sithi*
**syrup-** *v.* to suck
**thu•-** *v.* to put in one's mouth
**tu•- ~ ty•-** *v.* to feed (by putting food or drink into the mouth)
**wang•-** *v.* to bite a bit out of something, to take a mouthfull
**wang•kok-** *v.* to eat without using one's hands, with one's mouth

## 5.9 Cognition

**awan-** *vtr.* to forget
**bebe ra•a** *vtr.* to believe
**chanchi-** *vB.* to think (about/of)
**chanchichyp-** *v.* to suppose, imagine
**chol** *n.* idea
**dang•-** *vph.* to enter into a mental state
**khi-** *v.* to count
**mangsong-** *v.* to plan
**mykgythal** *n.* reality
**myksong-** *v.* to plan, to intend, to decide, to mean
**oltho ~ ortho** *n.* meaning
**porai- ~ pore-** *v.* to read; to study
**sai-** *v.* to write
**sap-** *vsec.* to know a skill
**sung** *n.* remembrance, thought, mind, brain, intelligence, spirit, life
**sungman- ~ suman** – *vgoal/v.* to remember
**syki- ~ ski-** *v.* to learn, to teach
**tho-** *v.* to compare
**tyng-** *v.* to know (a fact or person), to understand, to recognise
**tyngcheng-** *v.* to know first, to discover
**tyngwami** *n.* knowledge, understanding

## 5.10 Perception

**chai-** *v.* to look (at), to watch
**chaikhaw-** *v.* to spy (on), to peep
**chairura-** *vintr.* to look around you
**chairuru-** *v.* to look around
**chaithum-** *v.* to guard, to watch over
**natym-** *v.* to listen (to)
**na-** *v.* to hear
**nuk-** *v.* to see, to look like, to find
**sip-** *v.* to smell (to use one's nose to sense smells)
**mysepai chai-** *v.* to look/watch with one eye
**pyi•ram-** *v.* to feel for, to search by feeling
**thunuk-** *v.* to show

## 5.11 People

This section does not include the kinship terms. For kinship terms, see the Lexicon of Kinship Terms: Atong – English in §3.

### 5.11.1 Men, women and stages in life

**banthai** *n.* marriageable boy, bachelor
**bipha** *n.* lad, man, male
**bydyi** *n.* old person
**jyksai** *n.* married couple
**jyw• wa** *n.* parents
**jyw•bydyi** *n.* old woman, woman with children

**jyw•gaba** *n.* someone else's biological mother
**wa•gaba** *n.* someone else's biological father
**me•ama** *n.* married woman
**me•apha** *n.* married man
**morot** *n.* person, human, human being, man
**nawmyl** *n.* marriageable girl

### 5.11.2 Relationships between people

**baju** *n.* friend
**bakdong** *n.* a forbidden marriage
**bakdongmi sa• ~ bakdongmyng sa•** *n.* a child born out of a forbidden marriage.
**banthai** *n.* marriageable boy or man, bachelor, unmarried man
**bydyi badai** *n.* old couple
**chame ~ chamai** *n.* lover, sweetheart
**daldi** *n.* beloved person, love, darling
**gawi** *n.* female, marriageable girl, girl (unmarried)
**hijra** *n.* gay person, homosexual
**jat** *n.* tribe, race
**jonja** *n.* twin
**jora** *n.* partner, love (person), match in love
**jykmong ~ jykmongma** *n.* first wife of a man who has two wives
**jyknyi** *n.* widow, widower
**jykri ~ jykyryi** *n.* widow, widower
**jyktyi** *n.* second wife of a man who is already married

**jyw• wa** *n.* parents
**jyw•gaba** *n.* someone else's biological mother
**jyw•para** *n.* mother's house, mother's household
**kha•wa** *n.* lover
**khymgaba** *n.* spouse
**mahari** *n.* relatives (of the same clan), family (of the same clan), clan. Among the Garos, there are five maharis: Sangma, Marak, Momyn (Ha•chyk spelling Momin), Sira and Areng
**me•ama** *n.* married woman
**me•apha** *n.* married man
**mykchagaba** *n.* sweetheart, girl or boy that you fancy
**nawmyl** *n.* marriageable girl
**nokdang** *n.* the family that live together in one house
**sai** *n.* husband
**songsal** *n.* society
**wa•gaba** *n.* someone else's biological father

### 5.11.3 Profession/Function

See also §5.11.4, Religious people.
**babaji** *n.* fortune teller
**chaitumgaba** *n.* watchman
**chakol** *n.* servant
**chepgaba** *n.* prisoner

**chokida** *n.* warden
**dykyl** *n.* cannibal
**dylgaba** *n.* leader
**gun montyro man•ga(ba)** *n.* person who can control the spirits

**hetmadam** *n.* headmistress
**hetmastel** *n.* headmaster
**jokal** *n.* comic strip, cartoon, anime; a character from one of these categories
**khurutgaba** *n.* someone who performs an incantation to determine which spirit makes someone ill.
**kontrektyr** *n.* contractor
**loskor** *n.* highest rank in the system of customary law of the Garos, judge
**madam** *n.* female teacher
**mastel** *n.* male teacher
**mejistret** *n.* magistrate
**mistyri** *n.* mason, house builder and painter, vehicle repair man
**montyri** *n.* minister
**napit** *n.* hairdresser

**nokgaba** *n.* landlord, house owner, God
**nokhol ~ nokhor** *n.* slave
**nokma** *n.* village headman, rich man, respected man
**oja** *n.* medicine man, traditional herbal doctor
**opiser ~ ophiser** *n.* officer
**pha•lap** *n.* whore, prostitute
**rani** *n.* queen, also used to call one's daughter when she is a little child
**raja** *n.* king
**rakhigaba** *n.* caretaker
**sainokga(ba)** *n.* author
**sakhi** *n.* witness
**sipai** *n.* soldier
**ticher ~ tichyr** *n.* teacher
**saip ~ saep** *n.* European, white person, British military commander

## 5.11.4 Religious people
**kamal** *n.* priest
**kristan ~ kristen** *n.* Christian
**phadyr** *n.* Father (priest)
**songsyrek ~ songsarek** *n.* animism, an animist, pagan, heathen

## 5.11.5 Looks and afflictions
**aragong** *n.* a person who is too big for his age
**bimang** *n.* body, appearance
**chokrek** *n.* someone with a touting mouth
**dykyl** *n.* Khasi person (pejorative)
**ja•jol ~ ja•gol** *n.* person with long legs
**kala** *n.* deaf person
**kana** *n.* blind person
**khalbong** *n.* person who eats scandalously much
**khokalang** *n.* bold person
**khu•eng** *n.* person with a crooked or slant mouth
**khu•gri** *n.* someone who does not talk much, a quiet person
**khu•jylok** *n.* someone with an open mouth
**khu•ma** *n.* dumb person, someone who cannot speak
**kokalang** *n.* a bold person
**kykulwil** *n.* person who has an ear infection and has lost his balance

**menpart** *n.* most important or most salient person
**mykpeng mykpeng** *n.* name to call a cross-eyed person
**myn•dyluk** *n.* person without body hair (this word is used jokingly)
**nanggandai** *n.* naked person
**nanggodolong** *n.* naked person
**phebaw** *n.* person with a swollen cheek or swollen tonsils
**po•tolong** *n.* person with a naked chest
**raw•reng** *n.* someone who is slender and long
**ri•baw** *n.* person with one testicle bigger than the other
**sathup** *n.* sick person
**wagyleng ~ wagylok** *n.* person who is missing one or more teeth
**wakhol ~ wakholong** *n.* person who is missing one or more teeth

### 5.11.6 Persons with negative characteristics

**alsia ~ halsia** *n.* lazy person
**boba** *n.* crazy man, idiot, fool
**bobi** *n.* crazy woman, idiot, fool
**bobylawthok** *n.* fool
**boda** *n.* ignoramus, dumbass
**burbok ~ bulbok** *n.* idiot
**ga•tha** *n.* idiot
**gatha** *n.* fool, crazy person (masculine)
**gathi ~ ga•thi** *n.* fool, crazy person (feminine)
**jada** *n.* stupid person, idiot
**kongtoksi** *n.* used in the expression *halsia kongtoksi* 'lazy person (pejorative)'
**lapchagaba** *n.* a good-for-nothing
**nawang** *n.* retard, half-brain, fool, stupid, confused person
**jabyra** *n.* fool, crazy person
**gunda** *n.* criminal, brawler, fighter
**thol•am** *n.* liar
**sawthal** *n.* dirty person, person who never washes
**pokotia** *n.* freeloader, sponger, person who takes advantage of the kindness of others
**madong** *n.* somebody who has married a person from the same mahari as themselves.
**khalput** *n.* dirty person
**khuchylep** *n.* blabbermouth, someone who cannot keep secrets and talks a lot
**kanggal** *n.* poor person, pauper
**kaltyk** *n.* person who never washes himself/herself
**bawbyl chambyl** *khjyks*, *n.* enemy
**bawbyl** *n.* enemy
**tykhal** *n.* person who goes around eating in lots of other people houses

### 5.11.7 Babies, children

See also §5.6.2, Child bearing and raising.

**wa•ri ~ wa•ryi** *n.* child who lost its father
**sa•gyrai** *n.* child
**sa•** *n.* offspring
**sa•gyrai odek** *n.* baby
**odek** *n.* baby
**jyw•ri ~ jyw•ryi** *n.* child who lost its mother
**gogyrek** *n.* a baby with its neck bent sideways while it is being carried on the back
**babu** *n.* child or baby (used to address a small child or baby)
**sa•daiburung** *n.* child born out of an incestuous relationship

### 5.11.8 Dead people

**chyn-** *v.* to offer to the dead
**chywgyn**$_1$ *n.* festival of the dead
**chywgyn-**$_2$ *v.* to celebrate the festival of the dead
**gop-** *v.* to bury
**gopram** *n.* cemetery
**manggisi** *n.* corps, dead body
**mykrak-** *v.* to hold a wake
**sorot-** *v.* to hold a ceremony or celebrate in commemoration of a dead person one year after they died
**thyi-** *v.* to die

### 5.11.9 Nationality/Ethnicity

**Atong** *n.* Atong
**Banggal** *n.* Bengali, Bangladeshi
**Britis** *n.* British
**Garo** *n.* Garo
**Ha•chyk** *n.* Garo
**Holen** *n.* Dutchman / Dutchwoman
**Megam** *n.* Megam, Lyngam
**Nedyranmorot** *n.* Dutchman
**Nepal** *n.* Nepali
**phoren** *n.* white foreigner

## 5.11.10 Institutions
**sorkar** *n.* government
**phulis ~ pulis** *n.* police
**gobormen ~ golmen** *n.* government
**jyw•mrong solkari ~ jyw•myrong sorkar** *n.* central government

## 5.11.11 Groups of people
**bai•dam ~ baidam** *n.* some (people).
**darang ~ dyrang** *n.* people, anyone, everyone
**duma** *n.* crowd
**mondoli** *n.* The Church, congregation, church community
**tho•ma** *n.* group
**chasong** *n.* generation, era
**janggal** *n.* everybody, everything, all, all of them, all of it
**jinma** *n.* group, herd

## 5.11.12 Stages in life and death
**bia** *n.* wedding
**inggeech** *n.* engagement.
**janggi** *n.* life
**janggi khengwa** *n.* life
**khengwa** *n.* life
**pi•sa** *n.* childhood
**thyiwami ~ thyiwamyng** *n.* death

## 5.12 Human products

### 5.12.1 Clothes
**bagu** *n.* cloth for man worn around the waist
**bagukhawa** *n.* turban with a knot on the front side of the head
**cha•khop ~ ja•khop** *n.* shoe
**chakkhop** *n.* glove
**cheen** *n.* chain, zip fastener, zipper, zip
**chola** *n.* shirt
**dakmanda** *n.* long women's dress tied around the waist, skirt
**di•thap** *n.* diaper
**gamsa** *n.* a cloth
**genji** *n.* tank top *genji khung/jora ni* two tank tops
**happen** *n.* short pants, shorts
**hil** *n.* heel (of a shoe)
**jama** *n.* shirt
**juta** *n.* shoe
**kalai** *n.* loincloth
**kha•di** *n.* clothes
**khawkhuthuk** *n.* cloth for men worn around the head
**khawphyng** *n.* turban
**khophynga** *n.* cloth for women worn on the head with a knot at the back of the head
**longpen** *n.* a pair of pants, trousers, long pants, long trousers
**muja** *n.* sock
**pakyl** *n.* centre strap of a sandal
**phatsai** *n.* woman's dress
**ri•pan** *n.* a short dress that women wear around the waist
**sendel ~ sendyl** *n.* sandal
**seng•khi** *n.* traditional belt made of ivory beads
**sanglas** *n.* sun glasses
**tupi** *n.* cap, hat

### 5.12.2 Jewellery and makeup

**chaksan** *n.* bracelet
**hira** *n.* SUBST/ART. diamond
**jaksithem** *n.* ring
**jaksyl** *n.* bracelet
**lepstik** *n.* lipstick
**mukta** *n.* perl
**nakhungthek** *n.* nose piercing
**nakhalthek ~ nathek** *n.* earring
**pi•tyng** *n.* thread, necklace
**ryk** *n.* necklace
**syladyn** *n.* traditional necklace
**sylasyng** *n.* necklace

### 5.12.3 Tools

**ajip** *n.* fan
**apun** *n.* fishing hook
**apunkara** *n.* fishing line
**apunphong** *n.* fishing rod
**ba•sek** *n.* cloth in which to carry a baby on the body
**baton** *n.* button
**bawili ~ bawyli** *n.* pliers to put logs on the fire
**bek** *n.* bag
**belcha** *n.* spade
**betyri** *n.* battery
**biji** *n.* injection, injection needle
**bylet** *n.* razor blade
**cha•wek** *n.* broom to sweep outside the house
**chabi** *n.* key
**chachek** *n.* tea strainer
**chaka** *n.* wheel
**chamus** *n.* spoon
**chang•khui nagap ~ chang•khui kaldap** *n.* type of big knife with a wavy blade
**chang•kui ~ cheng•kui ~ chaw•kyi ~ chaw•ki** *n.* big knife with a curled blade used in the kitchen to prepare food, as well as in the field to cut plants and weeds
**chang•kuikatri** *n.* type of big knife with blade that has a rounded hook at the end
**chek** *n.* net, fishing net
**cheke** *n.* sieve.
**churi** *n.* knife
**dinggarai** *n.* fish trap
**dojanggre** *n.* a ladder with only one axis to which the runs are attached
**dokra** *n.* bag
**engkal ~ ingkal** *n.* handkerchief
**gan•thong** *n.* stick, handle (of knife etc.), stump (of a tree)
**guchung** *n.* ladder
**hen•** *n.* knot
**huksetgaba ~ huksetga** *n.* dustpan
**ja•ga** *n.* a trap
**janira** *n.* mirror
**janti** *n.* filter for rice beer
**jap** *n.* trap to drive away enemies
**jolpi** *n.* bamboo fish trap
**jomphol** *n.* hoist, crow bar, pry bar
**jonggi** *n.* type of fish trap that has an opening on the op with long spikes pointing inwards
**kabal ~ kabar** *n.* cover
**kak** *n.* lid
**kara** *n.* rope, vein
**kebyl** *n.* cable
**kensi ~ kesi** *n.* scissors
**kewal ~ khewal** *n.* a peddle, oar
**khalpak** *n.* belt of a basket
**khatchi** *n.* sickle
**khawkham** *n.* pillow
**khiil** *n.* nail (made from iron)
**khorat** *n.* a saw
**khorot** *n.* a saw
**khudal ~ kudal ~ wa•khu** *n.* hoe, chopper
**kilip ~ kylip** *n.* clip
**kombol** *n.* blanket
**kukuri** *n.* dagger, type of knife with a blade with an obtuse angle used to survive in the jungle
**kulal ~ kular ~ kural ~ kurar** *n.* axe
**lait** *n.* light
**musuri** *n.* mosquito net
**nabak** *n.* knot
**paila** *n.* scale (for weighing)
**paip** *n.* water pipe, water tube

**palak** *n.* bamboo spoon: piece of bamboo split in half and used to stir
**pen** *n.* pen
**pha•lak** *n.* piece of old cloth used to clean things
**phong** *n.* wooden handle of big knives, axes and spears
**phong•khal** *n.* stones to put a cooking pot on
**rong•sa** *n.* whetstone, flat stone for sharpening knives or edged tools
**rumal** *n.* head band
**sabun** *n.* soap
**saido** *n.* fishing line
**satha ~ sytha** *n.* umbrella
**sekari** *n.* pin lock
**sikol** *n.* a chain
**songsykhep** *n.* big pincher
**spun** *n.* spoon
**stulkhabar** *n.* tablecloth

**syldangkhep** *n.* big pliers to take pans off the fire
**sylkengkun** *n.* stick to drive a hoop
**symphak** *n.* type of blanket
**taiyr** *n.* tire
**tala** *n.* a lock
**tangka** *n.* money
**tangka poisa** *n.* money
**tawel** *n.* towel
**thali** *n.* plate (for eating) or its volume, plateful
**thiriphong** *n.* part of an elephant trap
**wa•gat** *n.* bamboo shoulder yoke
**wa•gydok** *n.* water pipe made of bamboo
**wa•tana** *n.* part of an elephant trap
**waiyr** *n.* wire
**wakeng** *n.* axe
**wal•** *n.* fire, torch.
**wal•byt** *n.* match (to make fire)
**wal•cham** *n.* bamboo torch

### 5.12.4 Weapons

**batdyl** *n.* slingshot
**bonduk ~ bondyk ~ byndyk** *n.* gun, shotgun
**danyl** *n.* shield
**darai** *n.* sword
**guthini ~ guthyni** *n.* spear, bamboo spear which is part of an elephant trap, walking stick
**jatha** *n.* a spear
**mongreng ~ mongyreng** *n.* very big knife on a long pole

**ostro** *n.* weapon
**patyl** *n.* slingshot
**sokhop** *n.* cover, sheath
**thiri** *n.* bow
**thirikun•** *n.* arrow
**wai•cheng** *n.* longest type of knife
**wai•seng** *n.* very big knife traditionally used to kill tigers and men

### 5.12.5 Instruments and related words

**bangsi** *n.* flute
**barat** *n.* lace that pulls the skin of a drum tight
**chigyryng** *n.* traditional snare instrument
**dama** *n.* drum of the Mandai people, a bit smaller than the Atong *khem*.
**dymchyrang** *n.* type of traditional snare instrument played by plucking
**kal** *n.* horn (traditional instrument)
**khem** *n.* big drum (traditional instrument) played during *chywgyn*

**raityng** *n.* line (to dry clothes on), washing line, clothes line
**rang** *n.* type of traditional brass drum or gong (instrument)
**rangkha** *n.* type of traditional metal gong or drum
**rangsyl** *n.* type of traditional metal gong or drum
**ymbyng** *n.* bamboo flute

### 5.12.6 Baskets and other receptacles

For a list of volume words, see Table 10.
**asok** *n.* type of basket
**baket ~ baltin** *n.* bucket
**boiom** *n.* jug
**bosta** *n.* big bag to transport things like rice and betel nut in
**botol** *n.* bottle or its volume, bottleful
**chokhoi** *n.* type of basket
**chongchang** *n.* bamboo bird cage
**dipot** *n.* teapot. *dipot thai• ni* two teapots
**dram** *n.* drum, barrel
**galon** *n.* ART/MSRE. jerry can
**gora** *n.* large earthen pot in which rice liquor (*chyw*) is made.
**gylas ~ gilas** *n.* glass or its volume, glassful. *cha gylas ni* two glasses of tea. *Gylas goi•tham bai•ok ge•thene.* He has broken three glasses.
**katha** *n.* type of basket
**khap** *n.* cup, teacup or its volume, cupful
**khugyri ~ koksi** *n.* type of basket
**kok** *n.* basket
**kok** *n.* type of basket
**kokbal** *n.* type of basket
**kokcheng** *n.* type of basket
**kokdam ~ koktang** *n.* type of basket
**koksep** *n.* type of basket
**korea** *n.* big metal pan
**maityk** *n.* pot for cooking rice

**nakamai ~ namakai** *n.* type of basket
**net** *n.* type of basket
**pai•ra ~ phai•ra** *n.* type of basket
**pawai** *n.* bowl to serve curry in or its volume, bowlful
**rongtyk** *n.* large clay pot to keep rice in, rice pot, rice barrel
**tannet** *n.* measure basket
**tannet** *n.* type of basket
**tawa** *n.* frying pan
**thali** *n.* plate (for eating) or its volume, plateful
**thyikhop** *n.* dried fruit in which water is stored for consumption
**thyk** *n.* pan for cooking rice
**tiup** *n.* tube
**tyibek** *n.* traditional bottle used to drink water out of and made of a dried vegetable also called *tyibek*
**tyigum** *n.* water container made of metal and shaped like a big vase used to store water in the kitchen. Its place in the house is in the *tyinok*.
**tyithai** *n.* water scoop made of a hollow, dried gourd.
**tykyw** *n.* water pot
**wa•sung** *n.* bamboo tube used as container, and used to cook *bering* in *wa•sung sung ni* two bamboo tubes

### 5.12.7 Bamboo mats

**dala** *n.* bamboo mat made of *wa•tyng* for drying *papol* or chillies in the sun, also called *damplak*
**dam** *n.* bamboo mat
**damdyl** *n.* bamboo mat that is used as the side of a house.
**damplak** *n.* bamboo mat, also called *dala*, made of *wa•tyng* for drying *papol* or chillies in the sun
**damthol** *n.* a rolled up mat

**dengdyl ~ beraberi** *n.* an open whickered type of bamboo mat used as fence of the balcony or veranda of a house
**taw•pachi ~ to•pachi** *n.* the triangular bamboo mat under the roof of a house between the *khyntyri*, *byrym* and *bylbang*
**wa•syl** *n.* green half of a strip of bamboo used to make rope (2) the outside of a bamboo tube

### 5.12.8 Business, trade and money

**badyng-** *vtr.* to trade, to deal in, to do business in, to peddle
**baki** *n.* credit
**bisyl** *n.* ART. coin
**chiwal-** *v.* to trade, to deal in, to do business in
**dam** *n.* price
**dorma** ~ **dolma** *n.* ART. salary
**gamchatga(ba)** *n.* value
**ha•chak** *n.* ART. wages
**hajira** *n.* ART. daily wages
**insuren** *n.* ART. insurance
**kam ~ gam** *n.* wealth, riches
**karaw** *n.* debt, obligation, trouble
**koros** *n.* expenses
**lap** *n.* profit, interest, gain, value
**lap-** *v.* to gain, to make profit, to be profitable
**man•dapami ~ mandapwami** *n.* profit, interest, gain
**phal-** *v.* to sell
**poisa** money
**ra•-** *v.* to buy
**samphat** *n.* ART. fee paid to a medicine man (oja) for his services
**tangka** *n.* money
**tanka poisa** *n.* money
**tika** *n.* ART. payment

### 5.12.9 Electronics

**ilektrisiti** *n.* electricity
**karen** *n.* electricity
**kemyra** *n.* camera
**keset ~ kheset** *n.* cassette, tape
**khori** *n.* watch
**kompiutyr ~ komputer** *n.* computer
**mobail** *n.* mobile phone
**motorajip** *n.* fan
**nokwek** *n.* broom
**otorewain** *n.* auto rewind
**philm ~ philym ~ philim** *n.* film, movie
**phone ~ phoon** *n.* telephone, telephone call
**pidio** *n.* video
**piktiyr** *n.* picture, photo
**redio** *n.* radio
**sidikeset** *n.* CD, compact disc
**skrin ~ sykrin** *n.* screen
**suis** *n.* switch
**suting** *n.* taking pictures, photo shooting
**telephon** *n.* telephone
**tenki** *n.* tank
**tibi** *n.* television
**wach** *n.* watch

### 5.12.10 Transportation and vehicles

**alukotar** *n.* ART. helicopter
**alupren** *n.* ART. aeroplane
**baik** *n.* ART. bike
**baisykyl** *n.* ART. bicycle
**bas** *n.* ART. bus
**chaka** *n.* ART. wheel
**dabogos** *n.* ART. skewer
**gari** *n.* ART. vehicle, car
**geer** *n.* ART. gear
**jahas** *n.* ART. ship
**koilagari** *n.* ART. coal truck
**lolal** *n.* ART. road roller
**ma•sugari** *n.* ART. bullock cart
**mobil** *n.* motor oil, engine oil
**oto** *n.* ART. auto rickshaw
**reel** *n.* ART. train, rail, stud of a fence
**reelgari** *n.* ART. train
**rens** *n.* ART. wrench
**taiyr** *n.* ART. tire

## 5.12.11 Building and creating

**ham-** *v.* to build, to construct
**chal-** *v.* to support
**chanpat-** *v.* to build a bamboo bridge
**ry-** *v.* to carve
**dok-** *v.* to unplug, to take apart, to disassemble, to unblock
**dok-** *v.* to weave
**khep-** *v.* to weave a bamboo mat
**thuk-** *v.* to enclose with a fence
**thuk-** *v.* to weave a bamboo mat
**rap-** *v.* to thatch, to roof
**wat-** *v.* to weave things from reed or bamboo, to make a mat or basket from bamboo or reed
**sak-** *v.* to make a rope by rubbing thread between one's hands

## 5.12.12 Buildings and other man-made structures

The reader is referred to photos 7–18 as illustrations of some of the words in this section.

**babylsi** *n.* kitchen
**bandaw ~ bando** *n.* tree house
**bilding** *n.* house built with masonry, building
**bo•rang** *n.* tree house
**dokhan** *n.* shop
**dolan** *n.* big building, big house
**dolong** *n.* bridge
**dylang ~ delang** *n.* spirit house
**gylja ~ gyljanok** *n.* church
**ha•banok** *n.* rice field house
**hospytyl** *n.* hospital
**jul** *n.* mine, coalmine
**nesynyl haiwe** *n.* national highway
**noga** *n.* tree house
**nok** *n.* house
**nokbanthai ~ nokphandai** *n.* bachelors' house
**nokthai** *n.* a small house separate from mother's house, small house next to the main house
**nol** *n.* ART. fence, fenced enclosure
**phaikana ~ paikhana ~ phaikhana** *n.* toilet
**pung** *n.* granary
**roal** *n.* primary school
**serek** *n.* balcony of a rice field house
**skul** *n.* school
**taw•nok** *n.* chicken coop
**taw•nok** *n.* chicken cove, chicken coop
**thuk-** *v.* to enclose with a fence
**bera** *n.* fence
**toilyt ~ toilet** *n.* toilet
**wa•dokolong** *n.* ART. water pipe made of bamboo
**waknol** *n.* pigsty
**waknol** *n.* pigsty

## 5.12.13 House structure

The reader is referred to photos 7–18 as illustrations of the words in this section.

**bangphak** *n.* vertical posts at the entrance of the bachelors' house between the floor and the horizontal beam above the entrance
**beraberi ~ dengdyl** *n.* loosely woven bamboo mat used as fence of the balcony or veranda of the house
**bylbang** *n.* tie beam, horizontal beam that runs over several *manjuri* and forms the base beam of the triangle of the roof
**byrym** *n.* inside rafter: beam that runs along the ridge board on the bottom of the roof on the inside of the roof and has the *kenchi* 'rafter' as its counterpart; together they form part of the support structure of the roof of a house
**chalgaba** *n.* a support
**dalni tatdepgaba** *n.* ramp of a door
**damdyl** *n.* bamboo mat that is used as the side of a house

**damdyl** *n.* bamboo mat that is used as the side of a house.

**dempharai** *n.* ART. lengthwise cut long bamboo strip used in the construction of a house

**do•khakhu** *n.* ART. carved, ornamented and colourfully painted king post of the bachelors' house above the entrance in between the tie beam (*bylbang*) and the peak of the roof

**engsyri** *n.* bamboo strip that runs on top of the bamboo floor of a house and has *wa•rap* as its counterpart underneath the floor to keep the bamboo strips that make up the floor in place

**eskrup** *n.* hinge

**gandai** *n.* big beam that forms the base of the house and rests on the *rong•thai*

**ges ~ kes** *n.* strut: beam that runs, as the long side of a rectangular triangle, between the *manjuri* and the *gandai* in the structure of a house

**han•dyng** *n.* beam that forms part of the base structure of a house perpendicular to the *gandai*

**jagybeng ~ jagebeng** *n.* beams that form the supporting structure of the floor of a house and on top of which the bamboo floor can be attached

**kenchi** *n.* outside rafter: big beam that runs from the ridge board to the bottom of the roof on the outside of the roof and that has *byrym* 'inside rafter' as its counterpart; together they form part of the support structure of the roof of a house

**kes ~ ges** *n.* strut: beam that runs, as the long side of a rectangular triangle, between the *manjuri* and the *gandai* in the structure of a house

**khelhi ~ kelki** *n.* window

**khiil** *n.* nail (made from iron), bamboo strip that is used to tie beams together

**khopja** *n.* hinge

**khyntyri** *n.* (1) the wooden beam running over the top of the roof of a house, comparable to a ridge board. (2) purlin (other speakers call purlins *wa•khaw*)

**manjuri ~ manjyri** *n.* supporting post for a house or other such structure, king stud

**nokchol** *n.* door, entrance

**nokhama** *n.* a supporting structure

**nokkhap** *n.* door

**nokkhung ~ nukkhung** *n.* roof

**noksam** *n.* wall of a house, the piece of ground where a house is built on.

**panchyreng** *n.* non-supporting horizontal beam that forms part of the structure of the side of a house and to which the *damdyl* can be attached

**parang** *n.* PLANT/ART. reed, thatch

**rochok ~ rotchok** *n.* ART. stud, vertical beam that forms part of the side of a house to which the *damdyl* can be attached

**rong•thai** *n.* base stone on which a house is built

**songkhamphek** *n.* forked branch or post

**tala** *n.* lock

**taw•pachi ~ to•pachi** *n.* the triangular bamboo mat under the roof of a house between the *khyntyri*, *byrym* and *bylbang*

**taw•pachi ~ to•pachi** *n.* ART. the triangular bamboo mat under the roof of a house between the *khyntyri*, *byrym* and *bylbang*

**tin** *n.* corrugated iron

**tota** *n.* plank

**wa•byrek** *n.* horizontal beam under the length of the roof underneath the *wa•khaw*. Together, the *wa•byrek* and *wa•khaw* form part of the structure of the roof of a house that keeps whatever covers the roof in its place.

**wa•khaw** *n.* purlin, horizontal beam on the outside over the length of the roof, that has the *wa•byrek* as its inside counterpart. Together, the *wa•khaw* and *wa•byrek* form part off the structure of the roof that keeps whatever covers the roof in its place Other speakers call a purlin a *khyntyri*.

**wa•rap** *n.* bamboo strip that runs underneath the bamboo floor of a house, and has *engsyri* as its counterpart on top of the floor to keep the bamboo strips that make up the floor in place

### 5.12.14 Furniture

**aina** *n.* mirror
**bangka ~ bangkha** *n.* fan
**choki ~ chuki** *n.* chair
**dakham** very small stool
**gadang** *n.* shelf
**jaljeng** *n.* cupboard
**janera** *n.* mirror
**mura** *n.* a stool
**palong** *n.* bed
**stulkhabar** *n.* tablecloth
**tebyl** *n.* table

### 5.12.15 Other man-made products

**bostu** *n.* thing, things
**dupliket** *n.* a fake
**kendyl ~ kendel** *n.* candle
**peking ~ pheking** *n.* luggage, packing

### 5.12.16 Materials and substances

**dekoresyn** *n.* decoration
**itha ~ ita** *n.* brick
**khul** *n.* pillow stuffing
**pawdyr** *n.* powder, baby powder
**simen** *n.* cement
**tin** *n.* corrugated iron sheet used to make roofs

## 5.13 The supernatural

### 5.13.1 Supernatural beings

**ajuju** ghost that is a person without legs or which looks like a monkey. It jumps from tree to tree and has a long tongue with which it can hit people, who then melt.
**Ambi Chakkhen** a liana (hanging root) that turns into an old lady at night. The old lady has very long arms and hands with very long nails. She will ask you to scratch her long arm and if you refuse, she will scratch you to death with her long nails.
**Ambi Jakbyryt** ghost that is only a hand. It scratches the back of someone walking in the dark.
**Babyra** supreme god
**Bandija•lang** the tree that lies across the beginning of the road to Balphakram and that acts like a gate for the spirits of the deceased. This tree is also called Dykhija•lang
**but** spirit that leads you astray
**dakal ~ takal** witch, demon
**Dykhija•lang** the tree that lies across the beginning of the road to Balphakram and that acts like a gate for the spirits of the deceased. This tree is also called Bandija•lang.
**Goira** the god of thunder. *goira kawa* the god of thunder shoots / the thunder roars *goira byl• tan•ok* the god of thunder has struck / lightening has struck
**Isol ~ Isor** God
**kynkongbang** woman ghost without back
**mangcha** a ghost that is a corps risen from the dead, a zombie
**marudyl** species of liana that turns into a ghost that makes a loud noise of rummaging through the jungle, breaking branches and trees. But when you look in the morning, there are no broken branches and trees. The noise is so loud and frightening, that when you are within a hundred metres of this ghost, you cannot sleep. When you cut a *marudyl*, a blood-like, red liquid comes out.
**matsadu** creature which is human during the day and becomes a tiger at night
**me•mang ~ mi•mang** ghost, spirit of a dead person
**me•mangkereng** skeleton ghost
**myte** deity, god

**Rabuga** god who created the world according to ancient religion
**Saljong** sun god
**sangkhyning ~ sangkhyni** mythical water dragon that lives in the Symsang river.
**Takgaba Rywgaba Phathigaba Ra•runggaba** supreme god
**tho•theng** forest creature that looks like a person with his feet pointing backwards, so that it looks like you are following his footsteps in the direction he is going, while in fact he was walking the other way.
**wai** spirit

## 5.13.2 Dreams, magic and supernatural practices and experiences

**jadu** *n.* magic
**jagat-** *v.* to experience that one's soul leaves one's body and temporarily enters an animal
**jasyri** *n.* the experience that one sees oneself in a different place while one is asleep as if one's soul leaves one's body
**jumang ~ jywmang** *n.* dream
**khurut ~ waikhurut-** *v.* to perform an incantation
**muni** *n.* a magic spell
**mynga-** *v.* to call upon someone or something
**nukcham-** *v.* to predict, to see into the future
**thabisi** *n.* amulet, antidote
**thama** *n.* divination

## 5.14 Animals

**mat** *n.* animal
**matburung ~ matpalyng** *n.* wild animal

### 5.14.1 Amphibians

**bengblok** *n.* toad, species of frog
**gandalak** *n.* species of frog which says *gagagagaga*
**kangkang** *n.* species of edible frog, green with black spots, which lives in caves and the hollows of stones at the side of a river
**lukchok ~ rukchok** *n.* species of frog
**lukpekpek ~ rukpek** *n.* species of frog
**lukwak ~ rukwak** *n.* toad
**me•mangkyi ~ mi•mangkyi** *n.* species of small frog that says *pekpekpekpek*
**na•luk** *n.* tadpole[12]
**ong ang** *n.* big edible frog that makes the sound *ong ang* (alternative spelling *ong•ang*)
**rukchok ~ lukchokchok** *n.* species of frog
**rukpek ~ lukpekpek** *n.* species of small frog which says *pekpekpekpek*
**rukwak ~ lukwak** *n.* toad

---

[12] This word is a compound with the word *na•* 'fish', just like the names of many other species of fish, and also the words *na•chan* 'firefly' and *na•cheng* 'river shrimp or river prawn'.

### 5.14.2 Arthropods (spiders, scorpions, ticks, fleas, lice, centipedes, millipedes)

**chengchengmachok** *n.* spider with a long stomach with yellow stripes which can be fried and eaten
**gawang ~ guwang** *n.* spider
**ha•mangkyrang** *n.* scorpion
**kaithuk** *n.* flea
**kakhirok** *n.* head lice, pubic lice, crabs
**khen•** *n.* river crab
**khen•jasyri** *n.* a river crab that is walking on the road
**khyryk** *n.* louse (plural: lice)
**manggywak** *n.* millipede[13]
**mangkhrang ~ mangkhram ~ mangkyrang ~ mankyrang** *n.* scorpion[14]
**na•cheng** *n.* river shrimp or river prawn[15]
**nasengkhet** *n.* tick
**nathyra** *n.* tick
**sanarai ~ sanyrai** *n.* centipede
**thik ~ kuythik** *n.* louse (on dogs)

### 5.14.3 Birds

**alabok** *n.* egret, cattle egret, white heron, *Bubulcus ibis*
**chanchora ~ chanchura** *n.* sparrow
**chebe** *n.* species of bird
**daw•-** *bound.* bird. This is the bound form of the word *taw•* 'chicken, bird' that appears in names of birds and compounds with the root bird in them
**daw•blok** *n.* bulbul bird
**daw•budok** *n.* lineated barbet, *Psilopogon lineatus* (species of bird)
**daw•chegydek** *n.* species of bird
**daw•gamdot** *n.* eagle
**daw•gep** *n.* duck
**daw•kha** *n.* black crow
**daw•kruha•sym** *n.* common emerald dove, Asian emerald dove, grey-capped emerald dove *Chalcophaps indica* (species of pigeon)
**daw•kyru** *n.* pigeon
**daw•phaw ~ do•pho** *n.* owl
**daw•phylgym** *n.* eagle
**daw•pynchyrep** *n.* common tailorbird, *Orthotomus sutorius*
**daw•reng** *n.* hawk, kite or falcon.
**daw•rugu** *n.* species of bird of prey
**daw•sik** *n.* parrot
**do•de** *n.* peacock
**do•pho ~ daw•phaw** *n.* owl
**gogylek** *n.* cock, rooster, cockerel
**hynggek** *n.* hornbill
**ma•rek** *n.* species of bird. When this bird sings, you know that someone will visit the village.
**mai•wek** *n.* species of bird that is believed to call every time somebody comes to visit the village
**moina** *n.* common myna (a species of bird), *Acridotheres tristis*
**pepylok** *n.* species of bird
**phylgym** *n.* eagle
**piong** *n.* species of bird
**saw•khyn ~ sawkun** *n.* vulture
**taw•** *n.* (1) bird (generic) (2) chicken
**taw•** *n.* chicken, bird
**taw•gylyk** *n.* species of jungle bird
**taw•khasi** *n.* capon, castrated rooster
**taw•pachi** *n.* swallow
**taw•palyng** *n.* jungle fowl
**taw•pynchyrep** *n.* common tailorbird, *Orthotomus sutorius*
**taw•reksyrup** *n.* banana bird (translated literally its name is 'banana-tree sucking bird')
**taw•sa•gyrai** *n.* chick

---

[13] Notice that the morph *mang* occurs in the words *mangkung* 'cricket', *manggywak* 'millipede', *mangka* 'species of fish' and *mangkhrang ~ mangkhram* 'scorpion'.
[14] See footnote at ARTHROPODS: *manggywak*.
[15] See footnote at AMPHIBIANS: *na•luk*.

## 5.14.4 Fish

**asynthalak** *n.* species of fish
**bangbol** *n.* species of fish
**bangganai** *n.* species of fish
**dynggyni** *n.* species of fish
**elong** *n.* species of fish
**era** *n.* species of fish
**galjak ~ kaljak** *n.* catfish
**kaljak ~ galjak** *n.* catfish
**khachol** *n.* species of fish[16]
**khadok** *n.* species of fish
**khagynyk** *n.* species of fish
**khaljong** *n.* species of fish
**khamynkhap** *n.* species of fish
**kharok** *n.* species of very small fish
**khuchia** *n.* species of fish
**khuchina** *n.* species of eel
**lathia** *n.* species of fish
**magal** *n.* species of fish
**mangka** *n.* species of fish[17]
**muchi** *n.* species of fish
**na•** *n.* fish (generic)
**na•garang** *n.* species of electric fish
**na•gungphel** *n.* species of fish
**na•jek** *n.* species of fish
**na•kha** *n.* species of fish
**na•lam** *n.* species of blue, purple river fish that tastes particularly good when prepared in a bamboo tube (see *bering-*)
**na•langtaupal** *n.* species of fish
**na•matsa** *n.* species of fish
**na•nyl** *n.* electric eel
**na•pat** *n.* species of fish
**na•phok** *n.* species of fish
**na•rong** *n.* species of fish
**na•ru** *n.* species of fish
**na•rym** *n.* species of fish
**na•rymkhu** *n.* species of fish
**na•sak** *n.* species of red fish
**na•wachak** *n.* species of fish
**na•wak** *n.* species of fish
**nawchak** *n.* species of fish
**phuthi** *n.* species of fish
**sengsyp** *n.* species of small fish
**serembut** *n.* species of fish
**singsip** *n.* species of fish
**synggi** *n.* species of fish

## 5.14.5 Insects

**achepchep** *n.* species of cricket
**aganggi gawrai** *n.* species of grasshopper
**aganggi** *n.* species of grasshopper
**aguk** *n.* grasshopper (generic)
**ambisuthyk** *n.* species of gold-coloured metallic beetle that flips itself back on its feet when it lies on its back
**asalchong•** *n.* species of black hairy caterpillar that lives on jackfruit trees
**begyri** *n.* cow fly
**bukylek** *n.* species of big grasshopper
**buna** *n.* big black and yellow flying insect
**butsa** *n.* species of big red ant
**byirakhem** *n.* species of bee
**che•et** *n.* species of green cricket
**chong•** *n.* insect, bug, lice
**chong•su** *n.* catererpillar (generic)
**chong•su** *n.* caterpillar
**di•but** *n.* dung beetle
**ganggawa** *n.* mosquito
**ganthai** *n.* species of cicada
**ganthirengreng** *n.* species of cicada that makes a very loud and high pitched sound
**gogak** *n.* beetle
**gogak** *n.* beetle (generic)
**gompara ~ gompyra** *n.* species of large, poisonous, black ant
**gongchit** *n.* stag beetle

---

[16] The generic word for fish in Lyngam (a.k.a. Lyngngam, Megam; Austroasiatic, Meghalaya, Northeast India and Bangladesh) is *kha*. The linguistic areas of Atong and Lyngam overlap (see van Breugel, 2015d).
[17] See footnote at ARTHROPODS: *manggywak*.

**gorweng ~ wengwang** *n.* species of cicada that makes the noise of a screaming baby or a woman being murdered
**gugyreng** *n.* species of grasshopper
**gugyrengsa•** *n.* species of bright-green grasshopper
**gukchepchep** *n.* grasshopper
**gukmadym** *n.* grasshopper
**ha•chepchep** *n.* grasshopper
**hangkyn** *n.* species of termite
**hangkyn raja** *n.* species of termite
**kabin** *n.* species of big black ant
**kal•tek ~ kal•thek** *n.* species of big red ant
**kalthek** *n.* species of big red ant
**khoryndachong** *n.* silkworm
**kyimang** *n.* fruit fly
**maijyk ~ maimijyk** *n.* dragonfly
**mangkung** *n.* cricket (not sure if this word is generic or not)[18]
**me•mangkereng** *n.* stick insect, walking stick, an insect from the order of *phasmadotea*
**me•mesi** *n.* flying insect
**melanggaw** *n.* poisonous red or black ant
**mongmachong•** *n.* caterpillar
**mongmachong•su** *n.* giant caterpillar
**na•chan** *n.* firefly[19]
**ne•** *n.* bee (generic)
**ne•kat** *n.* species of bee
**ne•wal** *n.* species of bee
**ong** *n.* wasp
**panchungchong•su** *n.* species of black hairy caterpillar that lives on jackfruit trees
**rongkhym** *n.* species of yellow beetle
**samalmaisirong** *n.* very small species of ant
**selu** *n.* cockroach
**seneng** *n.* species of red beetle with black dots, black wings with white tips, black legs with red joints and black antennae
**sot** *n.* species of very small fly that comes out in the evening and at night and cause itchiness
**sotmai** *n.* housefly
**taw•pak** *n.* bat, butterfly, moth
**thaba** *n.* bedbug
**thurung** *n.* species of flying insect that comes out after the first rain and fills the air in big swarms like a mist. They come out at the same time as the species of ant called *hang•kyn*
**wa•khal** *n.* grasshopper-like insect
**wa•mychym** *n.* fire fly
**wengwang ~ gorweng** *n.* species of cicada that makes the noise of a screaming baby or a woman being murdered

### 5.14.6 Mamals

**amak** *n.* macaque, monkey
**byira amanthong** *n.* jungle cat (the pattern on the skin of this cat is in the shape of an *aman* 'mortar')
**byira** *n.* cat
**bythyi** *n.* porcupine
**chyndyk** *n.* domestic water buffalo
**gada** *n.* donkey
**gondu** *n.* rhinoceros, rhino
**gorai ~ gore** *n.* horse
**hu•raw** *n.* gibbon, *Hylobates hoolock*
**husyring** *n.* rabbit
**kangguru** *n.* Kangaroo
**karat ~ ka•rat** *n.* squirrel
**kyi•** *n.* dog
**ma•su** *n.* cow
**ma•subolot** *n.* ox
**machok** *n.* species of large deer, maybe sambar deer
**magachak** *n.* barking deer
**makbul** *n.* bear
**matdam** *n.* otter
**matdi** *n.* wild water buffalo
**mathai** *n.* bachelor elephant, solitary male elephant
**matrong** *n.* jungle goat

---

[18] See footnote at ARTHROPODS: *manggywak*.
[19] See footnote at AMPHIBIANS: *na•luk*.

**matsa** *n.* tiger
**mes** *n.* sheep
**mongma ~ mungma** *n.* elephant
**mongmamathai ~ mungmamathai** *n.* bachelor elephant
**muchot** *n.* mouse, rat. *Abeknyng•chi muchotsa•gyrai mang byryi chepchap chepchap parawthokaidonga.* Inside the *abek* are four baby mice squeaking eek eek.
**pheru** *n.* fox
**purun** *n.* goat

**ranggorai** *n.* macaque. Monkey with a long tail, brown body and a red face.
**saphaw** *n.* rabbit
**sial** *n.* jackal
**singho** *n.* lion
**taw•pak** *n.* bat, butterfly, moth
**wak** *n.* pig
**wak** *n.* pig, pork
**waknok** *n.* domestic pig
**wakpalyng** *n.* wild pig

## 5.14.7 Reptiles

**baphe** *n.* species of large gecko
**dypyw** *n.* snake
**dypyw** *n.* snake
**dypywha•saw** *n.* species of small snake
**dypywkaram** *n.* species of black snake
**dypywkheng** *n.* species of green snake (possibly a species of pit viper)
**dypywnokma** *n.* anaconda
**dypywpoda** *n.* cobra
**gorial** *n.* crocodile
**ha•saw** *n.* species of black snake with red neck and head
**kangkylek** *n.* species of lizard with red neck, said to drink human blood
**katua ~ khatua** *n.* turtle, tortoise
**keko** *n.* species of large brown tokay gecko with narrow white stripes on its back and white-and-brown ringed tail and brown eyes

**khatua ~ katua** *n.* turtle, tortoise
**khu•sum ~ ku•sum** *n.* tortoise
**koke** *n.* species of large gecko that lives in trees
**lukchokchok** *n.* species of small gecko that creeps up the walls of houses at night[20]
**mejakbal** *n.* alligator
**noktapa** *n.* species of small gecko that creeps up the walls of houses at night
**pape** *n.* species of big, brown gecko
**pho•ot ~ phot** *n.* mythical black amphibian like a salamander
**phu•chul ~ phutsul** *n.* species of water monitor (lizard) that supposedly can eat humans, also called water dragon.
**thongmatchang** *n.* snake with many colours like the rainbow, when someone sees this snake, they know that someone in their family will die

## 5.14.8 Snails and clams

**chyhyl** *n.* species of snail
**echaluk** *n.* species of snail
**kapangsi** *n.* clam
**kapkung** *n.* snail
**sukrung ~ sykrung ~ sukyrung ~ sykurung** *n.* river snail

---

[20] This word has the morph *luk ~ ruk* in it, which also occurs as the first element in names for species of frogs. There is even a species of frog that is also called *rukchok ~ lukchokcho*, as can be seen in the list of amphibians.

### 5.14.9 Worms and leeches

**chygyl** *n.* species of worm that glows in the dark
**ha•mai** *n.* species of white earthworm
**khansyrui** *n.* earthworm
**kyrywkeng** *n.* parasitic worm that lives in the flesh of animals and humans. It is believed that when a *kyrywkeng* is crossing the road, and a pregnant woman steps over it, she will have a miscarriage.
**nadanggap** *n.* flat blood sucking parasite on humans and animals
**batro ~ u•chingrawri** *n.* species of brown leech that lives in the soil and mud
**gomga ~ gomgaba** *n.* leech
**u•chingrawri ~ batro** *n.* species of brown leech that lives in the soil and mud
**krimichong** *n.* parasitic worm that lives in the bowels
**uching ~ u•ching ~ ukching** *n.* leech

### 5.14.10 Animal body parts and products

**asu** *n.* fishbone, thorn
**basu** *n.* crown feathers of a bird
**bytym** *n.* fat, grease
**chengkhana** *n.* gills
**chun** *n.* trump
**chyngmat** *n.* comb of a rooster
**di•chongkhamai** *n.* cloaca
**di•chongkhanthyi** *n.* pygostyle
**di•mai** *n.* tail
**di•thom** *n.* gizzard.
**gangthai** fin *n.* (of fish)
**gawangsyryng** *n.* spider web
**kakmyn•** *n.* antenna (of insect), feeler
**karang** *n.* wing
**kaw•warai** *n.* gill
**kha•rongthai** *n.* chicken heart
**kha•thol** *n.* wattle (of a chicken)
**khen•khorong** *n.* crab's pincher, crab's claw
**khol** *n.* skin, hide, scale
**khoppalak ~ khoppylak** *n.* (1) the husky skin of an onion, garlic, corn etc. (2) skin or peel of fruit (3) eggshell
**khung** *n.* shell of a crab, tortoise etc., carapace
**koplak** *n.* egg shell
**korong ~ kyrong** *n.* horn (of an animal)
**mongmawa• ~ mungmawa•** *n.* elephant tusk
**myn•** *n.* body hair (of humans), fur (of animals)
**ne•katthup** *n.* hive, bee's nest
**phe•phong** *n.* floater organs of a fish
**suthul** *n.* comb of a rooster
**taw•di•mai** *n.* tail feathers
**taw•karang** *n.* bird's wing
**taw•kurung** *n.* the nest of a chicken
**taw•myn•** *n.* fluff, body feathers
**taw•puk** *n.* the innards of a chicken
**taw•tyi ~ taw•ti** *n.* egg
**thup** *n.* nest
**wa** tooth, tusk (of elephant)
**wakpuk** *n.* the innards of a pig

### 5.14.11 Animal behaviour

**bala-** *v.* to hold in the beak
**bam-** *v.* to brood, to sit on an egg
**del- ~ dyl-** *v.* to sting (of a bee etc.)
**kak-** *v.* to bite
**kakkhap-** *v.* to bite down on, to firmly hold between the teeth or in the beak
**khong•-** *v.* to bark
**pa•-** *v.* to perch
**paraw-** *v.* to call (of animal)
**sap-** *v.* to swoop down (of birds of prey)
**thangphytphyt-** *v.* to spatter, to splash
**thup-** *v.* to nest
**tyi•-** *v.* to lay an egg
**wyngwang-** *v.* to wag

## 5.14.12 Animal dwellings

**chirokhana** *n.* zoo
**chongchang** *n.* bird cage
**hang•khal** *n.* hole
**khurung** *n.* a chickens' nest
**kuti** *n.* cow shed
**ne•katthup** *n.* hive, bee's nest
**ringaba** *n.* place where domestic animals are kept, fishery
**rong•khal ~ rong•khol** *n.* underside of a stone, cave
**taw•gurung** *n.* nest
**taw•nok** *n.* chicken cove, chicken coop
**taw•thup** *n.* nest, bird's nest
**waknol** *n.* pigsty

## 5.15 Plants and trees

### 5.15.1 Plants

**balgyto•** *n.* orchid
**chachakphang ~ chaphang** *n.* teaplant
**chara** *n.* sapling
**chigi** *n.* species of plant
**chinkak** *n.* species of plant
**chywgundai** *n.* species of plant
**dachang** *n.* roselle, *Hibiscus sabdariffa*
**dala** *n.* young plant
**do•dokhichong** *n.* species of plant
**dukung** *n.* species of plant
**dymbyl** *n.* species of tree
**gegydek** *n.* species of plant
**golap** *n.* rose
**gylarong ~ siwi** *n.* species of sea bean or its pod
**gyrym** *n.* bush, patch
**ja•rytbok** *n.* species of plant
**jatram** *n.* species plant
**joba** *n.* Chinese rose
**khakhudyl** *n.* species of plant
**kokkylek ~ toktokkylek** *n.* species of plant
**kymkhalongphong** *n.* species of plant
**lekhaphul ~ getphul** *n.* bougainvillea
**mai•in** *n.* species of plant
**mai•kerep** *n.* species of plant
**maidu** *n.* species of plant
**marirang** *n.* species of plant
**matsamykhang** *n.* species of plant
**me•mangkoksi ~ mi•mangkoksi** *n.* pitcher plant
**mi•manggambyrai** *n.* species of plant
**mi•manggrai ~ mi•manggyrai** *n.* species of plant
**mochok** *n.* sapling
**mok** *n.* species of plant
**mongnal** *n.* lotus
**mykasyrep** *n.* species of plant
**myktoksi** *n.* species of plant
**pal** *n.* flower
**pan wa•** *n. khjyks*, *n.* plants
**panchan** *n.* species of plant
**pangyrym** *n.* jungle thicket
**penta** *n.* species of plant
**phakdemel** *n.* species of plant
**phalong ~ phalwang** *n.* species of plant
**raithai** *n.* species of tree
**re•koksi** *n.* species of plant
**rothop** *n.* species of plant
**ruda** *n.* species of cactus
**sam** *n.* weed, medicine
**samanggri ~ samanggyri** *n.* species of plant
**sambanggyri ~ sambanggri** *n.* species of plant
**sambarat** *n.* touch-me-not, *mimosa pudica*
**sampattar** *n.* species of plant
**samthai** *n.* species of plant
**samtokjang** *n.* species of plant
**saphairam** *n.* species of plant
**sarat** *n.* species of plant
**sok** *n.* the new young leaves of a plant (but not a tree) or vegetable, a shoot, sprout
**sosila** *n.* species of plant

**suksai** *n.* species of plant
**symgong** *n.* species of plant
**tha•makhu ~ tha•mykhu** *n.* tobacco
**thai• ~ thai** *n.* fruit
**thai•symphak** *n.* species of plant

**thai•thuka** *n.* species of plant
**thaikuka** *n.* species of plant
**thamat** *n.* species of plant
**toktokkylek ~ kokkylek** *n.* species of plant
**wasam** *n.* species of plant

## 5.15.2 Fruits, vegetables and leafy greens

**abong** *n.* corn, maize
**akyrudygyl ~ akrudygyl** *n.* pumpkin
**alu** *n.* potato
**anaros ~ kewa** *n.* pineapple
**bai•khop** *n.* species of bean
**biins** *n.* species bean
**bu•chot** *n.* mango
**bugyryk** *n.* species of vegetable
**chengcheng** *n.* tamarind
**chichot** *n.* dud jackfruit
**chin•thai** *n.* melon
**chinara** *n.* lemon
**daba** *n.* coconut
**dalim** *n.* pomegranate
**dengga** *n.* species of small leafy green
**diprin ~ dipyrin** *n.* species of vegetable
**dorai** *n.* lady's finger, okra ~ okro ~ ochro
**drakha** *n.* grape
**epyl** *n.* apple
**gakji** *n.* lemon
**galdai** *n.* star fruit, carambola, *averrhoa carambola*
**garu** *n.* mustard
**gomagundai** *n.* species of thick banana
**gomynda** *n.* pumpkin
**gomynthyri** *n.* species of vegetable
**gorothop** *n.* species of small leafy green
**ja•garu** *n.* species of vegetable
**ja•ryt** *n.* chilli pepper
**jamura** *n.* pomelo
**jingkha ~ jingka ~ sawel ~ sawyl** *n.* species of vegetable
**kaw** *n.* type of fruit
**khanmynchyw** *n.* species of edible shrub

**khirip ~ kyiryp** *n.* species of edible plant
**kobi** *n.* cabbage
**kolachita** *n.* bitter gourd, *Momordica charantia*
**komyla** *n.* orange
**kymkha** *n.* species of berry
**laisak** *n.* cabbage
**law** *n.* cucumber-like vegetable
**lechu** *n.* lychee
**mai•cheng** *n.* mezenga (edible shrub)
**mantaw** *n.* species of gourd
**mantawbylati** *n.* tomato
**mantawthai** *n.* type of vegetable
**mudu** *n.* papaya
**mula** *n.* white radish
**panchung** *n.* jackfruit
**pangkol** *n.* guava
**pangkywal** *n.* guava
**panthai** *n.* type of fruit
**phanthai** *n.* type of sour fruit
**phulkobi** *n.* cauliflower
**rekkun** *n.* banana flower
**rekthai** *n.* banana
**samchak** *n.* vegetable
**sawel ~ sawyl ~ jingka ~ jingkha** *n.* species of vegetable
**sengki** *n.* type of fruit
**sojana** *n.* species of long thin vegetable
**tha•gythyng** *n.* species of vegetable
**thai•gundai ~ thai•ma•thaigundai** *n.* species fruit
**thaw•jyw** *n.* type of fruit
**the•myt** *n.* cucumber
**tintyrin** *n.* tamarind

## 5.15.3 Beans

**biins** *n.* species of bean
**kha•rek** *n.* yardlong bean
**rekhep** *n.* species of huge beans
**siwi ~ gylarong** *n.* species of sea bean

## 5.15.4 Trees

**agychi** *n.* species of tree
**ambyrai** *n.* species of tree
**amu** *n.* species of tree
**aphubawbyl** *n.* species of tree
**aphut** *n.* species of tree
**asa** *n.* species of tree
**asakotoi** *n.* species of tree
**asel** *n.* species of tree
**atak** *n.* species of tree
**baksubipha** *n.* species of tree
**bebylokmai** *n.* species of tree
**buchotpan** *n.* mango tree
**chamchia** *n.* species of tree
**chara** *n.* sapling
**dalchini** *n.* species of tree
**daw•kumai** *n.* species of tree
**daw•mai•tak** *n.* species of tree
**egyro** *n.* species of tree
**gambiri** *n.* species of tree
**gamsili** *n.* species of tree
**goi** *n.* betel nut, areca nut (*Areca catechu*)
**goichara** *n.* a betel nut sapling, a young betel nut tree
**gungsynung** *n.* species of tree
**jambu** *n.* species of tree
**kawbipha** *n.* species of tree
**khawratcha** *n.* species of tree
**khymbal** *n.* species of tree
**khymkhalymphong** *n.* species of tree
**kulthuk ~ kyltyk ~ kyltuk** species of tree
**kuma** *n.* species of tree
**kymbal** *n.* species of tree
**mandal** *n.* species of tree, *erithrina superosastricta*
**maw** *n.* species of tree
**me•mangdanggai** *n.* species of tree
**monchara asu** *n.* species of tree
**palengma** *n. barebina-xariegata*
**pan** *n.* tree
**panchengrong** *n.* species of tree
**pandawsik** *n.* species of tree
**pangkollipa** *n.* species of tree
**panjyl** *n.* species of tree
**panmaikung** *n.* species of tree
**panmang** *n.* species of tree
**panmatha** *n.* species of tree
**panphek** *n.* sapling, young tree
**panrasun** *n.* species of tree
**panthjong** *n.* species of tree
**phalwang ~ phylwang** *n.* species of tree
**phe•ep ~ phep** *n.* banyan tree
**phe•epmisi** *n.* species of tree
**phylwang ~ phalwang** *n.* species of tree
**rabal** *n.* rubber tree
**rangrai** *n.* species of tree
**rek** *n.* banana tree
**rekphang** *n.* banana tree
**sa•ma** *n.* species of tree
**sadai** *n.* species of tree
**saigon** *n.* teak tree
**sakhap** *n.* species of tree
**sakhapnathyng** *n.* species of tree
**sidai** *n.* species of tree
**silongket** *n.* Shillong tree
**sisawkhyli** *n.* species of tree
**sisawmotgram** *n.* species of tree
**sokchuman ~ sokchuwan** *n.* species of tree
**sokjong** *n.* species of tree
**suit** *n.* species of tree
**suitbipha** *n.* species of tree
**suksyrui** *n.* species of tree of which the fruits can be eaten
**sympak** *n.* species of tree
**tai•symphak** *n.* species of tree
**taisympak** *n.* species of tree
**taw•pakcha** *n.* species of tree
**thai•thuka ~ thaikhungka** *n.* species of tree
**thaikhungka ~ thai•thuka** *n.* species of tree

### 5.15.5 Lianas
**durymytdyl** *n.* species of liana
**karydyl ~ kyrydyl** *n.* species of liana
**me•mangguchung** *n.* species of liana
**na•lamsusyrakdyl** *n.* species of liana
**raityng** *n.* rattan
**taw•cha•si** *n.* species of liana

### 5.15.6 Creepers
**budu** *n.* creeper
**mai•do** *n.* species of creeper
**megalaia** *n.* species of creeper
**rajami khu•symng** species of creeper
**saikhiribudu** *n.* species of creeper
**tyiribok** *n.* species of creeper poisonous to cows

### 5.15.7 Fungi and algae
**aphap** *n.* yeast
**chingchongphyrot ~ chomchomphyrot** *n.* species of white edible mushroom
**dumuta** *n.* species of edible mushroom
**mairugu ~ meringgu ~ meringgaw ~ merenggaw** *n.* mushroom (edible)
**panachol** *n.* mushroom (not edible)
**tainalap** *n.* algae
**tyinala** *n.* algae
**wa•chan** *n.* species of fungus that glows in the dark

### 5.15.8 Tubers, other edible roots and onions
**cheng•khu ~ cheng•khyw** *n.* ginger
**daw•kharasun** *n.* species of onion, crow onion,
**gajol** *n.* species of red carrot
**khambykthai** *n.* species of edible tuber
**khamphung** *n.* species of edible tuber
**khan** *n.* cassava, tapioca
**khansynen** *n.* species of edible tuber
**narot** *n.* species of edible tuber
**raidi** *n.* turmeric
**rasun** *n.* onion
**rasun tyisuk** *n.* spring onion
**ring** *n.* taro, species of edible tuber with green stems
**ringgong** *n.* species of plant that looks like *ring*, but is not edible.
**ringgythyng** *n.* species of plant that looks the same as *ring* but has black stems and is not edible.
**tha•malang ~ tha•mylang** *n.* sweet potato
**thanglang** *n.* species of tree
**tharam** *n.* species of tree
**tharamdaw•phit** *n.* species of tree

### 5.15.9 Grasses
**barangsi** *n.* species of grass
**gom** *n.* wheat
**jeng** *n.* broom plant
**ko•rot** *n.* sugarcane
**mai•wa** *n.* bamboo shoot
**maisi** *n.* millet
**narang** *n.* orange
**narykel ~ narykhel** *n.* coconut
**parang** *n.* reed, thatch
**rai** *n.* reed
**raima** *n.* cane
**samsai** *n.* low grass
**samsi** *n.* grass

### 5.15.10 Bamboo
**muluwa•** *n.* species of bamboo
**wa•** *n.* bamboo
**wa•da** *n.* species of bamboo
**wa•jong** *n.* species of bamboo
**wa•jongmagal** *n.* species of bamboo
**wa•kai** *n.* species of big bamboo

wa•khyntha *n.* species of bamboo
wa•lai *n.* species of bamboo
wa•puk *n.* the inside of a bamboo tube
wa•rung *n.* young bamboo
wa•thai *n.* species of bamboo smaller than wa•thyrai

wa•thaibok *n.* species of bamboo that is white from the ground a little up
wa•thyrai *n.* species of bamboo that grows one by one, not in a bush
warung *n.* young or immature bamboo

## 5.15.11  Bamboo parts

dempharai *n.* lengthwise cut long bamboo strip used in the construction of a house
engsyri *n.* bamboo strip that runs on top of the bamboo floor of a house and has *wa•rap* as its counterpart underneath the floor to keep the bamboo strips that make up the floor in place
jyw• *n.* a flattened bamboo used to make mats
kyryw *n.* thin strip of bamboo used to make rope, bamboo rope

wa•gylok *n.* a cut-off piece of bamboo
wa•khaw• *n.* one long half of a bamboo split lengthwise. *wa•khaw sa* one long half of a bamboo
wa•phuk *n.* white half of a strip of bamboo used to make rope
wa•tyng *n.* bamboo strip used to make baskets, and other woven utensils as well as rope

## 5.15.12  Rice

cha•gang *n.* bad rice that is thrown away in the husking process
cha•wek *n.* chaff
chaw- *v.* to winnow
ha•ba *n.* slash-and-burn field
mai *n.* rice
mai pylakgaba *n.* flowering rice
mai ga•- *v.* to thresh rice
maichek ~ maichyk *n.* cold rice
maidan *n.* new rice (just harvested)
maiguru *n.* species of rice from which beer is made for the chywgyn festival
maijyreng *n.* leftovers of cooked rice dried in the sun used to feed the pigs
maikhyt *n.* burnt rice
mainyl *n.* sticky rice
maipalak ~ maipylak *n.* stalks left over after threshing rice
maipal *n.* flowering rice, rice flower

maiphang *n.* paddy
mairong *n.* husked, uncooked rice
mairongkholnang ~ maikholnang *n.* unhusked rice
maisen *n.* sticky rice in a banana leaf
memaboro *n.* species of nice smelling rice
muri *n.* popped rice
phywra *n.* rice powder
pijyw *n.* (1) rice seeds for sowing, newly harvested rice (2) unhusked rice that is thrown away when cleaning a portion of rice before cooking it
rungkhut *n.* broken rice
ryng•chyw ~ ryngchyw ~ ryngchu *n.* flat-rice, flattened rice
saram *n.* dried rice grains
sithi *n.* fermented rice
sorok- *v.* to re-pound rice

## 5.15.13  Parts of plants and trees

borong *n.* cob
cha•dyl *n.* root, vein

cha•dylmorong *n.* main root of a tree
cha•dylsaphek *n.* small root

**cha•wek** *n.* chaff
**chachak** *n.* tea leaf
**-chak** *n.* leaf
**cheksi** *n.* stalk, twig
**chyw•** *n.* the new young leaves of a tree
**dala** *n.* branch of a tree not directly attached to the trunk
**dyl** *n.* root, vine
**gan•theng ~ ga•theng** *n.* the stem of a leaf or fruit
**gandi** *n.* a log
**gantheng** *n.* stalk
**japang ~ ja•phang** *n.* tree trunk, foot of a tree
**kokpylak** *n.* chaff
**kun•** *n.* a stick
**lapan** *n.* pan/paan leaf
**maikap** *n.* hay
**maikhol** *n.* the skin of a grain of rice
**maikhoppylak ~ maikhoppalak** *n.* stalks left over after threshing rice
**mawkhol** *n.* bark (of a tree)
**narykeltyi** *n.* coconut water
**pagawa** *n.* the white, spongy inside of a banana tree
**pakara ~ pakyra** *n.* stalk of a fruit
**pan** firewood
**pan** *n.* tree, firewood
**panbai** *n.* firewood
**panchak** *n.* leaf
**panchoka** *n.* small log
**panchong** *n.* tree trunk
**panchyksi** *n.* twig
**pandala** *n.* twig
**panju** *n.* firewood
**pankhol** *n.* bark (of a tree)
**pansok** *n.* terminal bud, end of a branch where the tree grows
**pantiki** *n.* wood chip
**panthong** *n.* wooden stick
**rochong** *n.* tree stump
**ros** *n.* juice, sap
**sok** *n.* the new young leaves of a plant (but not a tree) or vegetable, a shoot, sprout
**sun ~ sundul** *n.* tree trunk
**tokta** *n.* type of wood
**wa•cham** *n.* stubble, old rice stalk which is left over after harvesting the rice
**wa•cheksi** *n.* bamboo stalk or twig
**wa•phek** *n.* bamboo, branch, small bit of bamboo
**wa•thok** *n.* hollow bamboo stick
**weng•** *n.* node (of bamboo), joint

### 5.15.14 Seeds
**cha•ri** *n.* PLANT. seed for planting, paddy sprouts used for planting
**karan** *n.* PLANT. seed, kernel, fruit stone

### 5.15.15 Plant substances
**hanggal** *n.* charcoal, cinder
**laha** *n.* resin
**khaltyi** *n.* soda
**myn•tyi** *n.* resin; latex of jackfruit, thick fluid of various fruits
**tyi** *n.* juice

### 5.15.16 Verbs pertaining to trees, plants and fruit
**chogyp-** *v.* to break off and fall down (for branches and big leaves)
**khorop-** *v.* to break (only used for bamboo)
**nang-** *v.* to bear fruit
**pal-** *v.* to bloom
**phuruk-** *v.* (1) to become uprooted (2) to break (for plants and trees)
**rekhep-** *v.* to be dry (of plants)
**rydym-** *v.* to sprout leaves
**thymyn** *v.* to ripen

## 5.16 Places, spaces, position, direction

### 5.16.1 Places and spaces

**babylsi** *n.* kitchen
**badym** *n.* paddy field
**digi** *n.* well, ditch
**para** *n.* river junction
**bagan** *n.* garden
**bai** *n.* border
**bai•wak** *n.* land of broken-off rocks
**balaga** *n.* outside
**balpisa** *n.* a place to piss
**bari** *n.* garden, plantation
**basneng•thakgaba** *n.* bus sop
**bythyn** *n.* shade
**cha•kok** *n.* hollow between the roots of a tree
**cha•ma** *n.* lower side, downstream, bottom, below
**cha•masang** *n.* downstream
**cha•wekdam** *n.* place where the chaff is thrown after winnowing the rice
**chiakhkol** *n.* well, source
**chirokhana** *n.* zoo
**chokdeng** *n.* the end of a pointy object
**cholwat** *n.* a space
**chula** *n.* cooking space
**dainingrum** *n.* dining room
**dam** *n.* place
**dareng** *n.* edge
**digi** *n.* well, ditch
**din** *n.* bedroom
**disko** *n.* disco
**distrik** *n.* district
**dokhan** *n.* shop
**dykym** *n.* top, upside, upper side
**edres** address
**gesep ~ gysep ~ gisep** *n.* space, interval
**gopram** *n.* grave
**gycheng** *n.* side, near
**gythym ~ guthum ~ gythum** *n.* village
**ha•ba** *n.* slash-and-burn field, rice field
**ha•byreng** *n.* old rice field
**ha•byri** *n.* hill, mountain
**ha•dawak** *n.* lower side of a hill, low ground
**ha•gun** *n.* old plot of land in the *ha•ba*
**ha•gylsak ~ ha•gyrsak** the world, the earth

**ha•kha** *n.* hillslope, steep hill slope, upper side of a hill, high ground
**ha•khong** *n.* valley
**ha•khung** *n.* river bank
**ha•khyng** land that belongs to a rich person or *nokma*
**ha•pal** *n.* outside, field
**ha•rongrong** *n.* lower side of a hill, low ground
**ha•ryn** *n.* plot of land, parcel
**ha•song** *n.* village and surrounding lands, area that can comprise several *gythym ~ gythum ~ guthum* and the surrounding lands, country
**ha•thywkong** *n.* a puddle
**ha•tykylok** *n.* a puddle
**ha•wai** *n.* plain area
**hama** *n.* under, underneath, below, space between the floor or the base of something and the ground
**hang•khal** *n.* cave, hole
**hap** *n.* place
**ja•phang** *n.* foot of a tree
**jaboldam** *n.* garbage heap
**jagyra** *n.* right, right hand side, right hand
**jagysi** *n.* left, left hand side, let hand
**jaria** *n.* pathway
**jila** *n.* region
**jol** *n.* area
**kal** *n.* hole
**kantara** *n.* emptiness
**kep** *n.* cave
**khambai** *n.* top, upstream
**khambaisang** *n.* upstream
**khana** *n.* harbour
**klas** *n.* class
**kolani** *n.* colony
**kyndam** *n.* land behind the village
**kynsang** *n.* behind, later, after
**letrin** *n.* toilet
**ma•sutan•dam** *n.* place where cows are slaughtered
**matji ~ maji** *n.* middle, (in) between

**misyn** *n.* mission
**mykhang** *n.* front, face
**nalsasang** *n.* the other side
**nokhama** under, underneath, below, space between the floor or the base of something and the ground
**nokhap** *n.* level piece of land on which a house is built
**nokha•pal** *n.* outside
**noksuk** *n.* the side of an object that faces the wall
**nokweng** *n.* floor
**norok** *n.* hell
**nukhu** *n.* courtyard
**nygyl** *n.* market
**nyng•** *n.* inside
**opis ~ ophis** *n.* office
**otyk** *n.* bottom of a ravine or cliff
**palyng** *n.* jungle
**phoren** *n.* country of white foreigners
**ram** *n.* place, road, way, path
**ramga** *n.* the side of an object that faces a wall
**rewet** *n.* riverside, river bank
**rijap** *n.* forest reserve
**ringaba** *n.* place where domestic animals are kept, fishery
**robolpil ~ robolphil** *n.* football field
**rong•ka** *n.* cliff
**rong•khal ~ rong•khol** *n.* space under a stone, cave
**rugung** *n.* edge
**rygyn** *n.* side
**serek** *n.* surface
**sima** *n.* boundary, limit
**singsingkholong** *n.* deep hole in the ground
**song** *n.* village, area that can comprise several *gythym ~ gythum ~ guthum*
**songga** *n.* another village
**songmong** *n.* main village
**sorok** *n.* path, road, way
**syngsyngkhol** *n.* deep hole in the ground
**syngsyngkholong** *n.* deep hole in the ground
**tarang** *n.* layer
**tyichang** *n.* island
**tyigat** *n.* place in a river or at the end of a water pipe where the people get drinking water, take a bath and wash their clothes and dishes.
**tyisam** *n.* river bank, water's edge.
**wa•lung** *n.* place where stuff is burnt
**wal•mak** *n.* future rice field where the jungle has just been burnt
**wal•sam** *n.* fireside
**wari** *n.* deep place in the river where you can swim or take a bath

## 5.16.2 Position and positioning

**bam-** *v.* to bend one's head
**bamkhup ~ bangkhylok-** *v.* to bend one's head
**bangkhylok- ~ bamkhup-** *v.* to bend one's head
**bawen-** *v.* to be rolled up
**bythyw-** *v.* to block
**cha•choron- ~ cho•choron-** *vintr.* to squat
**cha•gywgyw-** *vintr.* to kneel down
**cha•ma** *n.* below, downstream
**cha•syrong-** *v.* to stretch one's leg
**chaksyrong-** *v.* to stretch one's arm
**chanchok-** *v.* to lean on
**chang•khet-** *v.* to be stuck
**chap-** *v.* to stand (be in standing position)
**chom-** *v.* to stack, to pile up
**chongchyron ~ choncholon-** *v.* to squat. *De•thengdo ramrygynchi di•etna chongchyronwa!* He squatted next to the road to shit!
**chu•ret** *v.* to hang down
**chugup-** *v.* to be on its side
**chygyp-** *v.* to fall face down on the ground
**da•rat-** *v.* to fall down (for persons)
**dan-** *v.* to spread out, to lay out (mats etc.)
**dandan-** *v.* to be pressed with one's back against something, to lean against something
**dap- ~ dep-** *v.* to be on top, to put on top
**dirikhap-** *v.* to catch
**dykymphak** *n.* above (side where the head is)
**dyngdai-** *v.* to dangle

gadaw- *v.* to lift one's chin up
gakgu- *v.* to nod one's head
ganang *v.* locative/existential verb, to exist, to be, there is/are, to have, to live
gom- *v.* to bend
gonggong- *v.* to bend over
gopjyrujyru- *v.* to look down with one's head bent down
gyching ~ giching *n.* vicinity, angle, inclination
hama *n.* below (underside)
hyn•- *v.* to give
jyw- *v.* to lie down (both the movement and the position), to sleep
jywbythyn- *v.* to lay on one's belly
jywbythyn- *v.* to lay on one's belly
jywdap- *v.* to lie on top of something
jywgebeng- *v.* to lie on one's side
jywkapkap- *v.* to lie on one's belly
jywkarang- *v.* to lie on one's back
kan•ol- *v.* to stretch
kapkap- *v.* to lie flat on one's belly
khambai *n.* on top of, upstream (top)
khang- *v.* to occupy
khawdam- *v.* to put down
khet *v.* to be stuck
khom•- *v.* to sit with one's head in one's lap and one's legs pulled up
khomchuk- *v.* to bend over
khyn *n.* behind (back)
matji *n.* between (middle)
mu•- *v.* to stay, to sit (be in sitting position), to sit down, to be at, to live somewhere, to keep *V*-ing (durative)

mu•chonchyron- ~ mu•choncholon- *v.* to squat
mu•dap- *v.* to sit on something
mu•peng- *v.* to sit and block someone's view
mu•si- *v.* to stay somewhere uncomfortable
mu•symbylek- ~ mu•symblek- *v.* to sit on the floor (with one's bum touching the floor)
mykhang *n.* front (face)
mykhang- *v.* to face
nang- *v.* to hang (down from/on)
pheng•chang- *v.* to hold something in front of something else
ping- *v.* to block the way
rygyn *n.* near, next to (side)
ryp- *v.* to dive, to be/stay under water, to be submerged
sep- *v.* to be stuck
sul- *v.* to stretch out, to stick out, to extend
syithai- ~ syithyi- ~ syithi- *v.* to hang.
sykdep- *v.* to press (with one's finger)
syket- *v.* to insert
takap- *v.* to stick
tam- *v.* to wait, to stop
tan- *v.* to put, to stop, to keep
tandap- *v.* to be on top, to cover
thom•- *v.* to make a heap
thumu•- *vtr.* to sit someone down (used for children)
wa•khel- *v.* to be stuck in one's teeth
wa•khelsep- *v.* to be stuck in one's teeth
wynget- *v.* to dangle

## 5.16.3 Directions of the compass

saknaram ~ salnyram *n.* east
salgypeng *n.* south
salgyro *n.* north
salniram *n.* west

## 5.16.4 Toponyms

This list provides the names of some toponyms in Atong spelling, i.e. names of villages, caves, rivers, hills, streams and other geographically important places. There are a lot more names of places in Atong, which are not listed here. Hopefully, future fieldwork research will make it possible to compile a more complete list of toponyms in Atong spelling. Where possible, coordinates are provided for places that can be found on Google Earth, and for places that cannot be found on Google Earth (yet), but of which the author knows the exact location. These coordinates can be read as follows. Alokphang's coordinates, 25.283825N, 90.668068E, can be read as 25°.28'38.25"N, 90°.66'80.68"E. Note that for all places with the word *gythym* 'village' in it, the word gythym can also be pronounced and written as *guthum* or *gythum*. Also note that the information provided by Google Earth about the location of some of these places at the time the of publication of this volume were found by the author to be inaccurate. Where possible, the author has provided the correct locations.

**Agrenggythym** 25.462087N, 90.461700E
**Alokphang** 25.283825N, 90.668068E
**Alokphang Buru Ha•wai**
**Alokphang Ningbrek**
**Badri Agreng**
**Badri Rongsa Ha•wai** 25.250706N, 90.432798E
**Badri Jaisyrugythym** 25.251485N, 90.463149E
**Badri Maidugythym** 25.234995N, 90.432484E
**Badri Rongdong** 25.231374N, 90.405058
**Badri Rongdyng Ha•wai** 25.240664N, 90.443061E
**Badri Rongsa Ha•wai** 25.251146N, 90.431625E
**Badri Wa•thaigythym** 25.235626N, 90.424952E
**Baghmara Bolsagre** 25.125567N, 90375792E
**Baghmara Konagythym**
**Baghmara Rangtokram**
**Baghmara Songdan**
**Baghmara Wa•kaisu**
**Balkhal** 25.162928N, 90.404182E
**Balphakram** 25.144430N, 90.491216E
**Balsriguthum** 25.495376N, 90.345932E
**Banglades** 23.410598N, 90.212279E
**Chanengtyikhal**
**Chidymak**
**Chondrosuk** 25.184462N, 90.453124E
**Dabatwari** 25.205848N, 90.410892E
**Dajong** 25.205882N, 90.41 27.75E
**Dalnenggythym**
**Dambuk Atong**
**Dangsa Ha•wai**
**Dapsi Nengrugythym**
**Dawel Agargythym**
**Dawel Rongragythym**
**Dawel Wakpangram**
**Dawylingchigythym**
**Do•renggo Wa•dachong** 25.221632N, 90.393776E
**Durakhal**
**Eksopus** 25.2133.16N, 90393083E
**Gandisunchok**
**Garaigytymh**
**Goirapatal** 25.204109N, 90.410705E
**Gongga Ha•dyng**
**Halwa Atong** 25.153595N, 90.455295E
**Jadi** 25.260484N, 90.432482E
**Kangkangkhal**
**Karamkhal**
**Khalu Bokchung**
**Khalu Garo Hills**
**Khalu Khasia**
**Khalu Kyndamgythym**
**Kharukhol**
**Kol India** 25.254448N, 90.434051E
**Kynchung Kasrigythym**
**Longlangwaidak**
**Ma•sigat** 25.140142N, 90.392447E
**Maidu Ha•wai**
**Me•mangmaisansa•ram**
**Mekalaia**
**Moradam**
**Nengrugythym** 25.230134N, 90.390204E
**Patalgythym** 25.265755N, 90.424355E
**Raiwak** 25.173842N, 90.400409E
**Raiwak Choklolgythym**

5 Semantic lexicon of verbs and nouns — **237**

Raiwak Chondolsu 25.182960N, 90.454536E
Raiwak Daw•kaparam
Raiwak Khambaipal 25.193833N, 90.435930E
Raiwak Malengmagythym
Raiwak Rongchegythym
Raiwak Rongraigythym
Rengchigythym
Rong•jyksai
Rongara 25.110472N, 90.471231E
Rongchek Ha•sym
Rongchekgre
Rongchekgre ~ Bokbakgythym
Rongchekgre Choklokgythym
Rongreng Pal 25.161334N, 90.420206E
Rongru Ha•sem 25.254030N, 90.270765E
Rongsu 25.212784N, 90.443195E
Rongsu Ha•sym
Rongthokgythym 25.271047N, 90.473110E
Ryngjywtyikhal
Sangkhyningkholwari
Senengkri
Sijyw 25.212821N, 90.394280E

Sijyw Arteka ~ Artika ~ Areteka 25.192409N, 90.404255E
Sijyw Damukgythym
Sijyw Duramong
Sijyw Ha•dura
Sijyw Songcham
Songdu 26.121897N, 90.362200E
Symsang 25.215500N, 90.430612E
Taw•pakkhal 25.210347N, 90.410282E
Tyihanggal
Tyinang
Tyipaityikhal 25.205969N, 90.38.0956E
Tyitykmak
Waimong Dalenggythym
Waimong Gonggrot 25.183152N, 90.492175E
Waimong Ha•bri 25.182695N, 90.443818E
Waimong Ha•gympal
Waimong Hangsapal
Waimong Maiduguthym 25.185987N, 90.505590E
Waimong Matchirampat

## 5.17 Shape

**bawen** *n.* circle
**gebeng** *n.* width, breadth
**jang•jot** *adj2.* biconcave, curved on both sides like the inner surface of a sphere, narrow in the middle.
**phak** *n.* side, half which is the result of a longitudinal section or a cut along the length.

**thong•** *n.* half which is the result of a cross section or a cut across the width or a cross-cut
**ympong** *adj2.* lopsided, convex, having a surface or boundary that curves or bulges outward, as the exterior of a sphere.

## 5.18 Time

See also the list of time words in PART 4, §20 as well as the lexica of Days of the Week and Months of the Year.

**dile** *n.* delay
**baji** *n.* o'clock
**tarik** *n.* date

## 5.19 Manipulation

**ak-** *v.* to pluck (leaves, fruit etc. not feathers), to pick (flowers)
**aset- ~ asyt-** *v.* to throw away, to dispose of
**baphai-** *vtr.* to drop
**bara-** *vtr.* to put in a hole, pan, *wa•sung*, bag etc.
**batkhynyng-** *vtr.* to smash
**batphai•-** *vtr.* to throw hard to break something
**batpyret-** *vtr.* to smash by throwing something to the ground
**bel•- ~ bil•-** *vtr.* to retract the foreskin from the glans penis
**benek-** *v.* to damage.
**bodol-** *vtr.* to change
**bul-** *vtr.* to dig up, to unearth, to stir
**but-** *v.* to squeeze in, to penetrate, to go inside a hole
**buthu-** *vtr.* to seal, to close a receptacle by putting something in the opening
**bychym-** *vtr.* to pull up/out
**bykot-** *vtr.* to unsheathe, to take out
**byl•-** *vtr.* to roll something into something
**bytjekjek-** *v.* to give short jerks, to draw or write scratchily
**bytjengjeng-** *v.* to pull jerkily
**bytphin-** *v.* to rewind
**bytphuruk-** *v.* to tear out with the roots
**bytphyrak-** *v.* to tear apart
**bytruru-** *v.* to drag something
**bytsorok-** *vtr.* to pull out
**bytsorok-** *vtr.* to pull out
**chel•-** *v.* to pry open
**chep- ~ chyp-** *v.* to close.
**chet-** *v.* to tear, to tear off (clothes paper etc.)
**chetpyrak-** *v.* to tear apart
**chit-** *v.* to tear, rip
**chok-** *v.* to scoop, serve up, dish up, dish out, to comb
**choka-** *v.* to take apart, to disassemble, to (be) torn, to cut off
**chokchok-** *v.* to sharpen (a pointy object)
**choket-** *v.* to scoop (for solid substances)
**chokset-** *v.* to scoop away
**chot-** *v.* to tear (off), to cu
**chuduk- ~ chyduk-** *v.* to turn upside down, to turn over
**chugup-** *v.* to cover with a lid
**chultet-** *v.* to shake off
**chyduk- ~ chuduk-** *v.* to turn upside down, to turn over
**chympyret-** *v.* to hit with one's fist, to crash head-on
**dap- ~ dep-** *v.* to press, keep together by force, pinch together, to pinch, to crush, to stack
**daw-** *v.* to open, to peel
**dem•-** *v.* to fold
**deng-** *v.* to untie
**dikirin-** *vtr.* to tear (clothes, paper etc.)
**diphing-** *vtr.* to fill
**doi•- ~ joi•-** *v.* to drag, to catch (by dragging a net through the water), to hold, to grasp, to scoop into a receptacle
**dol•romrom-** *v.* to roll up
**dolrorom-** *vtr.* to roll up
**dung-** *v.* to put something in something
**dypyleng-** *v.* to flatten, to make flat
**dyri- ~ diri-** *v.* to hold
**dyw-** *v.* to add
**ek-** *v.* to separate
**ga•-** *v.* to trample, to trod
**ga•dak-** *vtr.* to cut up, to cut into pieces, to cut up
**ga•dap-** *v.* to step on
**ga•dukduk-** *v.* to prod with one's legs or feet
**ga•jonong-** *v.* to trample on, to crush, destroy
**ga•jyret-** *v.* to crush with one's foot
**ga•kynyng-** *v.* to trample on, to crush, to destroy
**ga•phak-** *v.* to hit with one's foot while walking
**ga•phynek-** *v.* to stamp to death
**ga•pyret-** *v.* to stamp to death, to crush with one's foot
**ga•pyryw-** *v.* to stamp through something, to pierce by stamping
**ga•reret-** *v.* to tread on, to step on something
**ga•ryngreng-** *v.* to kick

## 5 Semantic lexicon of verbs and nouns — 239

**ga•thymbylong** *v.* to make a hole in a road or bridge by stamping
**ga•thyng-** *v.* to kick
**gat-** *v.* to put in/on, to load into/onto
**gatdap-** *v.* to stack, to put on top
**gogat-** *v.* to carry on the shoulders
**gop-** *v.* to bury, to hide, to cover up
**gyl-** *v.* to collect, to gather
**gyryp-** *v.* to cover
**gyryw-** *v.* to shake (an object that you can pick up, a non-fixed object)
**hot-** *v.* to extract, to draw a knife
**huk-** *v.* to sweep together
**jakhal-** *v.* to use
**jap-** *v.* to pile up
**jekjek-** *v.* to shake from side to side
**jingonget-** *v.* to shake
**jitymryma-** *vtr.* to roll something (for the purpose of transporting it)
**joi•- ~ doi•-** *v.* to drag, to catch (by dragging a net through the water), to hold, to grasp, to scoop into a receptacle
**jokset-** *v.* to drain
**jol-** *v.* to roll up
**jot-** *v.* to prod
**jothat-** *v.* to prod
**jotpyryw-** *v.* to pierce
**jul-** *v.* to jack up, to lift up, to dig up
**jumu-** *v.* to collect
**kak-** *v.* to close with a lid
**kakdep-** *v.* to bite on something
**kakpyret-** *v.* to crush by biting
**kap-** *v.* to catch
**kap-** *v.* to close
**kar-** *v.* to peel off
**kha-** *v.* to tie
**kha•-** *v.* to pour liquid into a jug
**kha•at-** *v.* to work with, to handle
**khabak-** *v.* to embrace, to grab firmly as in an embrace
**khai-** *v.* to carry on one's back with a strap tied around the head (like a basket that is carried on the back but that has a strap that is put around the head)
**khan•chot-** *v.* to cut
**khan•phyt-** *v.* to cut a solid object in half lengthwise

**khan•pyrak-** *v.* to cut a hollow object in half lengthwise
**khan•thong•-** *v.* to cut in half
**khan•tongthong•-** *v.* to cut up in pieces
**khasi-** *v.* to castrate, to remove the testicles
**khatdep-** *v.* to wrap, to wrap up, to fold
**khen-** *v.* to scratch
**khep- ~ khup ~ khyp-** *v.* to close, to cover, to spread out
**khep-** *v.* to pinch, to cut with scissors
**khet-** *v.* to tie the cloth in which you carry a baby
**khi•-** *v.* to hit (a target), to touch
**khok-** *v.* to remove (skin, bark, peel, dress etc.)
**kholthyrai-** *v.* to peel, to shed skin, to come off (of skin)
**khonok-** *v.* to search by feeling
**khuli- ~ kuli-** *v.* to open
**khyn-** *v.* to pick up, to gather, to collect
**khyw-** *v.* to drain, to shake out fluid
**matchirit-** *v.* to scratch, to be scratched
**mojekjek-** *v.* to shake (a fixed object)
**mot-** *v.* to shake a fixed object
**mu•pyret-** *v.* to crush by sitting on something
**myt-** *v.* to extinguish
**nat-** *v.* to scrub, to scour, to clean by scrubbing, to remove by scrubbing
**ni•et-** *v.* to switch off, to turn off
**nom•-** *v.* to loosen
**nong-** *v.* to apply, to put (on the skin or body, like a cream or medicine), to smear, to spread, to crush and smear
**okhynyng-** *v.* to break a round hollow object in half (crosswise)
**pai-** *v.* to carry by hand
**paitaw-** *v.* to lift up
**pan•pyrak-** *v.* to cut breadthwise
**peleng- ~ pel•eng-** *v.* to deflate
**pha•at- ~ pha•et-** *v.* to apply, to put on, to put on a wound, to apply to a wound
**phai•-** *vtr.* to break
**phai•thong-** *v.* to break a solid object in half (crosswise)
**phak-** *v.* to throw out, to empty
**phakset-** *v.* to throw away (for solid substances and things)
**phat-** *v.* to chuck away, to throw out

**phyn-** ~ **pyn-** *v.* to cover
**phyt-** *v.* to slice
**pirin-** *v.* to mix
**pyi•-** *v.* to touch, to grasp, to grab
**pyi•khap-** *v.* to catch with one's hands
**pyi•khep-** *v.* to hold firmly
**pyi•khyrep-** *v.* to crush with one's hand
**pyi•thyng-** *v.* to hold on to, to grasp
**pyn•-** *v.* to pack, to wrap up, to pack in a banana leaf
**pyndap-** *v.* to cover
**pyryw-** ~ **pyru-** *v.* to pierce, to make a hole in something
**pyt-** *v.* to wrap neatly as a present
**rai•byt-** *v.* to carry around
**rat-** *v.* to throw
**ratat-** *v.* to take out
**raw•-** *v.* to catch, to grasp
**riprip-** *v.* to rub
**rok-** *v.* to shave
**rokset-** *v.* to wipe off
**rong•dep-** *v.* to crush with a stone
**rongmyng-** *v.* to shuffle cards
**roprop-** *v.* to crumble
**ruchut-** ~ **ruchu-** *v.* to join, to connect
**ruru-** *v.* to make liquid come out
**sai-** *v.* to choose, to select, to elect
**saket-** *v.* to insert, to plug in
**sakrem-** *v.* to twist
**san-** *v.* to put in a bag
**sapset-** *v.* to drain
**sat-** *v.* to flush out
**sat-** *v.* to hit with a stick or bat, to cut with a sword
**satkhap-** *v.* to box, to slap
**satpyret-** *v.* to hit with the open hand
**sel•-** *v.* to pour into
**sep-** *v.* to wring, to squeeze out
**sepjyrot-** *v.* to wring
**si-** *v.* to peel
**si•-** *v.* to sharpen (a pointy object)
**si•wil-** *v.* to carve, to sharpen a pointy object
**sik-** *v.* to scratch, to pinch
**siksik-** *v.* to scrape, to rub
**sin-** ~ **sin•** ~ **syn•-** ~ **syn-** *v.* to lay, to lay out, to spread out on something

**sinthong•-** *v.* to cut/break in two pieces, to cut/break in half, to sever
**sintongtong-** *v.* to cut up in many pieces
**soal-** ~ **sual-** *v.* to divide, to share
**soksek-** *v.* to shake something without picking it up
**song-** *v.* to keep, to store
**song-** *v.* to set up post, to dig a hole and stick something in it so that it keeps standing up, to raise
**su•-** ~ **syw•-** *v.* to pound, to punch, to prod, to inject, to crush
**su•bylok-** *v.* to mash, to beat to a pulp
**su•pyrong-** *v.* to punch a hole through something
**su•that-** *v.* to prod, to poke
**suk-** *v.* to insert, to stitch
**sutuk-** *v.* to put over, to cover, to hide
**syk-** *v.* to insert, to be inserted, to press, to push
**sykrom-** *v.* to grasp someone
**sykup-** *v.* to fold
**sylet-** *v.* to make beautiful
**sym•-** *v.* to soak, to make wet
**taia-** *v.* to pull
**tan•-** *v.* to cut, to cut up, to chop, to chop up, to slay, to slaughter
**tan•chekchek-** *v.* to cut into small pieces
**tan•choleng-** *v.* to cut a piece out of something
**tan•pyrak-** *v.* to cut open
**tan•set-** *v.* to cut out, to cross out
**tan•thong-** *v.* to decapitate, cut off the head; to cut off
**tang•dap-** *v.* to splash on
**tankynyng-** *v.* to cut up in many pieces
**tap-** *v.* to hit, to beat-up
**tek-** *v.* to tie
**thabai•-** *vtr.* to break
**than•khoana-** *v.* to gut lengthwise
**thang-** *v.* to throw away with great force
**thap-** *v.* to beat, to beat up, to destroy
**tharai-** *v.* to change, to exchange, to swap
**thari-** *v.* to prepare, to arrange, to repair
**thatthongthong-** *v.* to tear to pieces
**the•met-** *v.* to fold

**thek-** *v.* to block off, to lock
**thek-** *v.* to insert
**thel•-** *v.* to tie
**them•-** *v.* to fold up (clothes, blankets etc.)
**themtaw-** *v.* to roll up
**thet-** *v.* to pull, to pull out
**thetchot-** *v.* to tear or break by pulling
**thojekjek-** *v.* to shake a fixed object
**thotphyret-** *v.* to smash by hitting against or on something
**thymyt-** *v.* to put out (fire), to switch off, to extinguish
**thyng-** *v.* to kick
**thyngel-** *v.* to tilt
**thyngpyret-** *v.* to kick
**thyp-** *v.* to throw (sidearm), to throw into
**thyrgyryw** *v.* to shake something large and unmovable
**tok-** *v.* to beat, to beat up, to grind, to crush, to play an instrument
**tokdepdep-** *v.* to crush, to grind
**tokgepgep-** *v.* to beat to a pulp
**tokhynyng-** *v.* to smash into pieces

**tokphyrong-** *v.* to take a powdered substance in the palm of one hand and softly tap on it with the other hand
**tokpyret-** *v.* to crush by hitting
**tokset-** *v.* to pull loose
**tokthong•-** *v.* to smash in half
**tong-** *v.* to copulate, to fuck
**tyt-** *v.* to pour
**ung-** *v.* to gather
**wai•-** *v.* to scoop (of liquid)
**waiset-** *v.* to drain a little bit of water, to scoop out water
**wat-** *v.* to switch on an electrical apparatus like a radio, TV, computer etc.
**watet-** *v.* to send (away), to post, to mail
**wen•- ~ wen-** *v.* to wind around, to wrap around, make as a coil
**wenphak-** *v.* to wind around
**wenwen- ~ winwin-** *v.* to wind around.
**wongong-** *v.* to stir
**wongwet-** *v.* to dangle
**wot- ~ wyt- ~ wen•- ~ wyn•-** *v.* to sharpen, to whet
**wungwung-** *v.* to stir
**wyiset-** *v.* to wipe off

## 5.20 Movement

**badai-** *vtr.* to cross beyond the limit, to pass a certain point
**bak-** *vtr.* to run after someone or something, to chase after
**bawen-** *v.* to move in a circle, to make a circle around something, to encircle
**byt-** *v.* to pull, to drag, to drive, to ride, to transport, to lead, to haul, to draw
**bytai-** *vtr.* to lead here, to bring here (by driving)
**bytganggang-** *v.* to drive a vehicle over a bumpy road
**cha•duk-** *v.* to bump
**cha•godot-** *v.* to stumble
**chagak-** *v.* to hit, to crash into
**chaw-** *v.* to go by boat, to row
**chaw•-** *v.* to float, to drown

**chingchoroi-** *v.* to swing from something
**dang•-** *v.* to enter, to go/come in
**dong•-** *v.* to arrive
**dongang-** *v.* to arrive (in the direction away from the deictic centre)
**dum-** *v.* to gather, to swarm
**duma-** *v.* to gather (of people)
**dung•-** *v.* to climb
**dyl-** *v.* to lead
**ga•khat-** *v.* to climb
**ga•sokhok-** *v.* to stumble
**guduk-** *v.* to wiggle, to budge, to move slightly, to be unstable, to wobble, to move unstably
**gul- ~ jul-** *v.* to walk through the jungle with difficulty.
**gylgyl-** *v.* to roam

**ho-** *v.* to jump
**hochorokchorok-** *v.* to jump like a deer
**hojokjok-** *v.* to jump up and down
**hongkhot-** *v.* to come out, to exit
**hopat-** *v.* to jump like taking a step, i.e. with one's legs apart, not with both legs together
**hung-** *v.* to swim
**jal-** *v.* to run away
**jalphakang-** *v.* to run from one side to the other
**jit-** ~ **jyt-** *v.* to move
**jok-** *v.* to escape, to come out
**jom•-** *v.* to sneak, to sneak up on somebody
**jul-** ~ **gul-** *v.* to walk through the jungle with difficulty.
**kynjung-** *v.* to make an about turn
**kyrurua** *vintr.* to roll (by itself)
**man-** *v.* to climb
**man-** *v.* to crawl, to creep
**manram-** *v.* to crawl
**nokphin-** *v.* to return home.
**parang-** *v.* to wander, to go astray
**pat-** *v.* to cross
**phak-** *v.* to gush out
**phekphek-** *v.* flipping and turning (like fish do on dry land or in a dammed-up fishing place)
**phok-** *v.* to lift up, to uproot
**poram-** *v.* to fly over
**pyw-** *v.* to fly, to flee
**pywtaw-** *v.* to jump over something
**rai•-** *v.* to go; to come
**rai•a-** *v.* to come
**rai•ganggang-** *v.* to go/drive/ride over things on a bumpy r
**rai•phak-** *v.* to go through; to hit with one's elbow while walk
**rai•sotwa** *n.* shortcut
**rai•wil-** *v.* to walk around something
**re•eng-** *v.* to go, to go away, to leave
**romrom-** *v.* to roll
**rongrong-** *v.* to slide over something
**rychup-** *v.* to fall on one's face, to fall head first.
**ryk-** *v.* to chase, to herd, to run to meet someone
**rymrym-** *v.* to roll
**ryngring-** *v.* to wiggle, to move back and forth, up and down
**ryt-** *v.* to pick up, to collect
**song•khot-** ~ **songkhot** *v.* to come out of a small opening or narrow space, to squeeze out of
**songkhel** *v.* to roll head first
**songrai-** ~ **songre-** *v.* to travel
**su•kherek-** *v.* to crash down
**sun-** *v.* to move, to shift
**sym-** *v.* to follow
**tat-** *v.* to drive in (as with a nail in wood)
**taw-** *v.* to go up, to ascend
**thang-** *v.* to come out with great force
**thangtaw-** *v.* to squirt out
**tharap-** *v.* to catch up with
**thin-** *v.* to climb a rope that is either vertically hung or horizontally strung
**thintaw-** *v.* to climb up
**thop-** *v.* to gang up on
**thorok-** *v.* to jump (down from/out of/into)
**thot-** *v.* to hit, to bump into something or against something
**thyl•-** *v.* to go very far
**wai-** *v.* to return, to go/come back
**waiphin** *n.* the return
**waiphin-** *v.* to go back, to return
**waisa** *n.* the going (to somewhere)
**wala-** *v.* to arrive at night, to be late so that it is already night
**wang•-** *v.* to turn, to wind
**wel•-** *v.* to turn left and right, to zigzag
**wil-** ~ **wyl-** *v.* to go down, to descend, to get off
**wylang-** *v.* to go down, descend
**wyngwet-** *v.* to swing, to move back and forth

## 5.21 Involuntary movement or change of state

**dangkhym-** ~ **dangthym-** *v.* to collapse (of a road or bridge), to go into a hole
**thang-** *v.* to fall down on
**thama-** *vtr.* to make lost
**sa•dap-** *v.* to spill
**thagal•-** *v.* to drop
**so•sorot-** *v.* to slip
**sel-** *v.* to leak
**pyi•ru-** *v.* to collapse
**pywgak-** *v.* to crash (in flight)
**rawsykot-** *v.* to slip out of the hand
**sat-** *v.* to spill
**pok-** *v.* to swell
**roprop-** *v.* to crumble
**pheret-** ~ **peret-** *v.* to split, to crack, to burst, to explode
**phingpyryt-** ~ **phingpurut-** *v.* to overflow
**phok-** *v.* to swell
**mykjyw-** *v.* to doze off
**phakphaklak-** *v.* to spill
**gangphu-** *v.* to swell, to blow up (like a chapatti on the fire)

**peret-** ~ **pheret-** *v.* to split, to crack, to burst, to explode
**ma-** *v.* to lose
**kanting-** *v.* to tear spontaneously
**chep-** *v.* to release contained air or water, to leak
**mimikakak-** *v.* to shake with laughter
**ga•sylek-** *v.* to sprain one's foot
**ga•syrot-** *v.* to slip and fall
**gal•-** *vintr.* to fall down
**gal•ruru-** *v.* scatter all over the place, to fall through something
**galat-** *v.* to fall
**gurum-** ~ **gyrum-** *v.* to collapse, to break off and fall down
**godot-** *v.* to bump
**bai•-** *vintr.* to break
**chokchok-** *v.* to drip out
**ga•ak-** *vsec.* to be compelled to, to be forced to
**jok-** *v.* to leak out, to jump because something startled you
**phet-** *v.* to swell up

## 5.22 Order

See also PART 4, §20, List of time words.
**dakanggaba** *n.* the first, first, the first time, previous
**jamkhamwa** *n.* the last one
**kynpha-** *v.* to be last, to be late
**las** *n.* the last one
**lasgaba** *n.* the last on, last
**phas** *n.* the first one
**phasgaba** *n.* first

## 5.23 Phase

**jam-** *vph.* to finish, to complete, to be complete, be total, to deplete, to use up, to succeed
**phet-** *v.* to arrive (at), to reach, to come out of the water, to emerge
**dang•-** *vph.* to start
**jam-** *vph.* to finish, to complete, to be complete, be total, to deplete, to use up, to succeed
**chu•sok-** *v.* to succeed
**machot-** *vph.* to finish, to end, to stop
**ha•bacheng-** *vB.* to start, to begin
**sok-** *v.* to succeed

## 5.24 Noise and sound

**chengchang bengchang** *n.* noise, racket
**khu•sylip ~ khu•sylyp-** *v.* to whistle
**kyryng-** *v.* to make noise, to make a sound
**kyrynggaba** *n.* sound, noise
**kyryngwa** *n.* sound, noise
**sawn** *n.* sound

## 5.25 Abstract

**aidia** *n.* idea, plan
**ain** *n.* custom, law, tradition
**atak-** *v.* to do what?
**bakdong** *n.* a forbidden marriage: a marriage between a man and a woman from the same *mahari*
**baki** *n.* credit
**baratwami** *n.* shame
**bewal** *n.* tradition, habit
**bimang** *n.* appearance
**bimung ~ bimyng** *n.* name
**byl** *n.* strength
**chasing** *n.* generation, era
**chin** *n.* a sign
**chol** *n.* idea, plan
**churu** *n.* very little food
**dam** *n.* price
**dosi** *n.* blame
**duk** *n.* sorrow
**dygri** *n.* degree
**gam ~ kam** *n.* wealth, riches, work matters, activity
**gamchat-** *v.* valuable, important
**gamchatgaba** *n.* value
**gawak** *n.* disease
**gebeng** *n.* width, breadth
**gremyr** *n.* grammar
**gyching ~ giching** *n.* angle, inclination
**ha•bachenggaba** *n.* (1) beginning (2) Genesis
**ha•gylsak ~ ha•gyrsak** *n.* everything, all
**histyri** *n.* history
**jagydok** *n.* strength
**jajyreng** *n.* confusion, danger
**jamkhamwa** *n.* the last one
**jaria** *n.* influence
**kan•-** *v.* to last
**karaw** *n.* (1) debt, obligation (2) trouble
**kethylik** *n.* Catholic
**kha•** *n.* fighting spirit
**koros** *n.* expenses
**kyrewami ~ kyriwami** *n.* danger
**lap** *n.* profit, interest, gain
**las** *n.* the last one
**lasgaba** *n.* the last one, last
**mamung ~ mamyng** *n.* nothing
**man** *n.* respect
**man•-** *vB.* to be able, to be possible, to be allowed to, to get, to obtain, to succeed
**man•dapwami** *n.* profit, interest, gain
**mejoryti** *n.* majority
**mondoli** *n.* the Church, congregation, church community
**mykgythal** *n.* reality
**nambal ~ nombol** *n.* number
**nang-** *v.* to need, to have to, must, to have to call someone by a certain term
**nemgyni** *n.* advantage, good fortune, good luck
**ni•- ~ nyi•-** *v.* negative locative/existential verb, to not exist, to not be
**oltho ~ ortho** *n.* meaning
**pang•wami** *n.* quantity, abundance
**phal** *n.* (1) share, shift of work (2) instead of
**phas** *n.* the first one
**phasgaba** *n.* first, the first one
**rokhom** *n.* shape, type

**rong** *n.* colour
**sak-** *vgoal.* to depend on
**somphi** *n.* riddle
**songsal** *n.* society
**songsyrek ~ songsarek** *n.* animism
**sotkat** *n.* (1) shortcut (2) short version
**spiit** *n.* speed
**sungchal-** *v.* to support (a structure)

**takbewal** *n.* tradition
**thik dong•- ~ dong-** *v.* to be correct
**thorom** *n.* (1) religion (2) denomination
**tiktik-** *v.* to make last
**tyngwami** *n.* knowledge, understanding
**wa•churek** *n.* capacity, capability
**ytyk-** *v.* to do like this/that

# PART 4: GRAMMATICAL LEXICA

This part of the book contains lists of lexemes and morphemes grouped together on the basis of grammatical criteria. The topics, and therefore the sections in this part of the book are organised alphabetically. The reader is referred to the Grammatical Compendium in PART 5, as well as van Breugel (2014) for grammatical information. Nouns are not listed here, but rather in the Semantic Lexicon of Verbs and Nouns in PART 3 of this book. Some lists of verbal subclasses of verb are presented here. However, the reader will not find lists of all transitive and intransitive verbs, because many verbs are ambitransitive, and the valency of many other verbs is not know with certainty. For more information on subclasses of verbs, the reader is referred to van Breugel (2014: Chapter 4).

## 1 Lists of adjectives of Type 1 and stative verbs

For grammatical information, see PART 5, §2. Table 5 presents a list of al recorded Type 1 adjectives organised according to semantic category (partly based on Dixon 2004), and alphabetically within each category.

Table 6 presents a list of stative verbs, which have not or not yet been identified as Type 1 adjectives (see PART 5, §2). This means that some or all of them might have the same properties as Type 1 adjectives, but that these properties are not attested to date. Future fieldwork will hopefully shed more light on the precise nature of the words in this list. The stative verbs in this list in Table 6 are organised according to semantic categories (partly based on Dixon 2004), and alphabetically within each category.

**Table 5:** Type 1 Adjectives.

<u>DIMENSION</u>
**chat-** bulky, thick
**chung-** big
**chyw•-** high, steep
**jaraw-** a long time
**kan•jot-** skinny
**myl-** small
**nyng•thyw-** deep
**phet-** thin
**raw•-** tall, long
**raw•reng-** slender and long
**sung•-** short (of time, person, thing)
**thyw•-** deep

<u>COLOUR</u>
**nak-** black
**sak-** red
**bok-** white

<u>VALUE</u>
**althu•-** easy
**baram-** rough
**bytym-** nicely smelling
**chaithawa-** beautiful
**cheng•-** light (not heavy)
**damrak-** expensive

## 1 Lists of adjectives of Type 1 and stative verbs

**Table 5** (continued)

ga•- good, nic (as a character trait or property of things)
ga•su- splendid, cool, terrific
han•tung- to be a dangerous place
jakhalthaw- very useful
man•dyk- physically difficult, complicated, troublesome
manam- to stink, to smell bad
nangchomot- important
nem- good
rongthala- clean
syl- beautiful, pretty
thal- clear, explicit
thaw- tasty

### Physical property
bu- to be sharp (of pointed things)
bui- murky, turbid
bylak- strong
chat- thick, bulky
chek- cold
chyng•- bright
chyrym- heavy
demdong- weak, soft
dumut- moulded
gyl- strong
ha•kha- *adj1.* very tight
jingjong- wiggly, unstable
ket- tight
kek- blunt (of pointed things)
kon• winding
kun- curly
kynkom- old and crooked
kyryng- tight
manak- dark (due to the absence of light)
mel•- fat (of people)
nom•- soft, weak, easy, tender
pa•- low, plain, flat, thin (of things)
pereng- straight
pyn•- dense, thick
rak- hard, strong, fast, difficult,
ran•- dry
rimyl- slippery

ronok- smooth
songrat- to be bent
sorong- straight
su•ut- damp
sym- sweet
thanthong- blunt (of pointed things)
tung- warm, hot
tyisi- wet
wekwak- to be very soft (like mud), sloppy

### Mental property
mythel- thankful, grateful
chyi•- tired, sleepy
neng•- tired (after making an effort)
seng- clever, intelligent

### Taste
beraw- to contain too much soda (MSG) (said about the taste of food)
jyryk- to have a nutty taste
ka•- ~ kha•- bitter (taste)
kha- salty
kha•sym- bitter-sweet
khyi- sour
kyisym- salty
parap- ~ pyrap to be (too) salty
thap- savoury, umami
symjin- to have the sour-sweet taste of a half-ripe fruit

### Position
jan•- far
nek- close, near

### Quantification
pang•- a lot, many

### Temporal
jaraw- (for) a long time

### Speed
jang- quick
tarak- quick, fast, swift

**Table 6:** Stative verbs not yet identified as Type 1 Adjectives.

### VALUE
**chuli-** to be useful
**khe- ~ ke-** to be proper, appropriate, suitable
**mili-** to be appropriate
**myngnang-** to be suitable
**sak-** to fit

### EXTENT
**bylong-** to be too much
**dai-** to be bigger, greater
**duga-** to be too much
**gusum-** to be spoiled (only used with meals)
**mykhal-** to be older than someone

### EXTENT/TEMPORAL
**dong•- ~ dong-** to be enough, to be sufficient, to be OK, to be convenient, to have passed, to be past

### TEMPORAL
**tharap-** to be on time

### POSITION
**chat-** to be fixed together (like a stapled pile of paper or a pile of wood etc.)
**thyk-** to be fixed sideways

### POSITION/ QUANTIFICATION
**heng•-** widely spaced, sparse

### QUANTIFICATION
**jel-** to be numerous

### PHYSICAL PROPERTY
**bu•chok-** to be sharp (of pointy objects)
**bythyw-** to be blocked
**chekjyrym-** to be coolish
**chenggang-** to be upright, to be erect
**cho•chep-** to be crumpled
**chogop-** fully bent but not touching the ground (used only with plants)
**chyng•chet-** to glitter
**dawel-** to be circular
**doksylok-** to be detached
**khalthyng-** to be nauseating

**khampyryw-** to have a hole in a cloth or paper as the result of burning
**khamthymbylong-** to have a hole in a road or bridge as the result of burning
**kung-** to be dammed up, to be enclosed by a dam or circle of stones
**kunremrem-** to curl, to be curly.
**kyp-** to fit tightly, to fit and close off
**kyryk-** to be clear, to be transparent
**mat-** to be sharp, to wound, to be able to wound, to cut, to be able to cut
**miniksuru-** to be flat-haired (of animals)
**mym•-** to be like a fist
**myn-** to be ripe, to be cooked, to be ready
**myn•sym-** to be hairy with small hairs
**nosto dong•-** to be damaged, to be defective
**ogynang- ~ oknak- ~ oknang- ~ okgynang-** to be pregnant
**pering-** to be straight
**phek-** to be drunk
**phing-** to be full
**phuk- ~ puk-** to be stuck
**phyryw-** to be hollow
**rekhep-** to be dry (of plants), to be wrinkled (of person)
**romthom-** to be spherical
**salam- ~ selem- ~ serem- ~ saram-** to break/tear easily, to be easily damaged
**saw-** to be rotten
**saw•myk-** to smell rotten, to smell foul
**saw•saw-** be able to cause a burning sensation
**sypsak-** to be scratched
**tan•pat-** to be oblique
**tengchypchyp-** to shine, to glitter.
**tuk-** overgrown, dense (of vegetation)
**tyikhyrep-** to be wrinkled because of being in the water for a long time

### MENTAL PROPERTY
**mal-** to be familiar, to be easy to deal with
**watbyrak-** to be shameless

## 2 List of adjectives of Type 2

A brief explanation of the difference between Type 1 and Type 2 adjectives is provided in PART 5, §2. The list presented here in Table 7 is the result of fieldwork conducted after the publication of van Breugel 2014 (see pp. 90–96). This list is updated and contains more words. All adjectives are organised according to semantic type (partially based on Dixon 2004), and alphabetically within each type.

**Table 7:** Type 2 Adjectives.

AGE
**bydyi** old (for persons)
**picham** old (of things)
**pidan** new

COLOUR
**ha•mangrong** brown
**khengchek** green, blue
**khengsyryk** dark green
**konglarong** orange
**pibok** white, unripe, very light green
**pinak** black
**pisak** red, blond
**rymyt** yellow, orange

VALUE
**ramram** ordinary, normal

PHYSICAL PROPERTY
**bangbang** empty
**bukalang** to have a hole in it (clothes)
**bykphyl** inside out
**byrymbyrym** multi-coloured
**ching•pheng** aslant, slant
**dymbrubru** shiny
**dymdam** naked
**gongdang** bent
**gopgylang** hollow
**gyching ~ giching** aslant, slant, diagonal, at an angle, inclined
**gythyng ~ githing ~ githyng** unripe, uncooked, raw
**jang•jot** biconcave, narrow in the middle
**jyngjang** dense
**karam** poisonous
**khingcheng** aslant, slant
**khurung** wanting to lay an egg
**kirin** torn (of cloth and paper)
**kompyl ~ kongpyl** bent
**nagok** deaf
**pelang ~ peleng ~ pylang ~ pyl•eng** flat
**phangphyl** upside down
**pikheng** alive
**pyryw ~ phyryw** to have a hole in it (of walls)
**rong•gyrym ~ rong•rymrym** being full of big rocks
**sorong** straight
**thokbyrang ~ thokbyrym** multi-coloured, many coloured
**thymbylong** to have a hole in it, with a hole in it, damaged (of roads, bridges and wooden planks)
**thywkhong** globular, protruding, bulging
**totyp** bent
**ympong** lopsided, convex, having a surface or boundary that curves or bulges outward, as the exterior of a sphere

MENTAL PROPERTY
**chalak** cunning, clever

BODILY CONDITION
**karan** thirsty
**tyikaran** thirsty

SIMILARITY
**baibai** the same
**dyngthang** different

**Table 7** (continued)

**sa•rong** of the same age
**ytykgaba** this kind of, like this, such

POSITION/ORDER
**abun** next, following, neighbouring, other, someone else
**bathan** lying on one's back
**byryp** lying on one's belly
**dyngdang** alone
**gapsan ~ hapsan** the same, together
**sul** next, neighbouring

SPEED
**kha•sin ~ khasin** slow
**kha•sin kadym** slow

TEMPORAL
**khengkhang** eternal

POSSESSIVE
**ginggang** having, with

# 3 List of adverbs

For grammatical information, see PART 5, §3. Table 8 is adapted from van Breugel (2014: 242).

**Table 8:** Adverbs sorted by denotation or function.

LOCATION IN SPACE
**byldyng byldang** all over the place
**repa chepa** 'in various places'
**phakwil phakwal ~ phakwyl phakwal** 'side by side'

LOCATION IN TIME
**bibyrokhon ~ bibykhoron** some day
**sirimynmyn** at the crack of dawn

FREQUENCY
**bichiba** never
**bichiba bichiba** sometimes, seldom
**bichiba** sometimes
**gasam gasam** seldom
**gisep gisep ~ gysep gysep** from time to time
**phangnan** ever, never (can be repeated for emphasis)
**wetanchian** every time

QUANTITY
**isykyn ~ iskyn** this much, so much, such

DURATION
**dykdyk** for a short while
**hawtyi** for some time

STATUS
**chong•mot ~ cho•mot** really
**bebe** truly
**beanbebe** truly

MANNER

**alamyla** somewhat, a little
**byk** suddenly
**bykbyk** quickly
**chacha** exactly
**chapchap** close together (as in a crowd)
**chup ~ chyp** fully dressed
**dymdym damdam** carelessly, disorderly
**jebadong** anyway, however it may be **jekhai** for example
**jenethene** somehow
**jetykyi** somehow
**jyrym jyrym** quietly
**kepleplep** stretched out flat on one's belly
**khasin ~ khasinsin** slowly
**kongken naken** zigzag
**kyryk kyryk** swiftly
**pyltawtaw ~ pywtawtaw** jerkingly over a rough road (can be repeated)
**rymrym** rolling down
**sykathang** carelessly, disorderly, in vain, for nothing
**syraksyrak** exactly, precisely
**thangguduk** suddenly **wetwet** quickly
**yndyn** simply, in vain, for free
**ytykyi** like this/that

INTENSIFICATION
**bylongen** very
**nemen** very

## 4 Lists of classifiers and volume words

For grammatical information on classifiers and volume words, see PART 5, §4. Table 9 presents an alphabetical list of all recorded classifiers in Atong. For most classifiers, some examples are given of the kind of words with which they are used. When the reader wants to know the classifier for a specific word, she/he is advised to search that word in the Atong-English Dictionary. If the information cannot be found, it means that the classifier was not recorded. The list of classifiers presented here is most probably not complete. Thus, when a classifier does not appear in this list, it does not mean that it does not exist, but simply that it has not been recorded. Future fieldwork on Atong will hopefully mend any shortcomings.

In this list, auto classifiers are labelled *(auto)*. Auto-classifiers are nouns that, when quantified, are directly followed by the quantifier, instead of needing an intervening classifier. For more information about classifiers, the reader is referred to van Breugel (2014: Chapter 12).

The most frequently used volume words are listed in Table 10. For a list of receptacles, see PART 3, §5.12.6.

**Table 9:** Classifiers.

**ali** classifier for small heaps or piles of things. *narang ali tham* 'three piles of oranges'

**bawang** length of the widely stretched arms and hands

**bek** repeater classifier for *abek*. *abek bek sa* 'one *abek*'

**bikha** classifier for surfaces of 80 by 80 *pit*. (A *pit* is the length of two fists and two thumbs when one joins the thumbs at the tip while making fists)

**byl•** *(auto)* stroke, blow

**bylsi** *(auto)* year

**cha•pha** a foot-length

**chak** classifier for leaves. *panchak chak sa* 'one leaf'

**chakwak** classifier for handfuls. *rong• chakwak chit sa* 'eleven handfuls of stones'

**ching** classifier for bamboo shoots. *mai•wa ~ maiwa• ching sa* 'one bamboo shoot'

**chok** classifier for bunches or small heaps. *ja•ryt chok sa* 'one small heap of chillies,' *rasunok chok sa* 'one bundle of spring onions'

**chol** classifier for ways, roads, paths and rivers. *tyikhal chol ni* 'two rivers', *ram chol tham* 'three roads, paths', *sorok chol byryi* 'four roads'

**chom•** classifier for small heaps of round fruits and vegetables the way they are presented at the market. *Narang chom•ni hyn•bone.* 'Give two little piles of oranges.'

**chong** classifier for iron nails. *khiil chong sa* 'one (iron) nail'

**dam** classifier for villages. *song dam ni* 'two villages'

**dol** *(auto)* group. *dol ni* 'two groups'

**dora** weight unit of 5 kg

**dot** classifier for long cylindrical things like logs (of wood), candles and bananas. *wa•dot sa* 'one culm of bamboo', *kendel dot sa* 'one candle', *pan dot sa* 'one log (of wood)'

**gasam** *(auto)* evening

**Table 9** (continued)

**gasamphang ~ gasamphak** *(auto)* afternoon, evening, later part of the day
**geng** classifier for long vegetables. *rasunok geng sa* 'one spring onion'
**goi•** residue classifier, used when something cannot be classified with one of the other classifiers
**grem** gram
**inchi** inch
**ja** *(auto)* month
**jora** classifier for things that occur in pairs. *sendel jora sa* 'one pair of sandals', *mykren jora sa* 'one pair of eyes'
**keji** kilogram
**kep** classifier for small flat things. *biskut kep sa* 'one biscuit'
**khabak** an armful
**khal** classifier for orifices, holes and caves. *hang•khal khal ni* 'two caves', *nakhungkhal khal ni* 'two nose holes' (*nakhungkhal ni* 'two nose holes' is also possible)
**khan** classifier for logboats. *rung khan ni* 'two boats'
**khantha** *(auto)* hour
**khap** classifier for flat piece of hard material like stone or metal. *so•rekhap khap ni* 'two pieces of mica'
**khap** classifier for hard and flat materials. *tota khap sa* 'one plank', *tin kahp sa* 'one sheet of corrugated iron', *damdyl khap sa* 'one bamboo mat used for the side of a house'
**khasot** classifier for bundles of things with stalks. *ra•sun khasot sa* 'one bundle of onions (which have stalks)'
**khatom** classifier for bagsful. *ra•sunok khatom sa* 'one bagful of spring onions'
**khaw** classifier for teeth. *wa khaw sa* 'one tooth'
**khaw•** classifier for teeth, planks, sheets of corrugated iron for roofs and flattened bamboos used to make mats (*jyw•*) when they are in a mat (*damdyl*). *damdyl khaw• sa* 'one *jyw•* of a *damdyl*', *wa khaw• ni* 'two teeth, two tusks', *tota khaw• tham* 'tree planks', *tin khaw• byryi* 'four sheets of corrugated iron'
**khung** classifier for flat things, clothes, written things and pictures, even when the pictures appear on a computer screen. *chiti khung ni* 'two letters', *longpen khung sa* 'one pair of trousers'
**khuru** length from the tip of the thumb to the tip of the middle finger when one puts one's hand down on the table on these points (Old English: span)
**kilomityr** kilometre
**kun•** classifier for culms and sticks. *parang kun• sa* 'one culm of thatch'
**lain** classifier for a collection of items lined up inside packets or on shelves
**lityr** litre
**mainym** length from the elbow to the tip of the fist
**manap** *(auto)* morning
**mang** classifier for animals, knives and other tools. *bythyi mang sa* 'one porcupine', *chaw•kyi ~ chang•kui mang sa* 'one big knife'
**minit ~ minyt** *(auto)* minute
**mityr** metre
**mon** weight unit of 40 kg
**myk** the length from the elbow to the tip of the middle finger; ell, cubit
**mym•** *(auto)* (1) fist (2) classifier for fists and classifier for things that are like a fist. *mym• ni* 'two fists'
**myng** classifier for spoken things, games and for the word *bostu* 'thing'. *khata myng ni* 'two words', *golpho myng ni* 'two stories', *git myng ni* 'two songs' *bostu myng tham* 'three things'
**myng•** classifier for humans
**nok** *(auto)* house
**nokkhung** *(auto)* roof
**nygyltyi** *(auto)* week
**pan** classifier for apparatus, appliances, mechanical and electrical things, cars, bikes, bicycles, mortars and umbrellas.

**Table 9** (continued)

redio pan sa 'one radio', satha pan sa 'one umbrella', gari pan sa 'one car', thep pan sa 'one tape', tibi pan sa 'one TV', asam pan tham 'three mortars'
**peket** classifier for packets. sigyret peket sa 'one packet of cigarettes'
**phak** classifier for parts of objects that are the result of a cut along the length or longitudinal cut. ang buchotaw phak tham kan•ni I will cut the mango in three pieces.
**phan** classifier for food packed in bundles in rai•chak 'big leaf used to pack food'
**phang** classifier for trees and flowers, culms and stalks. samsi phang sa 'one culm of grass' narang phang sa 'one orange tree'
**phat** classifier for clothes. ri•pan phat sa 'one traditional skirt'
**phek** classifier for branches of trees. dala phek sa 'a smaller but not very small branch of a tree and not directly derived from the trunk'
**phel ~ phul** classifier for baked things. barata phel sa 'one flatbread', biskut phel sa 'one biscuit'
**phong** classifier for cylindrical objects and for long sharp or pointy things
**piit** the length of two fists and two thumbs when one joins the thumbs at the tip while making fists
**rong** classifier for small round objects, money, small stones, seeds, stones in a game (when they have a value) and fruits, default classifier for counting for the sake of counting. narang rong sa 'one orange', tanka rong chyigyk 'ten rupees'
**sam** classifier for limbs: hands, arms, legs, feet, ears and tires. nakhal sam sa 'one ear', cha• sam sa 'one leg/foot', taiyr sam ni 'two tires'
**san** (auto) day
**sat** classifier for bundles. garu sat tham 'three bundles of mustard leaves'

**sekyn** (auto) second
**sentimityr** centimetre
**sung** classifier for hollow cylindrical objects or tubes. wa•sung sung tham 'three bamboo tubes'
**tang** classifier for koktang 'type of basket'. koktang tang ni 'two koktang'
**thai•** classifier for boxes and other receptacles. boiom thai• sa 'one jug', dipot thai• sa 'one teapot', khap thai• sa 'one cup'
**theng** classifier for pieces of food (especially meat). ma•surandai theng sa 'one piece of beef', khan theng sa 'a piece of tapioca'
**thep** classifier for heaps and small packets
**thom** classifier for things in heaps or piles. jyw• thom sa 'a pile of flattened bamboo used to make mats'
**thong•** classifier for cylindrical objects and for parts of objects cut across the width. betyri thong• byryi 'four batteries'
**thut ~ thun** classifier for big spherical things, stones, bricks, rocks, heads, hills, mountains and bars of soap. ha•byri thut sene 'seven hills/mountains' sabun thut sa 'one bar of soap', dykym thut sa 'one head', rong•thai thut tham 'three stones/rocks'
**tum** classifier for packets and places. peket tum sa 'one packet' hap tumbyisyk? 'How many places?'
**tung** classifier for things like bridges. dolong tung sa 'one bridge'
**tyi** replaces the word nygyltyi in the second part of an approximate quantification nygyltyi tham tyi byryi 'four or five weeks'
**tym** classifier for fields. ha•ba tym ni 'two dry-rice-and-vegetable fields'
**tyng** classifier for long thin things like ropes, chains and hair
**wal** (auto) night
**wen• ~ wet** (auto) time, turn

**Table 10:** Volume words.

| | |
|---|---|
| **botol** | a bottle or its volume |
| **gylas ~ gilas** | a glass or its volume |
| **hap** | half |
| **khap** | a cup or its volume |
| **pawai** | a bowl to serve curry in or its volume |
| **powa ~ pywa** | a bowl of rice or its volume |
| **thali** | a plate or its volume |
| **thothak** | a drop |

## 5 List of collocations

For grammatical information, see PART 5, §5. All recorded collocations are listed here according to their type, in Tables 11–14, and alphabetically within their type. When the members of a collocation appear juxtaposed, they are attested within the same phrase. This does not mean that they could not appear in different phrases or clause, but just that they are not attested used in those environments. When collocations are attested in different (but, as mentioned in PART 5, §5, always juxtaposed) phrases or clauses, their members are followed by an ellipsis. This does not mean that they could not appear in the same clause together, but just that they are not attested in that environment. The word class for each collocation is also provided.

**Table 11:** Synonymous collocations.

The two members of the collocation have similar meaning. Each of the constituent words can also be used independently.

**ain niam** *n.* laws, customs, traditions
**byl... jagydok...** *n.* strength
**dada... phaw•jong~phawjong...** *n.* elder brother
**gam jym** *n.* wealth, riches, fortune
**ha•sel... ha•mat...** *adv.* troublesome
**janggi khenwa** *n.* life
**naw... bai•...** *n.* blood relative
**naw... jai...** *v.* to scold
**nombok thyibok** *adj2.* unconscious, almost dead
**phet... dong•...** *v.* to arrive, to reach
**rasong... gal...** *n.* praise and pride
**tam•a toka** *v.* to play an instrument
**tangka poisa** *n.* money

**Table 12:** Associative collocations.

The two words are not synonymous, but have associated meanings.

---

**byl chak** *n.* strength
**byl... chak...** *n.* arm/hand
**byl... jagydok...** *n.,* strength
**jal... pyw...** *v.* to flee
**kana theka** *n* (< Indic). (1) have food and drink (liquor), food and drink (liquor) (2) have a drink (of liquor), liquor.
**pan wa•** *n.* plants, vegetation, plants and trees
**pan... wa•...** *n.* plants, vegetation, plants and trees
**re•eng... jok...** *v.* to go back
**re•eng... taw...** *v.* to go away, to leave
**rophyl... khele...** *v.* to joke around, to play around
**sa•a siwa** *v.* famine, starvation
**song... nok...** *n.* village

---

**Table 13:** Decorative collocations where only the second member is a decorative word.

---

**atak... adong...** *v.* what's happening? what's going on?
**bai• tyng** *n.* blood relative
**bai•maran chingmaran** *n.* two distant relatives
**bawbyl chambyl** *n.* enemy
**chak chok** *n.* hand
**chol chal** *n.* livelyhood, way to make a living
**ga•sokhok aksokhok** *adv.* stumbling
**gumuk gamak** *qtf.* altogether, in total
**jyw- siri-** *v.* to sleep
**nokwa• ha•chak** *n.* bamboo to build a house
**takwa rukwa** *n.* activities, customs, traditions
**thik thak** *adv.* exactly, precisely
**thop... tung...** *v.* to mob, to gather around

**Table 14:** Decorative collocations where both members are decorative words.

| | |
|---|---|
| **byldyng byldang** *adv.* | all over the place |
| **chuwil chuwal** *adv.* | spinning. |
| **chuwyng chuwang** *adv.* | with a spinning head, dizzily |
| **dymdym damdam** *adv.* | carelessly, just, any way |
| **phakwil phakwal ~ phakwyl phakwal** *adv.* | side by side |
| **repa chepa** *adv.* | in various places |

## 6 List of demonstratives

For grammatical information, see PART 5, §7. Table 15 is adapted from van Breugel (2014: 133).

**Table 15:** Demonstratives.

---
**ie ~ i-** proximal demonstrative 'this/these'
**ue ~ u-** distal demonstrative 'that/those'
**hawtyi** further away (and out of sight) 'there'
**hawe ~ haw-** remote demonstrative 'that/those way over there'
**hyiawe ~ hyiaw-** very remote demonstrative 'that/those way over there'
**hai•e ~ hai•-** generic proform demonstrative 'whatchamacallit, this/that ehm…'
**isykyn ~ iskyn** quantitative demonstrative 'this much'
---

## 7 List of discourse connectives

The functions of the discourse connectives in Table 16 are given in brackets after the translations. For grammatical information, see PART 5, §8.

**Table 16:** Discourse connectives.

**aro**  in addition, moreover (additive)
**uchi ~ uchie ~ uchian**  then (sequential)
**uchiba**  but then, but (sequential, contrastive)
**umigymynchi ~ umynggymynchi**  for that reason, therefore, that's why (reason)
**una**  then, therefore, because of that (sequential, reason)
**ytykchiba**  but, in that case (contrastive, conditional)
**ytykchido**  in that case, that being the case, but (conditional, contrastive)
**ytykma•chiba**  but (contrastive)
**ytykyimuan**  so then, having done that/this (sequential, pause filler)
**ytykyimyng ~ ytykyimu ~ ytykyimuna ~ ytykyimung ~ ytykyimungna**  so then, having done that/this (sequential, pause filler)
**ytykyisa**  therefore, that's why, then (reason, sequential)

# 8 List of event specifiers

For grammatical information, see PART 5, §9. Table 17, modified from van Breugel (2014: Table 74), presents an alphabetical list of event-specifier sufixes. The translation of an event specifier depends on the context in which it occurs. The translations offered here are those that fitted the recorded usages of these event specifiers. However, not everything people say is recorded; therefore, the translations offered here are not the only possible ones. Different contexts may require translations not offered here. Further research on Atong will hopefully increase the variety of translations for event specifiers. As mentioned in van Breugel (2014: 377), there are probably more event specifiers than the ones presented in this list.

**Table 17:** Event specifiers.

- **-a ~ ai** *V* towards
- **-an** still *V*-ing
- **-ang** *V* away, *V* affluently, *V* without holding back
- **-asol** really, verily, actually
- **-barai** always *V*
- **-bat** *V* even more, *V* most
- **-bebe** truly, verily
- **-bi** (intensifier suffix) very ADJECTIVE, *V* the most
- **-bongbong** *V* more than necessary, *V* in abundance (pejorative), *V* scandalously much
- **-bylok** *V* into pulp
- **-chai ~ -chyi** try to *V*
- **-chang** suddenly
- **-chap** *V* along with someone/something
- **-cheng** *V* first
- **-chep** *V* alone
- **-chichi** *V* with force, *V* into pieces
- **-chik ~ -chyk** *V* as long as you can
- **-chikchak** *V* in a swarm
- **-cho•mot ~ -chongmot** determinedly, certainly, definitely
- **-chyp** wastefully, unsuccessfully
- **-dam** truly
- **-damdam** *V* in different places
- **-dap** *V* and add, *V* on top, *V* on top of something
- **-dykdyk** about to *V*
- **-gachak** *V* until it is red hot
- **-gak** accidentally
- **-gat** *V* upon to, to start *V*-ing (inceptive)
- **-gorop** *V* together, *V* with a whole group
- **-jokjok** *V* up and down
- **-jol** (1) quickly (2) intensifier suffix
- **-joljol** very quickly
- **-jyryng** *V* daily, *V* all the time
- **-kham** *V* for a long time
- **-khan** *V* also, *V* as well, *V* in addition
- **-khaw** secretly, surreptitiously
- **-khelek** *V* for fun
- **-khep** firmly
- **-khynyng** *V* into small pieces
- **-khyrym** *V* in a group
- **-langlang** (intensifier suffix) very ADJECTIVE
- **-man ~**
- **-man•** already *V*-ed, resultative suffix
- **-mangmang** *V* only, just *V*, exclusively, continuously, *V* as best you can, barely *V*, *V* shoddily, poorly, negligently, not fully *V*-ed as it was supposed to be done
- **-mu•** keep *V*-ing
- **-nang** nicely, beautifully
- **-nap** *V* with all one's heart
- **-parang ~**
- **-pyrang** *V* without destination, *V* without goal, aimlessly
- **-pat** *V* across
- **-peng** to hinder/obstruct to *V*
- **-pha** (1) *V* also, *V* in addition, *V* along with, *V* together, *V* in total (2) please *V* (politeness suffix) (3) intensifier suffix

**Table 17** (continued)

-**phak** *V* in half lengthwise, *V* lengthwise, *V* and go through lengthwise, *V* by the side of something, *V* side by side
-**phet** detrimentally, *V* to one's detriment
-**phetphet** (1) repeatedly (2) intensifier suffix
-**phin ~ -phyn** (1) *V* backward, *V* back (2) intensifier suffix, over-*V*, *V* overtime, fully, obviously, fully, totally, completely
-**pyl** rapidly
-**pyrak** *V* and cut
-**pyryt** over-*V*
-**ram** inadvertently, unintentionally, fortuitously, *V* because of the situation
-**ramram** normally, naturally
-**rat** *V* downward
-**rawraw** continue to *V*, increasingly *V*
-**rong** usually, always
-**rongreng** *V* while spinning around
-**rum** *V* with a group, everyone/everything *V* (This suffix indicates that the action is done or undergone collectively as a group, at the same place and time. See also *-thok*.)
-**ruru** *V* more and more, *V* around, *V* all over the place
-**sak** appropriately
-**saw** (1) *V* and wait, expectantly, *V* for sure (2) intensifier suffix
-**sek** *V* and steal
-**seme** reluctantly
-**set ~ -syt** to *V* so as to dispose of something
-**si** uncomfortably
-**soso** *V* to/on the ground
-**sok ~ -soksok** to fail to *V*, to miss-*V*
-**sot** directly
-**su** (intensifier suffix) very ADJECTIVE
-**susa** competitively
-**sym** (1) *V* and follow; imitate someone's *V*-ing (2) intensifier suffix
-**symsym** continuously
-**syrang ~ -srang** (intensifier suffix) *V* completely, *V* wholly, *V* till the end, *V* very much

-**syret** wrongly, mistakenly
-**syruk** secretively
-**tat** compulsory *V*
-**taw** *V* upward
-**teng** (intensifier suffix) still too *V*
-**tengteng** still much too *V*
-**tham** barely *V*
-**thamak** barely *V*
-**that** *V* excessively
-**thel** surely *V*
-**thengtheng** still too *V*
-**thiri ~ -theri** *V* again, *V* back, reversely, *V* backward
-**thirithiri ~ -theritheri** *V* again and again
-**thok** everybody *V* (This suffix indicates that a group of individuals all do or undergo the same thing, but not necessarily together at the same place or at the same time. See also *-rum*.)
-**thong•** *V* in half, *V* crosswise, *V* and go through crosswise
-**thum** *V* on behalf of someone else, *V* for the benefit of someone else/something
-**thyl** *V* and avoid, *V* ahead
-**thylong** nicely
-**thyn** intensifier suffix
-**thyng ~ -thing** (1) only *V* (2) intensifier suffix
-**thyngthyng** only *V* and nothing else, continuously
-**tyn** as the leader, lead in *V*-ing
-**tyng** intensifier suffix
-**tyngtang** *V* all over the place
-**tyngtang** *V* all over the place
-**wenwen** *V* in circles
-**wil** *V* around
-**wilwil** *V* around and around
-**wyng** *V* with a swinging motion
-**wyngwang** *V* in a confused way

# 9 List of grammatical categories found in Atong and the morphemes associated with them

Table 18 presents an alphabetised list of grammatical categories found in Atong which are indicated by morphemes, together with the morphemes associated with these categories. These morphemes are suffixes (see van Breugel 2014: 369–375) and enclitics (see van Breugel 2014: 257 and 386). Note that this list does not contain descriptions of strategies other than morphological marking, which Atong may have to express certain grammatical categories. For more information on the morphemes associated with the grammatical categories listed here, as well as non-morphological strategies which limit the interlocutor's interpretation of an utterance in Atong, the reader is referred to van Breugel (2014). Moreover, the glossed and translated Atong texts with annotations in van Breugel (2019) illustrate how meaning in Atong arises in context.

**Table 18:** Grammatical categories and their morphemes.

| | |
|---|---|
| additive =ba | =edonga ~ =edong ~ =edok ~ =eronga ~ =erong ~ =erok |
| adverbial =ai ~ =e | |
| affirmation seeking =ne | emphatic mood =ba |
| affirmative mood =ba | epistemic mood =chym |
| alternative mood =sega ~ =siga | excessive degree -duga |
| approximation =darang ~ =dyrang | exclusive =tara |
| associative =para | factitive mood =wa ~ =a |
| attributive =gaba ~ =ga | focus =an |
| causative -et | frustrative mood =chym |
| change of state =ok ~=ak ~ =k | future (imperious/certain) mood =naka ~ =ka |
| comitative =mu ~ =myng ~ =mung | genitive =mi ~ =myng ~ =mung |
| comparative -khal | goal =na ~ ona |
| conative mood =chyi | identity =an |
| concomitant action =butung | imperative mood =bo |
| confirmative =mo | imperative emphasiser =to ~ =ta |
| contrastive topic (1) =e ~ =ai (2) =ba | imperious future mood =naka ~ =ka |
| customary aspect =a | implicative mood =chym |
| declarative mood (1) =te (2) =aro ~ =ro | incompletive status =khu |
| definiteness =aw ~ =taw | indefiniteness =ba |
| delimitative (1) =sa (2) =rara (3) =tara | individuation =aw ~ =taw |
| delimitative & mirative =mangmang | instrumental =sang |
| derelational =gaba ~ =ga | intensification (1) -bi (2) -phin (3) -syrang ~ -srang (4) -langlang (5) -saw (6) -teng (7) -su (8) -thyn (9) -tyng (10) -thyng (11) -pha (12) -jol (13) -phetphet (14) -sym (15) iskyn (16) ytykram•phinai |
| desiderative mood =na ~ =ona | |
| distributive =phek | |
| durative aspect =aidonga ~ =aidong ~ =aidok ~ =aronga ~ =arong ~ =arok ~ | |

**Table 18** (continued)

| | |
|---|---|
| interrogative mood  =ma | propriative[21]  =thang |
| irrealis  =chym | hearsay evidentiality  =no |
| irresultative  =chym | reciprocity  (1) =maran (2) -ruk |
| locative  (1) =chi (2) =sang | referentiality  =aw ~ =taw |
| mirative  =thai ~ =tyi ~ =syi | relational  =gaba ~ =ga |
| mobilitative  =sang | resultative status  -man• ~ -man |
| negative polarity  =cha | sequential  =myng ~ =mung ~ =mu ~ =mungna ~ muna |
| perlative  =tykyi | similative  =tykyi |
| personal pronoun plural  =tym | simplicitive status  =ari |
| plural number  =darang ~ =dyrang | speculative status  =khon |
| possessive  =myng ~ =mi ~ =mung | superlative degree  (1) =bat (2) =khal |
| privative  (1) =ri (2) =nyi ~ =ni | supposition  =chym |
| progressive aspect  =aidonga ~ =aidong ~ =aidok ~ =aronga ~ =arong ~ =arok ~ =edonga ~ =edong ~ =edok ~ =eronga ~ =erong ~ =erok | topic  (1) =do ~ =odo (given information) (2) =e (contrastive/new information) |
| prohibitive mood  (1) ta (2) =bai | uncertainty mood  =ni |

---

[21] The propriative enclitic marks a Possession, i.e. something that is possessed by someone. Contrastively, the genitive or possessive enclitic marks, among other things, a Possessor, i.e. someone or something possessing something.

# 10 List of idiomatic expressions

For grammatical information, see PART 5, §10.

**Table 19:** Idiomatic expressions.

| | |
|---|---|
| **balwa sak-** | to enjoy the wind |
| **biba jokgaba dam** | a wound |
| **chungai rai•cha** | I don't care. |
| **daikhalaisa-** | moreover. |
| **dong•arini ~ dongarini** | That's all right. That's okay. No worries. That will do. |
| **gol sa-** | to score a goal |
| **gumukan** | everybody, all (of them) |
| **ha• haw•ai sa•-** | to live from agriculture, to live from tilling the soil |
| **haw•ai kamai sa•-** | to work hard to survive |
| **haw•ai kamai sa•-** | to work hard to survive |
| **ja•ga sagaba** | enemy |
| **kata ~ khata ~ katha ~ khathajyksai** | collocation |
| **kata ~ khata ~ katha ~ khatha jyw•khynwa** | to tell long epic stories during the festival of chywgyn |
| **maidan syla thoka** | to celebrate the after-harvest festival |
| **mamung dong•cha** | It doesn't matter. |
| **man•ai sa-** | (1) to eat in great amounts (2) rich, wealthy |
| **mykren wa•thok song•phin-** | gazing in amazement |
| **ni•wa** | it's nothing, never you mind |
| **noga ~ nogaba** | so-called |
| **ra•ai sa-** | to marry off, to marry someone to someone else |
| **ronronok** | eyes almost closed because of tiredness |
| **sasyk sasyk tak-** | feeling unwell, feeling a small pain, feeling an urge |
| **te•ewrawraw ~ te•awrawraw** | (1) nowadays (2) up till now, still (3) not yet |
| **utyk udong** | let it be |
| **wa• sa•gaba** | though (of persons) |
| **warem sa•-** | to rust |
| **ytykkhal** | it doesn't matter |
| **ytykkram•phinai** | so (much) |

# 11 List of ideophones

For grammatical information, see PART 5, §11. The list presented here in Table 20 is a modified version of the one found in van Breugel (2014: 245–247). Note that this list is most probably incomplete. There are most probably many more ideophones in Atong that have not been recorded by the author. Future fieldwork will hopefully render this list more complete. Also note that the ideophones are presented here as they are recorded, and that different forms of the same ideophone may also exist. For example, the ideophone *thing thing thing* was recorded as such, i.e. with three iterations of *thing*. Atong speakers may very well also use *thing* and *thing thing*, even though these expressions were not recorded by the author.

**Table 20:** Ideophones.

### Animal sounds
**aak aak** caw, caw! the call of the crow
**baaa** Moo! the sound a cow makes
**bawbaw** woof! woof! the sound a dog makes
**budok budok** the call of the *daw•budok*
**chepchap chepchap ~ chepchep chepchap** squeak! the sound a mouse makes.
**chipchip** chirp! the call of a bird
**chutchut** squeak! the sound a mouse makes
**durrrmeme** eeeeee! the sound of a bleating goat
**gekgek** the call of the hornbill
**gepgep** quack, quack! the call of a duck
**kha** caw! the call of a black crow
**khrukhru** roo-koo! the call of a pigeon
**meee** meh-eh-eh! naa! the sound a goat makes
**mew** meow! the sound of a cat
**mmmm mmmm** screeeech! the call of an eagle.
**pekpek** croak! ribbit! the call of a frog
**wek** squeal! oink! the sound a pig makes
**ymbuuu** moo! the sound a cow makes

### Sounds made by humans
**•hm** grunt!
**•hmmmm** sigh! hmmm
**uph** oof! a puffing sound

### Sounds of the body
**phong** brap! blarp! the sound of someone farting
**gulgul galgal** growl! growling noise that the stomach makes

### Sounds having to do with falling
**chaw** splash! the sound of something plunging into the water.
**dam ~ dym** bam! thud! the sound of something heavy hitting the ground
**teng** plink! clang! plunk! ching ching! The sound of falling coins.
**thing thing thing** pling, pling, pling! clang, clang, clang! the sound of something falling made of metal
**wuuuuk** wooosh! the sound of something big falling down

### Hitting and beating sounds
**gyp** thunk!, tap!, bam! a hitting sound
**kak** slap! the sound of something hitting or slapping
**thap** hit! slap! a hitting sound
**thop** thok! thunk! a hitting sound:
**thup** thunk! slap! a beating sound

### Sounds having to do with movement
**kyryk** someone running, running and jumping like a deer

**Table 20** (continued)

| | |
|---|---|
| **thep gaw**  stamp, stamp! the sound of many animals stampeding through the forest | **them**  pow! bang! the sound of a gunshot |
| **thep thup**  clog clog. tap tap tap. | <u>ACTION IDEOPHONES</u> |
| **thup thup**  pitapat! the sound of footsteps | **krrrrr**  the sound of someone smoking viciously |
| <u>OTHER SOUNDS</u> | **wawa**  the sound of someone throwing something |
| **chyryt chyryt**  squirt, squirt! | |
| **thaw**  bang! pow! the sound of a gun firing | |

## 12 List of indefinite Proforms

For grammatical information, see PART 5, §12. Table 12 is adapted from van Breugel (2014: 172).

**Table 21:** Indefinite proforms.

| | |
|---|---|
| **atongba** | something |
| **bichiba** | somewhere, sometimes |
| **bimiba ~ bimyngba** | from somewhere |
| **bisangba** | to somewhere |
| **changba** | someone |
| **changgaba** | whoever |
| **darangba** | anybody, nobody, whoever |
| **je** | any, whichever, whatever |
| **jechiba** | anywhere, wherever |
| **jemi ~ jemyng** | any followed by a time noun in the locative |
| **jesangba** | to wherever |
| **jesykyn** | however much/many |

# 13 List of interjections

For grammatical information, see PART 5, §13. The list presented in Table 22 is a modified version of the one in van Breugel (2014: 247–249). This list is probably not complete. Future fieldwork will hopefully yield more data, with which this list can be rendered more complete.

**Table 22:** Interjections.

ACKNOWLEDGMENT
**o ~ ooo** oh (The Atong interjection is pronounced long, with rising intonation)

ANGER
**hyt** Hey! Used both as admonition and to repel someone or something
**hyt sala** Damn! You bastard!
**kha** General curse followed by family name, for example Kha Marak!
**sala** Damn! Idiot!
**tyi sala** Damn! You idiot!

ATTENTION SEEKING
**hu hu** Hello?
**o** hey!
**oi** Oi!

CALL OR REPEL ANIMALS
**ai ai ai** interjection to call a pig
**chui** interjection to repel a pig
**kutukutuk** interjection to call a dog
**ja** interjection to repel a cow
**khat** interjection to repel a dog
**tititi** interjection to call a chicken
**sa** interjection to repel a chicken
**pusipusi ~ puspus** interjection to call a cat
**sit** interjection to repel a cat

DEPRECIATION OR INDIGNATION
**bylongok** So stupid! Unbelievable!
**na** Nah!
**yis ~ hyits ~ hys ~ hyis ~ tyis ~ tys** What the...! Ugh! Yikes!

HESITATION OR PAUSE FILLING
**ba•** uhm, eh
**yh** uhm, eh
**yyy** uhm, eh

MUTUAL UNDERSTANDING AND CONCLUDING A CONVERSATION
**ba•** OK then
**de** OK then.
**ma•** Ok, Very well then.

REBUTTAL AND ADMONISHMENT
These can all be translated as 'Wow, wow, wow!', 'Wait a minute!' or 'Hey!'.
**ahyyy**
**hyw**
**hy•y•y•y**
**hy•y•y•yw**
**ho•o•o•o**

SELF-SATISFACTION
**ha** Ha!

SURPRISE
**ah** Oh!
**atyw** Huh?!
**ha•gylsak ~ ha•gyrsak** My goodness!
**hari ~ hare** Huh?
**ymyi** Huh?

SURPRISE, ASTONISHMENT AND ADMIRATION
All of these can be translated as 'wow!' or 'Wooow!'
**atyyyw**
**baaa**

**Table 22** (continued)

**baaapre**
**da•nang**

<u>SURPRISE, ASTONISHMENT, AMAZEMENT AND GRIEF</u>
All of these can be translated as 'Jeez!',
'Goodness!' Oh dear! or 'Huh?!'
**aia**
**aiaw**
**aiu**

# 14 List of interrogatives

For grammatical information, see PART 5, §14. Table 23 is adapted from van Breugel (2014: 161).

Table 23: Interrogatives.

| | |
|---|---|
| chang | who? |
| atong | what? |
| atongtykyi | why? how? |
| atongmai•nawhy? | for what purpose? |
| atakna ~ atana | why? for what purpose? |
| atakai ~ atykyi | how? |
| bie ~ bi- | which? where? |
| biskyn | how much, how many? |
| =byisyk | how much, how many? |
| biba | when? / in whatever place |
| bitykyi | by which way? |
| bichi | where? |
| bisang | to/from where? |
| bisangmi ~ bisangmyng | from where? |
| bimi ~ bimyng | from where? |
| bigaba | which? |
| INTERROGATIVE VERB | |
| atak- | to do what? |

# 15 List of particles

For grammatical information, see PART 5, §16.

**Table 24:** Particles.

| | |
|---|---|
| **chym** | irrealis |
| **khon** | speculative modality or status |
| **ma** | interrogative modality, or alternative conjunction |
| **mo** | confirmative tag |
| **ne** | affirmation seeking tag |

## 16  List of personal pronouns

For grammatical information, see PART 5, §17. The fourteen personal pronouns of Atong are listed in Table 25, according to their person and number, and are accompanied by their grammatical descriptions in brackets, and their English translations.

Table 25: Personal pronouns.

| Singular | | |
|---|---|---|
| ang ~ anga | (first person singular) | 'I/me' |
| nang• | (second person singular) | 'you' |
| na•a | (second person singular vocative) | 'you' |
| ge•theng ~ de•theng | (third person singular) | 'he/him/she/her/it' |
| phalthang | 'self' | |
| **Plural** | | |
| na•nang | (first person plural inclusive) | 'we/us (me and you)' |
| ning ~ ninga | (first person plural exclusive) | 'we/us (me and he/she/they)' |
| nang•tym | (second person plural) | 'you' |
| ge•thengtheng | (third person plural) | 'they/them' |
| itym | (third person plural proximate) | 'they/them' |
| utym ~ ytym | (third person plural distal) | 'they/them' |
| phalthangthang | 'selves' | |

## 17 List of postpositions

For grammatical information, see PART 5, §18.

**Table 26:** Positions with the markers of their complements.

| Marking of complement | Postpositions | Glosses |
|---|---|---|
| =mi ~ =myng (GEN) | dabat | 'since, from' |
| =chi=na (LOC=GOAL) | dabat | 'until' |
| NO MARKING | dabat | 'until' |
| =na (GOAL) | dakang | 'before' |
| =mi ~ =myng (GEN) | gymyn | 'because of, about' |
| =mi ~ =myng (GEN) | kynsang | 'after' |
| =mi ~ =myng (GEN) | thyl• | 'up to' |

# 18 List of proclauses

For grammatical information, see PART 5, §19. Table 27 is adapted from van Breugel (2014: 250).

**Table 27:** Proclauses.

| | |
|---|---|
| •mhm• ~ •mhmm• | That's right. |
| •mhm• ~ hm•m ~ m•m ~ •m ~ mm | I disagree. |
| am ~ ym | affirmative, okay, that's right |
| ha• | Take this. |
| hai | Let's go! Come on! |
| haida | I don't know. |
| ho•ong | I agree. That's right. |
| hy• ~ hy•y ~ yhy• | I disagree. |
| kyw | I am here! (answer to a search call) |
| tam•o ~ tam•aw | Wait! |

# 19 Lists of quantifiers

For grammatical information, see PART 5, §20. As mentioned in that section, two types of quantifiers can be distinguished in Atong, viz. numerals and non-numeral quantifiers. Within the numerals, there are the Atong numerals, presented in Table 28, the numerals borrwed from English, presented in Table 29, the numerals borrowed from Hindi, presented in Table 30. The non-numeral quantifiers are listed in Table 31.

Note that, as mentioned in PART 5, §20.1, the numerals *sa* 'one' *ni* 'two' and *tham* 'three' are bound morphemes. Therefore, in Table 28, these numerals are presented with the classifier *rong*.

The numerals above twenty are presented in different paragdigms according to their morphology. Some paradigms are a labeled modern or archaic, which refers to the status of their usage, which is explained in PART 5, §20.1.

**Table 28:** Atong numerals.

| | | | |
|---|---|---|---|
| 1 | rong sa | 11 | chit sa |
| 2 | rong ni | 12 | chi ni |
| 3 | rong tham | 13 | chi tham |
| 4 | byryi | 14 | chi byri |
| 5 | banga | 15 | charanga ~ chi banga |
| 6 | korok | 16 | chi dok |
| 7 | sene | 17 | chi sene ~ chi syni |
| 8 | chatgyk | 18 | chi chat |
| 9 | chykhyw | 19 | chi sykhu |
| 10 | chyigyk ~ chek | 20 | kholgyk ~ kholgryk ~ khol ~ rum•sa (archaic) |

| | MODERN | MODERN | ARCHAIC | ARCHAIC |
|---|---|---|---|---|
| 21 | kholgyk sa | kholgryk sa | khole rong sa | khole rong sa |
| 22 | kholgyk ni | kholgryk ni | khole rong ni | khole rong ni |
| 23 | kholgyk tham | kholgryk tham | khole rong tham | khole rong tham |
| 24 | kholgyk byryi | kholgryk byryi | khole byryi | khole rong byryi |
| 25 | kholgyk banga | kholgryk banga | khole banga | khole rong banga |
| 26 | kholgyk korok | kholgryk korok | khole korok | khole rong korok |
| 27 | kholgyk sene | kholgryk sene | khole sene | khole rong sene |
| 28 | kholgyk chatgyk | kholgryk chatgyk | khole chatgyk | khole rong chatgyk |
| 29 | kholgyk chykhyw | kholgryk chykhyw | khole chykhyw | khole rong chykhyw |

**Table 28** (continued)

| | ARCHAIC (vigesimal 1) |
|---|---|
| 21 | rum• sa rong sa |
| 22 | rum• sa rong ni |
| 23 | rum• sa rong tham |
| 24 | rum• sa byryi |
| 25 | rum• sa banga |
| 26 | rum• sa korok |
| 27 | rum• sa sene |
| 28 | rum• sa chatgyk |
| 29 | rum• sa chykhyw |

| | MODERN | ARCHAIC | ARCHAIC (vigesimal 1) |
|---|---|---|---|
| 30 | kholachi | kholechyi | rum•sa chyigyk |
| 31 | kholachi sa | khole chit sa | rum•sa chit sa |
| 32 | kholachi ni | khole chi ni | rum•sa chi ni |
| 33 | kholachi tham | khole chi tham | rum•sa chi tham |
| 34 | kholachi byri | khole chi byri | rum•sa chi byri |
| 35 | kholachi banga | khole charanga | rum•sa charanga |
| 36 | kholachi dok | khole chi dok | rum•sa chi dok |
| 37 | kholachi sene | khole chi sene | rum•sa chi sene |
| 38 | kholachi chat | khole chi chat | rum•sa chi chat |
| 39 | kholachi sykhyw | khole chi sykhu | rum•sa chi sykhu |

| | MODERN | ARCHAIC (vigesimal 1) |
|---|---|---|
| 40 | sot byri | rum• ni |
| 41 | sot byri sa | rum• ni rong sa |
| 42 | sot byri ni | rum• ni rong ni |
| 43 | sot byri tham | rum• ni rong tham |
| 44 | sot byri byryi | rum• ni byryi |
| 45 | sot byri banga | rum• ni banga |
| 46 | sot byri korok | rum• ni korok |
| 47 | sot byri sene | rum• ni sene |
| 48 | sot byri chatgyk | rum• ni chatgyk |
| 49 | sot byri chykhyw | rum• ni chykhyw |
| 50 | sot bonga | rum• ni chyigyk |
| 51 | sot bonga sa | rum• ni chit sa |
| 52 | sot bonga ni | rum• ni chi ni |

**Table 28** (continued)

| | | | |
|---|---|---|---|
| 53 | sot bonga tham | rum• ni chi tham | |
| 54 | sot bonga byryi | rum• ni chi byri | |
| 55 | sot bonga banga | rum• ni charanga | |
| 56 | sot bonga korok | rum• ni chi dok | |
| 57 | sot bonga sene | rum• ni chi sene | |
| 58 | sot bonga chatgyk | rum• ni chi chat | |
| 59 | sot bonga chykhyw | rum• ni chi sykhu | |
| 60 | sot dok | rum• tham | |
| 61 | sot dok sa | rum• tham rong sa | |
| 62 | sot dok ni | rum• tham rong ni | |
| 63 | sot dok tham | rum• tham rong tham | |
| 64 | sot dok byryi | rum• tham byryi | |
| 65 | sot dok banga | rum• tham banga | |
| 66 | sot dok korok | rum• tham korok | |
| 67 | sot dok sene | rum• tham sene | |
| 68 | sot dok chatgyk | rum• tham chatgyk | |
| 69 | sot dok chykhyw | rum• tham chykhyw | |
| 70 | sot sene ~ sot syni | rum• tham chyigyk | |
| 71 | sot sene sa | rum• tham chit sa | |
| 72 | sot sene ni | rum• tham chi ni | |
| 73 | sot sene tham | rum• tham chi tham | |
| 74 | sot sene byri | rum• tham chi byri | |
| 75 | sot sene banga | rum• tham charanga | |
| 76 | sot sene korok | rum• tham chi dok | |
| 77 | sot sene sene | rum• tham chi sene | |
| 78 | sot sene chatgyk | rum• tham chi chat | |
| 79 | sot sene chykhyw | rum• tham chi sykhu | |
| | <u>MODERN</u> | <u>ARCHAIC</u> (vigesimal 1) | <u>ARCHAIC</u> (vigesimal 2) |
| 80 | sot chet | rum• byryi | khol chang byryi |
| 81 | sot chet sa | rum• byryi rong sa | khol chang byryi rong sa |
| 82 | sot chet ni | rum• byryi rong ni | khol chang byryi rong ni |
| 83 | sot chet tham | rum• byryi rong tham | khol chang byryi rong tham |
| 84 | sot chet byryi | rum• byryi rong byryi | khol chang byryi rong byryi |
| 85 | sot chet banga | rum• byryi banga | khol chang byryi banga |
| 86 | sot chet korok | rum• byryi korok | khol chang byryi korok |
| 87 | sot chet sene | rum• byryi sene | khol chang byryi sene |

Table 28 (continued)

| | | | |
|---|---|---|---|
| 88 | sot chet chatgyk | rum• byryi chatgyk | khol chang byryi chatgyk |
| 89 | sot chet chykhyw | rum• byryi chykhyw | khol chang byryi chykhyw |
| 90 | sot sykhu | rum• byryi chyigyk | khol chang byryi chyigyk |
| 91 | sot sykhu sa | rum• byryi chit sa | khol chang byryi chit sa |
| 92 | sot sykhu ni | rum• byryi chi ni | khol chang byryi chi ni |
| 93 | sot sykhu tham | rum• byryi chi tham | khol chang byryi chi tham |
| 94 | sot sykhu byryi | rum• byryi chi byri | khol chang byryi chi bri |
| 95 | sot sykhu banga | rum• byryi charanga | khol chang byryi charanga |
| 96 | sot sykhu korok | rum• byryi chidok | khol chang byryi chi dok |
| 97 | sot sykhu sene | rum• byryi chi sene | khol chang byryi chi sene |
| 98 | sot sykhu chatgyk | rum• byryi chi chat | khol chang byryi chi chat |
| 99 | sot sykhu chykhyw | rum• byryi chi sykhu | khol chang byryi chi sykhu |
| | MODERN | ARCHAIC (vigesimal 1) | |
| 100 | raja sa | rum• banga | |
| 101 | raja sa rong sa | rum• banga rong sa | |
| 102 | raja sa rong ni | rum• banga rong ni | |
| 103 | raja sa rong tham | rum• banga rong tham | |
| 104 | raja sa rong byryi | rum• banga rong byryi | |
| 105 | raja sa rong banga | rum• banga rong banga | |
| | etc. | etc. | |
| 115 | raja sa charanga ~ raja sa chi bonga | | |
| | etc. | | |
| 200 | raja ni | | |
| 300 | raja tham | | |
| | etc. | | |
| 1,000 | hajal ~ hajal sa | | |
| 1,101 | hajal sa raja sa rong sa | | |
| | etc. | | |
| 2,000 | hajal ni | | |
| | etc. | | |
| 7,895 | hajal sene raja chatgyk khol chang byryi charanga ~ | | |
| | hajal sene raja chatgyk sot sykhu banga | | |
| 1,00,000 | lak ~ lak sa | | |
| 2,50,675 | lak kholgyk banga raja korok sot syni banga ~ | | |
| | lak khole banga raja korok rum• tham banga | | |

**Table 28** (continued)

| | |
|---|---|
| 1,00,00,000 | kror ~ kror sa |
| 15,59,81,733 | kror charanga lak sot bonga chykhyw hajal khol chang byryi rong sa raja sene kholachi tham ~ kror charanga lak rum• ni chi sykhu hajal rum• byryi rong sa raja sene khole chi tham |

**Table 29:** Numerals borrowed from English.

| | | | | |
|---|---|---|---|---|
| 0 | jero | | 10 | ten |
| 1 | wan | | 11 | ileben |
| 2 | tu | | 12 | twelp |
| 3 | tri | | 13 | tyrtin |
| 4 | por | | 14 | portin |
| 5 | paip | | 15 | piptin |
| 6 | siks ~ sik | | 16 | sikstin ~ siktin |
| 7 | seben | | 17 | sebentin |
| 8 | eet | | 18 | eetin |
| 9 | nain | | 19 | naintin |

| | |
|---|---|
| 20 | twenti |
| 21 | twenti wan |
| 30 | tyrti |
| 40 | porti |
| 50 | pipti |
| 60 | sikti |
| 70 | sebenti |
| 80 | eeti |
| 90 | nainti |
| 100 | wan handryt |
| 1000 | wan tawsyn |
| 1000000 | wan milion |

## 19 Lists of quantifiers

**Table 30:** Numerals borrowed from Hindi.

|   | Atong | Hindi |    | Atong | Hindi |
|---|-------|-------|----|-------|-------|
| 1 | eek | एक | 7 | saat | सात |
| 2 | do | दो | 8 | aat | आठ |
| 3 | tiin | तीन | 9 | no | नौ |
| 4 | cha ~ char | चार | 10 | dys ~ das | दस |
| 5 | panch | पाँच | 11 | giara | ग्यारह |
| 6 | che | छः | 12 | bara | बारह |

**Table 31:** Non-numeral quantifiers.

| | |
|---|---|
| =byisyk | how much/many? (interrogative quantifier enclitic) |
| =phek | each (distributive enclitic) |
| =gumuk | all, whole, everything, everyone |
| =khakhet ~ =khakhet | all, whole, everything, everyone |

## 20 List of time words

For grammatical information, see PART 5, §21.[22] Table 32 is adapted from van Breugel (2014: 155).

**Table 32:** Time words.

| NON-DEICTIC | |
|---|---|
| bylsi | year |
| gasam | evening |
| gasamphang ~ gasamphak | afternoon, evening, later part of the day |
| ja | month |
| khantha | hour |
| manap | morning |
| minit ~ minyt | minute |
| nygyltyi | week |
| san | day |
| sanmaji | midday, noon |
| sekyn | second |
| wal | night |
| walsymsym | twilight, dusk |
| **DEICTIC** | |
| dakang | before, in the past, earlier |
| kynsang | later, after |
| teraka | last year |
| maja | the day before yesterday or longer ago |
| myia | yesterday |
| taija | last night |
| tai•nep | this morning |
| tai•sa | a moment ago, just now, a little while ago |
| te•ew | now |
| tai•ni | today |
| tarai | this year |
| te•en | later but still today |
| hampyi | later today, in the evening |
| hanep | tomorrow |
| cheknai | the day after tomorrow |
| hambun | later but not today, in the far future |
| naija | next year |

---

[22] The word *sirimynmyn* 'at the crack of dawn' is mentioned as a non-deictic time word in van Breugel (2014: 155), but has since been reclassified as an adverb (see §3 in this part of the book).

# 21 List of intransitive and transitive verbal pairs

The transitive verbs in Table 33 are all derived from their intransitive counterparts by a fossilised morphological prosses involving a prefix, except for the pair *bai•-/phai•-* 'to break'. Note also that the word *pyleng* was only recorded as a Type 2 adjective. Its transitive counterpart is a verb. For more details, the reader is referred to van Breugel (2014: 82–83).

**Table 33:** Intransitive and transitive verb pairs (adapted from van Breugel 2014: 83–84).

| INTRANSITIVE | TRANSITIVE |
| --- | --- |
| **ba•-** to carry a child (in a cloth on the body) | **thaba•-** to make someone carry a child (in a cloth on the body) |
| **bai•-** to break | **thabai•-** to break |
| **bai•- ~ bai** to break | **phai•-** to break, to translate |
| **barat-** to be ashamed | **thabarat-** to make ashamed |
| **bejaw-** to feel the sensation of being tickled | **thebejaw- ~ thebajaw-** to tickle someone |
| **gal•-** to fall | **thagal•-** to drop |
| **kung-** to be dammed up by a circle of stones | **dukung-** to dam up water by making a circle of stones in the river |
| **ma-** to lose | **thama-** to make lost |
| **myt-** to extinguish, to go out (of fire) | **thymyt-** to extinguish, to put out (of fire) |
| **mimi-** to smile | **thimimi-** to make someone smile |
| **myn-** to be ripe | **thymyn-** to ripen |
| **nuk-** to see | **thunuk-** to show |
| **sa-** to wake up | **thasa-** to wake somebody up |
| **khyp- ~ khup-** to put clothes on, to cover, | **dykhyp- ~ dukhup-** to dress someone |
| **kyryng-** to make noise | **dykyryng-** to make noise on purpose |
| **kyryi-** to fear | **dykyryi-** to threaten |
| **khep-** to cry | **dekhep-** to make someone cry |
| **thyi-** to die | **dythyi-** to kill |
| **kirin-** to tear (of clothes, paper etc.) | **dikirin-** to tear (clothes, paper etc.) |
| **phing-** to be full | **diphing-** to fill |
| **pyleng** (*adj2*) flat | **dypyleng-** to flatten |

## 22 List of verbs of emotion and interaction

For grammatical information, see PART 5, §15. The list in Table 34 contains all recorded verbs of emotion and interaction, which are marked *vgoal* in the Atong-English dictionay.

**Table 34:** Verbs of emotion and interaction.

| | |
|---|---|
| bam- | obey |
| barat- | be ashamed of |
| chanpheng- | defend |
| jajyreng- | worry about |
| kha•dang- | care about/for |
| kha•gal- | love |
| kha•phak- | miss |
| kyryi- | fear, be afraid of |
| pai- | support |
| rakhi- | to protect against |
| sa•nyl- ~ sa•nal- | be jealous of |
| sak- | depend on |
| sari- | shun, ignore |
| sungman- | remember, think of |
| symsak- | care about/for |

# PART 5: COMPENDIUM OF ATONG GRAMMAR

The purpose of this compendium is to provide succinct information about the Atong word classes, collocations, idiomatic listed in this volume, so that the reader can use it without van Breugel (2014) immediately at hand. Some new insights into the grammar are presented here as well, complementing or correcting van Breugel (2014). After the overview of Atong word classes in Section 1, all subsequent topics are presented alphabetically.

## 1 Overview of the word classes of Atong

Word classes and their sub-classes are defined and described in detail in van Breugel (2014: Chapters 3–15). Lists of members of the various word classes, are presented in PART 3 of this book, together with succinct descriptions of the word classes. What is presented here is a brief overview of the word classes in Atong, All word-class distinctions are made on the basis of grammatical properties.

The Atong lexicon can be divided into two basic categories, viz. words that can occur as the head of the predicate of a clause, and words that cannot. All the words that can occur as a predicate head together form the Predicative Word Classes. All other words belong to the Non-Predicative Word Classes.

The Predicative Word Classes can be subdivided into the classes listed in Table 35, whereas the Non-Predicative Word classes are listed in Table 36. In both tables, the specific chapter or section in van Breugel (2014) where the word classes are described is provided for each word class. For more information on the predicate in Atong, the reader is referred to van Breugel (2014: Chapter 20).

**Table 35:** The Predicative word classes.

| | |
|---|---|
| Classifiers (Chapter 12) | Time Words (§8.6) |
| Demonstratives (§8.2) | Type 1 Adjectives (§5.2) |
| Interrogatives (Chapter 9) | Type 2 Adjectives (§5.3) |
| Nouns (Chapter 6) | Verbs (§4.1) |
| Personal Pronouns (§8.5) | |

**Table 36:** The non-predicative word classes.

| | |
|---|---|
| Adverbs (§15.1) | Particles (§22.11–13, §23.3.4) |
| Discourse Connectives (Chapter 14) | Postpositions (Chapter 13) |
| Indefinite Proforms (Chapter 10) | Proclauses (§15.5) |
| Ideophones (§15.3) | Quantifiers (Chapter 11) |
| Intensifiers (§15.1) | Conjunctions (§15.2) |
| Interjections (§15.4) | The Prohibitive Word *ta* (§23.3.5vi) |
| Numerals (Chapter 11) | |

## 2 Adjectives

There are two types of adjectives in Atong, viz. Type 1 and Type 2. Type 1 adjectives, listed in Table 5, are a subclass of stative verbs (van Breugel, 2014: 92), i.e. verbs which denote a state of being. Thus, Type 1 adjectives are a predicative word class (see van Breugel 2014: Chapter 3). Type 2 adjectives, listed in Table 7, are also a predicative word class, but with different properties than Type 1 adjectives. The different semantic categories of Type 1 and Type 2 adjectives are presented in Table 37.

**Table 37:** The semantic categories found in Type 1 and Type 2 adjectives.

| TYPE 1 ADJECTIVES | TYPE 2 ADJECTIVES |
|---|---|
| Colour | Age |
| Dimension | Bodily Condition |
| Mental Property | Colour |
| Physical Property | Mental Property |
| Position | Physical Property |
| Quantification | Position/Order |
| Speed | Possessive |
| Taste | Similarity |
| Temporal | Speed |
| Value | Temporal |
| | Value |

Type 1 adjectives are distinguished from other stative verbs (listed in Table 6) based on the different semantic effects the change-of-state enclitic =*ok* ~ =*ak* (COS) can have on the members of these word classes. When the change-of-state enclitic =*ok* ~ =*ak* occurs on predicative words that are not Type 1 adjectives, the change of state marker can only be interpreted as indicating a change of state, i.e. something that was not the case before, is now the case. Examples of this interpretation are *walangok* (night-away=COS) 'it has become night', and *re•engok* 'has/have left'. However, on Type 1 adjectives, the change of state marker can either be interpreted as indicating a change of state, or as an intensifier. Which interpretation is required depends on the context in which the Type 1 adjective occurs. For example, *syl=ok* (beautiful=COS) can mean either 'has become beautiful', in the Change of state interpretation, or 'very beautiful', in the Intensifier interpretation.

There are two more ways in which Type 1 adjectives differ from Type 2 adjectives, both having to do with their ability or inability to fulfil certain grammatical functions. First, Type 1 adjectives need to take predicative enclitics to function as predicates, whereas Type 2 adjectives can function as predicates without such enclitics, i.e. in their root form (see van Breugel 2014: Chapter 22 on predicative enclitics). In the sentence *nang•=mi nok syl=a* (2SG=GEN house beautiful=CUST) 'your house is beautiful', the Type 1 adjective *syl* is used with the customary aspect predicate enclitic *=a* (CUST). It would be ungrammatical to use the adjective without this or any other predicate enclitic, e.g. *\*nang•mi nok syl*. Type 2 adjectives can be predicates without such enclitics, as does the Type 2 adjective *thymbylong* 'have holes in it', as in *ie ram thymbylong* (PROX road have.holes.in.it) 'this road has holes in it'.

The fact that predicate enclitics can only be attached to the Type 2 adjective and not the noun phrase *ie ram* is proof of the fact that *thymbylong* is indeed the predicate, e.g. *ie ram thymbylong=an=cha* (PROX road have.holes.in.it.=FOC=NEG) 'This road does not have holes in it'. In this example, the predicate enclitics *=an* (FOC) and *=cha* (NEG) are attached to the predicate *thymbylong*. Even when the noun phrase is moved to the position for given information, i.e. after the predicate, would the predicate enclitics still be on the word *thymbylong*, e.g. *thymbylongancha ie ram* (have.holes.in.it.=FOC =NEG PROX road) '[it] does not have holes in it, this road'.

Second, Type 1 adjectives cannot function as modifiers within a noun phrase without being marked by the attributive enclitic *=gaba ~ =ga*,[23] e.g. *Nilam=do gari syl=gaba ra•=wa* (Nilam=TOP beautiful=<u>ATTR</u> car buy=FACT) 'Nilam bought a beatiful car'. In this example, the head of the noun phrase, *gari* 'car' is modified by the Type 1 adjective *syl-* 'beautiful' with the help of the attributive enclitic *=gaba*.

---

[23] See van Breugel (2014: 297–306) and van Breugel (2010) on attributive clauses and the enclitic *=gaba ~ =ga*.

Contrastively, Type 2 adjectives can function as adnominal modifiers without this enclitic, i.e. in their root form, e.g. *Nilam=do gari pidan ra•wa* (Nilam=TOP car new buy=FACT) 'Nilam bought a new car'. The head of the noun phrase, *gari* 'car', is modified by the Type 2 adjective *pidan* 'new' without the help of any morphological marking.

Note that the both Type 1 and 2 adjectives, when they are being used as modifiers within a noun phrase, can occur on either side of the head without any difference in meaning.

## 3 Adverbs

Adverbs are listed in Table 8. Adverbs modify following predicates. Adverbs can be divided into semantic types based on their denotation or function. The following semantic types can be discerned in Atong: Location in Space, Location in Time, Frequency, Quantity, Duration, Status, Manner and Intensification.

There is no derivational morphology applicable to adverbs. Only very few instances of adverbs hosting a phrasal enclitic are attested in the recorded materials. Adverbs can consist of one phonological word, e.g. *dykdyk* 'for a while', or two phonological words, e.g. *byldyng byldang* 'all over the place'. Adverbs consisting of more than two phonological words are not attested. If an adverb consists of two phonological words, these cannot be separated by extraneous materials, but function as one grammatical words. Adverbs are the only word class where Type 4 collocations (see §5 in this part of the book) can be found, in which both members are decorative words, without any denotation of their own, e.g. *byldyng byldang*.

## 4 Classifiers and volume words

### 4.1 Classifiers

Classifiers are listed in PART 4, §4. A classifier can be defined as any word that can occur in the classifier slot in a quantifier phrase. A quantifier phrase (QP) is a phrase that consists of a classifier and a quantifier, in that order (see van Breugel 2014: Chapter 11 for more details). A quantifier phrase either precedes or follows a noun within a nominal clump.[24] The function of a quantifier phrase is, as the

---

[24] Van Breugel, (2014: 98) states that "[t]he Nominal Clump is a referential unit that consists of one or more phrases, none of which is obligatory and 'each is perfectly capable of occurring

name suggests, to quantify a referent. Quantification means telling how much of something there is, or how many there are. The quantified referent can be explicitly stated by a noun phrase (NP) within the nominal clump, as in (2), or it can be stated more vaguely just by the quantifier phrase itself, as in (3). In the latter case, the classifier provides some information, although not very precise, about the category of thing the quantified referent belongs to.

In example (2), the quantified referent is expressed by the noun phrase, which consists only its head noun, *sa•* 'child'. The quantifier phrase follows the noun phrase, and consists of the classifier for humans, *myng•*, and the quantifier *sene* 'seven'. Together, the noun phrase and the quantifier phrase make up the nominal clump.

(2) *Gorialchie sa• myng• sene ganangno.*

|  |  |  | NOMINAL CLUMP | | | | |
|---|---|---|---|---|---|---|---|
| Gorial | =chi | =e | [*sa•*]<sub>NP</sub> | [*myng•* | +*sene*]<sub>QP</sub> | *ganang* | =*no.* |
| crocodile | =LOC | =CT | child | CLF:HUMANS | +seven | exist | =QUOT |

'The crocodile had seven children, it is said.'

In example (3), reference to what is quantified is only provided by the classifier in the quantifier phrase. All the information the listener gets is that the referents are human. The actual referent of the quantifier phrase in the example is one of the six children of a single mother. The referent himself is never explicitly referred to in the story. the listener (or reader) can infer that he exists, because explicit reference to a group of sons, of which he is a member, is made at the beginning of the story, which is about a hundred sentences earlier than the one in example (3) (see van Breugel 2019: 220–264).[25] This proves very clearly that quantifier phrases without accompanying noun phrases are used to refer in Atong.

---

alone' (Haiman 2011: 141). The order of the elements within the nominal clump is not entirely fixed.

Since all the phrases in the Clump can occur on their own and refer on their own, it is not possible to say what the head of the nominal clump is. It is possible to say what the heads of the phrases are, namely either deictic words or the words that denote the entity that the phrase is about, which may be overt or may have to be inferred." All phrases within the nominal clump have the same referent, i.e. they co-refer.

[25] Atong thus has four ways to deal with topical referents, i.e. referents which are considered given information (also called information within the pragmatic presupposition). A clause can lack a referential expression altogether, or mention the topical referent the end of the clause, after the predicate (see van Breugel 2014: 420 for both these strategies, see van Breugel 2010 for details on omission of referential expressions). Alternatively, topical referents can be expressed

(3)  *Myng• sa tem! kawoknotyi!*
    [*Myng•*        +*sa*]<sub>QP</sub>   *them!*       *kaw*    =*ok*    =*no*    =*tyi*.
    CLF:HUMANS   +one         IDEO:pow   shoot   =COS    =QUOT   =MIR
    'One of them, pow! shot [at it], it is said, to [our] surprise.'

Some classifiers are used with many different nouns, while others are only used with a few, or even one specific noun. Thus, the specificity of the referent of the quantifier phrase depends on the specificity of the classifier. The classifier *myng•*, for example, can be used with any noun denoting a human being, and even anthropomorphised animals in fables, while the classifier *ching* is only used with the noun *mai•wa* 'bamboo shoot'. More general classifiers are thus more context dependent than more specific ones when it comes to retrieving a possible referent when one is used without an accompanying noun phrase.

A quantifier phrase can either precede or follow the noun phrase it quantifies, without a difference in meaning. In the majority of recorded cases, the quantifier follows the noun phrase, and this construction is therefore considered unmarked. The order NP-QP is more rare, and therefore considered marked. When the quantifier phrase precedes the noun phrase, the referent of the noun phrase is always considered known information, which could also have been omitted, and the focus of the construction lies on the quantifier phrase, which conveys the new information. When the quantifier follows the noun phrase, either all the information in both phrases is new, or the focus lies on the noun phrase, or on the quantifier phrase, depending on the context. Focus can also be obtained by extra stress on the element that needs to be in focus.

In the first sentence of example (4), the quantifier phrase (QP) follows the noun phrase (NP) it quantifies, which is the unmarked scenario. This is the opening sentence of a story, and all the information in the sentence is considered to be new. When the information that the crocodile has many children is well established, the second time those children are quantified, in Sentence 3, the quantifier phrase precedes the noun phrase. The referent of the noun phrase *nang• sa•* 'your children' is known information. What is new, and therefore in focus in the marked construction QP-NP, is its quantification. It would also have been possible to omit the noun phrase altogether.

---

clause initially, marked by the topic enclitic =*do* (TOP). Finally, when the topical referent is quantified, reference may only be made by means of a classifier in a quantifier phrase.

(4) Te•ewba gorialchie sa• myng• sene ganangno. Pherue balokno: "[...], nang•chido sa• pang•ate. Angna myng• tham nang• sa•aw poraikhalna watetboto."

Sentence 1.  Te•ew   =ba   gorial      =chi   =e
             now     =CT   crocodile   =LOC   =CT

[sa•]_NP  [myng•          +sene]_QP  ganang   =no.
child     CLF:HUMAns      +seven     exist    =QUOT

Sentence 2.  Pheru   =e    bal    =ok     =no:
             fox     =CT   say    =COS    =QUOT

nang•   =chi   =do    sa•     pang•        =a      =te.
2SG     =LOC   =TOP   child   be.many      =CUST   =DECL

Sentence 3.  "Ang   =na
             1SG    =GOAL

NOMINAL CLUMP
[myng•         +tham]_QP   [nang•   sa•]_NP   =aw
CLF:HUMANS     3            2SG     child     =ACC

porai   -khal    =na     watet   =bo    =to."
study   -MORE   =GOAL    send    =IMP   =IMPEMPH

'Now, the crocodile had seven children, it is said. So then, a fox came, it is said. The fox spoke, it is said: "Hey, friend, you have many children! Send me three of your children to me to study!"'

The quantified noun phrase and the quantifier phrase can be marked separately by phrasal enclitics, which proves that they are separate phrases. In (5), the quantified noun phrase *ang ma•su* is marked by the accusative enclitic *=aw*, while the quantifier phrase *mang ni* is unmarked. In (6), the quantified noun phrase *jyk* is marked by the phrasal enclitic *=ba*, while the quantifier phrase is unmarked.

(5) Uchie: "Ang ma•suaw mang ni ra•wa, ..."
    Uchie:  "[Ang   ma•su]_NP   =aw    [mang          +ni]_QP   ra•    =wa,
    then    1SG     cow         =ACC   CLF:ANIMALS    +two      get    =get
    'Then: "[He] bought my two cows [, and...]'

(6) Ytykyimyng jykba myng• ni khymanoro.
    Ytykyimyng   [jyk]_NP   =ba       [myng•        +ni]_QP
    so.then      spouse     =ADD/CT   CLF:HUMANS    +two

    khym    =a       =no      =ro.
    marry   =FACT    =QUOT    =DECL
    'So then, [he] is married to two wives, it is said.'

Contrastive to the examples above are cases where the nominal clump is marked as a whole, which means that the marking occurs at the end of the referential expression. Illustrative of this way of marking are Sentence 3 in example (4), where the nominal clump *myng• tham nang• sa•* is marked with the accusative enclitic =*aw*, whose host is the head of the quantified noun phrase, i.e. the noun *sa•* 'child'. Another example is (7), where the nominal clump *ge•theng ma•su mang byryi* is marked by the accusative suffix, which attaches to the quantifier phrase.

(7) *ge•theng ma•su mang byryiaw*

| NOMINAL CLUMP | | | | |
|---|---|---|---|---|
| [ge•theng | ma•su]$_{NP}$ | [mang | +byryi]$_{QP}$ | =aw |
| 3SG | COW | CLF:ANIMALS | +four | =ACC |

The classifier or auto-classifier and the numeral form one phonological word.[26] Usually, in the recorded data, a single quantified noun and the following quantifier phrase form one phonological word together. This happens both when the quantified noun has one syllable, as in (8), or two syllables, as in (9). In (8), the quantified noun is the monosyllabic word *song* 'village'. The stress is on the classifier in this case. In (9), the quantified noun is the bisyllabic word *chyn•thai* 'melon'. The stress is on the numeral *ni* 'two' in this case. Note that stress assignment in Atong is not fully understood by the author, and more research is needed to provide a good description of it.

(8) *song dam sachi.*

| PHONOLOGICAL WORD | | | |
|---|---|---|---|
| [song]$_{NP}$ | +[dam | +sa]$_{QP}$ | =chi. |
| village | +CLF:VILLAGES | +one | =LOC |

'in a village'

(9) *chyn•thai rong ni*

| PHONOLOGICAL WORD | | |
|---|---|---|
| [chyn•thai]$_{NP}$ | +[rong[27] | +ni]$_{QP}$ |
| melon | +CLF:ROUND.THINGS | +two |

'two melons'

---

[26] If a quantified noun precedes the quantifier phrase, the stress of this phonological word is usually on the last syllable, but can occur on any other syllable as well, for reasons not fully understood by the author. More fieldwork research is needed here.

[27] IIn van Breugel (2019: 78, Sentence 49), the plus sign was unfortunately erroneously omitted before *rong* and before its gloss.

When the quantified noun is more than two syllables, it may constitute a separate phonological word from the following quantifier phrase, as is illustrated in example (10). In that example, the compound noun *mongmawa* is one phonological word, and the quantifier phrase is another

(10)  *mongmawa dora byryi*

| PHONOLOGICAL WORD | PHONOLOGICAL WORD |
|---|---|
| [*mongma+ wa*]<sub>NP</sub> | [*dora +byryi*]<sub>QP</sub> |
| elephant+ tooth | 5kg +four |

'the twenty kilogram weighing elephant tusks'

However, in Atong orthography, the quantified noun, the classifier and the numeral are written as separate orthographic words, as can be seen in the first tier of each example (see also van Breugel 2015a). Only non-numeral quantifiers (see §20.2 in this part of the book) are written together with the preceding classifier.

An auto-classifier is a noun that can be quantified without the help of a separate classifier. The syntagmatic construction is NOUN<sub>AUTO-CLASSIFIER</sub>+NUMERAL, e.g. *bylsi sa* (year one) 'one year'. When the quantified noun is an auto-classifier, it is part of the quantifier phrase. The reason for the this syntactic classification is the fact that quantified auto-classifiers cannot host phrasal enclitics. Any enclitic necessary to mark the quantified auto-classifier gets attached to the end of the quantifier phrase, e.g. [*san+sa*]<sub>QP</sub>=*chi* (day+one=LOC) 'one day', where the locative enclitic =*chi* comes after the numeral *sa*, at the end of the quantifier phrase, in which *san* 'day' is the auto-classifier.

## 4.2 Volume words

Volume words are lexemes, which are used to refer to both receptacles and their volumes. When used to refer to a receptacle, these words behave like regular nouns, which can be used as the head of a noun phrase, can be possessed or otherwise modified, or quantified with the use of a classifier (either *thai•* or *goi•*).

When volume words are used to quantify the volume of something, they are used in the classifier slot of a quantifier phrase (see previous section), e.g. [*cha*]<sub>NP</sub> [*khap+tham*]<sub>QP</sub> (tea cupful+three) 'three cups of tea'. The most frequently used volume words are listed in PART 4, §4, Table 10. For a list of receptacles, see PART 3, §5.12.6.

## 5 Collocations

A list of all recorded collocations organised by type can be found in, Section 5 of this book. Collocations are not treated in *A grammar of Atong* (van Breugel 2014), but are discussed in some detail in van Breugel (2019: 484–485), where rigorous systematic analysis and description was first attempted.

Collocations are words that occur in pairs, either juxtaposed in the same phrase, as in example (11), or in different but juxtaposed phrases, as in (12), or in different but juxtaposed clauses, as in (13). The members of collocations are not chosen at random, but rather by convention, i.e. when two word form a collocation, it is two particular words that are usually used together. Except for the decorative words in decorative collocations (treated below), the words in other types of collocation can also be used separately as single words. The semantic relationships between them fall into four types, which are outlined below.

(11)   <u>gam</u> jymaw khaiaimu
       COLLOCATION
       [*gam*   *jym*]$_{PHRASE}$   =*aw*   *khai*                    =*ai*    =*mu*
       wealth  riches              =ACC   carry.from.forehead       =ADV    =SEQ
       '[he] carried the wealth and riches'

(12)   <u>bai•</u>na <u>tyng</u>na baratai
       NOMINAL CLUMP
       FIRST MEMBER              SECOND MEMBER
       [*bai•*]$_{PHRASE.\#1}$   =*na*    [*tyng*]$_{PHRASE.\#2}$   =*na*    *barat*       =*ai*
       blood.relative            =GOAL    DECO                      =GOAL    be.ashamed    =ADV
       '[he] shied away from his blood relatives, and [...]'

(13)   <u>byl</u> neng•chiba <u>chak</u> neng•chiba
       FIRST MEMBER
       [*byl*            *neng•*    =*chi*   =*ba*]$_{CLAUSE.1}$
       muscle            tired      =LOC    =CT

       SECOND MEMBER
       [*chak*           *neng•*    =*chi*   =*ba*]$_{CLAUSE.2}$
       hand/arm          tired      =LOC    =CT
       'if your muscles are tired, if your arms/hands are tired'

Collocations in Atong differ from compounds both morphologically and phonologically. Compounds consist of two or more roots, which together form one grammatical and phonological word. The two members of a collocation do not enter into such a tight bond, but remain separate grammatical and phonological words. Hence, the members of a collocation are written as two separate orthographic words (i.e. with a space in between them), e.g. *bagan bari* (garden garden) 'garden'; whereas the members of a compound are written as one orthographic word (without space between them), e.g. *achuambi* (*achu+ambi*) 'ancestors'. For more information on stress in Atong, the reader is referred to van Breugel (2014: 37–38 and 2019: 17–19). There are four different types of collocations, which will be treated one by one below.

Type 1, synonymous collocations, are those where the two members of the collocation have similar meanings, e.g. *gam jym*, which both mean 'wealth/riches', as in example (11). In this type of collocation each of the constituent words can also be used independently.

Type 2, associative collocations, are those where the two words are not synonymous, but have associated meanings, e.g. *song* 'village' and *nok* 'house', as in *song=mi nok=mi morot* (village=GEN house=GEN person) 'villagers'. These words have associated meanings, because a village consists of houses. In this type of collocation, each of the members can also be used independently.

Types 3 and 4 are both types of so-called decorative collocations. Type 3, consists of collocations where the first member can occur as an independent word, but the second member is purely decorative, e.g. *bawbyl chambyl* (enemy DECO) 'enemy'. Only one collocation where the first word is decorative, and the second word can be used independently is attested in Atong, viz. *phakwil phakwal ~ phakwyl phakwal* 'side by side'. In this collocation, the second member means 'armpit'. However, this could just be a coincidence due to decorative vowel change. Type 4, consists of collocations where both members are decorative, e.g. *repa chepa* (DECO DECO) 'in various places'.

In certain rare cases – the author has unfortunately forgotten which ones – the purely decorative nature of the second member of a Type 3 collocation could not be established with absolute certainty during fieldwork research. This means that these some collocations listed as Type 3 could actually belong to Type 1 or Type 2. Perhaps what is decorative to some speakers is not to others. Perhaps a suitable translation for certain words in collocations could not be found, and therefore it was said that these words were decorative. Therefore, rather than categorically marking collocations as belonging to a certain type or other, the more generic labelling as *kathajyksai* was chosen.

## 6 Conjunctions

Conjunctions link phrases within a clause, or clauses within a sentence. There are two conjunctions in Atong, viz. the alternative conjunction *ma* 'or' (treated in PART 5, §16) and the additive conjunction *aro* 'and'.

## 7 Demonstratives

Demonstratives are listed in PART 4, §6. Demonstratives are deictic words used to indicate if something is closer or further away in space, or to make a noun phrase definite. There are seven demonstratives in Atong, presented in the list. The first five of them indicate different degree of remoteness, i.e. indications of how close or far away things are. English only has two degrees of remoteness, expressed by the words *this*, *these*, for close objects, and *that*, *those*, for far objects. This difference makes it impossible to translate the Atong demonstratives one to one into English, and more descriptive translations have to be attempted.

The demonstrative *hai•e ~ hai•-* is used as a pause filler, when the speaker cannot think of another word to use in its place. The quantitative demonstrative *isykyn ~ iskyn* is used to answer the question *biskyn?* 'how much?'. This word also functions as an intensifier,[28] in which case it would be translated as 'so', or 'such a', e.g. *iskyn chunga* 'so big', or *iskyn madam* 'such a (female) teacher'.

Except for *hawtyi* and *isykyn ~ iskyn*, the demonstratives have a free and a bound allomorph. The free form occurs without any enclitics, e.g. *ie rong•* (PROX rock) 'this/these rock(s)'; while the bound form is used with enclitics, e.g. *i=chi* (PROX=LOCATIVE) 'here'.

Demonstratives can be used as adnominal modifiers, as in (14), or as pronouns, as in (15). In (14), the distal demonstrative *u* modifies the noun *song* 'village'. When used as pronouns, demonstratives are referential phrases on their own, i.e. demonstrative phrases, which can take phrasal enclitics. In (15), the distal demonstrative *u* functions as a pronoun, is the head of its own phrase, and is host to the accusative phrasal enclitic *=aw*.

Demonstrative phrases often co-occur with co-referring noun phrases within the same nominal clump (see van Breugel 2014: 2014:98–103), carrying the same phrasal enclitics, which is illustrated in (16). In that example, both the demonstrative *u* (DST) and the noun *tangka* 'money' are hosts of the accusative phrasal

---

[28] Intensifiers are words that make the meaning of the following word stronger.

enclitic =*aw*. Both the demonstrative phrase and the noun phrase have the same referent.

Demonstratives are always the left-most element in a nominal clump, and thus always precede the head of a noun phrase, whether they are used as modifiers or pronouns.

(14) *ue song dam niaw Songma Songgyni Khychu Badri nowaai.*

| NOUN PHRASE | | | | |
|---|---|---|---|---|
| MODIFIER | HEAD | | | |
| *ue* | *song* | *dam* | +*ni* | =*aw.* |
| DST | village | CLF:PLACES | +two | =ACC |

| *Songma* | *Songgyni* | *Khychu* | *Badri* | *no* | =*wa* | =*ai.* |
|---|---|---|---|---|---|---|
| Songma | Songgyni | Khychu | Badri | say | =FACT | =POS |

'Those two villages, were called Songma Songgyni Khychu Badri.'

(15) *Aro ṳawan ajot noaiba mynga.*

| *Aro* | *u* | =*aw* | =*an* | *ajot* | *no* | =*ai* | =*ba* | *myng* | =*a.* |
|---|---|---|---|---|---|---|---|---|---|
| and | DST | =ACC | =FOC | ajot | say | =ADV | =EMPH | call.a.name | =CUST |

'And it's that, which is called *ajot*.'

(16) *Ge•thengdo ṳaw tangkaaw ra•akno.*

| *Ge•theng* | =*do* | *u* | =*aw* | *tangka* | =*aw* | *ra•* | =*ak* | =*no.* |
|---|---|---|---|---|---|---|---|---|
| 3SG | =TOP | DST | =ACC | money | =ACC | take | =COS | =QUOT |

'He took that money, it is said.'

When a demonstrative and a noun both occur in a nominal clump, it often happens that it is the demonstrative which hosts a phrasal enclitic, and not the last element of the noun phrase. This phenomenon is illustrated in (17), where the demonstrative *u* (DST) is accusative marked, and the following noun *rupekbisi* 'frog poison' is unmarked. More often than not, in sentences in which this construction occurs, the demonstrative and what follows belong to the same prosodic unit. This is not, however, evidence that the demonstrative and what follows belong to the same phrase. Thus, this phenomenon blurs the distinction between the modifying and pronominal usages of demonstratives. The reason for the existence of this construction is not fully understood by the author, and merits further investigation.

(17) Na•dyrangdo uaw rukpekbisi ryngaimu gumukan thyithokoknowa.

| Na• | =dyrang | =do | u | =aw | rukpek+ | bisi | ryng | =ai | =mu |
|---|---|---|---|---|---|---|---|---|---|
| fish | =PL | =TOP | DST | =ACC | frog+ | poison | drink | =ADV | =SEQ |

| gumukan | thyi | -thok | =ok | =nowa. |
|---|---|---|---|---|
| all.of.them | die | -ALL.OF.THEM | =COS | =QUOT |

'The fish, having drunk the frog poison, all died, it is said.'

For more information on demonstratives, the reader is referred to van Breugel (2014: 133–142).

## 8 Discourse connectives

Discourse connectives are listed in PART 4, §7. They usually indicate the relationships between stretches of text, i.e. between sentences, or between multiple sentences or paragraphs. Prosodically, discourse connectives can either belong to the end of a sentence, or to the beginning of one, depending on the speaker's preference. The following type of connection are found to be expressed by discourse connectives in Atong: sequential, contrastive, conditional and reason. For a detailed description of discourse connectives, the reader is referred to van Breugel (2014: Chapter 14).

## 9 Event specifiers

Event specifiers, listed in PART 4, §8, are derivational suffixes which can occur on words whose roots belong to the predicative word class (see van Breugel 2014: 65, 369). In the vast majority of recorded occurrence however, event specifiers occur on verbs, which are a sub-class of predicative. Event specifiers provide extra information about how the event or state denoted by the predicative root or stem comes about, is carried out, accomplished or persisting through time, as well as who are involved in the situation and in what way. For example, in the word natym-thok=bo (listen-everybody=IMP) 'everybody listen', we see the event specifier -thok 'everybody V'. The analysis and translation of the word natymthokbo are as follows: natym-thok=bo 'listen everybody', where natym 'listen' is the root, followed by the event specifier -thok 'everybody V' and the imperative enclitic =bo. Similarly, in the word sa•ronga 'usually eat', we find the root sa• 'eat', the event specifier -rong 'usually' and the customary aspect enclitic =a. For more information about event specifiers, the reader is referred to van Breugel (2014: 376–385).

## 10 Idiomatic expressions

Idiomatic expressions are listed in PART 4, §10. Idiomatic expressions are words, phrases and clauses, whose meaning cannot be gauged from their composition. In other words, even if you understand all the parts of the expression, its meaning is different than the sum of its parts. For example, the word *ytykram•phinai* can be analysed as *ytyk-ram•-phin=ai* (do.like.that-INADVERTENTLY-TOTALLY=ADV). However, the word functions as an intensifier to Type 1 adjectives as in example (18).

(18)  *Ang nang•na ykykramphinai kha•gala.*
      Ang    nang•  =na    ytykram•phinai   kha•gal   =a.
      1SG    you    =GOAL  so.very.much     love      =CUST
      'I love you so very much'

An example of an idiomatic phrase is *biba jokgaba dam*. This phrase can be analysed as *biba jok=gaba dam* (vapour leak.out=ATTR place). Yet, the translation of this expression is 'wound'. An example of an idiomatic clause is *chungai rai•cha*, which is analysed as *chung=ai rai•=cha* (big=ADV go/come=NEG). Nevertheless, this clause means 'I don't care'.

## 11 Ideophones

Ideophones are listed in Table 20. Ideophones are expressions that "evoke a vivid impression of a sensation" (Matthews 1997). Most of the recorded Atong ideophones are onomatopoeic, which means that the sound of the word is closely associated with the sound that is referred to with that word. For example, when we want to express the fact that someone is experiencing pain in Atong, we can use the word *aiaaa* 'ouch', which sounds like the sound made by someone who is in pain. Other words can also be onomatopoeic. For example, the frog species *ong ang* is named after the sound the frogs make. There are also a few ideophones which evoke actions rather than sounds. These ones are called 'action ideophones' (exemplified in van Breugel, 2014: 247).

The following categories of ideophones are recorded: animal sounds, sounds made by humans, sounds of the body, sounds having to do with falling, hitting and beating sounds, sounds having to do with movement, other sounds and action ideophones. For more information on ideophones, the reader is referred to van Breugel (2014: 245–247).

## 12 Indefinite proforms

Indefinite proform are listed in PART 4, §12. They are words that are used to refer to non-identifiable referents. The only indefinite-proform root is *je* 'any, whichever, whatever'. The other forms listed in Table 21 are formed by way of the indefinite phrasal enclitic =*ba*. For a description of all twelve Atong proforms, the reader is referred to van Breugel (2014: Chapter 10).

## 13 Interjections

Interjections are listed in PART 4, §13. They are words which are used to express a range of different functions, viz. to express emotion on the part of the speaker, such as surprise, astonishment, amazement, admiration, grief, anger, depreciation, indignation, or self-satisfaction. Interjections are also used for a speaker to seek attention, to call or repel animals, to express acknowledgement of what someone else has said or done, to rebut someone else's remarks or actions, or to simply fill a pause, when thinking about what to say or when hesitating to say something.

## 14 Interrogatives

Interrogatives are listed in PART 4, §14. Interrogatives, or questions words, are used to question referents, places, manners, time, quantities etc. The interrogative verb is used to ask questions about actions, or things happening. Question words form interrogative phrases that can take phrasal enclitics (see van Breugel 2014: Chapter 17). The allomorph *bie* of the interrogative *bie ~ bi* 'which? where?' cannot host enclitics, whereas the allomorph *bi* cannot occur without an enclitic, e.g. *bisang* 'to/from where' or *bichina* 'until where?'. The interrogative =*byisyk* 'how much/many' is not a question "word", but rather an interrogative quantifier enclitic, which can be used on classifiers, including auto-classifiers (see van Breugel 2014: Chapter 12). The interrogative verb can be a predicate, and take predicate enclitics (see van Breugel 2014: Chapters 20 and 22).

 The interrogatives *biskyn* 'how much/many?' and the derived interrogative *bisang* 'to/from where?' are the only members of their word class which can take the change of state predicative enclitic =*ok* (see van Breugel 2014: Chapter 23). Thus, these interrogatives can form causes on their own, viz. *biskynok* 'how much is it/are they?' and *bisangok* 'where did something/someone go?', or with and argument, e.g. *Ranus bisangok?* 'Where did Ranus go?'. For more information on interrogatives, the reader is referred to van Breugel (2014: Chapter 9).

## 15 Nouns and verbs

Nouns and verbs are listed in the semantic lexica in PART 4 of this book. The precise intricacies of nouns and verbs, being far too great, are outside the scope of this book. What is presented in this section is a very concise overview of the differences between nouns and verbs in Atong for the purposes of academic rigour and completion.

In Atong, nouns and verbs are predicative word classes, which means that they can function as the head of a predicate. A predicate is defined as "a clump consisting of one or more phrases, which denotes a state or event and is the locus of the illocutionary force" (van Breugel 2014: 349). The words belonging to the Predicative Word Classes can be thought of as existing somewhere on a cline from most prototypically predicative to least prototypically predicative.

The most prototypically predicative words are those that can occur as predicate head in all clause types, with the maximum amount of predicative morphology and with arguments. Words that are limited in any of these properties are said to be less prototypically predicative. It is the denotation of a word, which determines where it sits on the predicativity cline.

Verbs are the most prototypically predicative words. Nominal roots can only appear in certain clause types, and are very limited in the amount of predicate morphology they can occur with, which is of course due to the semantic incompatibility between the root and a lot of the predicate morphology. Most words that are normally used to refer can be used to mean 'X' (as participant) or 'be X' (as predicate). They are somewhere at the lower end of the prototype cline. In between these two are words like *ajip* 'fan', that can be used to refer to a thing or to denote an event or state and can thus express all verbal categories like causative, aspect, modality, negation etc. and all nominal categories like semantic roles and referentiality. The least prototypically predicative words are the personal pronouns, classifiers and interrogatives.

Some examples of nominal predicates are in order. Nouns can head predicates of both independent and dependent clauses. Example (19) is illustrative of a noun functioning as the head of a predicate of an independent clause. The phrase of which *bai•siga* 'friend' is the head is the host of the irrealis predicative enclitic =*chym*, marking the phrase as a predicate. In example (20), the noun *sa•gyrai* 'child' is the head of its phrase, which is marked by the factitive and goal predicative enclitics =*wa* and =*na* respectively, turning the noun phrase into a predicate.

The clause sa•gyraiwana is a Reason Clause, which is a type of dependent clause (see van
   Breugel 2014: 468–471).

In both these examples, it would be possible to add a modifier to the nominal heads of these predicative phrases, as in example (21). In that example, an attributive clause (AC) is added to the phrase.[29] Modification by an attributive clause (see van Breugel 2010, 2014: 298–303) is a strictly nominal property. The possibility of modification means that the noun still has nominal properties, even though the phrase of which it is the head functions as a predicate.

(19)  Ge•thengtheng balrukbutungchi bai•segathangmaranchym.
      Ge•thengtheng   bal     -ruk     =butung    =chi
      3PL             speak   -RECP    =WHILE     =LOC

      <u>PREDICATE</u>
      [bai•sega]      =thang   =maran    =chym.
      friend          =OWN     =RECP     =IRR
      'When they were speaking to each other, they were no longer friends.'

(20)  Sa•gyraiwana kymchawa.
      <u>PREDICATE</u>
      [sa•gyrai]   =wa     =na      kym     =cha    =wa
      child        =FACT   =GOAL    marry   =NEG    =FACT
      'Because [she's] a child, [I] will not marry her.'

(21)  Sa•gyrai mylgabawana kymchawa.
      <u>PREDICATE</u>
                    <u>AC</u>
      [[sa•gyrai]<sub>HEAD</sub>  [myl]    =gaba]   =wa    =na      kym     =cha    =wa
      child               small    =ATTR    =FACT  =GOAL    marry   =NEG    =FACT
      'Because [she's] a small child, [I] will not marry her.'

For more details on nouns and nominal predicates, the reader is referred to van Breugel (2014: Chapters 6 and 20) For more details on verbs and sub-categories of verbs, the reader is referred to van Breugel (2014: Chapter 4). In van Breugel

---

[29] An attributive clause is a clause functioning as a modifier of the head of a noun phrase. For a clause to be able to do this, its predicate has to be marked with the attributive enclitic =gaba ~ =ga, whose allomorphs are in free variation (see van Breugel 2010).

(2014: 68) Table 20 is particularly useful for details on the properties on both verbs and nouns.[30]

# 16 Particles

The particles are listed in PART 4, §15. Particles in Atong are defined in van Breugel (2014: 389) as "phonological words without denotation". Thus, particles can form words on their own in a sentence, with their own stress, i.e. particles are phonological words. However, particles cannot be used to refer to a thing, or a state of being, or an action, like nouns or verbs can. Particles are written as separate orthographic words in the Atong writing system.

The speculative modality or status particle *khon*, for example, indicates that whatever the speaker says might be the case. The conversation in (22) (adapted from van Breugel 2014: 409) is illustrative of the use of this particle.

(22)  Speaker 1: *Nang ama nygylsang re•engwama?*
      Speaker 1:  Nang   ama     nygyl    =sang   re•eng    =wa     =ma?
                  2SG    mother  market   =MOB    go.away   =FACT   =Q
                  'Has your mother gone to the market?'

      Speaker 2: *Ho•ong, khon.*
      Speaker 2:  Ho•ong   [PAUSE]   khon.
                  yes                SPEC
                  'Yes, maybe.'

The same morphemes that can be particles can also be used as enclitics. When this is the case, the morphemes form part of a larger phonological word, and are written together with the rest of the word they belong to, e.g. *mungmakhonte*, which is analysed as *mungma=khon=te* (elephant=SPEC=DECL) 'It might be an elephant, I'm telling you!'. In this example, the morpheme *khon* is an enclitic.

---

[30] Even though the Table 20 in van Breugel (214: 68) indicates that, just like nouns, verb and Type 1 & 2 adjectives can form compounds with nouns, this does not mean that, in noun-plus-verb compounds, the function of the compounded noun is one of modification of the verb. Verbs, as head of a predicate, for example, may phonologically form compounds when occurring with a prototypically associated noun. The noun is always optional, i.e. the verb can also be used without it. Most importantly, the noun does not modify the verb, i.e. the meaning of the compound is the same as the meaning of the simplex form. Examples are *tyi+hunga* (water+swim) 'swim' and *nakhal+na* 'ear+hear' (see van Breugel 2014: 355–358). The function of the noun is just to make the construction more explicit than when it would just be the verb.

## 17 Personal pronouns

Personal pronouns are listed in PART 4, §16. Personal pronouns are referent-tracking devices, which may be used when the referent in a clause is within the pragmatic presupposition, which means that the speaker considers the listener to know the identity of the referent, so that a full noun phrase can be omitted. However, in Atong, pronouns are in no case obligatory. Van Breugel (2014: 416) states that "[i]n Atong, arguments are not obligatorily expressed, which most often leaves it up to inference on the listener's part as to which entities need to be conceived to achieve the greatest amount of relevance of an utterance in a certain context." The presence or absence of arguments in a clause, has no consequences for the grammaticality of the clause in Atong.

The first person singular pronouns each have two allomorphs: *ang* ~ *anga* (1SG) and *ning* ~ *ninga* The allomorphs *ang* (1SG) and *ning* (1PL) can take enclitics, like for example =*aw* (ACC) → *nang•aw*, or =*ba* (AFF) → *ningba*. On the contrary, the allomorphs *anga* and *ninga* cannot take enclitics. As a consequence, *anga* and *ninga* are used when their referents are the Topic of the clause, i.e. the person the clause is about, or when their information status is that of 'given information', i.e. information that the listener is supposed to already know. In addition, the allomorphs *ang* and *ning* can be unmarked Possessors, e.g. *ang nok* 'my house' or *ning nok* 'our house', while *anga* and *ninga* cannot.

The first person plural makes an inclusive/exclusive distinction. The first person plural inclusive, *na•nang*, means 'me and you'. This pronoun is used when the speaker speaks on behalf of him/herself and the person or persons he or she speaks to. The first person plural exclusive, *ning*, means 'me and he or she or they'. This pronoun is used when the speaker speaks on behalf of him/herself and someone else, but not the addressee.

The third person singular and plural each have several forms. The words *ge•theng* ~ *de•theng* (different speakers favour one form over the other) and *ge•thengtheng* are almost exclusively used for human referents. The plural form *de•thengtheng* is not attested, but it might exist. The other third person forms, *itym* and *utym* ~ *ytym*, are used for both human and non-human referents. The different forms (allomorphs) *utym* and *ytym* are in free variation. Apart from the personal pronoun *ge•theng* ~ *de•theng*, the demonstratives *ue* and *ie* can also be used to refer to third persons, both human and non-human. For more information about personal pronouns, the reader is referred to van Breugel (2014: 145–155).

# 18 Postpositions

Postpositions, listed in PART 4, §17, are words that occur immediately after a phrase. This phrase is termed the *complement* of the postposition. Different postpositions take different types of complements. The post positions require the preceding complement to carry a specific semantic-role enclitic. Postpositions can be phonological words on their own, or be cliticised to their complement. For more information about postpositions, the reader is referred to van Breugel (2014: Chapter 13).

# 19 Proclauses

Proclauses, listed in PART 4, §18, are different from the interjections treated in §13 (in this part of the book),[31] because proclauses need to be interpreted as expressing polarity,[32] modality,[33] and sometimes even person and number of the speaker, while interjections do not express any of these categories. For example, the proclauses *hai* 'come on!/Let's go!' and *ha•* "take it!' are imperatives, and express positive polarity. The proclause *haida* 'I don't know' expresses negative polarity, indicative mood, and first person singular. *Kyw* expresses positive polarity and first person singular. Another example is the proclause *hai* 'Come on! Let's go!', which is an adhortative expression and therefore expresses modality.

In addition, some proclauses can be followed by predicate enclitics (see van Breugel 2014: Chapter 22), which is another property that interjections do not have. The affirmative proclause *ho•ong*, for example, can take the irrealis enclitic

---

[31] As mentioned in §13 of this part of the book, interjections are used to express emotion on the part of the speaker, such as surprise, astonishment, amazement, admiration, grief, anger, depreciation, indignation, or self-satisfaction. Interjections are also used for a speaker to seek attention, to call or repel animals, to express acknowledgement of what someone else has said or done, to rebut someone else's remarks or actions, or to simply fill a pause, when thinking about what to say or when hesitating to say something.

[32] In Atong, when somebody asks *Mai sa•akma?* 'Have you eaten?', we can answer positively, like this: *Sa•ak*, or we can answer negatively, like this: *Sa•khucha*. The enclitic =*cha* in the negative answer shows negative polarity. When =*cha* is not expressed, the answer is positive. We can also say that the answer without =*cha* has positive polarity. Proclauses do not need =*cha* to have negative polarity. For example, the proclause *haida* 'I don't know' has negative polarity without the use of =*cha*.

[33] The words *tam•o* or *tam•aw* and *hai* 'Let's go! Come on!' are used to make a suggestion or tell someone what to do; while *am* or *ym* is used to make a statement. Suggestions, commands and statements are all examples of modalities.

=*chym*, viz. *ho•ongchym* 'yes, supposedly' or the speculative enclitic =*khon*, viz. *ho•ongkhon* 'yes, maybe'. For more information about proclauses, the reader is referred to van Breugel (2014: 249–253).

## 20 Quantifiers

Quantifiers indicate a precise or approximate quantity, i.e. how much there of something there is. There are two categories of quantifiers, viz. numbers, which are called numerals in linguistic jargon, and other quantifiers. The numerals are discussed in §20.1, and the non-numeral quantifiers in §20.2.

### 20.1 Numerals

Numerals are used in three different ways: counting, quantification and enumeration. Counting happens when a speaker uses numerals in a sequence, without quantifying anything. This happens for example when teaching someone how to count, or when playing hide-and-seek, e.g. when someone counts to one hundred: *rong+sa, rong+ni, rong+tham, byryi, banga ... raja+sa* (CLF:ROUND.THINGS+one, CLF:ROUND.THINGS+two, CLF:ROUND.THINGS+three, four, five ... hundred+one) 'one, two, three, four, five ... one hundred'. Quantification, as was mentioned above, is used to talk about how much of something there is, e.g. *Morot myng•+ni ganang* (person CLF:HUMANS two exist) 'There are two people', or *parata phel+ni sa•=bo* (flatbread CLF:FLAT.THINGS+two eat=IMP) 'eat two pieces of flatbread'. Enumeration is sequential quantification. In other words, it happens when a speaker counts several things one after the other, e.g. *myng•+sa, myng•+ni, myng•+tham, myng•+byryi, myng•+banga, myng•+banga=khu=a* (CLF:HUMANS+one, CLF:HUMANS+two, CLF:HUMANS+three, CLF:HUMANS+four, CLF:HUMANS+five, CLF:HUMANS+five=INCOM=CUST) 'there are still one, two, three, four, five, six people'.

Atongs count with traditional Atong numbers, numbers borrowed from Standard Garo, numbers borrowed from English, and numbers borrowed from Hindi. English and Hindi numbers are usually used to count English and Hindi loan words referring to objects associated with modern day life introduced into the Atong society through Hindi or English. For example, when telling the time, Hindi numbers are used for the hours, while English numbers are used for the minutes, e.g. *giara baji twelp minit* (eleven hours twelve minutes) 'twelve minutes past eleven'. For more information on borrowed numerals, and more examples of what is counted with them, see van Breugel (2014: 192–195).

As can be seen in Table 28, Atong has several set or paradigms of numbers above 19. In all sets of numbers above 19, the ones labelled 'modern' are the ones in current use; the others are archaic in Badri and Siju, where most of the fieldwork was conducted. Only people old enough to be grandparents seemed to know them, when the author was there.

All modern sets of numbers, and some of the archaic ones, follow the decimal system. As is indicated in Table 28, within the archaic numerals, two sets follow the vigesimal system. The first vigesimal set starts at twenty, *rum• sa* (twenty one), and the second one starts at eighty, *khol chang byryi* (twenty times four) 'eighty'.

The Atong numerals *sa* 'one', *ni* 'two' and *tham* 'three' are bound morphemes. This means that they never occur by themselves, but are always accompanied by a classifier (see also van Breugel 2014: Chapter 12) or by an auto-classifying noun (see §4 in this part of the book).[34] When counting, this classifier is usually *rong*, the classifier also used when quantifying round things. This classifier is also the classifier used in the list of numerals in Table 28. When quantifying other things, this classifier needs to be replaced by an appropriate alternative. Thus, when saying how many people there are, the classifier *myng•* should be used, e.g. *morot+myng•+khole myng•+sa* (person+CLF:HUMANS+twenty CLF:HUMANS+one) 'twenty-one people'.

Other bound numerals are *raja ~ ratja* 'hundred' and *hajal* 'thousand'. These numerals cannot occur without a following multiplier numeral, i.e. *raja+sa* (hundred+one) 'one hundred', *hajal+tham* (thousand+three) 'three thousand'.

Some words do not need classifiers to be quantified, e.g. *san* 'day', as in *san ni* 'two days'. These words are called auto-classifiers. The numeral always follows the classifier or noun, e.g. *morot myng•+ni* (person CLF:HUMANS+two) 'two people', or *rasun khasot+byryi* (onion CLF:BUNDLES+four) 'four bundles of onions', *san charanga* 'fifteen days'. Apart from numerals, classifiers can be followed by some other quantifiers, listed below.

All recorded numerals are single phonological words. Their spelling rules are outlined in van Breugel (2015a). This said, the pronunciation of the very long numerals 7,895; 2,50,675 and 15,59,81,733 presented in Table 28 has not been recoded. It is entirely likely that these long numerals contain more than one phonological word in their pronunciation. More fieldwork is necessary to test this hypothesis. Despite being written separately in the spelling system, a classifier or

---

[34] The phenomenon of the numerals *one*, *two* and *three* being bound morphemes is also recorded for Hakhun Tangsa (Boro 2017: 115, 178), which, like Atong, belongs to the Sal or Bodo-Konyak-Jinghpaw group of the Tibeto-Burman language family (Burling 1983, 2003a).

auto-classifier and the following numeral together form one phonological word (see also van Breugel 2019: 23).

For a detailed description of quantifiers, the reader is referred to van Breugel (2014: Chapter 11). For details on how to spell Atong numbers, see van Breugel (2015a: Chapter 7).

### 20.2 Non-numeral quantifiers

There are two non-numeral quantifiers which can occur in the quantifier slot after a classifier or auto-classifier, viz. the interrogative quantifier =*byisek*, and the distributive quantifier =*phek*. These non-numeral quantifiers are enclitics, i.e. they form single phonological words with the preceding classifier or noun. The quantifiers =*gumuk*, which is an enclitic, and *khakhet*, which is not attested as enclitic, are only attested as quantifiers on noun phrases. Enclitic quantifiers are written together with the classifier or noun they follow, as is the usual spelling rule for words containing enclitics.[35] Examples of these four quantifiers can be found in van Breugel (2014: 196–198).

The Type 2 adjective *abun* 'other' is the only other non-numeral morpheme that can occur in the quantifier slot after a classifier. In its single recorded occurrence in this position, *abun* and the preceding classifier form one phonological word, with the primary stress on the second syllable of *abun* viz. [mən² a'bun]: *myng•+abun boba* (CLF:HUMANS+other crazy.person) 'another crazy person' (see van Breugel 2014: 197).

## 21 Time words

Time words are listed in PART 4, §20. As mentioned in van Breugel (2014: 155), "[t]here are two types of time words: non-deictic and deictic ones. The non-deictic ones denote periods of time while the deictic ones denote moments in time relative to the time of speaking." For a detailed description of time words, the reader is referred to van Breugel (2014: 155–160). For the days of the week and months of the year, see PART 3, §§1 and 2 respectively.

Van Breugel (2014: 155) mentions that non-deictic time words cannot function as predicate head, which would set them apart from predicative words.

---

[35] The general spelling rule is thus that numerals are written as separate words in the spelling, while non-numeral quantifiers are written together with the preceding word.

However, this statement needs to be corrected. Non-deictic time words do belong to the predicative word classes, and share propositional and syntactic properties with nouns.

First, non-deictic time words can function as the head of a predicate. For example, in the clause *wal-ang=aidok* (night-away=COS) 'It's getting night', the word *wal* 'night' occurs as the predicate. It is the operator enclitics which indicate that *wal* is the predicate of the clause (see van Breugel 2014: 349).[36] This time-word predicate can even have an argument, viz. *san* 'day' in *san wal=ok* (day night=COS) 'The day has become night'.

Second, just like other nouns, non-deictic time words can be arguments or adjuncts in clauses without taking any derivational morphology. Also, like other nouns, they can be quantified, and pluralised, e.g. (23). When non-deictic time words are quantified, they all function as auto-classifiers, which means that they are directly followed by a quantifier (see §20 in this part of the book), instead of needing an intervening classifier (see §4 on this part of the volume).

(23)  pang•a bylsi<u>darang</u>mi kynsangan
   pang•  =a[37]   bylsi   =<u>darang</u>   =mi    kynsang
   many   =CUST    year    =PL              =GEN   later
   'many years later'

In addition, like nouns, the non-deictic time words can be possessed, e.g. tai•ni nang•mi sa. (today 2SG=GEN day) 'today is your day'. In that example, the second person singular nang• (2SG) is the Possessor. Moreover, non-deictic time words can be modified like nouns, e.g. san sagaba (day one=ATTR) 'the first day'. However, a property of nouns that non-deictic time words lack is being used as modifiers to other nouns.[38]

---

**36** Operator enclitics "modify the clause and its parts" (Van Valin & LaPolla 1997: 40). They are listed in Table 76 in van Breugel (2014: 388). There are operator enclitics indicating illocutionary force, aspect, modality, status, negation and evidentiality.
**37** The stative verb *pang•* is the only recorded one that can modify a noun while marked by the customary aspect enclitic =*a* (CUST). Other stative verbs, like all other verbs, need to be marked by the attributive enclitic =*gaba* ~ =*ga* (ATTR) to modify nouns within a noun phrase, e.g. *nem=gaba bylsi* (good=ATTR year) (a/the) good year(s). (See van Breugel 2010 and 2014: 298–306 for more information about attributive clauses.)
**38** Except for the compound san+maji (day+middle) 'middle of the day', which is itself a non-deictic time word. The second member of this compound is the Indic loanword maji, most probably from Hindi मध्य /madhya/ 'middle'.

All these observations about the properties of non-deictic time words lead to the conclusion that they are a predicative word class, and that they share most of their properties with nouns.

## 22 Verbs of emotion and interaction

Verbs of emotion and interaction, listed in PART 4, §22, need only two participants to complement their meanings, viz. an Emotor or Agent and a Target. The Emotor is the referent who has an emotion for someone. The agent is a "wilful, purposeful instigator of an action" (Van Valin & LaPolla 1997). The referent to whom the emotion or interaction is directed is the Target. The Emotor and Agent are unmarked or their semantic roles, while the Target is marked by the goal enclitic =na (see van Breugel 2014: 75–76 and 272–273).

In example (24), the verb of emotion is kha•gal- 'to love'. The Emotor is the first person singular, ang (1SG), which is unmarked for semantic role. The Target is the second person singular, nang• (2SG), which is marked for its semantic role by the goal enclitic =na (GOAL).

(24) Ang nang•na kha•gala.

| EMOTOR | TARGET | | | |
|---|---|---|---|---|
| Ang | nang• | =na | kha•gal | =a. |
| 1SG | 2SG | =GOAL | love | =CUST |

'I love you.'

In example (25), the Agent is the unmarked first person plural exclusive, ning (1PL/EXCL). The Target is mongma 'elephant', marked by the goal enclitic.

(25) Ning mongmana rakhiaronga.

| AGENT | TARGET | | | |
|---|---|---|---|---|
| Ning | mongma | =na | rakhi | =aronga. |
| 1PL/EXCL | elephant | =GOAL | guard.against | =PROG |

'We are guarding against elephants.'

# Appendix of Photos

The photos presented in this appendix serve as illustrations to words in the Atong-English Dictionary. Most pictures in this appendix were taken by the author, except for the ones in which he himself is depicted, which were taken by Mr Samrat N. Marak on behalf of the author, and Photo's 124 and 125. The author extends his gratitude to Dr Erik de Maaker for his permission to use his work in Photo's 124 and 125.

**Photo 1:** Using an *abek* 'hollow spoon' to scoop up liquor (van Breugel 2019: 549).

**Photo 2:** Making *chywbok* 'white rice beer' from *sithi* by adding water to it (van Breugel 2014: 633).

**Photo 3:** Sandish drinking *chywbok* 'white rice beer' from an *abek* (van Breugel 2019: 549).

**Photo 4:** *Gora* 'large earthen pots in which rice liquor (*chyw*) is made'.

**Photo 5:** *Aphap* 'yeast' on a bed of leaves hanging from a hook from the ceiling of a house.

**Photo 6:** Clumps of *aphap* 'yeast' in a pot.

**Photo 7:** *Nok* 'bamboo house' (left) and *bablysi* 'kitchen' (right) with roofs of *parang* 'reed'.

**Photo 8:** *Bandaw ~ bando ~ bo•rang ~ noga* 'tree house'.

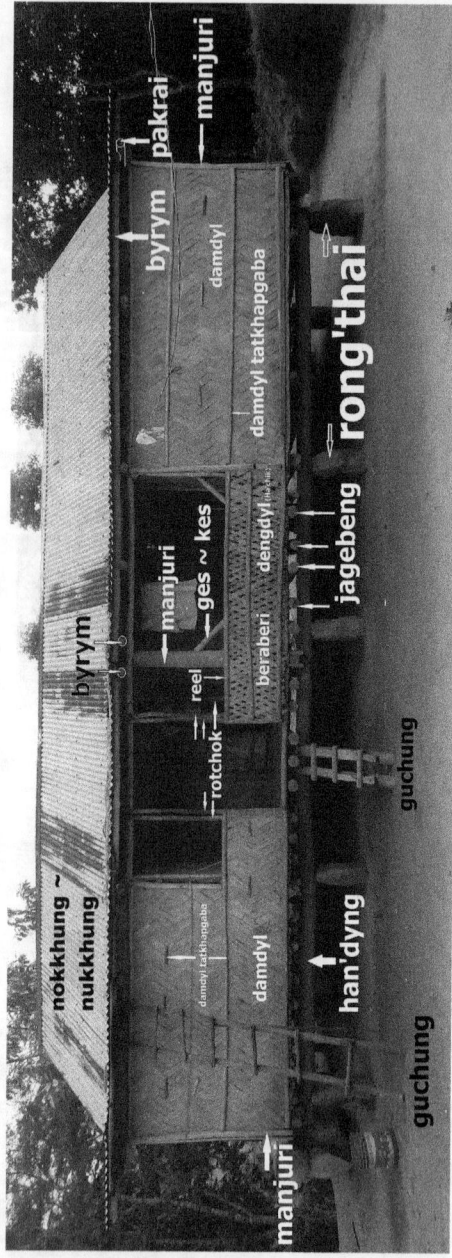

**Photo 9:** Parts of the house (1), front of the house with a roof of *tin* 'corrugated iron' (adapted from van Breugel 2014: 634).

**Photo 10:** Parts of the house (2), side of the house (adapted from van Breugel 2014: 635).

**Photo 11:** Parts of the house (3), *pannok* 'a wood stock house' (van Breugel 2014: 635).

**Photo 12:** *Bilding* 'a house built with masonry'.

**Photo 13:** *Pung* 'granary', side and frontal views (van Breugel 2019: 533).

**Photo 14:** Front of the *nokbanthai* 'bachelors' house' of Siju with the *do•khakhu* and *phu•chul* ornaments (van Breugel 2014: 648).

**Photo 15:** *Akan* and *gan•chang* 'racks hanging above the cooking fire'.

**Photo 16:** *Tyinok* 'utensil storage'.

**Photo 17:** Radia D. Sira *chula ryphigaba* 'plastering the cooking place'.

**Photo 18:** *Chula* 'cooking place'.

**Photo 19:** *Mai pylakgaba* 'flowering rice'.

**Photo 20:** *Ha•khamgaba* 'burnt soil'.

**Photo 21:** *Ha•banok* 'rice-field house' in *ha•ba* 'rice field' in Badri (van Breugel 2014: 643).

**Photo 22:** Walsrang M. Sangma and Tambu M. Sangma in the *ha•byreng* 'old rice field' in Badri (van Breugel 2014: 640).

**Photo 23:** *Badym* 'paddy field' (van Breugel 2019: 534).

**Photo 24:** Waimong Ha•bri with its characteristic camel-hump shape as seen from the road between Siju and Baghmara (van Breugel 2019: 531).

**Photo 25:** Waimong Ha•byri and the rice fields of Rongsu seen from a rice field high above the Symsang river in Badri (van Breugel 2019: 531).

Appendix of Photos —— 325

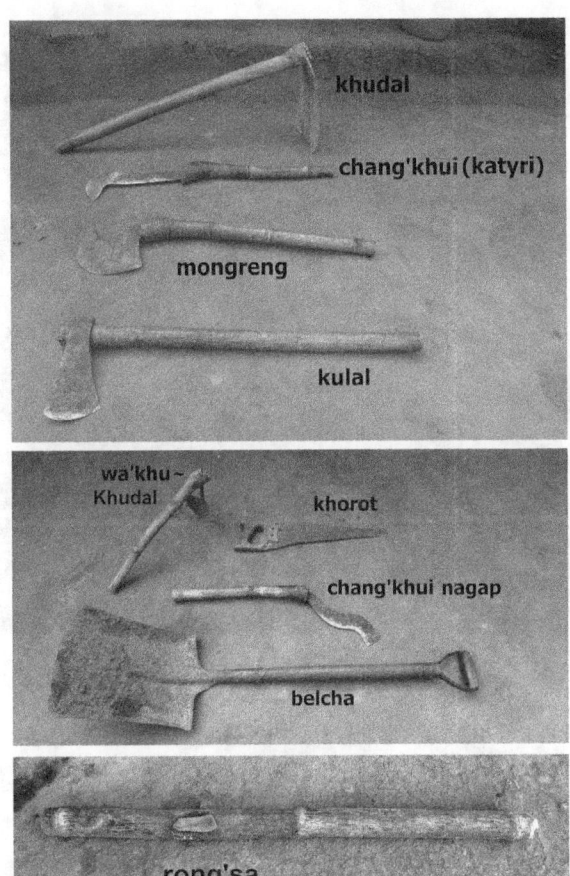

**Photo 26:** Tools (van Breugel 2014: 636).

**Photo 27:** Baskets (1) (van Breugel 2014: 633).

**326** — Appendix of Photos

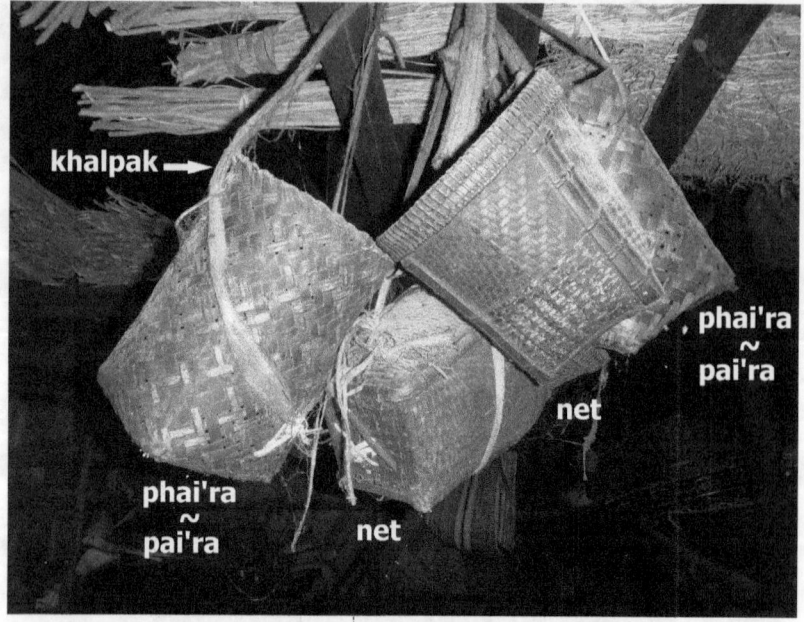

**Photo 28:** Baskets (2) (van Breugel 2014: 637).

**Photo 29:** Chicken in a *koksep* to be sold at the market (van Breugel 2014: 637).

**Photo 30:** Inside a *pung* 'granary' in Badri Maidugythym, baskets and a heap of unhusked rice as well as other food supplies are kept (adapted from van Breugel 2019: 534).

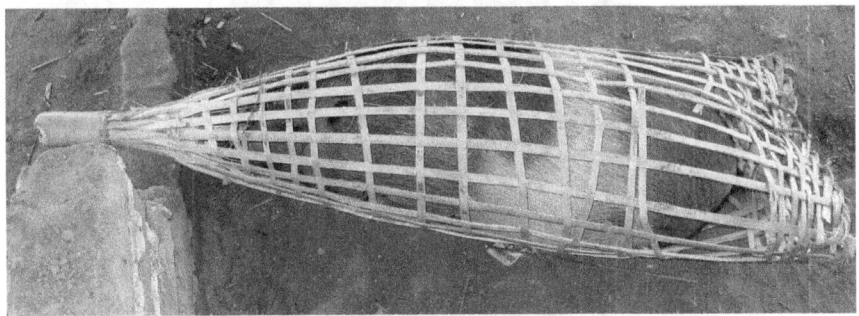

**Photo 31:** A pig trapped in an *asok* (adapted from van Breugel 2014: 638).

**Photo 32:** *Dinggarai* 'fish trap' (two different types).

**Photo 33:** *Sal•tareng* 'broom'.

Appendix of Photos — **329**

**Photo 34:** *Nokwek ~ nogek* 'broom plant' drying in a field.

**Photo 35:** *Khai-* 'to carry on one's back with a strap tied around the head', Jupina and Anarutha M. Sangma.

**Photo 36:** Jister Sangma stirring food with a *Palak* 'bamboo spoon'.

**Photo 37:** *Dakham* 'very small wooden stool'.

**Photo 38:** *Mura* 'stool'.

**Photo 39:** Fishermen in a *rung* 'logboat' at Dabatwari (van Breugel 2014: 638).

**Photo 40:** *Asam* 'mortar' and *aman* 'pestle' (van Breugel 2019: 540).

**Photo 41:** Nisajyw• holding an *awan* 'winnowing basket', and *mairongkholnang* 'unhusked rice' lying on a *damplak* or *dala* 'bamboo mat'.

**Photo 42:** *Ja•ryt* 'chillies', *mairongkhol* 'unhusked rice' and leafy greens drying in the sun on mambo mats called *damplak* or *dala*.

**Photo 43:** *Tyigatchi natgaba* 'doing the dishes at a *tyigat*. (1) *korea* 'wok' (2) *tyigum* 'waterpot' (3) *tyksyl ~ dyksyl* 'metal cooling pot' (4) *paip* 'water pipe' (5) *sorea ~ soraia* 'tub'.

**Photo 44:** *Edin M. Sangma wakpuk sytgaba.* 'Edin M. Sangma is cleaning pig intestines by rinsing them and pulling them over a stick.'.

**Photo 45:** *Rai•chak* 'big leaf used to pack food'.

**Photo 46:** *Rai•chak* 'big leaves used to pack food' are used to make bundles of food, tied with *wa•tyng* 'bamboo strip'.

**Photo 47:** Eating food from *rai•chak* 'big leaf used to pack food'.

**Photo 48:** *Dachang*.

**Photo 49:** *Law*.

Appendix of Photos —— 337

**Photo 50:** *Gomynda*.

**Photo 51:** *Sawel ~ sawyl ~ jingka ~ jingkha* (left) and *garuthai* (right).

**Photo 52:** *Sojana*.

**Photo 53:** *Dorai*.

Appendix of Photos — **339**

**Photo 54:** *Maiguru*.

**Photo 55:** *Rothop*.

**Photo 56:** *Ja•ryt* (left), *kha•rek* (middle), *kolachita* (left).

**Photo 57:** *Kha•rek*.

Appendix of Photos — 341

**Photo 58:** *Garu*.

**Photo 59:** *The•myt*.

**Photo 60:** *Khanmynchyw*.

**Photo 61:** *Suksai*.

Appendix of Photos — 343

**Photo 62:** *Ring* with big leaves and green stems.

**Photo 63:** *Ringgythyng* with big leaves and black stems.

**Photo 64:** *Man*.

**Photo 65:** *Toktokkylek ~ kokkylek*.

**Photo 66:** *Me•mangguchung*.

Appendix of Photos — **345**

**Photo 67:** *Sambarat*, leaves open (left) and leaves closed after touching (right).

**Photo 68:** *Wa•da*.

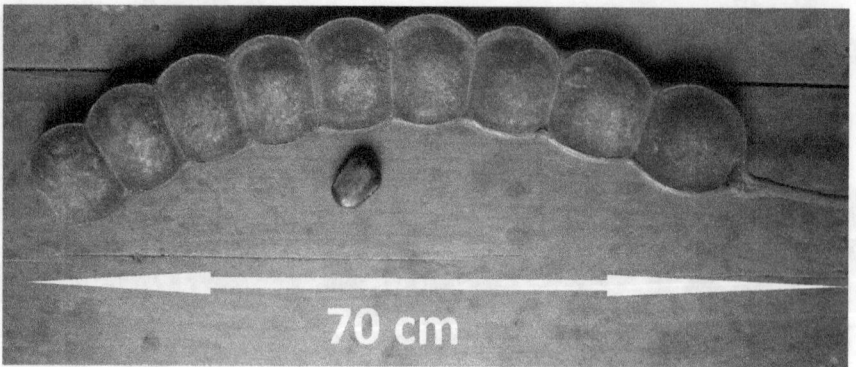

**Photo 69:** *Gyralong* or *siwi* and its seed.

**Photo 70:** *Ruda*.

Appendix of Photos — 347

**Photo 71:** *Rajami khu•symang ~ rajamyng khu•symang.*

**Photo 72:** *Ambyrai* with fruit.

**Photo 73:** *Ambyrai* tree.

**Photo 74:** *Ambyraichak* 'leaves of the *ambyrai* tree'.

Appendix of Photos — 349

**Photo 75:** *kyrydyl ~ karydyl*, which changes into *Ambi Chakkhen* at night (van Breugel 2019: 548).

**Photo 76:** *Symgong*.

**Photo 77:** *Joba*.

**Photo 78:** *Sosila* (right) and its fruit (left).

**Photo 79:** *Wa•jongmagar ~ wa•jongmagal.*

**Photo 80:** Samrat N. Marak holding the beans of a *kyltyk ~ kulthuk ~ kyltuk.*

**Photo 81:** *Lekhaphul ~ getphul.*

**Photo 82:** *Thai•gundai ~ thai•ma•thaigundai.*

**Photo 83:** *Aganggi.*

Appendix of Photos — 353

**Photo 84:** *Aganggi gawrai ~ ne•balang.*

**Photo 85:** *Ambisuthyk.*

**Photo 86:** *Gugyreng.*

**Photo 87:** *Gukchepchep.*

Appendix of Photos — 355

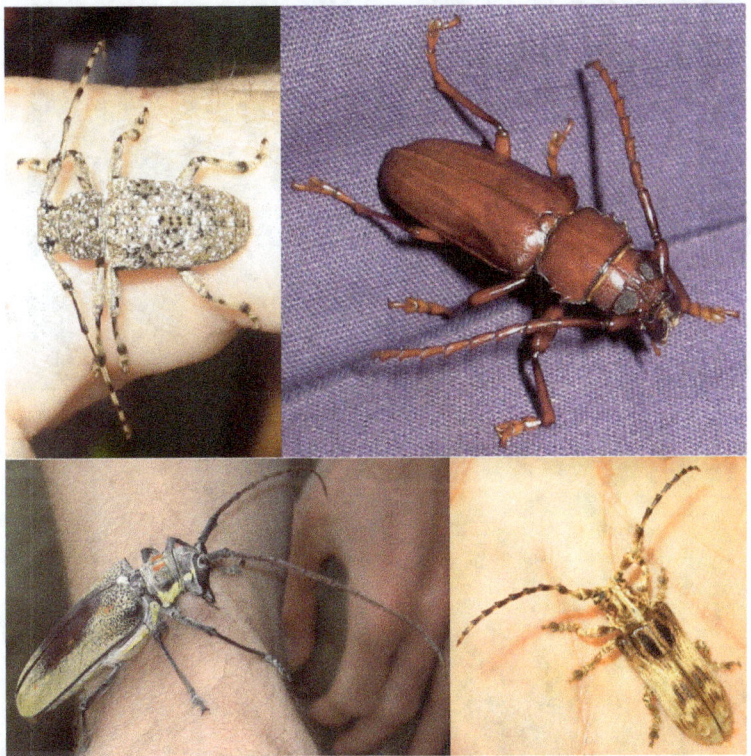

**Photo 88:** *Gogak* 'beetle' (different species).

**Photo 89:** *Gukmadym*.

**Photo 90:** *Seneng*.

**Photo 91:** *Ha•bong*.

**Photo 92:** *Hangkyn raja*.

**Photo 93:** *Chengchengmachok* (van Breugel 2019: 548).

**Photo 94:** *Gawanghu•raw*.

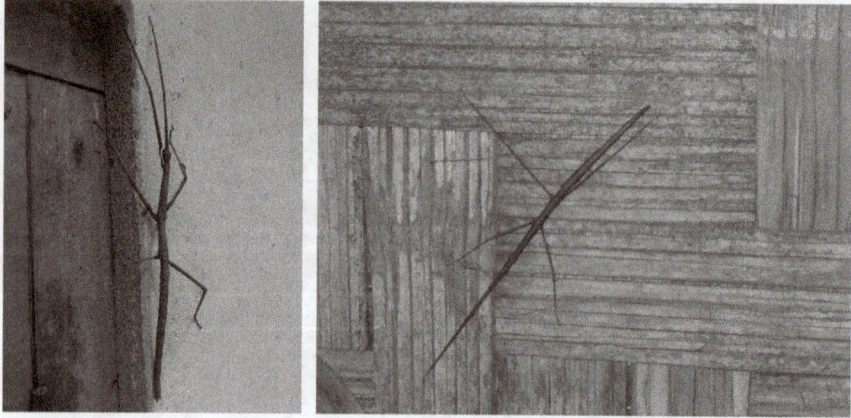

**Photo 95:** *Me•mangkereng* (two different species).

**Photo 96:** *Sanyrai*.

Appendix of Photos — **359**

**Photo 97:** *Manggywak*.

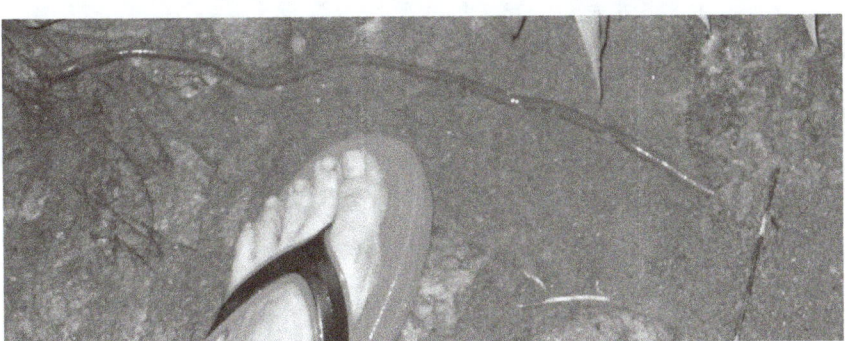

**Photo 98:** *Khansyrui* compared to the size of a foot.

**Photo 99:** *Panchungchong•su* on Samrat's face.

**Photo 100:** Upper fish: *muchi*, lower fish: *galjak*.

**Photo 101:** *Khamynkhap.*

**Photo 102:** *Na•rymkhu*.

**Photo 103:** (1) *era* (2) *na•lam* (3) *na•rong* (4) *khen•*.

**Photo 104:** *Lukwak ~ rukwak*.

**Photo 105:** *Na•luk* (two different species).

**Photo 106:** *Phu•chul*.

**Photo 107:** Type of *ne•katthup* (van Breugel 2019: 536).

**Photo 108:** *Ne•kat thupaidonga* 'bees at their nest'.

**Photo 109:** *Keko*.

**Photo 110:** *Dypywkheng*.

**Photo 111:** *Kangkylek*.

**Photo 112:** *Daw•pynchyrep ~ taw•pynchyrep*, common tailorbird, *Orthotomus sutorius*.

**Photo 113:** *Moina*.

**Photo 114:** *Ryk*.

**Photo 115:** *Daw•rigi ~ daw•rygi ~ daw•rugoi.*

**Photo 116:** *Dymchyrang* played by Winchipa (van Breugel 2019: 542).

**Photo 117:** *Chigyryng* played by Winchipa.

**Photo 118:** Two men carrying *khem* 'traditional drums' to the Tyisam Atong Festival in Baghmara (van Breugel 2019: 541).

**Photo 119:** Tontonwa• blowing a *kal*.

**Photo 120:** Different types of *rang* 'brass drum' (van Breugel 2019: 535).

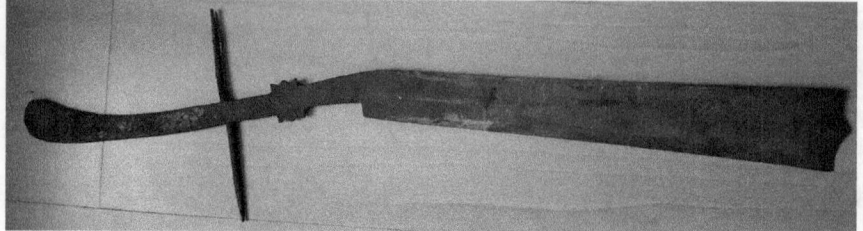

**Photo 121:** *Darai* 'sword'.

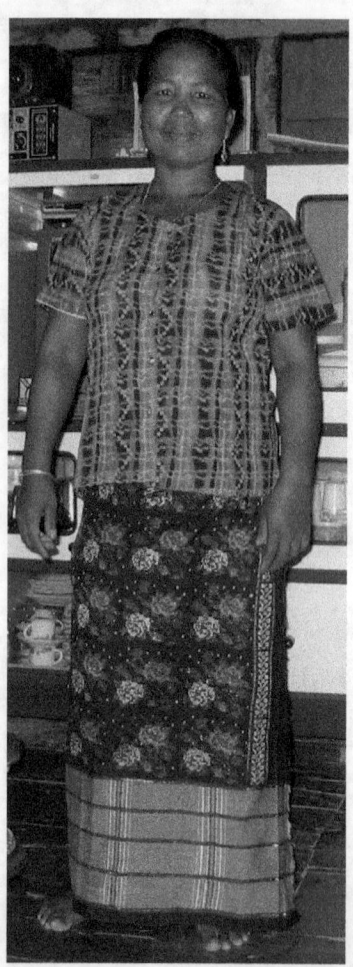

**Photo 122:** Miss Jamila M Sangma wearing a *dakmanda* 'long dress' (van Breugel 2019: 361).

**Photo 123:** *Plindar R. Marak sa•aw ba•aidonggaba.* 'Plindar R. Marak carrying his child in a cloth on his body.'.

**Photo 124:** *Seinomu Bairyckmu Tyipaityikhalchi wa•rok phakwengaidong* 'Seino and Bairyck on a bamboo raft in the Tyipai River'.

**Photo 125:** Goirapatal.

**Photo 126:** *Delang* of the Ambengs (1). (Courtesy of Erik de Maaker)

Appendix of Photos — 373

**Photo 127:** *Delang* of the Ambengs (2). (Courtesy of Erik de Maaker)

It was unfortunately impossible for the author to find a photograph of a *delang* or *dylang* made by the Atongs. Photo's 126 and 127 are presented instead, to show the reader what such a structure can look like. The Ambengs, like the Atongs, belong to the Garo Tribe.

# References

Boro, Krishna. 2017. *A grammar of Hakhun Tangsa*. PhD dissertation. Graduate School of the University of Oregon.
Breugel, Seino van. 2009a. *Atongmorot balgaba golpho, First edition*. Tura: Tura Book Room.
Breugel, Seino van. 2009b. *Atong-English Dictionary*, 1st edn. Tura: Tura Book Room.
Breugel, Seino van. 2010. No common argument, no extraction, no gap: Attributive clauses in Atong and beyond. *Studies in Language* 34(3). 493–531.
Breugel, Seino van. 2014. *A grammar of Atong*. Leiden, Boston: Brill.
Breugel, Seino van. 2015a. *Atong spelling guide*. https://www.academia.edu/15882391/Atong_Spelling_Guide (accessed 30 April 2020).
Breugel, Seino van. 2015b. *Atongmorot balgaba golpho, Second edition*. https://www.academia.edu/487053/Atongmorot_Balgaba_Golpho (accessed 30 April 2020).
Breugel, Seino van 2015c Atong-English dictionary, $2^{nd}$ edition. https://www.academia.edu/487044/Atong_English_Dictionary (accessed 30 April 2020).
Breugel, Seino van. 2015d. Journey to the Lyngams: a people of Meghalaya, Northeast India. *Humanities Journal* 22(1). Bangkok: Kasetsart University. 252–290.
Breugel, Seino van. 2016. Review of Jose, U.V. & Horsing Kholar at al. 2014. *Tiwa-English Dictionary: with English-Tiwa index*. LTBA 38(2). 324–327.
Breugel, Seino van. 2019. *Atong texts: Glossed, translated and annotated narratives in a Tibeto-Burman language of Meghalaya, Northeast India*. Leiden, Boston: Brill.
Breugel, Seino van. 2020. New perspectives on Atong kinship terms. In: Q. Marak & Sarit K. Chaudhuri (eds.), The *cultural heritage of Meghalaya*. Oxford, New York: Routledge. 205–250.
Burling, Robbins. 1961. *A Garo grammar* (Deccan College Monograph Series 25). Poona: Deccan College Post Graduate and Research Institute.
Burling, Robbins. 1992. Garo as a minimal tone language. *Linguistics of the Tibeto-Burman Area* 15(2). 331–51.
Burling, Robbins. 1983. The Sal Languages. *Linguistics of the Tibeto-Burman Area* 7(2). 1–31.
Burling, Robbins. 2003a. The Tibeto-Burman Languages of Northeastern India. In G. Thurgood & R. J. LaPolla (eds.), *The Sino-Tibetan Languages*, 169–192. London and New York: Routledge.
Burling, Robbins. 2003b. Garo. In G. Thurgood & R. J. LaPolla (eds.), *The Sino-Tibetan languages*, 387–400. London and New York: Routledge.
Burling, Robbins. 2004a. *The Language of the Modhupur Mandi (Garo), Vol. II: the lexicon*. Ann Arbor: University of Michigan. Online: http://archiv.ub.uni-heidelberg.de/savifadok/31/1/ModhupurVol2.pdf.
Burling, Robbins. 2004b. *The language of the Modhupur Mandi (Garo), Vol. III: Glossary*. Ann Arbor: University of Michigan.
Clackson, James. 2007. *Indo-European linguistics: An introduction*. Cambridge, New York, Melbourne, Madrid, Cape Town, Singapore, São Paulo: Cambridge University Press.
Dixon, R.M.W. 2004. Adjective classes in Typological perspective. In: R.M.W. Dixon and Alexandra Y. Aikhenvald (eds.). *Adjective classes: A cross-linguistic typology*, 1–9. Oxford University Press.

# References

Dixon, R.M.W. 2006. Complement clauses and complementation strategies in typological perspective. In: R.M.W. Dixon and Alexandra Y. Aikhenvald (eds.). *Complementation*, 1–48. Oxford University Press.

Eberhard, David M., Gary F. Simons, and Charles D. Fennig (eds.). 2020. *Ethnologue: Languages of the World. Twenty-third edition*. Dallas, Texas: SIL International. Online version: http://www.ethnologue.com. Accessed: 26 June 2020.

Haiman, John. 2011. *A grammar of Cambodian Khmer*. Amsterdam, Philadelphia: John Benjamins Publishing Company.

Jose, U.V., Horsing Kholar et al. 2014. *Tiwa-English Dictionary: with English-Tiwa Index*. Shillong: Don Bosco Centre for Indigenous Cultures.

Ladefoged, Peter. 2000. *A course in phonetics*, fourth edition. Thomson Wordsworth.

Matthews, P.H. 1997. *Concise dictionary of linguistics*. Oxford: Oxford University Press.

Members of the Garo Mission of the American Baptist Missionary Union. 1905. *English-Garo dictionary*. Shillong: Garo Mission, American Baptist Missionary Union.

Nengminza, D.S. 1978. *The School Dictionary Garo to English*. Tura: Sushil Kr. Das.

Nengminza, D.S. 2001. *The School Dictionary Garo to English*. Tura: Sushil Kr. Das.

Van Valin, Robert D. & Randy J. LaPolla. 1997. *Syntax: Structure, meaning and function*. Cambridge University Press.

# Index

allomorphy 4
apostrophe XXII
aspirated stop XIX
Atong
– closest linguistic relatives XVII
– language family XVII
– location XVII
– population XVI
attributive clause 302
Austroasiatic 223
auto-classifier 293, 307, 309

bilingualism XVI
Boro K. 307
bound morpheme 3, 4, 307
bullet XXII
Burling R. XXII, 4, 194, 307

Clackson J. 9
classifier (*See also* auto-classifier) 8, 9
cline 301
collocation 2
compound 221, 294, 295, 303
context 304
context dependent 290
counting 306

decorative words 288, 294, 295
demonstrative phrase 296
denotation 288
diphthong XXI
Dixon R.M.W. XXV, 246, 249

Eberhard D.M. et al. XVI
ellipsis 2
enclitic 3, 263
entry 1, 2, 5, 7
enumeration 306
example sentence with glosses 140, 178, 287, 288, 289, 290, 291, 292, 293, 294, 297, 298, 299, 302, 303, 309, 310
expressions 139

fieldwork . (*See also research*)
– conducted 307
– future 236, 246, 252, 266, 269
– more is necessary/needed 22, 93, 120, 292, 295, 307
– was not possible XV

Garo
– Standard XXII, 4, 73, 115, 131, 306
– Tribe 4
glide XX, XXI
glottal stop XXI, XXII
glottalisation XIX, XX, XXI, XXII
glottalised syllable XXII
grammatical word 288, 295
grammaticality 304

Haiman J. 289
headword 2, 3, 135
homonymy 8
homophony 8

inclusive/exclusive distinction 304
Indic 9
inference 304
infinitive 4
information status 290, 304
Intensifiers 296

Jose U.V et al. 194

*kathajyksai* 2

label 5
Ladefoged P. XXI
lexicographic tradition 4
loanvowels XX
loanwords XX, 9, 10, 186, 309
Lyngam 223

markedness 290
Matthews P.H. 299
Megam. *See* Lyngam 223

Members of the Garo Mission  4
modality  305
modification  302, 303
mutual intelligibility  XVI

Nengminza D.S.  4
nominal clump  288, 296, 297
nominal predicate  301, 302

onomatopoeic  299
operator enclitic  309
orthographic word  293, 295, 303, 307, 308

phonemes  XIX
phonological word  288, 292, 293, 295, 305, 307, 308
phonology  XIX
polarity  305
polysemy  138
population. *See* Atong  XVI
pragmatic presupposition  304
predicate  301
predicative word classes  285, 301, 308
predicativity cline  301
prestige language  XXII, 4
Primary-B verbs  XXV
prosody  297, 298

quantification  289, 290, 306
quantifier phrase  288

*raka*  XXI, XXII
referent *or* reference  289, 290, 304
– specificity  290
relevance  304
research. (*See also* fieldwork)
– further  261
– more is necessary/needed  85, 292, 297

scientific name  6
semantic
– domain  1
– extension  8
– incompatibility  301
– interpretations (different)  286
social media  XV
spelling
– rules  307, 308
– system  XVIII
– variation  4
stress  XX, 292, 295, 308
sub-entry  1, 2
suffix  3, 263
– derivational  298
syllable  XXI
syllable structure  XXI
synonymy  139, 295
syntagmatic construction  136

taboo  12
Topic  304
topical referents  289
transitivity (*see also* valancy)  4
translation  135, 136, 137, 261, 295, 296

valency (*see also* transitivity)  246
van Breugel S.  XVIII, XIX, XXI, XXII, XXV, XXVII, 3, 4, 116, 135, 137, 140, 178, 187, 194, 223, 246, 249, 252, 261, 263, 266, 269, 282, 283, 285, 286, 287, 288, 289, 292, 293, 294, 295, 296, 298, 299, 300, 301, 302, 303, 304, 305, 306, 307, 308, 309, 310
Van Valin R.D. & R.J. LaPolla  309, 310
verb  4
voiceless bilabial nasal  XXII
vowel off-glide sequence  XXI

word class  1

www.ingramcontent.com/pod-product-compliance
Lightning Source LLC
Chambersburg PA
CBHW061927220426
43662CB00012B/1827